ORGANIC MECHANISMS

ORGANIC MECHANISMS

Reactions, Methodology, and Biological Applications

Second Edition

XIAOPING SUN
University of Charleston
Charleston, West Virginia

Registered Office
John Wiley & Sons, Inc., 111 River Street, Hoboken, NJ 07030, USA

Editorial Office
111 River Street, Hoboken, NJ 07030, USA

For details of our global editorial offices, customer services, and more information about Wiley products visit us at www.wiley.com.

Wiley also publishes its books in a variety of electronic formats and by print-on-demand. Some content that appears in standard print versions of this book may not be available in other formats.

Library of Congress Cataloging-in-Publication Data
Names: Sun, Xiaoping, 1960– author.
Title: Organic mechanisms : reactions, methodology, and biological
    applications / Xiaoping Sun, University of Charleston, Charleston, West Virginia.
Description: Second edition. | Hoboken, NJ : Wiley, 2021. | Includes index.
Identifiers: LCCN 2020020410 (print) | LCCN 2020020411 (ebook) | ISBN
    9781119618829 (cloth) | ISBN 9781119618843 (adobe pdf) | ISBN
    9781119618867 (epub)
Subjects: LCSH: Organic reaction mechanisms.
Classification: LCC QD502.5 .S86 2021 (print) | LCC QD502.5 (ebook) | DDC
    547/.139–dc23
LC record available at https://lccn.loc.gov/2020020410
LC ebook record available at https://lccn.loc.gov/2020020411

Cover image: Light against rippled background (Digital) © Chad Baker/Getty Images,
Organic mechanism diagram Courtesy of Xiaoping Sun
Cover design by Wiley

Set in 10/12 Times SPi Global, Pondicherry, India

Printed in the United States of America

SKY10022983_120420

CONTENTS

PREFACE

The first edition of this book, *Organic Mechanisms: Reactions, Methodology, and Biological Applications*, was published in July 2013. In January 2019, Mr. Jonathan Rose, Senior Editor of Wiley in Hoboken, New Jersey, asked me to consider writing the second edition of the book. We both realized that because 6 years had passed since the publication of the first edition, it would be necessary to update the book with a new addition to meet the increasing needs for various upper-level college students, instructors, and practicing chemists. I came up with a detailed proposal regarding how to modify the book and what new information will be added to the second edition. The proposal received very favorable peer reviews and was approved by Wiley in March 2019. Then I started writing the new edition. It took me 11 months to finish writing the second edition of the book. Meanwhile, I was teaching full-time as well.

Since 1998, I have been on faculty at two higher educational institutions, West Virginia University Institute of Technology (1998–2001) and University of Charleston (2001–present). My current and previous research interests have much emphasis on reaction mechanisms. When I teach upper-level organic chemistry and biochemistry courses, the lectures are also strongly mechanisms based. In the past 20 years or so, my teaching and research have been reinforcing each other. Now, I wish to share my experiences with other colleagues, students, and different scientific workers in the chemical community by publishing the second edition of this book. Much new information, particularly many more new examples of biological applications and green chemistry methodology which encompass important organic reaction mechanisms, has been included in the updated edition.

The second edition of the book consists of 10 chapters. It starts with reviews of various fundamental physicochemical principles (Chapter 1), which are essential for studying organic mechanisms. General principles of green chemistry methods and enzymatic catalysis have been added to Chapter 1 in the new edition. Then, each of the following chapters is devoted to one major class of organic reactions. Thorough discussions on various reaction mechanisms are presented in the book in a very good detail, and a sophisticated and readily understandable manner. Special attention has been paid to mechanisms of different organic functionalization processes, such as methodology of aliphatic C—H bond activation and functionalization, charge-transfer aromatic nitration and recently developed chemistry of aromatic compounds, and cycloaddition of alkenes to 1,3-dipolar-like molecules. Substantial efforts have been made in demonstrating direct applications of organic mechanisms in elucidating sophisticated biological and biochemical processes and designing organic synthesis. This can be seen throughout all the chapters and reflects a remarkable feature of the book. In the second edition, many new examples of cycloaddition reactions, including those conducted in water representing green chemistry methodology, have been added in Chapter 4. Some new electrophilic aromatic substitution reactions developed in the recent 10 years or so have been added in Chapter 5. More biological applications in enzymatic reactions involving key organic mechanisms have been added in Chapters 6 and 8. Chapter 7 (Eliminations) has been essentially rewritten to cover many more interesting and challenging elimination reaction mechanisms and applications.

To facilitate teaching and learning, a Solutions Manual and PowerPoint slides of all the figures will be provided by the author and will be available for professors on the book's page on www.wiley.com\go\Sun\OrgMech_2e. I greatly appreciate all the constructive comments given by peer-reviewers. The reviewers' comments have helped the author tremendously in improving the book.

I would like to take this opportunity to dedicate my book to Dr. Jack Passmore, my former PhD supervisor, and to my organic chemistry and biochemistry students at University of Charleston. The book is also dedicated to my wife Cindy and my son Oliver.

XIAOPING SUN, PHD

Professor of Chemistry
Charleston, West Virginia
June 2020

FIRST EDITION PREFACE

In Summer 2010, I was contacted by Mr. Jonathan Rose, a senior editor of Wiley in Hoboken, New Jersey, for book review and possibly writing an organic mechanisms-based textbook. We both agreed that a new book in this subject will meet the increasing needs of various upper-level college students, instructors, and practicing chemists. I took up the challenge and came up with a detailed proposal regarding the contents, style, and features of the book. The proposal received favorable peer reviews and was approved by Wiley in December 2010. Then I started the writing process. It took me 18 months to finish writing the book. Meanwhile, I was teaching full-time as well.

Since 1998, I have been on faculty at two higher educational institutions, West Virginia University Institute of Technology (1998–2001) and University of Charleston (2001–present). My current and previous research interests have much emphasis on reaction mechanisms. When I teach upper-level organic chemistry and biochemistry courses, the lectures are also strongly mechanisms based. In the past dozen years or so, my teaching and research have been reinforcing each other. Now I wish to share my experiences with other colleagues, students, and different scientific workers in the chemical community by publishing this book.

The book consists of 10 chapters. It starts with reviews of various fundamental physicochemical principles (Chapter 1) which are essential for studying organic mechanisms. Then each of the following chapters is devoted to one major class of organic reactions. Thorough discussions on various reaction mechanisms are presented in the book in a very good detail, and a sophisticated and readily understandable manner. Special attention has been paid to mechanisms of different organic functionalization processes, such as methodology of aliphatic C—H bond activation and functionalization, charge-transfer aromatic nitration and recently developed chemistry of aromatic compounds, and cycloaddition of alkenes to 1,3-dipolar-like molecules. Substantial efforts have been made in demonstrating direct applications of organic mechanisms in elucidating sophisticated biological and biochemical processes and designing organic synthesis. This can be seen throughout all the chapters and reflects a remarkable feature of the book. To facilitate teaching and learning, a Solutions Manual and PowerPoint slides of all the figures will be provided by the author and will be available for professors on a companion website for adopting instructors. I greatly appreciate all the constructive comments given by seven peer-reviewers on the initial manuscript. The reviewers' comments have helped the author tremendously in improving the book.

I would like to take this opportunity to dedicate my book to Dr. Jack Passmore, my former PhD supervisor, and to my organic chemistry and biochemistry students at University of Charleston. The book is also dedicated to my wife Cindy and my son Oliver.

XIAOPING SUN, PHD

Professor of Chemistry
Charleston, West Virginia

ABOUT THE COMPANION WEBSITE

This book is accompanied by a companion website:
www.wiley.com\go\Sun\OrgMech_2e

The website includes:
- Solution Manual
- PPT

1

FUNDAMENTAL PRINCIPLES

1.1 REACTION MECHANISMS AND THEIR IMPORTANCE

The microscopic steps in a chemical reaction which reflect how the reactant molecules interact (collide) with each other to lead to the formation of the product molecules are defined as **mechanism** of the reaction. The mechanism of a reaction reveals detailed process of bond breaking in reactants and bond formation in products. It is a microscopic view of a chemical reaction at molecular, atomic, and/or even electronic level.

The structure of most organic compounds is well established by X-ray crystallography and various spectroscopic methods with the accuracy of measurement in bond distances and angles being the nearest to 0.01 Å and 1°, respectively. Only **effective molecular collisions**, the collisions of the molecules with sufficient energy that take place in appropriate orientations, lead to chemical reactions. The extent of a chemical reaction (chemical equilibrium) is determined by the changes in thermodynamic state functions including enthalpy (ΔH), entropy (ΔS), and free energy (ΔG). The combination of kinetic and thermodynamic studies, quantum mechanical calculations, and geometry and electronic structure-based molecular modeling has been employed to reveal mechanisms of various organic chemical reactions.

Organic Mechanisms: Reactions, Methodology, and Biological Applications,
Second Edition. Xiaoping Sun.
© 2021 John Wiley & Sons, Inc. Published 2021 by John Wiley & Sons, Inc.
Companion website: www.wiley.com/go/Sun/OrgMech_2e

Reaction mechanisms play very important roles in the study of organic chemistry. The importance of mechanisms not only lies in that they facilitate an understanding of various chemical phenomena but also that mechanisms can provide guidelines for exploring new chemistry and developing new synthetic methods for various useful substances, drugs, and materials. In this regard, mechanistic studies will allow *synthetic chemists* to vary reaction conditions, temperatures, and proportions of chemical reagents to maximize yields of targeted pure products. For *industrial chemists*, mechanistic knowledge allows the prediction of new reagents and reaction conditions which may affect desired transformations. It also allows optimization of yields, reducing the costs on raw materials and waste disposals. It provides a tool for the chemists to make reactions occur in their desired ways and manufacture the ideal products. For *biochemists* and *medicinal chemists*, the microscopic view of organic reactions can help them better understand how the metabolic processes in living organisms work at molecular level, how diseases affect metabolism, and how to develop appropriate drug molecules to assist or prevent particular biochemical reactions [1].

Overall, the goal of this book is to tie reaction mechanisms, synthetic and green chemistry methodology, and biochemical applications together to form an integrated picture of organic chemistry. While the book emphasizes mechanistic aspects of organic reactions, it is a practical textbook presenting the synthetic perspective about organic reaction mechanisms appealing to senior undergraduate-level and graduate-level students. The book provides a useful guide for how to analyze, understand, approach, and solve the problems of organic reactions with the help of mechanistic studies.

In this chapter, fundamental principles that are required for studies and understanding of organic reaction mechanisms are briefly reviewed. These principles include basic theories on chemical kinetics, transition states, thermodynamics, and atomic and molecular orbitals.

1.2   ELEMENTARY (CONCERTED) AND STEPWISE REACTIONS

Some chemical reactions only involve *one microscopic step*. In these reactions, the effective molecular collision, the collision of reactant molecules with sufficient energy in appropriate orientation, leads to *simultaneous* breaking of old bonds in reactants and formation of new bonds in products. This type of reactions is defined as **elementary (or concerted) reactions**. An elementary (concerted) reaction proceeds via **a single transition state**. The transition state is a short-lived (transient) activated complex in which the old bonds are being partially broken and new bonds are being partially formed concurrently. It possesses the maximum energy level (in the free energy term) in the reaction profile (energy profile).

Many other chemical reactions involve *many microscopic elementary (concerted) steps* in the course of the overall reactions. These reactions are defined as **stepwise (or multistep) reactions**. A stepwise reaction proceeds via **more than one transition state**. Each microscopic concerted step proceeds through one transition state, giving a distinct product which is referred to as an **intermediate**. Each intermediate formed in the course of a stepwise reaction is metastable and usually highly reactive, possessing a relatively high energy level. Once formed, the intermediate undergoes a subsequent reaction eventually leading to the formation of the final product.

Figure 1.1 shows reaction profiles for concerted and stepwise reactions using examples of S_N2 and S_N1 reactions, respectively [1]. In a concerted reaction such as the S_N2 reaction of bromomethane (CH_3Br) with hydroxide (OH^-) (Fig. 1.1a), as the reactant molecules start colliding effectively, namely that OH^- approaches (attacks) the carbon atom in CH_3Br from the opposite side of the –Br group, formation of a new bond (the O—C bond) and breaking of an old bond (the C—Br bond) occur **simultaneously**. At the same time, the hydrogen atoms in CH_3Br move gradually from the left side toward the right side. The reaction proceeds via a single transition state (activated complex) in which the old C—Br bond is being partially broken, coincident with the partial formation of a new O—C bond. The hydrogen atoms have moved to the "middle," forming a roughly trigonal-planar configuration. The transition state possesses the maximum energy level in the reaction profile. It is short-lived and highly reactive. As the reaction further progresses, the transition state collapses (dissociates) spontaneously to lead to full breaking of the old C—Br bond in the reactant and concurrent complete formation of the new O—C bond in the product. Simultaneously, the hydrogen atoms move to the right side. The overall process is a one-step transformation. The extent of the reaction is determined by the difference in free energy (ΔG) between reactants and products.

In contrast to a concerted reaction, a stepwise reaction proceeds via more than one transition state. It consists of two or more elementary (concerted) steps (Fig. 1.1b), and distinct reactive intermediate(s) is formed in the course of the reaction [1]. The S_N1 reaction of 2-bromo-2-methylpropane (Me_3CBr) in Figure 1.1b demonstrates the general feature of a stepwise reaction. The first step is the dissociation of Me_3CBr to a reactive carbocation Me_3C^+ intermediate. In the second step, Me_3C^+ reacts with water to give a tertiary alcohol product (via a hydronium $Me_3C–OH_2^+$ which is very

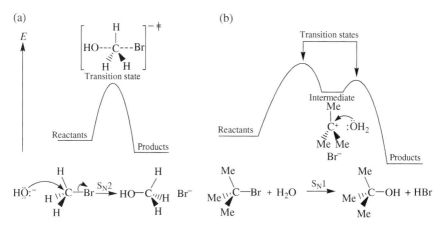

FIGURE 1.1   Reaction profiles for a concerted S_N2 reaction (a) and a stepwise S_N1 reaction (b).

often omitted as its subsequent deprotonation to $Me_3C–OH$ is spontaneous and very rapid, and nearly simultaneous). Each concerted step proceeds via a transition state. The extent of a stepwise reaction is also determined by the difference in free energy (ΔG) between reactants and products, while the overall reaction rate is dictated by the relative stability of the reactive intermediate(s).

Whether a chemical reaction is concerted or stepwise is determined by geometry and electronic structure of reactant and product molecules and reaction conditions. In many cases, the mechanism is predictable. In the individual chapters of this book, we will study the various types of concerted and stepwise reactions and the specific conditions which make them happen.

1.3   MOLECULARITY

The number of molecules contained in the transition state of a *concerted reaction* is called **molecularity** of the reaction. Clearly, the molecularity is determined by the number of reactant molecules that are involved in the mechanism (microscopic step) of a concerted reaction.

1.3.1   Unimolecular Reactions

The microscopic steps of many concerted chemical reactions only involve a single reactant molecule. Such a concerted reaction whose mechanism only involves *one reactant molecule* is defined as a **unimolecular reaction**. It is generalized as follows (Eq. 1.1):

$$A \longrightarrow \underset{\text{Activated}}{A^*} \longrightarrow P \tag{1.1}$$

A, A∗, and P represent a reactant molecule, an activated reactant molecule (transition state), and a product molecule, respectively. In a unimolecular reaction, a reactant molecule can possibly gain energy and then is activated by several means, including collision of the reactant molecule with a solvent molecule or with the wall of the reactor, thermally induced vibration of the reactant molecule, and photochemical excitation of the reactant molecule. After the molecule is activated, some simultaneous bond-breaking and bond-formation processes will take place in A∗ intramolecularly. As a result, the reactant molecule A will be transformed into one or more product molecules. Common examples of unimolecular reactions are thermal or photochemical dissociation of a halogen molecule (Reaction 1.2) and intramolecular ring-opening and ring-closure reactions (Reaction 1.3).

$$X{-}X \xrightarrow[\text{or } \Delta]{h\nu} 2\,X^{\cdot} \qquad (X = Cl,\ Br,\ or\ I) \tag{1.2}$$

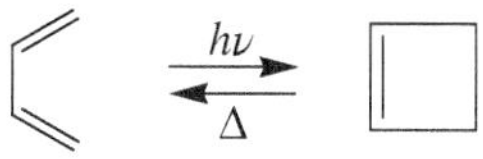

$$\text{(1.3)}$$

1.3.2 Bimolecular Reactions

For most of concerted chemical reactions, their microscopic steps (mechanisms) involve effective collisions between two reactant molecules. Such a concerted reaction that is effected by collision of *two reactant molecules* to directly lead to the formation of products is defined as a **bimolecular reaction**. A bimolecular reaction can be effected by collision of two molecules of a same compound (Eq. 1.4) or two molecules of different compounds (Eq. 1.5).

$$A + A \longrightarrow \underset{\text{Activated}}{A_2{}^*} \longrightarrow P \tag{1.4}$$

$$A + B \longrightarrow \underset{\text{Activated}}{[AB]^*} \longrightarrow P \tag{1.5}$$

As a result, simultaneous bond-breaking and bond-formation take place within the activated complex (transition state) A_2* or $[AB]*$. This leads to spontaneous collapse of the activated complex (transition state) giving product molecules. Common examples of bimolecular reactions are thermal decomposition of hydrogen iodide (HI) to elemental iodine (I_2) and hydrogen (H_2) (Reaction 1.6), the S_N2 reaction of hydroxide with bromomethane (Reaction 1.7), and Diels–Alder reaction of 1,3-butadiene and ethylene (Reaction 1.8).

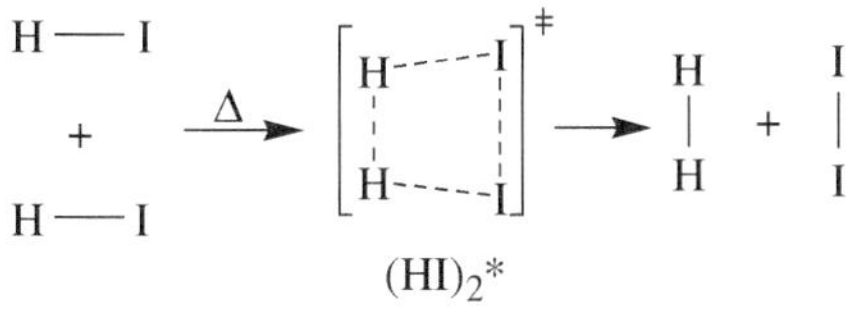

$$\tag{1.6}$$

$$HO^- + CH_3{-}Br \longrightarrow \left[HO\text{----}CH_3\text{----}Br\right]^{-\ddagger} \longrightarrow HO{-}CH_3 + Br^- \tag{1.7}$$

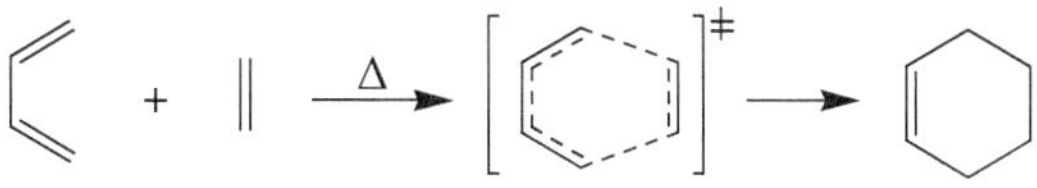

$$\tag{1.8}$$

Almost all the concerted processes in organic reactions are either unimolecular or bimolecular steps.

1.4   KINETICS

1.4.1   Rate-Laws for Elementary (Concerted) Reactions

For elementary reactions, the reaction orders are consistent with the molecularity. A unimolecular reaction is the first-order in the reactant and a bimolecular reaction has a second-order rate law.

Unimolecular reactions   A unimolecular reaction (Eq. 1.1: A ➔ P) follows the first-order rate law as shown in Equation 1.9

$$-\frac{d[A]}{dt} = k[A] \tag{1.9}$$

where k is the **rate constant** (with the typical unit of s^{-1}) for the reaction, and it is independent of the concentration of the reactant. The rate constant is the quantitative measure of how fast the reaction proceeds at a certain temperature.

Rearranging Equation 1.9 leads to

$$\frac{d[A]}{[A]} = -kdt \tag{1.10}$$

Integrating Equation 1.10 on both sides and applying the boundary condition $t = 0$, $[A] = [A]_0$ (initial concentration), we have

$$\int_{[A]_0}^{[A]} \frac{d[A]}{[A]} = -\int_0^t kdt \tag{1.11}$$

From Equation 1.11, we have $\ln[A] - \ln[A]_0 = -kt$
Therefore,

$$\ln[A] = -kt + \ln[A]_0 \quad \text{or} \quad [A] = [A]_0 e^{-kt} \tag{1.12}$$

Equation 1.12 is the integrated rate law for a unimolecular reaction.

The half-life ($t_{1/2}$) of reactant A (the time required for conversion of one-half of the reactant to the product, i.e., when $t = t_{1/2}$, $[A] = \frac{1}{2}[A]_0$) can be solved from Equation 1.12 as follows:

$$\ln\left(1/2[A]_0\right) = -kt_{1/2} + \ln[A]_0$$

Therefore,

$$t_{1/2} = \frac{\ln 2}{k} \tag{1.13}$$

Equation 1.13 shows that the half-life of a substance that undergoes first-order decay is inversely proportional to the rate constant and independent of the initial concentration.

Bimolecular reactions   A bimolecular reaction that involves two reactant molecules of the same compound (Eq. 1.4: $2A \rightarrow P$) follows the second-order rate law as shown below:

$$-\frac{d[A]}{dt} = k[A]^2 \tag{1.14}$$

where k is the rate constant (with the typical unit of $M^{-1}s^{-1}$) for the reaction.
Rearranging Equation 1.14 leads to

$$\frac{d[A]}{[A]^2} = -k dt \tag{1.15}$$

Integrating Equation 1.15 on both sides and applying the boundary condition $t = 0$, $[A] = [A]_0$ (initial concentration), we have

$$\int_{[A]_0}^{[A]} \frac{d[A]}{[A]^2} = -\int_0^t k dt \tag{1.16}$$

From Equation 1.16, we have

$$\frac{1}{[A]} = kt + \frac{1}{[A]_0} \tag{1.17}$$

Equation 1.17 is the integrated rate law for a bimolecular reaction involving two molecules from the same compound.
A bimolecular reaction that involves two reactant molecules of different compounds (Eq. 1.5: $A + B \rightarrow P$) also follows the second-order rate law (first-order in each of the reactants) as shown in Equation 1.18.

$$-\frac{d[A]}{dt} = -\frac{d[B]}{dt} = \frac{d[P]}{dt} = k[A][B] \tag{1.18}$$

Assume that at a given time t, the molar concentration of the product P is x. Therefore, the molar concentrations of reactants A and B are $[A] = [A]_0 - x$ and $[B] = [B]_0 - x$, respectively. $[A]_0$ and $[B]_0$ are initial concentrations of reactants A and B, respectively.

From Equation 1.18, we have

$$\frac{dx}{dt} = k\left([A]_0 - x\right)\left([B]_0 - x\right) \tag{1.19}$$

If the quantities of the two reactants A and B are in stoichiometric ratio ($[A]_0 = [B]_0$), Equation 1.19 becomes

$$\frac{dx}{dt} = k\left([A]_0 - x\right)^2 \tag{1.20}$$

Rearranging Equation 1.20 leads to Equation 1.21.

$$\frac{dx}{\left([A]_0 - x\right)} = k\,dt \tag{1.21}$$

Integrating Equation 1.21 on both sides and applying the boundary condition $t = 0$, $x = 0$, we have

$$\int_0^x \frac{dx}{\left([A]_0 - x\right)} = \int_0^t k\,dt \tag{1.22}$$

From Equation 1.22, we have

$$\frac{1}{[A]_0 - x} = kt + \frac{1}{[A]_0} \tag{1.23}$$

Since $[A] = [A]_0 - x$, Equation 1.23 becomes

$$\frac{1}{[A]_0} = kt + \frac{1}{[A]_0} \qquad \text{(the same as 1.17)}$$

If the reactants A and B have different initial concentrations, Equation 1.19 becomes

$$\frac{dx}{\left([A]_0 - x\right)\left([B]_0 - x\right)} = k\,dt \tag{1.24}$$

Integrating Equation 1.24 on both sides and applying the boundary condition $t = 0$, $x = 0$, we have

$$\int_0^x \frac{dx}{([A]_0 - x)([B]_0 - x)} = \int_0^t k\,dt \tag{1.25}$$

From Equation 1.25, we have

$$\ln \frac{[A]_0 - x}{[B]_0 - x} = k([A]_0 - [B]_0)t + \ln \frac{[A]_0}{[B]_0} \tag{1.26}$$

Since $[A] = [A]_0 - x$ and $[B] = [B]_0 - x$, Equation 1.26 becomes

$$\ln \frac{[A]}{[B]} = k([A]_0 - [B]_0)t + \ln \frac{[A]_0}{[B]_0} \tag{1.27}$$

Equation 1.27 represents the integrated rate law for a bimolecular reaction involving two different reactant molecules with different initial concentrations.

If one of the reactants (such as B) in Equation 1.5 (the bimolecular reaction: A + B → P) is in large excess (typically 10–20-folds, i.e., $[B]_0/[A]_0 = 10$–20), the change in molar concentration of reactant B in the course of the reaction can be neglected ($[B] \sim [B]_0$) [2]. The rate law (Eq. 1.18) becomes

$$-d[A]/dt = k[A][B] = k[B]_0[A]$$

Let $k' = k[B]_0$ (the observed rate constant). We have

$$-d[A]/dt = k'[A]$$

The reaction becomes pseudo first order. The integrated rate law is

$$\ln [A] = -k't + \ln [A]_0$$

1.4.2 Reactive Intermediates and the Steady-State Assumption

First, let us consider a reaction that consists of two consecutive irreversible *unimolecular processes* as shown in Reaction 1.28.

$$X \xrightarrow{k_1} Y \xrightarrow{k_2} Z \tag{1.28}$$

X is the reactant. Z is the product. Y is a reactive intermediate. k_1 and k_2 are rate constants for the two unimolecular processes. In order to determine the way in which the concentrations of the substances change over time, the rate equation for each of the substances is written down as follows (Eq. 1.29–1.31) [2]:

$$-\frac{d[X]}{dt} = k_1[X] \tag{1.29}$$

$$\frac{d[Y]}{dt} = k_1[X] - k_2[Y] \tag{1.30}$$

$$\frac{d[Z]}{dt} = k_2[Y] \tag{1.31}$$

Equation 1.30 shows the net rate of increase in the intermediate Y, which is equal to the rate of its formation ($k_1[X]$) minus the rate of its disappearance ($k_2[Y]$). Equation 1.31 shows the rate of formation of the product Z. Since Z is produced only from the k_2 step which is a unimolecular process, the rate equation for Z is first order in Y.

It is tedious to obtain the accurate solutions of the above simultaneous differential equations. Appropriate approximations may be employed to ease the situation [2].

In most of the stepwise organic reactions, the intermediates (such as radicals and carbocations) possess high energies and are unstable and highly reactive. Therefore, the formation of such an intermediate is usually relatively slow (with a high E_a), while the subsequent transformation the intermediate experiences is relatively fast (with a low E_a). Mathematically, k_1 is much smaller than k_2 ($k_1 \ll k_2$). As a result, *the concentration of the reactive intermediate (such as Y in Reaction 1.28) remains at a low level and it essentially does not change in the course of the overall reaction.* This is referred to as the **steady-state approximation**. The changes in concentrations of the reactant, intermediate, and product over time for Reaction 1.28 are illustrated in Figure 1.2. It shows that the intermediate Y in Reaction 1.28 remains in a steady-state (an essentially constant low concentration) in the course of the overall reaction, formulated as

$$\frac{d[Y]}{dt} = 0 \tag{1.32}$$

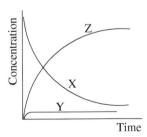

Time

FIGURE 1.2   The changes in concentrations of the reactant (X), intermediate (Y), and product (Z) over time for Reaction 1.28. The intermediate Y is shown to remain in a steady-state ($d[Y]/dt = 0$) in the course of the overall reaction.

In general, the steady-state approximation is applicable to all types of reaction intermediates in organic chemistry. Equation 1.32 is the mathematical form of the steady-state assumption.

With the help of the steady-state approximation, the dependence of concentrations of all the substances in Reaction 1.28 on time can be obtained readily [2].

Integration of Equation 1.29 leads to Equation 1.33 (c.f. Eqs 1.9–1.12).

$$[X] = [X]_0 e^{-k_1 t} \qquad (1.33)$$

$[X]_0$ is the initial concentration of X.

From Equation 1.30 (rate equation for Y) and Equation 1.32 (steady-state assumption for Y), we have

$$[Y] = \frac{k_1}{k_2}[X] \qquad (1.34)$$

Substituting Equation 1.33 for Equation 1.34, we have

$$[Y] = \frac{k_1}{k_2}[X]_0 e^{-k_1 t} \qquad (1.35)$$

On the basis of the stoichiometry for Reaction 1.28, the initial concentration of the reactant X can be formulated as

$$[X]_0 = [X] + [Y] + [Z]$$

Therefore,

$$[Z] = [X]_0 - [X] - [Y] \qquad (1.36)$$

Combination of Equations 1.33, 1.35, and 1.36 gives Equation 1.37.

$$[Z] = [X]_0 \left\{ 1 - \left(1 + \frac{k_1}{k_2} \right) e^{-k_1 t} \right\} \qquad (1.37)$$

Equations 1.33, 1.35, and 1.37 show the dependence of concentrations of all the substances in Reaction 1.28 on time [2].

Substituting Equation 1.34 (derived from the steady-state approximation for Y) for Equation 1.31 gives Equation 1.38.

$$\frac{d[Z]}{dt} = k_1[X] \qquad (1.38)$$

Comparing Equations 1.29 and 1.38 indicates that rate for consumption of the reactant X is approximately equal to rate for the formation of the product Z (Eq. 1.39).

$$-\frac{d[X]}{dt} = \frac{d[Z]}{dt} \tag{1.39}$$

1.4.3   Rate-Laws for Stepwise Reactions

Let us use the following consecutive reaction (Reaction 1.40) that involves both reversible and irreversible elementary processes to demonstrate the general procedure for obtaining rate laws for stepwise reactions [3]:

$$X \underset{k_{-1}}{\overset{k_1}{\rightleftharpoons}} Y \xrightarrow{k_2} Z \tag{1.40}$$

The rate (r) for the overall reaction can be expressed as an increase in concentration of the product (Z) per unit time (Eq. 1.41):

$$r = \frac{d[Z]}{dt} = k_2[Y] \tag{1.41}$$

Since Z is produced only from the k_2 step which is a unimolecular process, the rate equation for Z is first order in Y.

The steady-state assumption is applied to the intermediate Y, and its rate equation is written as follows:

$$\frac{d[Y]}{dt} = 0 = k_1[X] - k_{-1}[Y] - k_2[Y] \tag{1.42}$$

From Equation 1.42, we have

$$[Y] = \frac{k_1}{k_{-1} + k_2}[X] \tag{1.43}$$

Substituting Equation 1.43 for Equation 1.41 leads to Equation 1.44, the rate law for Reaction 1.40.

$$r = \frac{k_1 k_2}{k_{-1} + k_2}[X] = k_{obs}[X] \tag{1.44}$$

where $k_{obs} = k_1 k_2/(k_{-1} + k_2)$ is the observed rate constant.

There are several limiting situations for such a stepwise process [3]. If $k_2 \gg k_{-1}$ (the intermediate Y is converted to the product Z much faster than going back to the reactant X), Equation 1.44 can be simplified to

$$r = k_1[X] \qquad (\text{for } k_2 \gg k_{-1}).$$

In this case, the first step of Reaction 1.40 is the rate-determining step and actually irreversible.

If $k_2 \ll k_{-1}$ (the intermediate Y is converted to the product Z much more slowly than going back to the reactant X), there is a fast preequilibrium between the reactant X and the intermediate Y before the product Z is formed. In this case, the Equation 1.44 can be simplified to

$$r = (k_1/k_{-1})k_2[X] = K_{eq}k_2[X] \qquad (\text{for } k_2 \ll k_{-1}),$$

where $K_{eq} = (k_1/k_{-1})$ is the equilibrium constant for the fast preequilibrium between X and Y ($K_{eq} = [Y]/[X]$). Therefore, $r = k_2[Y]$, and the second k_2 step is the rate-determining step. Since the fast preequilibrium between X and Y is established prior to the formation of the product, the steady-state assumption is not necessary if $k_2 \ll k_{-1}$.

If the values of k_2 and k_{-1} are comparable, the full steady-state assumption is needed to establish the rate equation as shown in Equation 1.44.

1.5 THERMODYNAMICS

1.5.1 Enthalpy, Entropy, and Free Energy

Enthalpy (H), entropy (S), and free energy (G) are all thermodynamic state functions. Enthalpy (H) is defined as the sum of internal energy (U) and the product of pressure (P) and volume (V), formulated as

$$H = U + PV \tag{1.45}$$

From Equation 1.45, the change in enthalpy (ΔH) can be calculated as

$$\Delta H = \Delta U + \Delta(PV) \tag{1.46}$$

At constant pressure (P), Equation 1.46 becomes

$$\Delta H = \Delta U + P\Delta V = \Delta U - w$$

According to the first law of thermodynamics, $q_P = \Delta U - w$ (heat).

Therefore,

$$\Delta H = q_P \tag{1.47}$$

The physical meaning of Equation 1.47 is that *the enthalpy change in a process (including a chemical reaction) at constant pressure is equal to the heat evolved.* Since most of the organic reactions are conducted at constant pressure, the reaction heat can be calculated on the basis of the enthalpy change for the reaction.

Entropy (S) is considered as the **degree of disorder**. In thermodynamics, the infinitesimal change in entropy (dS) is defined as the reversible heat (dq_{rev}) divided by the absolute temperature (T), formulated as

$$dS = dq_{rev}/T$$

For a finite change in state,

$$\Delta S = \int \frac{dq_{rev}}{T} \tag{1.48}$$

Free energy (G) is defined as

$$G = H - TS$$

At constant temperature and pressure, the change in free energy (ΔG) can be calculated as

$$\Delta G = \Delta H - T\Delta S \tag{1.49}$$

1.5.2   Reversible and Irreversible Reactions

In general, chemical reactions in thermodynamics can be classified as two types, reversible and irreversible reactions. An **irreversible reaction** is such a reaction that proceeds only in *one direction*. As a result, **the reactant is converted to the product completely (100%) in the end of the reaction.** In contrast, a **reversible reaction** is such a reaction that can proceed to both forward and backward directions. In other words, there is an interconversion between the reactants and the products in a reversible reaction. As a result, **all the reactants and the products coexist in the end of the reaction, and the conversion is incomplete.**

The reversibility of a chemical reaction can be judged by the second law of thermodynamics. Originally, the second law is stated based on the entropy criterion as follows: *A process (including a chemical reaction) is reversible if the universal entropy change (ΔS_{UNIV}) associated to the process is zero; and a process is* *irreversible if the universal entropy change (ΔS_{UNIV}) associated to the process is positive (greater than zero). $\Delta S_{UNIV} = \Delta S + \Delta S_{SURR}$, the sum of the entropy change in the system (ΔS) and the entropy change in surroundings (ΔS_{SURR}).*

Since it is difficult to calculate the entropy change in surroundings (ΔS_{SURR}), very often the free energy criterion is used to judge reversibility for any processes that take place at constant temperature and pressure. By employing the free energy change (ΔG) in a system, the second law can be modified as: *At constant temperature and pressure, a process (including a chemical reaction) is irreversible (spontaneous) if the free energy change (ΔG) of the process is negative ($\Delta G < 0$), a process is reversible (at equilibrium) if the free energy change (ΔG) of the process is zero ($\Delta G = 0$), and a process is nonspontaneous if the free energy change (ΔG) of the process is positive ($\Delta G > 0$).* The free energy criterion is widely used in organic chemistry because most of the organic reactions are conducted in open systems at constant temperature and pressure.

According to Equation 1.49, both enthalpy (ΔH) and entropy (ΔS) effects need to be considered when judging reversibility of a reaction using the free energy criterion. The spontaneity of a reaction is favored by a negative enthalpy ($\Delta H < 0$, exothermic) or a positive entropy ($\Delta S > 0$, increase in disorder), while a positive enthalpy ($\Delta H > 0$, endothermic) or a negative entropy ($\Delta S < 0$, decrease in disorder) works against a reaction. Variations in temperature (T) can change the extent of the entropy effect ($T\Delta S$), and therefore they affect the reversibility accordingly. High temperatures favor reactions with a positive entropy change ($\Delta S > 0$), and low temperatures favor reactions with a negative entropy change ($\Delta S < 0$). The effects of enthalpy and entropy on reversibility of the chemical reactions conducted at constant temperature and pressure are summarized in Figure 1.3.

1.5.3   Chemical Equilibrium

Many organic reactions are reversible, namely that the conversion of the reactant to the product is incomplete. When the rate of the forward process is equal to the rate of

$$\Delta G \; = \; \Delta H \; - \; T\Delta S$$

(−)	(+)	($\Delta G < 0$, always spontaneous)
(−)	(−)	(spontaneity depends on temperature)
(−)	(−)	(spontaneity depends on temperature)
(+)	(−)	($\Delta G > 0$, always nonspontaneous)

FIGURE 1.3   The effects of enthalpy and entropy on reversibility of the chemical reactions conducted at constant temperature and pressure.

the backward process for a reversible reaction, the concentrations of all the reactants and products cease to change, and the reaction has reached a **dynamic equilibrium**.

While the rate constant of a reaction serves as the quantitative measure of how fast the reaction proceeds (Section 1.4), **the equilibrium constant (K)** is used as a quantitative measure for the extent of a reversible reaction, which is defined as follows:

$$aA + bB = cC + dD$$
$$K = [C]^c[D]^d/[A]^a[B]^b \tag{1.50}$$

Equation 1.50 represents a balanced chemical equation for a reversible reaction (concerted or stepwise). A, B, C, and D represent chemical formulas of different substances (reactants or products). a, b, c, and d represent the corresponding stoichiometric coefficients. [A], [B], [C], and [D] represent the molar concentrations of the species A, B, C, and D, respectively. At a certain given temperature, the K value remains constant, and it is independent of the concentrations of any reactants or products.

The equilibrium constant expression indicates that *having one of the reactants (such as B) in excess can increase the percentage of conversion of the other reactant (such as A) to the products. On the other hand, removal of one product (decrease in its concentration) from the reaction system can also increase the percentage of the conversion of the reactants to the products.* In the case that one reactant is in very large excess, the conversion of the other reactant (limiting reagent) can be essentially complete (~100%). Therefore, a reversible reaction has been essentially converted to an irreversible reaction.

In organic chemistry, the strategy of using a certain reactant in excess is employed for many reversible reactions to increase the product yields. For example, most of the acid–catalyzed esterification reactions (Reaction 1.51) have the equilibrium constants $K \sim 5$–10.

$$RCO_2H + H'OH = RCO_2R' + HOH \quad K \sim 5 - 10 \tag{1.51}$$

When the carboxylic acid (RCO_2H) and the alcohol ($R'OH$) are used in 1:1 molar ratio, the conversion of the reactants to the products is 70–75%. If $R'OH$ is used in 10-folds of excess, the conversion of RCO_2H (limiting reactant) to the ester product will be ~99%. In this case, the reversible reaction has been almost transformed into an irreversible reaction. For some esterification reactions, the essential quantitative conversion of the reactants to the ester product can also be obtained by removal of water from the reaction system once it is formed.

The relationship between the equilibrium constant (K) and standard free energy ($\Delta G°$) is formulated as

$$\Delta G° = -RT \ln K \tag{1.52}$$

Substituting Equation 1.49 for Equation 1.52 leads to

$$\Delta H° - T\Delta S° = -RT \ln K$$

Therefore,

$$\ln K = -(\Delta H^\circ/RT) + \Delta S^\circ/R \qquad (1.53)$$

Equation 1.53 describes the dependence of the equilibrium constant on temperature. Very often, for an exothermic reaction, the standard enthalpy $\Delta H^\circ < 0$ ($-\Delta H^\circ/R$ is positive). The equilibrium constant (K) decreases as a function of the temperature (T). On the other hand, an endothermic reaction often has $\Delta H^\circ > 0$ ($-\Delta H^\circ/R$ is negative). The equilibrium constant (K) increases as a function of the temperature (T). Therefore, in many cases (exceptions exist) high temperatures facilitate endothermic reactions and low temperatures facilitate exothermic reactions.

Equation 1.53 also shows that plot of $\ln K$ versus $1/T$ defines a straight line, from which both the standard enthalpy (ΔH°) and standard entropy (ΔS°) can be obtained from the slope and intercept, respectively, for an unknown reaction.

$$\text{Slope} = -\Delta H^\circ/R \quad \Delta H^\circ = -\text{Slope} \times R$$

$$\text{Intercept} = \Delta S^\circ/R \quad \Delta S^\circ = -\text{Intercept} \times R$$

1.6 THE TRANSITION STATE

1.6.1 The Transition State and Activation Energy

The transition state is the structural bridge that links reactants and products for any *concerted chemical reactions*. In order for the reactant molecules to collide effectively giving the products, they must overcome an energy barrier, which is called **activation energy (E_a)** (Fig. 1.4). This is true for both exergonic ($\Delta G < 0$, Fig. 1.4a) and endergonic ($\Delta G > 0$, Fig. 1.4b) reactions. The activation energy (E_a) is a **free energy term**, which includes contributions from both enthalpy and entropy. The state in which the reaction system reaches a maximum energy level in the energy profile (Fig. 1.4a or b) is called **transition state**. It is also referred to as **activated complex** in which reorganization of the atoms in the reactant molecules are taking

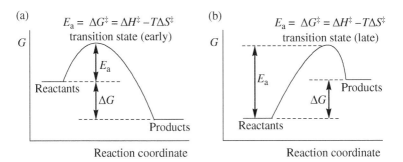

FIGURE 1.4 Early transition state (a) and late transition state (b).

place such that some old bonds are being partially broken, coincident with the partial formation of some new bonds. The transition state (activated complex) is highly energetic and therefore, it is in general very unstable and short-lived [with the half-life being in the order of picosecond (10^{-12} s) for many reactions]. Once formed, it rapidly collapses (dissociates) spontaneously. As a result, the old bonds in the reactants are fully broken, and simultaneously, the new bonds are completely formed, giving the final stable products in the end of the reaction.

In a concerted reaction (Fig. 1.4), the activation energy (E_a) is equal to the free energy of the transition state ($E_a = \Delta G^{\ddagger} = \Delta H^{\ddagger} - T\Delta S^{\ddagger}$). $\Delta G^{\ddagger}$, $\Delta H^{\ddagger}$, and $\Delta S^{\ddagger}$ are the difference in free energy, enthalpy, and entropy values, respectively, between the transition state and the reactants ($\Delta G^{\ddagger} = G_{TS} - G_{reactants}$, etc.), and they are referred to as **free energy, enthalpy, and entropy of the transition state**. In the case that the entropy ($\Delta S^{\ddagger}$) effect is small and can be neglected, the activation energy is then approximately equal to the activation enthalpy ($E_a = \Delta H^{\ddagger}$). In the individual chapters, we will see that this approximation is very often valid and works well for many reactions. The energy level (in the free energy term) that the transition state possesses is the quantitative measure for the stability of the transition state. A fast reaction requires a relatively stable transition state (with a relatively low energy level). A reaction is getting slower as the stability of the transition state decreases (with an increase in its energy level). The quantitative relationship between the rate constant (k) of a reaction (concerted) and the activation energy (E_a) is described by the well-known Arrehnius equation (Eq. 1.54):

$$k = \mathrm{Ae}^{-Ea/RT} \left(E_a = \Delta G^{\ddagger} = \Delta H^{\ddagger} - T\Delta S^{\ddagger} \right) \tag{1.54}$$

1.6.2   The Hammond Postulate

The exact structure of a transition state is in general not measurable due to its instability. It is in-between the structures of reactants and products, which can be qualitatively predicted and described by the **Hammond postulate** [3]. It states that *the structure of the transition state for a concerted reaction resembles (closer to) the species (reactant or product) to which it is most similar in energy*. According to the Hammond postulate, if a concerted reaction is exergonic (or exothermic if entropy of the reaction ΔS is small and negligible) (Fig. 1.4a), the energy of the transition state is most similar to that of the reactant. Therefore, the structure of the transition state resembles the reactant. Such a transition state is called **early transition state**. If a concerted reaction is endergonic (or endothermic if entropy of the reaction ΔS is small and negligible) (Fig. 1.4b), the energy of the transition state is most similar to that of the product. Therefore, the structure of the transition state resembles the product. Such a transition state is called **late transition state**.

Figure 1.5 shows examples of exothermic ($\Delta H < 0$) and endothermic ($\Delta H > 0$) S_N2 reactions (concerted). Usually, the entropies of simple S_N2 reactions are small relative to the reaction enthalpies, and they are negligible. According to the Hammond

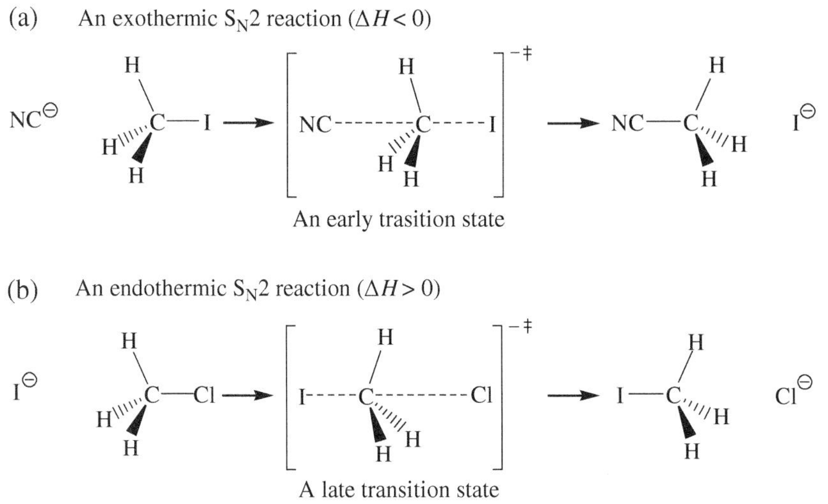

(a)   An exothermic S_N2 reaction ($\Delta H < 0$)

An early trasition state

(b)   An endothermic S_N2 reaction ($\Delta H > 0$)

A late transition state

FIGURE 1.5   The S_N2 reactions that proceed via (a) an early transition state and (b) a late transition state.

postulate, the exothermic S_N2 reaction has an early transition state whose structure resembles the reactants (Fig. 1.5a), while the endothermic S_N2 reaction has a late transition state whose structure resembles the products (Fig. 1.5b). Although a simple S_N2 reaction has a small entropy, the entropy of its transition state ($\Delta S^{\ddagger}$) possesses a relatively large negative value. This is because in the transition state two reactant molecules combine loosely resulting in decrease in entropy. For the exothermic S_N2 reaction (Fig. 1.5a), since the reaction has an early transition state which resembles the reactants, both breaking of the old C—I bond and formation of the new C—C bond in the transition state have only proceeded to a small extent. This gives rise to a small $\Delta H^{\ddagger}$ value. Therefore, the major contributor to the activation energy is the entropy effect ($E_a \approx -T\Delta S^{\ddagger}$). On the other hand, the endothermic S_N2 reaction (Fig. 1.5b) has a late transition state which resembles the products. Breaking of the old C—Cl bond and formation of the new C—I bond in the transition state are close to completion ($\Delta H^{\ddagger} \approx \Delta H$). As a result, both enthalpy ($\Delta H^{\ddagger}$) and entropy ($\Delta S^{\ddagger}$) have substantial contributions to the activation energy ($E_a = \Delta H^{\ddagger} - T\Delta S^{\ddagger}$). Their combination makes the E_a value greater than the reaction enthalpy (ΔH).

1.6.3   The Bell–Evans–Polanyi Principle

For similar concerted reactions that take place at a certain given temperature, the activation energy (E_a) can be directly correlated to the reaction enthalpy (ΔH) as follows (the **Bell–Evans–Polanyi principle**):

$$E_a = c_1 \Delta H + c_2 \tag{1.55}$$

where c_1 and c_2 are positive constants.

Equation 1.55 is applicable to both endothermic and exothermic reactions and can be employed to analyze the relative activation energies for similar reactions based on their reaction enthalpies. For endothermic reactions ($\Delta H > 0$, Fig. 1.4b), the greater the energy gap (ΔH) between the reactant and product (the more endothermic), the higher is the activation energy E_a. For exothermic reactions ($\Delta H < 0$, Fig. 1.4a), the greater the energy gap (the absolute value of ΔH) between the reactant and product (the more exothermic), the smaller is the activation energy E_a. The greater energy gap in an exothermic reaction makes the $c_1\Delta H$ more negative. As a result, the E_a becomes smaller. Using this principle, one can directly relate the kinetics (activation energy) to the thermodynamics (enthalpy of the reaction) for similar chemical reactions.

1.7   ELECTRONIC EFFECTS AND HAMMETT EQUATION

1.7.1   Electronic Effects of Substituents

Substituents on aromatic (phenyl) rings can increase or decrease the electron density in the ring by donating or withdrawing electrons. This alters rates and/or equilibrium constants for the reactions occurring in the ring and in the side group attached to the ring. Such electronic effects of substituents can be quantitatively defined on the basis of their influence on the acidity (dissociation constant) of *para*- or *meta*-substituted benzoic acid (XC_6H_4COOH) relative to benzoic acid (C_6H_5COOH) (Eq. 1.56):

$$\lg\left(\frac{K_A}{K_H}\right) = \sigma \quad \text{(Hammett substituent constant)}$$

(1.56)

In Equation 1.56, K_A and K_H are the acid dissociation constants of XC_6H_4COOH and C_6H_5COOH, respectively. lg is the common logarithm (10-based logarithm). The σ value defined in the equation is called **Hammett substituent constant** for a given substituent $-X$ at *para*- or *meta*-position. The *para*- or *meta*-substituted benzoic acids (XC_6H_4COOH) with different $-X$ groups are in general commercially available or easy to synthesize. The pK_A ($-\lg K_A$) value for each XC_6H_4COOH is numerically equal to pH of a solution containing equalmolar concentrations of the acid and the sodium salt of the conjugate base and can be readily determined experimentally. Therefore, the σ constants for various substituents can be obtained readily [1].

The electron-donating and electron-withdrawing effects of various substituents originate from combination of inductive and conjugation effects. In general, an electron-withdrawing group (EWG) on the phenyl ring of $XC_6H_4COO^-$, such as $-NO_2$ (nitro), $-C(O)R$ (acyl), $-SO_3H$ (sulfonic acid), and $-CN$ (cyanide), lowers the electron density in the ring facilitating delocalization of the negative charge of the carboxylate to the ring. This stabilizes the $XC_6H_4COO^-$ anion, resulting in an increase in K_A relative to K_H. Therefore, **the σ constants for EWGs are usually positive (σ > 0 for EWGs)**. On the other hand, an electron-donating group (EDG) on the phenyl ring of $XC_6H_4COO^-$, such as –R (alkyl), –OH (hydroxyl), –OR (alkoxy), and –NH$_2$ (amino), enhances the electron density in the ring to disfavor the delocalization of the negative charge of the carboxylate. This destabilizes the $XC_6H_4COO^-$ anion, giving rise to decrease in K_A relative to K_H. Therefore, **the σ constants for EDGs are usually negative (σ < 0 for EDGs)**. The substituents on the *ortho*-position of the phenyl rings often produce steric hindrance on the side group next to them. Therefore, the electronic effects of the *ortho*-substituents are usually not considered.

For a given substituent (EWG or EDG), the extent of its electronic effects on the side group (–COO$^-$) of $XC_6H_4COO^-$ is different when the substituent is placed on the *para*-position and on the *meta*-position. Therefore, the σ constants for the substituent on *para*-position (σ_{para}) and on *meta*-position (σ_{meta}) are different. For example, the electron withdrawing effects of –NO$_2$ on both its *para*- and *meta*-carbons are mainly due to its conjugation effect. The conjugation effect of a *para*–NO$_2$ to the side group –COO$^-$ is stronger than a *meta*–NO$_2$. Therefore, the σ_{para} (0.81) is greater than the σ_{meta} (0.71) for –NO$_2$ (σ > 0 for EWGs) [1]. This is also true for some other EWGs (–CN, –CF$_3$, and –CO$_2$Me) whose major electronic effects on both their *para*- and *meta*-carbons are conjugation effects, and we have $\sigma_{para} = 0.70$ and $\sigma_{meta} = 0.62$ for –CN; $\sigma_{para} = 0.53$ and $\sigma_{meta} = 0.46$ for –CF$_3$; and $\sigma_{para} = 0.44$ and $\sigma_{meta} = 0.35$ for –CO$_2$Me [1]. For EDGs whose major electronic effects are conjugation effects, the absolute value of σ_{para} is greater than the absolute vale of σ_{meta} (σ < 0 for EDGs). For example, we have $\sigma_{para} = -0.14$ and $\sigma_{meta} = -0.06$ for –CH$_3$; and $\sigma_{para} = -0.32$ and $\sigma_{meta} = -0.10$ for –N(CH$_3$)$_2$ [1].

1.7.2 Hammett Equation

For various aromatic compounds (substituted benzenes), the substituents have substantial influence on rates and equilibrium for their reactions occurring on the phenyl ring as well as on the side group attached to the ring (Eq. 1.57):

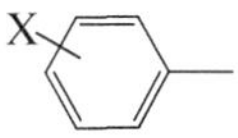

X— (side group) ⟶ A certain reaction

$$\lg\left(\frac{k_A}{k_H}\right) = \rho\sigma \quad \text{(Hammett equation)}$$

(1.57)

In Equation 1.57, the chemical structure shows a general substituted benzene with –X being an EWG or EDG at the *para*-position or *meta*-position to the side group. A reaction may occur on the side group (typically on the first atom of the group connecting to the ring) with the aromatic ring intact. A reaction may also occur on the ring with the side group intact. k_A and k_H are the rate constants for a certain reaction of the compound with and without the substituent, respectively. lg is the common logarithm (10-based logarithm). The σ value is the Hammett substituent constant of $-X$ as defined in Equation 1.56. It has been found that for various reactions occurring on the phenyl ring and on the side group, as the substituent $(-X)$ is altered (σ changes accordingly), the plot of $\lg(k_A/k_H)$ versus the substituent constant σ gives a straight line, showing that the $\lg(k_A/k_H)$ value is directly proportional to σ of the substituent and is independent of its structure. The slope (ρ) for the $\lg(k_A/k_H)$ versus σ plot is characteristic of different reactions. The quantitative relationship between $\lg(k_A/k_H)$ and the substituent constant σ in Equation 1.57 is referred to as the **Hammett equation**. For a given reaction that occurs on the phenyl ring or on the side group, ρ (rho) in the Hammett equation is a constant.

For the reactions which develop a positive charge (or destroy a negative charge) in the transition state of the rate-determining step, ρ in the Hammett equation is a negative constant ($\rho < 0$). An EWG ($\sigma > 0$) will destabilize the transition state carrying a positive charge, making the $k_A/k_H < 1$ [$\lg(k_A/k_H) < 0$]. An EDG ($\sigma < 0$) will stabilize the transition state carrying a positive charge, making the $k_A/k_H > 1$ [$\lg(k_A/k_H) > 0$]. Both EWG and EDG require a negative ρ constant to maintain the Hammett equation. **For the reactions which develop a negative charge (or destroy a positive charge) in the transition state of the rate-determining step, ρ in the Hammett equation is a positive constant ($\rho > 0$).** An EWG ($\sigma > 0$) will stabilize the transition state carrying a negative charge, making the $k_A/k_H > 1$ [$\lg(k_A/k_H) > 0$]. An EDG ($\sigma < 0$) will destabilize the transition state carrying a negative charge, making the $k_A/k_H < 1$ [$\lg(k_A/k_H) < 0$]. Both EWG and EDG require a positive ρ constant to maintain the Hammett equation.

When a reaction occurs on the aromatic ring to develop a positive or a negative charge in the transition state, the influence of a substituent (EWG or EDG) on the reaction rate is usually stronger than that on a reaction taking place in the side group. Therefore, the absolute value of the ρ constant for a reaction occurring on the aromatic ring is often greater than the absolute value of the ρ constant for a reaction occurring on the side group (exceptions exist). All this is illustrated by the reactions in Figure 1.6, with all the ρ constants taken from Ref. [1].

Figure 1.6a is an electrophilic aromatic substitution reaction (occurring on the aromatic ring) via an arenium cation (a positive charge is developed in the transition state). The substituent can strongly interact with the positive charge in the transition state formed in the ring, giving rise to a big negative ρ constant ($\rho = -12.1$) for the reaction. Figure 1.6b is a nucleophilic aromatic substitution reaction (occurring on the aromatic ring) via an intermediate Meisenheimer anion (a negative charge is developed in the transition state). The substituent can strongly interact with the negative charge in the transition state, giving rise to a positive ρ constant ($\rho = 3.9$) for the reaction. On the other hand, Figure 1.6c reaction takes place at the first atom of a side chain attaching to the aromatic ring and the ring is intact in the reaction. A carbocation is formed in the course of the reaction, thus a positive charge is developed in the transition state. The ρ constant for this reaction is negative ($\rho = -4.54$). Its absolute value is much smaller than that of the ρ constant for Reaction (a) occurring in the ring. As the substituent is distanced from the reactive center, its electronic effect on the reaction is

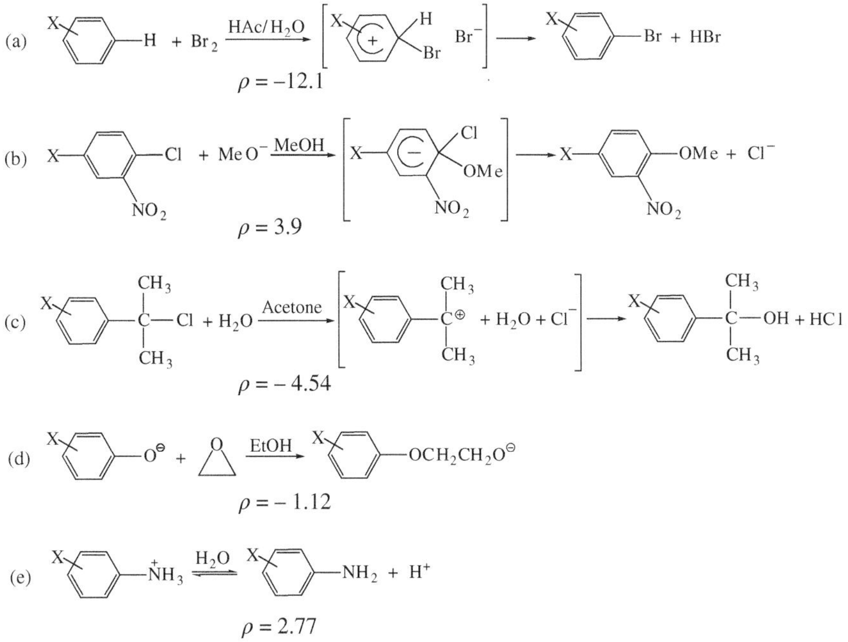

(a)

$\rho = -12.1$

(b)

$\rho = 3.9$

(c)

$\rho = -4.54$

(d)

$\rho = -1.12$

(e)

$\rho = 2.77$

FIGURE 1.6 The ρ constants for various reactions of substituted benzenes. (a) Electrophilic aromatic substitution, (b) Nucleophilic aromatic substitution, (c) Hydrolysis, (d) S_N2 reaction, and (e) Dissociation.

getting smaller. Figure 1.6d reaction occurs in the side group, with a negative charge in the phenoxide oxygen destroyed (equivalent to development of a positive charge). Therefore, the reaction has a small negative ρ constant ($\rho = -1.12$). Its absolute value is substantially smaller than those of the ρ constants for Reactions (a) and (b), indicating weaker electronic effects of substituents on the side groups relative to the effects on the ring. Figure 1.6e reaction also occurs on the side group, with a positive charge in the nitrogen atom destroyed (equivalent to development of a negative charge) in the course of the reaction. For this reaction, the Hammett equation can be established using the acid dissociation constants of $XC_6H_4NH_3^+$, formulated as $\lg(K_A/K_H) = \rho\sigma$, where K_A and K_H are dissociation constants of $XC_6H_4NH_3^+$ and $C_6H_5NH_3^+$, respectively. Since the reaction occurs on the side group and destroys a positive charge, it has a small positive ρ constant ($\rho = 2.77$).

1.8   THE MOLECULAR ORBITAL THEORY

1.8.1   Formation of Molecular Orbitals from Atomic Orbitals

The microscopic particles such as electrons possess dual properties, which are particle-like behavior and wave-like behavior. The latter can be quantitatively characterized by the **wavefunction (ψ)**, which is a function of the space coordinates (x, y, z

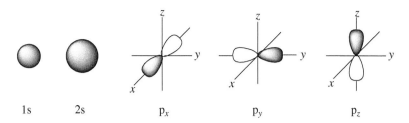

FIGURE 1.7   The shapes of the s and p orbitals in the three-dimensional space.

in three dimensions) of an electron. The one-electron wavefunction in an atom is called **atomic orbital (AO)**. The square of a wavefunction (ψ^2) is the probability of finding an electron (also called electron density). The atomic orbitals in the valence shells of the atoms of main group elements include **s** and **p** orbitals. Their shapes in the three-dimensional space are illustrated in Figure 1.7.

Studying the behavior of fundamental particles in chemistry must eventually go beyond the classical laws. It requires that chemical bonding in molecules be explained as a superposition phenomenon of electron wavefunctions. When two atoms (such as hydrogen atoms) approach each other, their valence electrons will start interacting. This makes the wavefunctions (atomic orbitals) of the interacting atoms superimpose (overlap). Mathematically, such a superposition phenomenon (also called **orbital overlap**) can be expressed in terms of the **linear combinations of atomic orbitals (LCAOs)** leading to a set of new wavefunctions in a molecule, called **molecular orbitals (MOs)** which are shown in Equations 1.58 and 1.59 [2].

$$\Phi_1 = c_{11}\psi_1 + c_{12}\psi_2 \tag{1.58}$$

$$\Phi_2 = c_{21}\psi_1 + c_{22}\psi_2 \tag{1.59}$$

ψ_1 and ψ_2 represent atomic orbitals of the two approaching atoms 1 and 2, respectively. c_{11}, c_{12}, c_{21}, and c_{22} are constants (positive, zero, or negative). Φ_1 and Φ_2 are the resulting molecular orbitals from linear combinations of ψ_1 and ψ_2. By the nature, the molecular orbitals are one-electron wavefunctions. However, they can approximately characterize the behavior of electrons in a many-electron molecule. In principle, *the number of molecular orbitals formed is equal to the number of participating atomic orbitals which overlap in a molecule*. In other words, the participating atomic orbitals can combine linearly in different ways. The number of LCAOs is equal to the number of the atomic orbitals.

Figure 1.8 illustrates how an H_2 molecule is formed from two H atoms. When two H atoms approach to one another, their 1s orbitals ($1s_A$ and $1s_B$) overlap giving two molecular orbitals σ_{1s} and σ_{1s}^* through the following linear combinations of $1s_A$ and $1s_B$ [2].

$$\sigma_{1s} = c_1 1s_A + c_2 1s_B$$

$$\sigma_{1s}^* = c_1' 1s_A - c_2' 1s_B$$

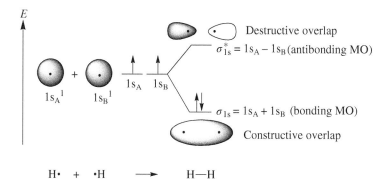

FIGURE 1.8   Formation of the hydrogen molecule (H_2) from two hydrogen (H) atoms.

Since $1s_A$ and $1s_B$ are identical, their contributions to each of the MOs (σ_{1s} and σ_{1s}^*) should be equal. Therefore, we have $c_1 = c_2 = c$ (>0) and $c_1' = c_2' = c'$ (>0).

In order to normalize the molecular orbital σ_{1s}, the following integral must have the value unity

$$\int (\sigma_{1s})^2 d\tau = 1,$$

where $d\tau$ is the volume factor.

Therefore,

$$\int (c1s_A + c1s_B)^2 d\tau = 1$$

$$c^2 \left[\int (1s_A)^2 d\tau + 2 \int (1s_A 1s_B) d\tau + \int (1s_B)^2 d\tau \right] = 1$$

Since the wavefunction of the 1s orbital is normalized, we have

$$\int (1s_A)^2 d\tau = \int (1s_B)^2 d\tau] = 1$$

The term $S = \int (1s_A 1s_B) d\tau$ is referred to as the **overlap integral**. Therefore, we have

$$c^2 [1 + 2S + 1] = 1$$

$$c = 1/[2(1 + S)]^{1/2}$$

Similarly, by normalizing σ_{1s}^*, we can obtain

$$c' = 1/[2(1 - S)]^{1/2}$$

Therefore, we have

$$\sigma_{1s} = 1/[2(1 + S)]^{1/2} (1s_A + 1s_B) \qquad (1.60)$$

$$\sigma_{1s}{}^* = 1/[2(1 - S)]^{1/2} (1s_A - 1s_B) \qquad (1.61)$$

The overlap integral S is determined by the internuclear distance. At equilibrium H—H bond distance, the electron density of σ_{1s} in the midregion of the bond is maximum, while the electron density of $\sigma_{1s}{}^*$ in the midregion of the bond is zero. Therefore, σ_{1s} is called **bonding molecular orbital**. It is formed by constructive interaction (overlap) of two atomic orbitals and is responsible for the formation of the H—H σ bond. $\sigma_{1s}{}^*$ is called **antibonding molecular orbital**. It is formed by destructive interaction (overlap) of two atomic orbitals and is responsible for dissociation of the H—H bond. Since each of the 1s orbitals makes the same contribution to the bonding σ_{1s} and antibonding $\sigma_{1s}{}^*$ MOs, the coefficients $1/[2(1 + S)]^{1/2}$ and $1/[2(1 - S)]^{1/2}$ are often omitted when writing the LCAOs. Therefore, the bonding and antibonding MOs in H_2 can be simply written as $\sigma_{1s} = 1s_A + 1s_B$ and $\sigma_{1s}{}^* = 1s_A - 1s_B$.

The diatomic halogen X_2 (X = F, Cl, Br, or I) molecules are among fundamental main group molecules. Bonding in these molecules, usually represented by F_2, is described in Figure 1.9 using MO theory. As two F atoms come together, the two single electrons (in p_z orbitals) interact resulting in constructive and destructive orbital overlaps (LCAOs) in the line connecting the two nuclei, giving rise to the formation of the bonding MO ($\sigma_{2p} = 2p_{z,A} + 2p_{z,B}$) and antibonding MO ($\sigma_{2p}{}^* = 2p_{z,A} - 2p_{z,B}$), respectively.

In the category of nonmetallic main group elements, a π bond is formed by overlap of two p orbitals (LCAOs) in sideways (Fig. 1.10). If both p orbitals are identical (such as p orbitals in a C=C π bond), each of the p orbitals has the same contribution to the bonding (π_p) and antibonding ($\pi_p{}^*$) MOs (Fig. 1.10a). They are formed by constructive and destructive sideway orbital overlaps, respectively: $\pi_p = p_1 + p_2$ (fused lobes due to a positive linear combination—constructive orbital overlap)

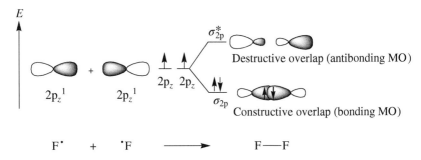

FIGURE 1.9 Formation of the fluorine molecule (F_2) from two fluorine (F) atoms.

(a)

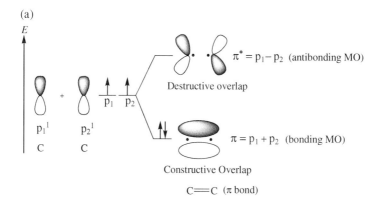

$\pi^* = p_1 - p_2$ (antibonding MO)

Destructive overlap

$\pi = p_1 + p_2$ (bonding MO)

Constructive Overlap

C$=$C (π bond)

(b)

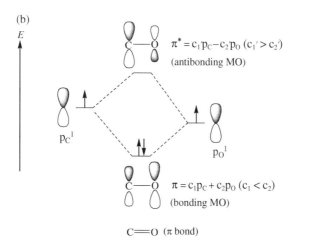

$\pi^* = c_1'p_C - c_2'p_O \; (c_1' > c_2')$

(antibonding MO)

$\pi = c_1 p_C + c_2 p_O \; (c_1 < c_2)$

(bonding MO)

C$=$O (π bond)

FIGURE 1.10 Formation of (a) the C=C π bond from two equivalent p orbitals and (b) the C=O π bond from two nonequivalent p orbitals.

and $\pi_p^* = p_1 - p_2$ (separated lobes due to a negative linear combination—destructive orbital overlap). If the two p orbitals are from atoms of different elements (such as p orbitals in a C=O π bond), the contribution of each p orbital to the bonding (π_p) and antibonding (π_p^*) MOs is different (Fig. 1.10b). Usually, the p orbital in the more electronegative atom has greater contribution to the bonding MO (π_p), and the p orbital in the less electronegative atom has greater contribution to the antibonding MO (π_p^*). In the C=O π bond, the bonding (π_p) and antibonding (π_p^*) MOs can be expressed as

$$\pi_p = c_1 p_C + c_2 p_O \; (0 < c_1 < c_2)$$

$$\pi_p^* = c_1' p_C - c_2' p_O \; (c_1' > c_2' > 0)$$

The above equations show that for the formation of π_p, the p orbital in oxygen (more electronegative) makes a greater contribution than does the p orbital in carbon (less electronegative). For the formation of $\pi_p{}^*$, the p orbital in carbon (less electronegative) makes a greater contribution than does the p orbital in oxygen (more electronegative). In each case, the bonding π_p MO is responsible for the formation of a π bond, and antibonding orbital $\pi_p{}^*$ is responsible for dissociation of the π bond.

When more than two p orbitals overlap sideways, it results in the formation of a **conjugate π bond**. Similar to the separate π bonds (formed by sideway overlap of two p orbitals), a conjugate π bond consists of series of MOs formed by linear combinations of the contributing p orbitals. The number of constituent MOs is equal to the number of contributing p orbitals. For example, the conjugate π-bond of allyl radical ($CH_2=CHCH_2{}^{\cdot}$), formed by sideway overlap of three p orbitals in the carbon atoms, consists of the following three MOs (Fig. 1.11a):

$$\psi_1 = p_1 + p_2 + p_3 \ (\text{bonding})$$

$$\psi_2 = p_1 - p_3 \ (\text{nonbonding})$$

$$\psi_3 = p_1 - p_2 + p_3 \ (\text{antibonding})$$

The conjugate π-bond of 1,3-butadiene ($CH_2=CHCH=CH_2$), formed by sideway overlap of four p orbitals in the carbon atoms, consists of the following four MOs (Fig. 1.11b):

$$\psi_1 = p_1 + p_2 + p_3 + p_4 \ (\text{bonding})$$

$$\psi_2 = p_1 + p_2 - p_3 - p_4 \ (\text{bonding})$$

$$\psi_3 = p_1 - p_2 - p_3 + p_4 \ (\text{antibonding})$$

$$\psi_4 = p_1 - p_2 + p_3 - p_4 \ (\text{antibonding})$$

In each of the molecules, since all the p orbitals are from carbon atoms, their contributions to each of the MOs are equal.

1.8.2   Molecular Orbital Diagrams

When two AOs combine linearly (overlap) to form MOs (Figs 1.8–1.10), the bonding MO formed by positive LCAO has lower energy than each of the starting AOs, while the antibonding MO formed by negative LCAO possesses higher energy than the AOs [4, 5]. As a result, the electrons from the starting AOs flow into the lower-energy-level bonding MO upon the formation of the molecule and the antibonding MO with a higher energy level remains empty. The overall energy decreases. The diagrams showing correlations of AOs and the resulting MOs and their relative energy levels are called **molecular orbital diagrams**. Figs 1.8–1.10 show the MO diagrams for the formation of H_2 from hydrogen atoms, the formation of F_2

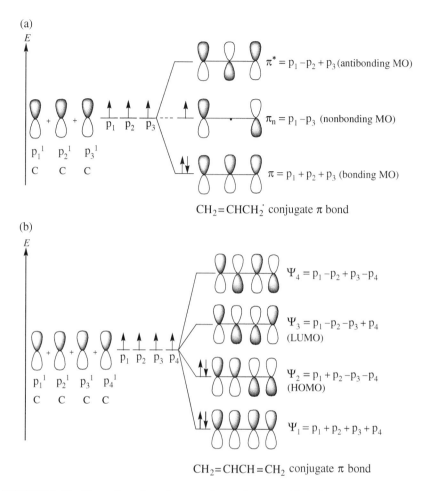

FIGURE 1.11   Formation of conjugate π bonds from p orbitals in (a) the allyl radical $(CH_2=CHCH_2\cdot)$ and (b) the 1,3-butadiene $(CH_2=CHCH=CH_2)$ molecule.

from fluorine atoms, and the formation of a π-bond from p orbitals. All the MO diagrams clearly indicate the energy gains (driving forces) for the formation of a molecule (or a bond) from individual atoms. Figure 1.11 shows MO diagrams for the formation of a conjugate π-bond from p orbitals. The MO diagrams also indicate that any MOs whose energies are lower than those of the starting AOs are bonding orbitals, responsible for the formation of the molecule. If an MO has the same energy as that of a starting AO, it is a nonbonding orbital which does not make any contribution to the formation of the molecule. Any MOs whose energies are higher than those of the starting AOs are antibonding orbitals, responsible for the dissociation of the molecule if they are populated with electrons [5].

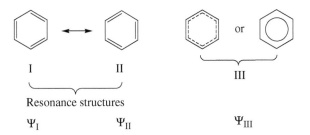

I   II

III

Resonance structures

Ψ_I Ψ_{II} Ψ_{III}

FIGURE 1.12   Resonance stabilization of benzene.

1.8.3   Resonance Stabilization

By the nature, **resonance stabilization** is a result of electron delocalization in a molecule, which leads to decrease in energy and stabilizes the molecule. It typically occurs in the p_π systems. First, we will use the benzene molecule as an example to demonstrate the nature of resonance stabilization.

Figure 1.12 shows two valid Lewis structures (I and II) for benzene [4]. Each of them has alternating double bond and single bond in the ring structure. Since six π electrons in the molecule are delocalized along the ring via the sideway overlap of six p orbitals forming a conjugate π bond and all the C—C bonds are of equal length, neither of the Lewis structures alone can truly describe the geometry and the real electronic configuration of the benzene molecule. Due to delocalization of the π electrons, the real electronic configuration can be considered resonating between the Lewis structures I and II, which are called **resonance structures**. In other words, the real structure and electronic configuration of benzene contain characters of both resonance structures and can be thought the average of the two. The bond order of each of the C—C bonds is 1.5, which is between a single bond (b.o. = 1) and a double bond (b.o. = 2). This way of characterization of benzene using two resonance structures I and II is consistent with the consequence of the π electrons delocalization as described by structure III. Therefore, the real wavefunction of the benzene molecule Ψ_{III} can be formulated as the linear combination of the wavefunctions of structures I and II (Ψ_I and Ψ_{II}, respectively) [4]:

$$\Psi_{III} = a\Psi_I + b\Psi_{II} \quad (a = b).$$

Since the structures I and II are equivalent, the contributions of both Lewis structures to the real structure of the benzene molecule should be equal. Therefore, we have a = b. Due to the electron delocalization, the real structure III has lower energy than that of structure I or II. Such stabilization by electron delocalization is called **resonance stabilization**.

Figure 1.13 shows that the real structure of the carbonyl (C=O) group can be characterized by two resonance structures A and B owing to delocalization of the π electrons in the C=O bond domain. Each of the structures A and B makes certain

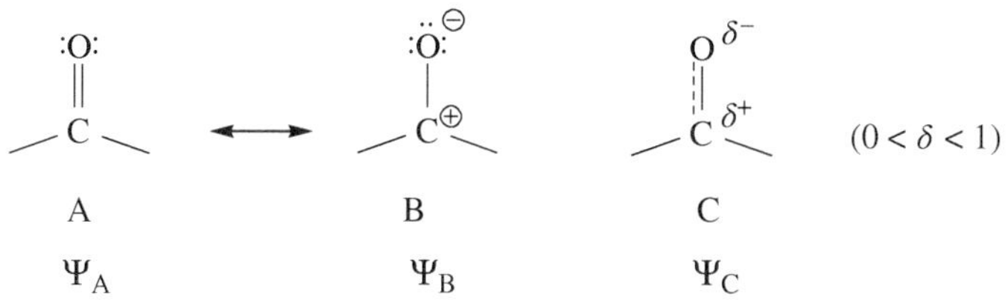

FIGURE 1.13   Possible resonance structures for the carbonyl (C=O) group.

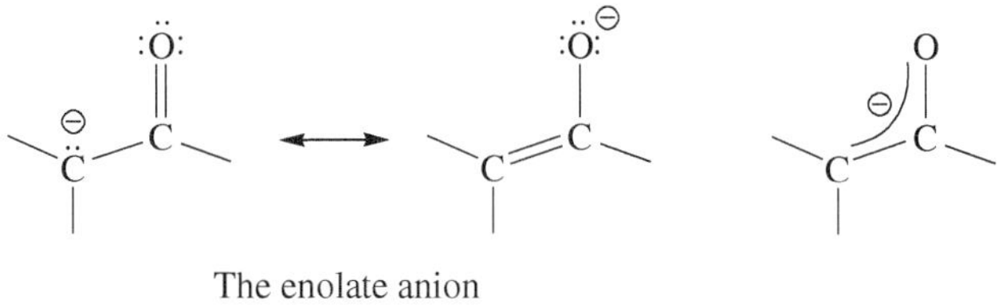

The enolate anion

FIGURE 1.14   Resonance stabilization of the anolate anion.

contribution to the real structure of the carbonyl group, consistent with its overall bond polarity as shown in structure C. The real wavefunction of carbonyl Ψ_C can be formulated as the linear combination of the wavefunctions of structures A and B (Ψ_A and Ψ_B, respectively).

$$\Psi_C = a\Psi_A + b\Psi_B \quad (a > b).$$

Due to a charge separation, the structure B possesses a higher energy than does the structure A. The contributions of structures A and B to the real structure of carbonyl are not the same. *In general, a resonance structure possessing a lower energy has greater contribution than that which possesses a higher energy* [4]. Therefore, the structure A is the major contributing Lewis structure to carbonyl, while the structure B only makes a minor contribution.

Similarly, due to delocalization of the negative charge via a conjugate π bond, an enolate anion resonates between two anionic structures as shown in Figure 1.14. As a result, each of the oxygen and α-carbon atoms is partially negatively charged, giving rise to the resonance stabilization. Further discussions on the structure of enolate and its reactivity will be presented in Chapter 9.

Since bonding electrons in a σ-bond are also delocalized in the bonded atoms, a compound that only contains σ-bonds can also be characterized by different Lewis structures (resonance structures). Figure 1.15a shows that the σ bond in hydrogen chloride (Ψ_{HCl}) contains both covalent ($\Psi_{covalent}$) and ionic (Ψ_{ionic}) characters [5], which can be expressed mathematically as

$$\Psi_{HCl} = a\Psi_{covalent} + b\Psi_{ionic} \quad (a > b)$$

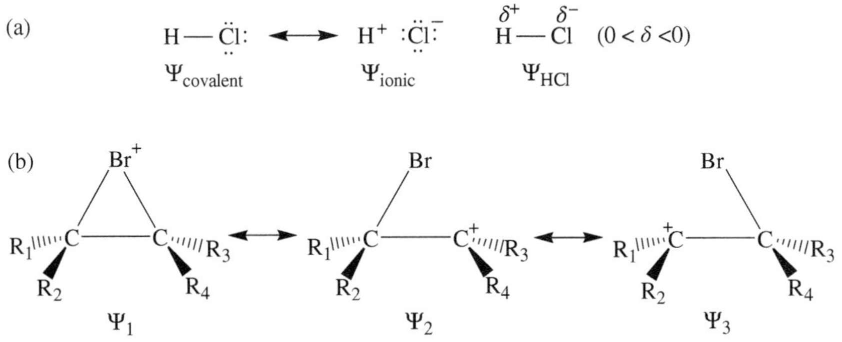

(a)

$$H \!-\! \ddot{C}l\!: \quad \longleftrightarrow \quad H^+ \ :\!\ddot{\underset{..}{C}l}\!:^- \qquad \overset{\delta^+}{H}\!-\!\overset{\delta^-}{Cl} \quad (0<\delta<0)$$

$$\Psi_{covalent} \qquad\qquad \Psi_{ionic} \qquad\qquad \Psi_{HCl}$$

(b)

FIGURE 1.15   Possible resonance structures for (a) hydrogen chloride (HCl) and (b) a brominium cation.

The covalent contributor (HCl) possesses lower energy. Thus, it is more important than the ionic contributor (H^+Cl^-). In other words, the true structure of HCl is neither pure covalent nor pure ionic. It resonates between the two structures, consistent with the observed bond polarity.

Figure 1.15b describes the nature of a brominium ion, the intermediate of electrophilic bromination of an alkene (see Chapter 3 for more details). The positive charge can be delocalized to all the three cyclic atoms giving rise to three contributing resonance structures. Overall, the real wavefunction of the species (Ψ) can be expressed in terms of linear combination of the wavefunctions of the individual resonance structures as shown below:

$$\Psi = a\Psi_1 + b\Psi_2 + c\,\Psi_3$$

The relative importance of the contributing resonance structures depends on the nature of the R groups. We will have more discussions on this situation in the individual chapters.

1.8.4   Frontier Molecular Orbitals

In a molecule, AOs of the constituent atoms overlap (LCAOs) giving rise to the formation of a set of MOs. As a result, all the electrons will move in the entire molecule. Some MOs with lower energies will be filled (occupied) by the electrons. Other MOs with higher energies will remain empty (unoccupied). The highest occupied molecular orbital (HOMO) and lowest unoccupied molecular orbital (LUMO) in a molecular are particularly important, and they are referred to as **frontier molecular orbitals (FMOs)**. In H_2, there are only two MOs (Fig. 1.8). Both of them, σ_{1s} (HOMO) and σ_{1s}^* (LUMO), are FMOs. In the C=C double bond, there are one σ-bond and one π-bond (Fig. 1.10a). The FMOs are π_p (HOMO) and π_p^*

(LUMO). In 1,3-butadiene ($CH_2=CHCH=CH_2$), the FMOs are ψ_2 (HOMO) and ψ_3 (LUMO) [Fig. 1.11b]. *In general, only the FMOs (HOMO and LUMO) participate in reactions and the other orbitals (lower-lying or upper-lying MOs) remain approximately intact during a chemical reaction* [6]. Many concerted reactions are effected by interaction of the HOMO of one reactant with the LUMO of the other. This requires that the reacting FMOs must have symmetry-match [1]. This case will be further addressed in the individual chapters, particularly in Chapter 4. In addition, when the energy of a photon ($h\nu$) at certain wavelength (frequency) matches the HOMO–LUMO energy gap, the photon can be absorbed by the molecule resulting in electronic transition from HOMO to LUMO. This type of absorption is common to many compounds whose molecules contain π-bonds. π (HOMO)–$\pi*$ (LUMO) transition occurs when the light with matching energy is absorbed. The transition gives rise to certain colors for the compounds containing π-bonds. In addition, the transition makes both HOMO and LUMO singly occupied. As a result, photochemical reactions may become possible via the singly occupied molecular orbitals (SOMOs) (Chapter 4).

1.9   ELECTROPHILES/NUCLEOPHILES VERSUS ACIDS/BASES

In organic chemistry, any chemical species that function as acceptors of an electron pair (2e) from another species are termed **electrophiles**. In contrast, any species that function as donors of an electron pair (2e) to an electrophile are termed **nucleophiles**. In inorganic chemistry, the electron-pair acceptors (electrophiles) are called **Lewis acids**. The electron-pair donors (nucleophiles) are called **Lewis bases**. Therefore, by the nature, electrophiles and Lewis acids are equivalent terms, and they are used to describe the same type of chemical species. Nucleophiles and Lewis bases are another set of equivalent terms, and they are used to describe another same type of chemical species.

Electrophiles are electron-deficient species so that in chemical reactions they can accept a pair of electrons from a nucleophile to form a covalent bond. In contrast, nucleophiles are electron-rich in order to be able to donate a pair of electrons to an electrophile to form a covalent bond (Eq. 1.62) [1, 6].

$$E^+ \quad + \quad :Nu^- \quad \longrightarrow \quad E\text{——}Nu \qquad (1.62)$$

Electrophile   Nucleophile   A covalent bond
(e-pair acceptor) (e-pair donor)

An electrophile can be a cation (positively charged) or a neutral molecule. In general, in order for a neutral molecule to be an electrophile, the molecule usually should contain a polar covalent bond, with the reactive center (the atom that accepts an electron pair) being partially positively charged. Alternatively, if an atom in a neutral molecule contains an empty orbital (typically, a p orbital), the molecule can also be an electrophile, with the reactive center being on the atom that contains the empty orbital. A nucleophile can be an anion (negatively charged) or a neutral molecule. In general, a nucleophile contains at least one lone-pair of electrons which is relatively active and able to be donated to an electrophile forming a covalent bond (Eq. 1.62). In some types of molecules, a bonding electron-pair including σ-bond and π-bond electrons can also be donated to an electrophile forming a covalent bond.

Now let us go over briefly some common types of electrophiles and nucleophiles. In Chapters 6 and 7 on individual types of reactions, we will present more intensive discussions on electrophiles and nucleophiles.

1.9.1   Common Electrophiles

The first type of common electrophiles we will study is **carbocations**. A carbocation is a species that contains a positively charged trivalent carbon atom possessing a trigonal plannar structure. The simplest carbocation is, in principle, the methyl (CH_3^+) cation (Fig. 1.16). CH_3^+ has not been identified experimentally presumably due to its extremely high instability. It is, however, employed as a prototype from which all the carbocations are derived by replacing one or more hydrogen atoms with different groups. The central carbon atom in CH_3^+ is sp^2 hybridized, with an empty p orbital holding a positive charge and perpendicular to the trigonal plane defined by the three sp^2 orbitals. Each C–H bond is formed by the 1s-sp^2 overlap. When one hydrogen atom in CH_3^+ is replaced by an alkyl (R) group, a primary (1°) carbocation RCH_2^+ is formed. Replacement of the second and third hydrogen atoms with alkyl groups results in a secondary (2°) carbocation R_2CH^+ and a tertiary (3°) carbocation R_3C^+, respectively (Fig. 1.16). In general, all types of carbocations are energetic. Thus, they are usually very unstable and highly reactive. Most of them act as strong electrophiles primarily owing to the active empty p orbital in the central carbon atom which has highly strong tendency to accept a pair of electrons from a nucleophile (almost any types of nucleophiles).

The relative stability for different types of carbocations increases in the order of methyl cation CH_3^+, 1° carbocation RCH_2^+, 2° carbocation R_2CH^+, and a 3°

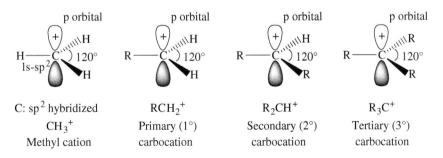

FIGURE 1.16   Structure of different types of carbocations.

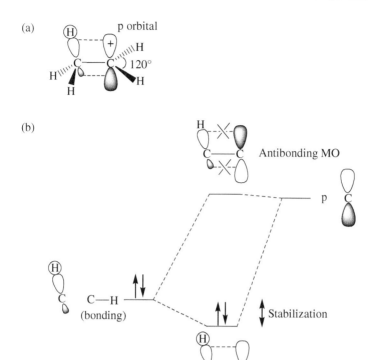

FIGURE 1.17   (a) Overlap of a C—H bond of the methyl group in the ethyl cation $(CH_3CH_2^+)$ with one lobe of the empty p orbital in the carbocation (hyperconjugation) and (b) linear combination of the C—H bonding orbital with the empty p orbital giving rise to formation of bonding and antibonding molecular orbitals.

carbocation R_3C^+. This is mainly attributed to the **hyperconjugation effect** of the C—H bonds in the alkyl groups, which can be well explained using the ethyl $CH_3CH_2^+$ cation (a primary carbocation) (Fig. 1.17a). In $CH_3CH_2^+$ one of the C—H bonds in the methyl group can overlap in sideway with one lobe of the unhybridized p orbital (hyperconjugation effect) in the primary CH_2 carbon. It results in delocalization of the positive charge into the C—H bond domain. In addition, the positively charged CH_2 carbon (sp^2 hybridized) attracts electrons from the CH_3 carbon (sp^3 hybridized) through the C—C σ-bond (inductive effect) also leading to delocalization of the positive charge. The charge delocalization lowers energy of the carbocation. The hyperconjugation effect is further demonstrated in Figure 1.17b using a MO model. The interactions (linear combinations) of a low-energy C—H bond and a high-energy p orbital lead to the formation of a bonding MO (with the energy level lower than that of the C—H bonding orbital) and an antibonding MO (with the energy level essentially the same as that of the p orbital). The difference in energy between the C—H bond and the bonding MO represents stabilization of the carbocation (decrease in energy) by a methyl group.

In $(CH_3)_2CH^+$ (a secondary carbocation), two C—H bonds (each from one methyl group) can overlap simultaneously with one lobe of the unhybridized p orbital in the secondary CH carbon. In $(CH_3)_3C^+$ (a tertiary carbocation), three C—H bonds (each from one methyl group) can overlap simultaneously with one lobe of the unhybridized p orbital in the tertiary carbon. As a result, the increase in number of the C—H bonds overlapping with the unhybridized p orbital (hyperconjugation effects) makes the positive charge delocalize to greater domains and further lowers the energies of the carbocations. In addition, the inductive effects through the methyl–C^+ σ bonds are getting more appreciable as the number of methyl groups on the positive carbon increases. This also makes the positive charge delocalize to greater domains and further lowers the energies of the carbocations.

When unsaturated groups such as vinyl and phenyl are attached to a positively charged carbon, the carbocations are greatly stabilized. As a result, the stability of allylic cation $(CH_3=CHCH_2^+)$ and benzylic cation $(PhCH_2^+)$ is even higher than a regular tertiary carbocation such as $(CH_3)_3C^+$. The stabilization is due to large **conjugation effects** of the unsaturated groups. In each of the allylic and benzylic cations, the positively charged empty p orbital overlaps in sideways with the π bond of the unsaturated group (conjugation effect), which delocalizes the positive charge to the vinyl or phenyl group and lowers energy of the cation.

Usually, a carbocation is produced by dissociation of a tertiary or secondary haloalkane, which in turn is attacked by a nucleophile resulting in an S_N1 reaction (Reaction 1.63) [1, 6].

$$R_3C-X \underset{Slow}{\overset{-X^-}{\rightleftharpoons}} R_3C^+ \xrightarrow[Fast]{:Nu^-} R_3C-Nu \qquad (1.63)$$

Dissociation of the haloalkane to a carbocation can also be facilitated by a Friedel–Crafts catalyst, such as $AlCl_3$ [1, 6]:

$$R_3C-X + AlCl_3 \rightleftharpoons R_3C^+ (AlXCl_3)^-$$

Another common route for generation of a carbocation is the electrophilic addition of a hydrogen halide to an alkene (Reaction 1.64) [1, 6].

$$CH_3CH=CH_2 + H-X \xrightarrow{Slow} CH_3\overset{+}{C}HCH_3 + X^- \xrightarrow{Fast} \underset{CH_3CHCH_3}{\overset{X}{|}} \qquad (1.64)$$

Once formed, the intermediate carbocation reacts fast with the halide (a nucleophile) to give a haloalkane addition product.

The concentrations of the carbocations in Reactions 1.63 and 1.64 are extremely low, and their existence cannot be detected by common spectroscopic methods. However, in certain conditions, a secondary or a tertiary carbocation can be stabilized and identified experimentally. For example, 2-chloromethane dissociates in the medium of antimony pentafluoride (SbF_5, a very strong Lewis acid) to give isopropyl cation $(CH_3)_2CH^+$ (Reaction 1.65) [1, 7].

$$(CH_3)_2CH\text{---}Cl \ + \ SbF_5 \ \xrightarrow{\ SbF_5\ } \ (CH_3)_2\overset{+}{C}H \ (SbClF_5)^- \tag{1.65}$$

2-Chloropropane   Excess   Isopropyl
                                          cation

Due to the extremely weak basicity (nucleophilicity) of the $(SbClF_5)^-$ anion, the concentration of $(CH_3)_2CH^+$ can be enhanced to a sufficient level in SbF_5 so that the carbocation has been identified by ^1H NMR spectroscopy. The NMR spectrum shows a one-proton (the CH proton) signal at $\delta = 13.5$ ppm, split into septets by six methyl protons. The $-CH_3$ proton appears at $\delta = 5.1$ ppm. The very low-field signal of the CH proton (13.5 ppm) is consistent with the strong deshielding effect of the positive charge on the secondary CH carbon [1, 7]. Another NMR-characterized carbocation is the secondary diphenylmethyl cation Ph_2CH^+, which is greatly stabilized by two aromatic groups, and this makes the experimental iden- tification of the carbocation practically possible. In contrast to the secondary and tertiary carbocations, primary carbocations (e.g., the ethyl $CH_3CH_2^+$ ion) are usually not observable experimentally due to their high instability. This is comparable to the above-mentioned situation for the methyl CH_3^+ cation.

The neutral haloalkane molecules RX (X = Cl, Br, or I) function as another com- mon type of electrophiles owing to polarity of the carbon–halogen bond. The slightly positively charged carbon in the carbon–halogen bond of a primary or sec- ondary haloalkane can be attacked by a strong nucleophile (usually, an anion) effect- ing an S_N2 reaction (Reaction 1.66).

$$Nu\!:^- \ + \ R\text{---}X \ \longrightarrow \ Nu\text{---}R \ + \ X^- \tag{1.66}$$

A carbonyl group (C=O) functions as an electrophile, with the reactive (electro- philic) center being the slightly positively charged carbon atom as described earlier in Section 1.8.3 (Fig. 1.13). As a result, a ketone (or an aldehyde) can undergo a nucleophilic addition with a strong nucleophile as shown in Reaction 1.67:

$$Nu\!: \ + \ \underset{R_1}{\overset{O}{\underset{}{\overset{\|}{C}}}}R_2 \ \longrightarrow \ \underset{R_1}{\overset{O^-}{\underset{Nu}{\overset{|}{C}}}}R_2 \tag{1.67}$$

A Bronsted (protic) acid (a proton donor) can always accept a pair of electrons from nucleophiles via the proton attached to the acid molecule. Therefore, the Bronsted acid is also a Lewis acid or an electrophile. For example, a hydrogen halide (H–X) molecule functions as an electrophile in Reaction 1.64, with the reactive cen- ter being the electrophilic proton. In addition, a protic acid can protonate the carbonyl oxygen, which is nucleophilic because of the lone pairs of electrons (Reaction 1.68):

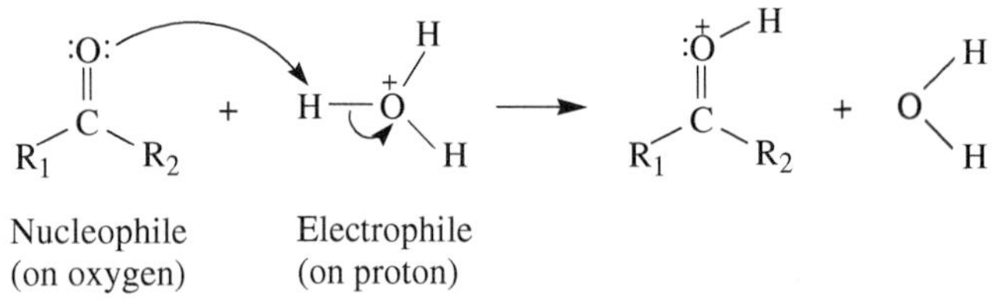

Nucleophile	Electrophile
(on oxygen)	(on proton)

$$(1.68)$$

Some strong Lewis acids such as $AlCl_3$, BF_3, and BH_3 (generated by dissociation of B_2H_6) are also strong electrophiles commonly used in many organic reactions. The electrophilicity of these species lies in the active empty p orbitals in the central aluminum and boron atoms. The functions of the compounds will be discussed extensively in Chapters 3 and 5.

1.9.2   Common Nucleophiles

Many anions that contain a lone pair of electrons (e.g., OH^-, OR^-, RCO_2^-, HS^-, Br^-, and CN^-) are good (strong) nucleophiles. On the other hand, many electrically neutral molecules that contain a lone pair of electrons (e.g., H_2O, ROH, and RCO_2H) act as poor (weak) nucleophiles.

When we determine the relative strength of nucleophiles, consideration is focused on the reactivity for the nucleophiles toward electrophilic carbon atoms of organic substrates. The general guideline is *The greater the electron density does the species have on its reactive center and the larger is the size of the reactive center, the more nucleophilic is the species.* On the basis of this guideline, we have the following general rules:

(1) An anion is stronger in nucleophilicity than a neutral molecule given that the reactive (nucleophilic) centers for both the anion and neutral molecule are the atoms of the same element. This is because in general an anion, due to its negative charge, has greater electron density than does a neutral molecule. It can be readily seen by comparison of hydroxide (OH^-, a strong nucleophile) and water (H_2O, a weak nucleophile) and by comparison of an alkoxide (RO^-, a strong nucleophile) and an alcohol (ROH, a weak nucleophile).

Hydroxide	Water	Alkoxide	Alcohol

For both pairs of species, the reactive centers are in oxygen atoms. The above Lewis structures show that each of the anions has three lone pairs of electrons in the oxygen atom, while the corresponding neutral molecule has two lone pairs of electrons in the oxygen atom. Clearly, for each pair of the species, the anion has greater electron density in the reactive center than the corresponding neutral molecule, giving rise to stronger nucleophilicity.

(2) For the same group of elements, the atomic radii increase from the top to bottom. As a result, for species containing the atoms of the same group of elements as reactive centers and having the same number of electron pairs in the reactive centers, their nucleophilicity increases as one moves from the top to the bottom along the group of elements. For example, the nucleophilicity of halides increases in the order of fluoride (F^-), chloride (Cl^-), bromide (Br^-), and iodide (I^-). All of them have four electron pairs. Another example is that the nucleophilicity of hydrogen sulfide (SH^-) is stronger than hydroxide (OH^-). For both of them, the central atoms contain three lone pairs of electrons.

Some molecules do not contain any lone pairs of electrons. However, a bonding electron pair in these molecules may be donated to an electrophile making them nucleophilic. In this category, compounds containing a C–M [M = Li or MgX (X = Cl, Br, or I)] bond are strong nucleophiles because the C—M bonding electron pair is active and has strong tendency to be donated to an electrophilic center. For example, a Grignard reagent undergoes nucleophilic addition to a carbonyl group (a carbon electrophile) as shown below:

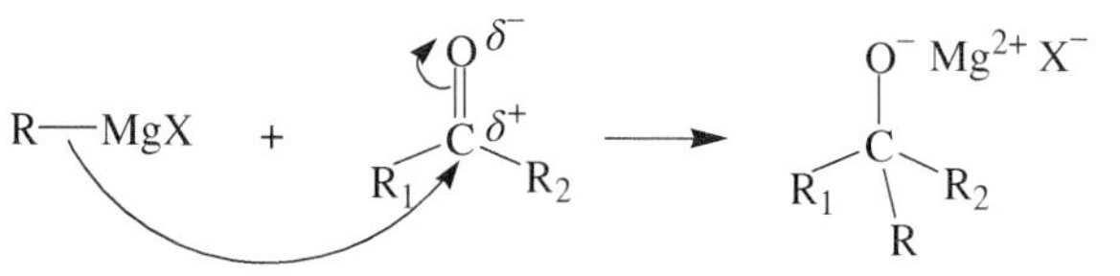

The C=C π bond in alkenes is nucleophilic. They are characterized by electrophilic addition reactions (e.g., Reaction 1.64). The aromatic rings are nucleophilic due to the activity of the conjugate π electrons. Therefore, arenes undergo extensive electrophilic substitution reactions as illustrated below:

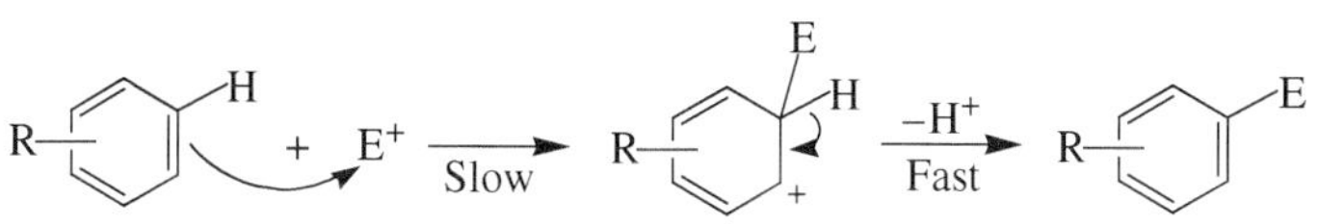

1.10   ISOTOPE LABELING

The enrichment of specific isotope of an element for an atom in a compound molecule is called **isotope labeling**. It is a very useful technique for studying kinetics and mechanisms for organic reactions. For example, let us first consider the acid-catalyzed hydrolysis of esters such as ethyl acetate (Reaction 1.69) [1]

$$
\underset{\substack{\text{O} \\ \parallel}}{\text{Me}-\text{C}-\text{O}-\text{Et}} + \text{H}_2\text{O} \xrightarrow{\text{H}^+} \underset{\substack{\text{O} \\ \parallel}}{\text{Me}-\text{C}-\text{OH}} + \text{EtOH} \tag{1.69}
$$

From the overall reaction, it is not clear what bond, the acyl-oxygen bond or the alkyl-oxygen bond, is broken. Cleavage of either bond could lead to the formation of the products.

In order to establish the mechanism for the reaction, the oxygen atom of the alkoxide group (–OEt) in the ester is labeled (enriched) with the oxygen-18 isotope (^{18}O) and the hydrolysis of the ^{18}O-labeld ester is conducted in the same condition (Reaction 1.70):

$$
\underset{\substack{\uparrow \\ \text{Bond breaking}}}{\underset{\substack{\text{O} \\ \parallel}}{\text{Me}-\text{C}-^{18}\text{O}-\text{Et}}} + \text{H}_2\text{O} \xrightarrow{\text{H}^+} \underset{\substack{\text{O} \\ \parallel}}{\text{Me}-\text{C}-\text{OH}} + \text{Et}^{18}\text{OH} \tag{1.70}
$$

$$
\xcancel{\longrightarrow} \underset{\substack{\text{O} \\ \parallel}}{\text{Me}-\text{C}-^{18}\text{OH}} + \text{EtOH}
$$

Et^{18}OH is identified from the products by mass spectrometry, but EtOH is not found. MeCO^{18}OH is not found either [1]. The results unambiguously show that the acyl-oxygen bond is broken in the reaction giving the observed products. However, the alkyl-oxygen bond remains intact, as cleavage of the alkyl-oxygen bond would lead to MeCO^{18}OH, but it is not observed. On the basis of the oxygen-18 isotope labeling experiment and other supporting data, the reaction mechanism is established and described in Figure 1.18. The acyl–oxygen bond cleavage, unambiguously identified from the isotope labeling experiment, is accomplished via the combination of a nucleophilic addition of water and a subsequent elimination of Et^{18}OH.

Very often, isotope labeling on a compound may alter the rate constants for certain reactions of the compound. This is referred to as the **kinetic isotope effects** [1, 3]. In principle, a heavier isotope of an element has a lower zero-point energy. Therefore, a bond to a heavier isotope (such as C—D bond) possesses a lower energy level than a bond to a lighter isotope (such as C—H bond) (Fig. 1.19). More energy is required to break a bond to a heavier isotope (such as C—D bond) than a bond to a lighter isotope (such as C—H bond). Since breaking of a C—H/C—D bond has a late transition state, the difference in energies of the transition states for breaking these bonds is much smaller than the difference in energy levels of the C—H and C—D bonds. Thus, the activation energy for breaking a C—D bond ($E_{a,D}$) is greater than that for breaking a C—H bond ($E_{a,H}$) (Fig. 1.19). This means the activation energy for a reaction that involves breaking a bond formed on a heavier isotope (such as C—D bond) is greater, corresponding to a smaller rate constant at a given temperature, if (*only if*) breaking of the bond is the rate-determining step for the studied reaction. If breaking of this bond occurs prior to or after the rate-determining step, the kinetic isotope effect is very minor. This effect is most remarkable for substitution of deuterium (D) for hydrogen (H) because the ratio in masses between these

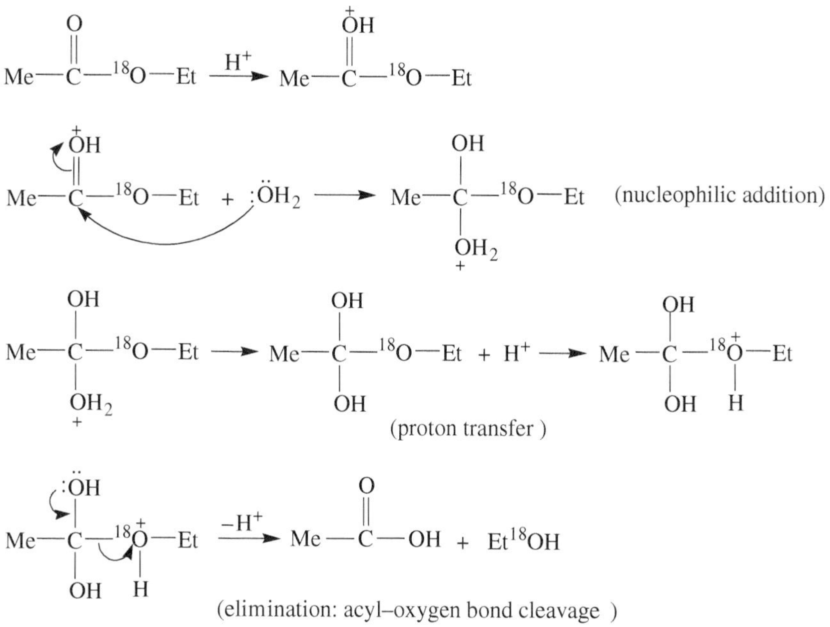

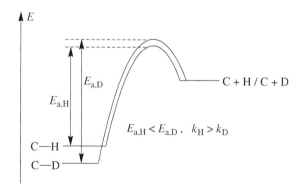

FIGURE 1.18   Reaction mechanism for acid-catalyzed hydrolysis of the oxygen-18 isotope labeled ethyl acetate.

FIGURE 1.19   Energetics for C—H and C—D (deuterium) bonds.

two isotopes ($D/H = 2$) is larger than those for any other pairs of isotopes. *In general, if a bond to hydrogen (such as C–H) or deuterium (such as C–D) is being broken in the rate-determining step of a reaction, the ratio of k_H (rate constant for the reaction involving breaking the bond to hydrogen, such as C—H bond) to k_D (rate constant for the reaction involving breaking the bond to deuterium, such C—D bond), k_H/k_D,* *is typically 2–8.* This is termed **kinetic deuterium isotope effect.** If the k_H/k_D values are found between 1 and 1.5, the corresponding bond breaking does not occur in the rate-determining step.

Kinetic deuterium isotope effect is often used in studies of the C—H bond functionalization. For example, the radical bromination of toluene ($PhCH_3$) to benzyl bromide ($PhCH_2Br$) has the kinetic isotope effect $k_H/k_D \sim 5$ [3]. This shows that breaking of the C—H bond in the methyl group of toluene by a bromine radical (Reaction 1.71) is the rate-determining step for the overall bromination reaction:

$$PhCH_2\text{—}H + Br\cdot \longrightarrow \left[PhCH_2\text{-----}H\text{---}Br\right]^{\ddagger} \longrightarrow PhCH_2^{\cdot} + H\text{—}Br \quad (1.71)$$
$$\text{Transition state}$$

The mechanism of radical halogenations of alkanes will be discussed extensively in Chapter 2.

The electrophilic nitration of benzene by acetyl nitrate ($CH_3CO_2NO_2$) involves substitution of the aromatic C—H bond by a nitro ($-NO_2$) group (Reaction 1.72) [3]:

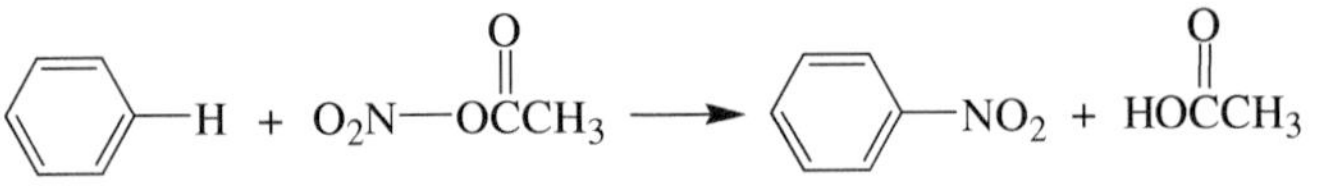

$$(1.72)$$

Overall, the reaction has the second-order rate law, first order in benzene and first order in acetyl nitrate (rate = $k[C_6H_6][CH_3CO_2NO_2]$). Study of the kinetic isotope effect using fully deuterated benzene C_6D_6 shows $k_H/k_D = 1$ [3]. The data indicate that the C—H bond cleavage is not involved in the rate-determining step. Instead, the rate-determining step is the electrophilic attack of the nitrating agent on benzene, which is supported by the observed overall second-order rate law. Mechanism for the reaction is shown in Figure 1.20. The aromatic nitration and related electrophilic substitution reactions of arenes will be studied in Chapter 5.

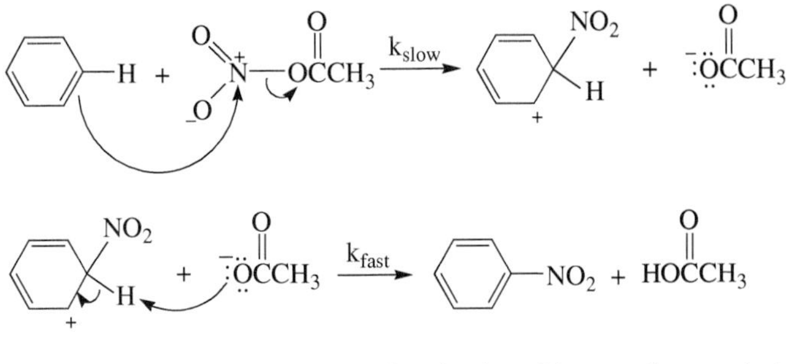

FIGURE 1.20   Reaction mechanism for nitration of benzene by acetyl nitrate.

1.11   ENZYMES: BIOLOGICAL CATALYSTS

Biochemistry is the study of life at the molecular level. Among main types of bio-molecules are carbohydrates (including monosaccharides, disaccharides, and poly-saccharides), lipids (including triacylglycerols, phospholipids, and steroids), proteins, and nucleotides (e.g., ATP) and nucleic acids (including RNA and DNA). Most of the biochemical reactions that occur *in vivo* (in the living systems such as animals and plants) are catalyzed by **enzymes**, which are also called biolog-ical catalysts. Almost all the enzymes are protein molecules which are composed of α-amino acid residues linked by peptide bonds. There are six types of enzymes clas-sified on the basis of the types of reactions that the enzymes catalyze. They are *oxi-doreductases* (the enzymes catalyzing oxidation–reduction reactions), *transferases* (the enzymes catalyzing group transfers, namely to substitute one functional group for another), *lyases* (the enzymes catalyzing elimination of two groups from adjacent carbon atoms to form a C=C double bond), *hydrolases* (the enzymes catalyzing hydrolysis of biomolecules), *isomerases* (the enzymes catalyzing isomerization reactions), and *ligases* (the enzymes catalyzing bond formation coupled with ATP hydrolysis) [8].

An enzyme contains an **active site** which is the catalytic center where a biochem-ical reaction takes place. The active site is usually composed of functional groups from side chains of some α-amino acid residues in the protein (enzyme) molecule. The biological reactant molecule(s) [often termed substrate(s), one or more] enters the structure of the enzyme and *loosely binds* to the enzyme usually by hydrogen bonding, hydrophobic interactions, and/or ionic bonding to form an intermediate **enzyme–substrate complex** (ES complex). Then various chemical processes take place in the ES complex to lead to the formation of the product. Finally, the product departs from structure of the enzyme to regenerate a free enzyme molecule. The enzyme catalyzed biochemical reactions are generalized as follows (Eq. 1.73):

$$S + E \rightleftharpoons ES \longrightarrow P + E \qquad (1.73)$$

In Equation 1.73, S and E represent the biological substrate and enzyme, respec-tively. ES is the intermediate enzyme–substrate complex. P is the reaction product. The formation of the ES complex is usually reversible, while the conversion of ES to the final product is irreversible.

An enzyme catalyzes a biochemical reaction by lowering activation energy of the reaction. On the basis of the types of interactions between the enzyme and the biological substrate in the transition state of a biochemical reaction, there are in gen-eral three types of catalytic mechanisms.

(1) *The acid–base catalysis*. Hydrogen bonds are formed in the transition state, which stabilize the transition state (decrease the level of its free energy) to lower the activation energy ($\Delta G^{\ddagger}$). As a result, the reactivity of the substrate is enhanced and the reaction goes faster.

(2) *The metal–ion catalysis.* The catalytic center is a metal ion which is attached to functional groups of some side chains of α-amino acid residues of the enzyme. A coordination bond between the substrate and the metal ion is formed (or partially formed) in the transition state to stabilize the transition state and activate the substrate.

(3) *The covalent catalysis.* A covalent bond between the substrate and the enzyme is formed (or partially formed) in the active site in the transition state to stabilize the transition state and activate the substrate.

A remarkable feature for enzymatic reactions is that the enzyme and substrate are both structurally and electronically complementary (also called key-to-lock model). As a result, the catalytic efficiency is extremely high. An enzymatic reaction is often 10^6–10^{12} times as fast as the uncatalyzed reaction [8]. In addition, most of the enzymatic reactions are highly selective and stereospecific due to the enzyme–substrate complementarity.

Many common enzymatic reactions follow mechanisms of the acid–base catalysis. The catalysis is usually concerted and achieved by *proton transfer* between the enzyme and substrate. The general principle for this type of catalysis can be illustrated using a generalized keto–enol tautomerization (Fig. 1.21).

Figure 1.21a shows an uncatalyzed, concerted keto–enol tautomerization taking place via a single transition state. The transition state is greatly destabilized (with a high level in free energy) by the formation of partial electric charges in different atoms. Figure 1.21b and c show acid- and base-catalyzed tautomerization, respectively. The acid H–A and base :B⁻ represent acidic and basic functional groups, respectively, in the active sites of enzymes. In the acid catalysis (Fig. 1.21b), a hydrogen bond is formed in the transition state on the carbonyl of the keto substrate to stabilize the partial negative charge on oxygen (via the electrostatic attraction to the partial positive charge on the HA proton). This stabilization leads to decrease in the energy level (in the free energy term) of the transition state and the activation energy ($\Delta G^{\ddagger}$) is lowered. In the base catalysis (Fig. 1.21c), a partial deprotonation on α-carbon by the basic group occurs in the transition state. As a result, the partial positive charge developed on the α-hydrogen of the keto substrate is stabilized (via the electrostatic attraction to the negative charge in the basic group B⁻). This stabilization leads to decrease in the free energy level of the transition state and the activation energy is lowered. In both acid- and base-catalyzed tautomerization (Fig. 1.21b and c), the formations of the enol and regeneration of the free enzymes from the transition states (TS1 and TS2) proceed via additional proton transfer steps. However, these steps are spontaneous and fast and do not affect the activation energies. The comparison of energetics for the uncatalyzed tautomerization and acid- and base-catalyzed tautomerization is demonstrated in Figure 1.21d.

Very often, enzymatic catalysis can also be achieved by splitting an uncatalyzed mechanistic step with a high activation energy into multiple catalyzed microscopic steps with lower activation energies. This way of catalysis can be demonstrated by

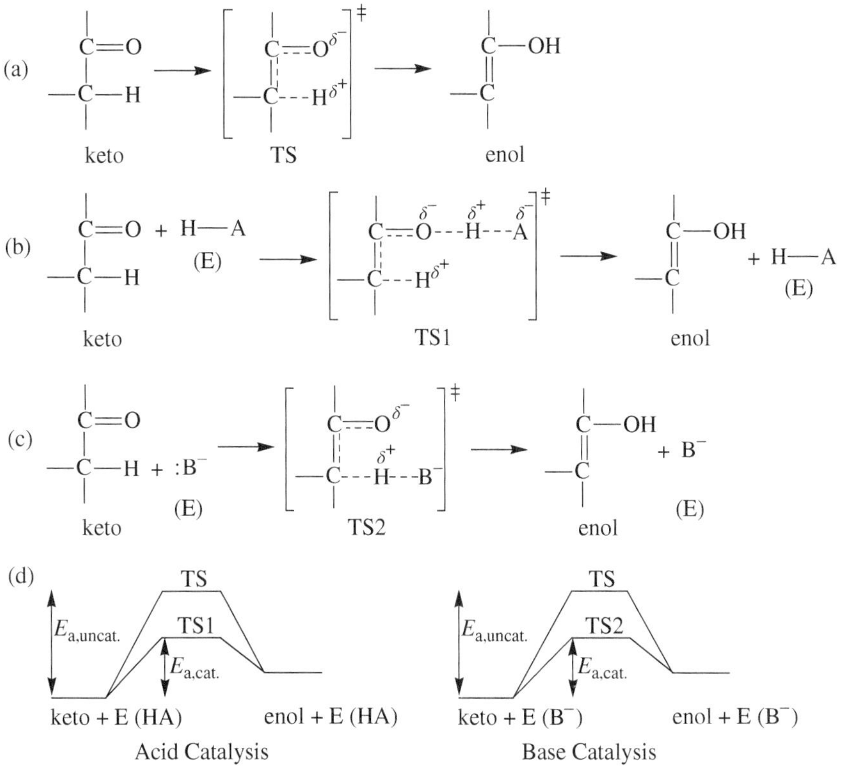

FIGURE 1.21   Acid–base catalysis for enzymatic reactions. (a) Uncatalyzed concerted keto–enol tautomerization; (b) The acid-catalyzed mechanism for the keto–enol tautomerization; (c) The base-catalyzed mechanism for the keto–enol tautomerization; and (d) Comparison of energetics for uncatalyzed and acid- or base-catalyzed keto–enol tautomerization.

using *carbonic anhydrase*, the enzyme that catalyzes the reaction of carbon dioxide (CO_2) with water (H_2O) giving bicarbonate (HCO_3^-) in human blood [8]. When CO_2 produced from respiration is transferred into the venous blood, it combines with water in the blood and this reaction happens. By virtue, the reaction of CO_2 with H_2O is a simple inorganic chemical reaction. When it happens in the venous blood and is catalyzed by carbonic anhydrase, it becomes a biochemical reaction.

The catalytic center of carbonic anhydrase is a sp^3-hybridized zinc ion (Zn^{2+}) connecting to three imidazole rings (Im) of histidine residues in the enzyme's polypeptide chain [8]. The unoccupied (vacant) sp^3 orbital in Zn^{2+} is strongly electrophilic. The transition state (TS) of the uncatalyzed reaction of CO_2 with H_2O is highly destabilized (with a high level of free energy) due to the formation of partial electric charges (Fig. 1.22a). In the presence of carbonic anhydrase, the reaction pathway is altered such that the first step of the reaction is that H_2O (hydrogen

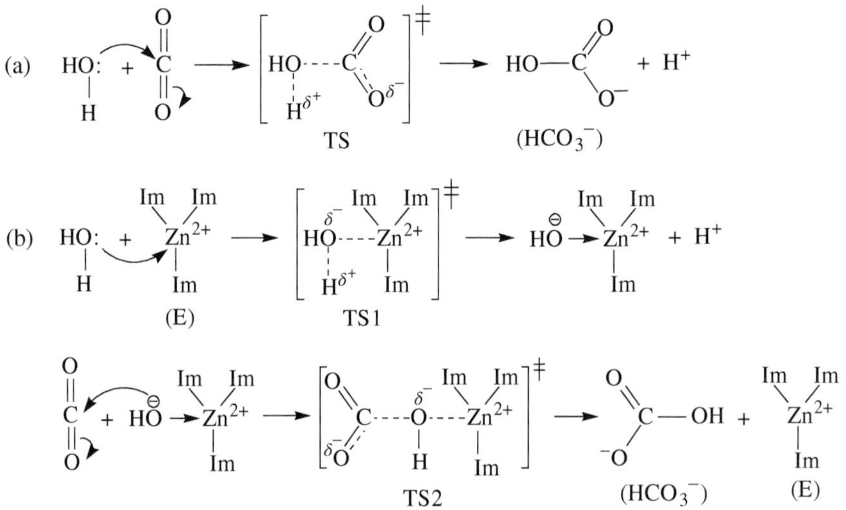

FIGURE 1.22 (a) Mechanism for the concerted reaction of H_2O and CO_2 giving HCO_3^- and (b) Mechanism for the enzyme (carbonic anhydrase) catalyzed stepwise reaction of H_2O and CO_2 giving HCO_3^-.

bonded to the enzyme) coordinates to the electrophilic Zn^{2+} center in the enzyme to lead to the formation of a nucleophilic hydroxide (OH^-) attached to Zn^{2+} (Fig. 1.22b) [8]. The transition state (TS1) is substantially stabilized, relative to that (TS) of the uncatalyzed reaction, by the partially formed $O{\ldots}Zn^{2+}$ coordination bond. Then the strongly nucleophilic OH^- in the Zn^{2+} center attacks CO_2 to bring about a nucleophilic addition with a low-level transition state (TS2), giving HCO_3^- and regenerating a free enzyme (Fig. 1.22b) [8]. In the *in vivo* reaction, the proton by-product combines with hemoglobin in the blood. The comparison of energetics and mechanisms for the uncatalyzed and enzyme catalyzed reactions of CO_2 with H_2O is demonstrated in Figure 1.23.

Many biochemical reactions follow some fundamental organic reaction mechanisms demonstrated in this book. Various biological applications of the mechanisms are discussed in all the individual chapters.

1.12 THE GREEN CHEMISTRY METHODOLOGY

Traditionally, synthesis of organic compounds has been performed very commonly in various organic solvents to make homogeneous reactions occur because many organic reactants are hydrophobic (water insoluble), but lipophilic (soluble in

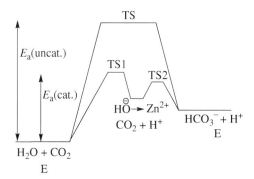

FIGURE 1.23 Comparison of energetics for the concerted and the enzyme (carbonic anhydrase) catalyzed stepwise reactions of H_2O and CO_2 giving HCO_3^-.

"oil"—the organic media). Organic solvents are in general hazardous substances. Utilization of large amounts of toxic organic solvents in synthetic reactions has created a big burden to the environment. The ideal green synthesis of organic compounds should avoid utilization of any solvents or only include environmentally benign solvents in the reactions, among which the most convenient, most natural, and cheapest is water. Conducting organic reactions in (or on) water will minimize the difficulty in the waste disposal, greatly reduce the pollution of the environment with hazardous materials, and ease the product work-up and separation. This protocol has received much attention from organic chemists in the past 15 years [9, 10]. Many synthetic reactions performed in water heterogeneously have been found substantially faster than the reactions traditionally carried out in organic solvents [9–12]. Effective and efficient organic synthesis in water represents a major aspect of methodology in the current green chemistry.

Research by vibrational sum frequency (VSF) spectroscopy has shown that on the hydrophobic interface between water and an organic layer, such as CCl_4 and hexane, the intermolecular hydrogen bonding between water molecules is substantially weaker than the hydrogen bonds of water in the liquid–vapor interface and inside the liquid phase of water. This is indicated by a blue shift of the vibrational bands for hydrogen bonded water in H_2O/CCl_4 and H_2O/hexane interfaces relative to that for the water in the liquid–vapor interface [13]. This weakening in hydrogen bonding allows water molecules in the interface to move more freely and gives rise to strong orientation effects of water on the hydrophobic interface, such that the O—H bond in some water molecules (1 in 4 H_2O molecules) on the hydrophobic interface penetrates into the organic layer [9, 14]. In the cases of the heterogeneous organic reactions performed in water, if a reactant molecule can act as a hydrogen bond acceptor, the OH bond in water which penetrates into the organic layer can form a strong hydrogen bond with the reactant molecule and more importantly, **form a strong** hydrogen bond with the transition state of the reaction to stabilize the transition state and lower the activation energy [9, 14]. In addition, upon vigorous agitation of the reaction mixtures in water, the bulk organic phase can be separated into aggregates (droplets) by water. Formation of aggregates increases surface area of the reactants and therefore, enhances their energy to lower the activation energy as well [9]. As a result, the reactions that take place on the hydrophobic interface (heterogeneous reactions) become much faster than the homogenous reactions occurring in organic solvents or in neat reactants [9, 14]. The hydrophobic effects are summarized in Figure 1.24. This way water is not only used as a green solvent, but it also acts as an effective "catalyst" to speed up the reactions.

Figure 1.24a illustrates that a free OH bond in a water molecule of the hydrophobic interface partially penetrates into the organic layer and is hydrogen bonded to the transition state (via a hydrogen bond acceptor) of a reaction taking place in the organic phase. This water molecule is also hydrogen bonded to other water molecules (denoted by W) inside the water phase. The hydrogen in the free OH bond acts as a nearly bare proton. Therefore, the hydrogen bond formed on the transition state is strong, greatly stabilizing the transition state (Fig. 1.24b). $[TS]_O^{\ddagger}$ and $[TS]_W^{\ddagger}$ represent the transition states for the reactions taking place in organic media (homogeneous) and in hydrophobic water/organic interface (heterogeneous), respectively. $E_{a,O}$ and $E_{a,W}$ are the corresponding activation energies for the homogeneous and heterogeneous reactions ($E_{a,W} < E_{a,O}$), respectively. Both aggregation of the organic reactants and especially, the hydrogen-bond stabilization for the transition state result in decrease in the activation energies of the organic reactions on the

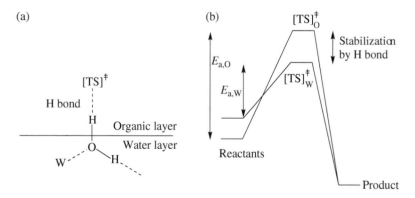

FIGURE 1.24   Hydrophobic effects on organic reactions. (a) The intermolecular hydrogen bond between water and transition state of reactions taking place in the hydrophobic interface, and (b) comparison of the energetics for the homogeneous reactions ($E_{a,O}$ and $[TS]_O^{\ddagger}$) in organic media and heterogeneous reactions ($E_{a,W}$ and $[TS]_W^{\ddagger}$) in the water–organic interface.

hydrophobic interface and greatly speed up the reactions relative to those performed in organic media.

Many types of organic reactions, such as Diels–Alder cycloadditions and Claisen rearrangements, have been carried out in water with fast speeds and good yields [9–12]:

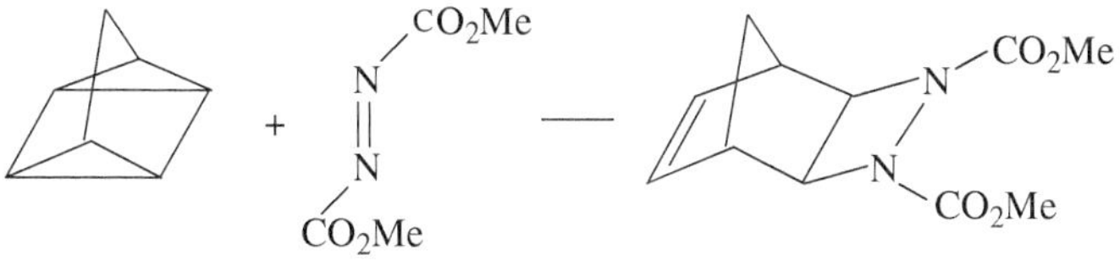

$$(1.74)$$

For example, it only takes 10 min for Reaction 1.74 to complete in water as the reaction mixture is vigorously agitated at 23 °C. When the reaction is performed by neat reactants with no solvent, the completion time is 48 h. It takes more than 120 h for the same reaction to complete in toluene homogeneously [12].

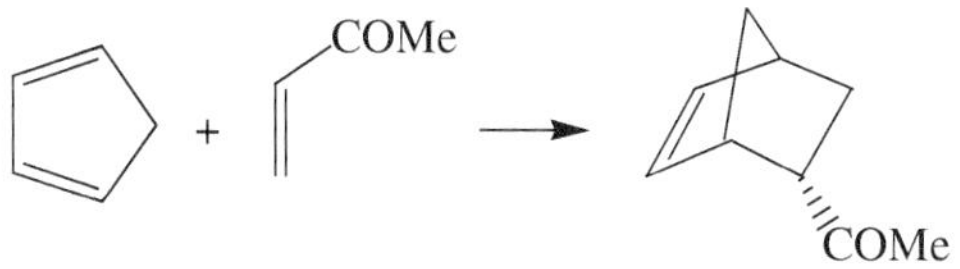

$$(1.75)$$

The rate constant for Reaction 1.75 performed in water (k_{water}) is many times greater than the rate constants for the reaction performed in organic solvents ($k_{solvent}$). Research has shown that the $k_{water}/k_{solvent}$ ratios at room temperature for isooctane, acetonitrile, and methanol are 740, 290, and 58, respectively [9].

More examples and mechanistic details of organic reactions carried out in water will be discussed in individual chapters (Chapters 4, 8, and 10) in the sections of green chemistry methods and applications.

PROBLEMS

1.1 Dibenzylmercury ($(PhCH_2)_2Hg$) contains linear C—Hg—C bonds. At 140 °C the compound undergoes thermal decomposition to give 1,2-diphenylethane $PhCH_2CH_2Ph$ and mercury. Suggest a concerted mechanism and a stepwise mechanism for this reaction.

1.2 The following are two alternative mechanisms for the $AlCl_3$ catalyzed Friedel–Crafts reaction of benzene with chloroethane. Show qualitative reaction profiles for both mechanisms. Which mechanism is more plausible? Explain.

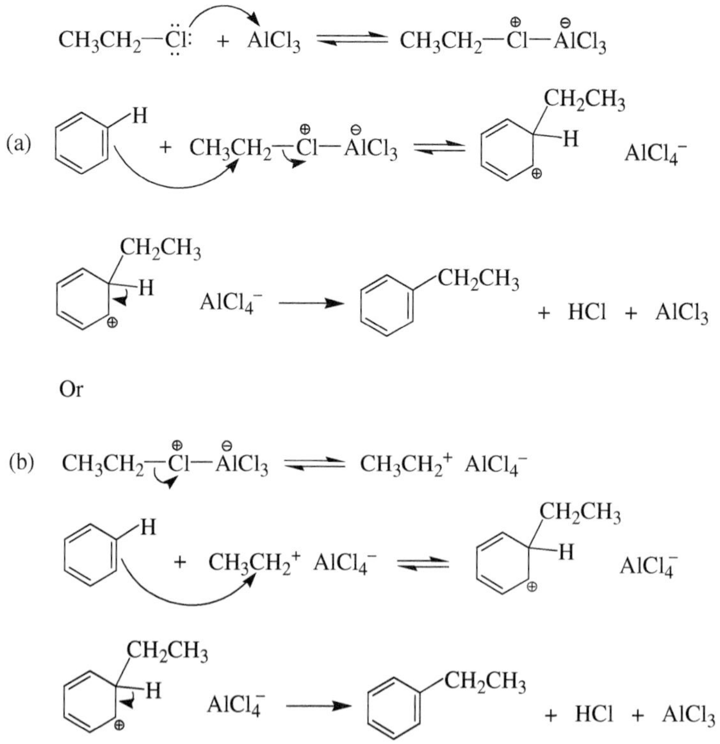

$$CH_3CH_2-\overset{..}{\underset{..}{Cl}}: \;+\; AlCl_3 \;\rightleftharpoons\; CH_3CH_2-\overset{\oplus}{Cl}-\overset{\ominus}{AlCl_3}$$

(a)

$$C_6H_5\text{-H} \;+\; CH_3CH_2-\overset{\oplus}{Cl}-\overset{\ominus}{AlCl_3} \;\rightleftharpoons\; [\text{arenium: } CH_2CH_3, H]^{\oplus} \quad AlCl_4^-$$

$$[\text{arenium: } CH_2CH_3, H]^{\oplus}\; AlCl_4^- \;\longrightarrow\; C_6H_5-CH_2CH_3 \;+\; HCl \;+\; AlCl_3$$

Or

(b)

$$CH_3CH_2-\overset{\oplus}{Cl}-\overset{\ominus}{AlCl_3} \;\rightleftharpoons\; CH_3CH_2^+ \; AlCl_4^-$$

$$C_6H_5\text{-H} \;+\; CH_3CH_2^+\; AlCl_4^- \;\rightleftharpoons\; [\text{arenium: } CH_2CH_3, H]^{\oplus} \quad AlCl_4^-$$

$$[\text{arenium: } CH_2CH_3, H]^{\oplus}\; AlCl_4^- \;\longrightarrow\; C_6H_5-CH_2CH_3 \;+\; HCl \;+\; AlCl_3$$

1.3 The decomposition of bromoethane in gaseous phase to ethene and hydrogen bromide follows a first-order rate law. Suggest a possible concerted and a stepwise mechanisms which are consistent with the rate law.

1.4 The rate of nitration of benzene in a mixture of nitric acid and sulfuric acid does not depend on the concentration of benzene. What conclusion can be drawn from this? Show qualitative energy profile for the stepwise reaction.

1.5 The mechanism for the autoxidation of hydrocarbons by oxygen is shown below:

$$\text{Initiator} \xrightarrow{\;k_i\;} 2\,R^{\cdot}$$

$$R^{\cdot} + O{=}O \xrightarrow[\text{Fast}]{\;k_{p1}\;} R-O-O^{\cdot}$$

$$R-O-O^{\cdot} + H-R \xrightarrow[\text{Slow}]{\;k_{p2}\;} R-O-O-H + R^{\cdot}$$

$$2\,R-O-O^{\cdot} \xrightarrow{\;k_t\;} R-O-O-R + O{=}O$$

Apply the steady-state approximation to work out the expected kinetics for the reaction.

1.6 For the S_N1 solvolysis of the tertiary chloroalkanes R–Cl in ethanol at 25°C, the relative rates for three different chloroalkanes are 1, 1.2, and 18.4 for R = Me_3C, $^tBuMe_2C$, and $^tBu_2MeC$, respectively. Is the difference due to electronic or steric effects? Explain. Draw qualitative energy profiles for the S_N1 solvolysis of the three chloroalkanes.

1.7 For the following each pair of anions, compare their basicity and nucleophilicity. Account for the difference.

(1) OH^- and SH^- (2) CH_3COO^- and $C_6H_5COO^-$

1.8 Which of the following S_N2 reactions goes faster? Explain by kinetic isotope effect.

(1) $CH_3CH_2CH_2CH_2CH_2OH$ + HBr $\longrightarrow$ $CH_3CH_2CH_2CH_2CH_2Br$ + H_2O

  1-Pentanol (regular oxygen)

(2) $CH_3CH_2CH_2CH_2CH_2{}^{18}OH$ + HBr $\longrightarrow$ $CH_3CH_2CH_2CH_2CH_2Br$ + $H_2{}^{18}O$

  1-Pentanol (oxygen-18 labelled)

REFERENCES

1. Jackson, R. A. *Mechanisms in Organic Reactions*, The Royal Society of Chemistry, Cambridge, UK (2004).
2. Silbey, R. J.; Alterty, R. A.; Bawendi, M. G. *Physical Chemistry*, 4th ed., Wiley, Danvers, MA, USA (2005).
3. Hoffman, R. V. *Organic Chemistry: An Intermediate Text*, 2nd ed., Wiley, Hoboken, NJ, USA (2004).
4. Huheey, J. E.; Keiter, E. A.; Keiter, R. L. *Inorganic Chemistry: Principles of Structure and Reactivity*, 4th ed., Harper Collins, New York, NY, USA (1993).
5. Sun, X. An Integrated Approach to the Lewis Model, Valence Bond Theory, and Molecular Orbital Theory: A New Model for Simple Molecular Orbitals and a Quicker Way of Learning Covalent Bonding in General Chemistry. *Chem. Educator*, 2007, *12*, 331–334.
6. Fox, M. A.; Whitesell, J. K. *Core Organic Chemistry*, 2nd ed., Jones and Bartlett, Sudbury, MA, USA (1997).
7. Olah, G. A. 100 Years of Carbocations and Their Significance in Chemistry. *J. Org. Chem.* 2001, *66*, 5943–5957.
8. Voet, D.; Voet, J. G.; Pratt, C. W. *Fundamentals of Biochemistry*, 5th ed., John Wiley & Sons, Inc., Hoboken, NJ, USA (2016).

9. Butler, R. N.; Coyne, A. G. Water: Nature's Reaction Enforcer–Comparative Effects for Organic Synthesis "In-Water" and "On-Water". *Chem. Rev.* 2010, *110,* 6302–6337.

10. Simon, M.-O.; Li, C.-J. Green Chemistry Oriented Organic Synthesis in Water. *Chem. Soc. Rev.* 2012, *41*, 1415–1427.

11. Narayan, S.; Muldoon, J.; Finn, M. G.; Fokin, V. V.; Kolb, H. C.; Barry Sharpless, K. "On Water": Unique Reactivity of Organic Compounds in Aqueous Suspension. *Angew. Chem. Int. Ed.* 2005, *44*, 3275 –3279.

12. Klijn, J. E.; Engberts, J. B. F. N. Fast Reactions "On Water". *Nature,* 2005, *435*, 746–747.

13. Scatena, L. F.; Brown, M. G.; Richard, G. L. Water at Hydrophobic Surfaces: Weak Hydrogen Bonding and Strong Orientation Effects. *Science,* 2001, *292*, 908–912.

14. Jung, Y.; Marcus, R. A. On the Theory of Organic Catalysis "On Water". *J. Am. Chem. Soc.* 2007, *129*, 5492–5502.

2

THE ALIPHATIC C—H BOND FUNCTIONALIZATION

Alkanes make up about 50% of petroleum. They are a primary source of fuel and also constitute a significant fraction of the carbon skeleton for organic synthesis. Therefore, functionalization of the alkane (aliphatic) C—H bonds (the chemical process in which the hydrogen atom is replaced by a functional group) is among the most important and most useful aspects in organic chemistry.

Saturated hydrocarbons (alkanes and cycloalkanes) are in general chemically inactive, which is often attributed to their high C—H bond dissociation energies (typically 90–100 kcal/mol) and very low acidity (for C_{sp^3}–H pK_a = 45–60) [1]. Elucidating the conditions necessary for activation (cleavage) of the C—H bond and its subsequent transformation into other bonds (functionalization) represents both a fundamental and a practical challenge to organic chemists.

Although C—H bonds of the sp^3 hybridized carbons are very strong and more difficult to cleave than other types of linkages, alkanes are not completely inert. Several general methods have been developed to effectively activate and functionalize the aliphatic C—H bond with appropriate selectivity. The most effective common method for the C—H bond functionalization is photochemical and high-temperature thermal halogenations (chlorination and bromination) of alkanes giving haloalkanes (alkyl halides). The reactions take place via reactive intermediate halogen and alkyl radicals. The second general method is activation (cleavage) and subsequent functionalization (typically oxidation) of the C—H bond by various transition metal complexes. Usually, the reactions take place via an intermediate agostic bond formed

Organic Mechanisms: Reactions, Methodology, and Biological Applications,
Second Edition. Xiaoping Sun.
© 2021 John Wiley & Sons, Inc. Published 2021 by John Wiley & Sons, Inc.
Companion website: www.wiley.com/go/Sun/OrgMech_2e

between C–H and the metal center, resulting in cleavage of the strong C—H bond and formation of a relatively weak carbon–metal bond. The carbon–metal bond very often can be transformed into other bonds in carbon by functionalizing agents, completing the functionalization process. A third general method has involved usage of superacids, which can effectively protonate certain inorganic functionalizing agents or the alkane substrates and make the alkane C—H bond functionalization possibly occur. Other methods involve some key electrophilic inorganic cations to activate the alkane C—H bonds. In addition, biochemical methodology has been developed to selectively activate and functionalize C—H bonds in some alkanes by enzymes found in some living organisms.

In this chapter, detailed discussions of mechanisms and synthetic utility for the aforementioned general methods are presented.

2.1 ALKYL RADICALS: BONDING AND THEIR RELATIVE STABILITY

Halogenation of alkanes, which takes place via reactive intermediate alkyl radicals, represents a major common method which has been employed for effective functionalization of the alkane C—H bonds. Understanding of the nature of alkyl radicals is essential for proper study of the reaction mechanism.

In principle, an alkyl radical (R·) can be formed by **homolytic cleavage** of a C—H bond in an alkane (R–H) molecule: The pair of bonding electrons is separated. One electron is transferred to the carbon atom forming an alkyl R· radical. The other electron is transferred to the hydrogen atom forming H·.

$$R \overset{\frown}{-} H \longrightarrow R· + H·$$

The energy level of the alkyl R· radical is determined by the standard enthalpy ($\Delta H°$) for the homolytic C—H bond cleavage (bond dissociation energy, BDE). Therefore, the BDE can serve as a quantitative measure for stability of the alkyl radical, namely, that **the smaller the BDE, the more stable is the R· radical.** The following formulas show the BDEs associated to various alkyl radicals including methyl, ethyl (a primary radical), isopropyl (a secondary radical), and *tert*-butyl (a tertiary radical) [2]:

$$H_3C - H \rightarrow CH_3· + H·   \quad \Delta H°(BDE) = 104\,kcal/mol$$
$$CH_3CH_2 - H \rightarrow CH_3CH_2· + H·   \quad \Delta H°(BDE) = 98\,kcal/mol$$
$$(CH_3)_2CH - H \rightarrow (CH_3)_2CH· + H·   \quad \Delta H°(BDE) = 95\,kcal/mol$$
$$(CH_3)_3C - H \rightarrow (CH_3)_3C· + H·   \quad \Delta H°(BDE) = 92\,kcal/mol$$

In each of the radicals, the carbon atom in which the unpaired electron resides is sp^2-hybridized. The unpaired electron is held in an unhybridized p orbital which is

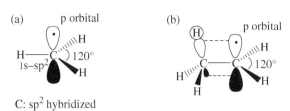

FIGURE 2.1   (a) Structure of and bonding in the methyl radical and (b) bonding and hyperconjugation effect in the ethyl radical.

perpendicular to the trigonal plane defined by the three sp^2 orbitals. Bonding in $CH_3\cdot$ is illustrated in Figure 2.1a. Adding alkyl (methyl) groups to the carbon which holds the unpaired electron reduces BDEs associated to the radicals and therefore, the stability of the radicals is enhanced. This is attributed to the hyperconjugation effect of the C—H bonds in the methyl groups, which can be well explained using the ethyl $CH_3CH_2\cdot$ radical as an example (Fig. 2.1b).

In $CH_3CH_2\cdot$, one of the C—H bonds in the methyl group can overlap in sideway with a lobe of the unhybridized p orbital in the primary CH_2 carbon. Such a sideway overlap of a σ-bonding orbital (including C—H and C—C bonds) with one, and only one, lobe of a p orbital in its adjacent atom is called **hyperconjugation effect**. In this case, the hyperconjugation effect results in delocalization of the unpaired electron into the C—H bond domain and energy of the radical is lowered, giving rise to a decrease in BDE by 6 kcal/mol relative to that for $CH_3\cdot$. The hyperconjugation effect

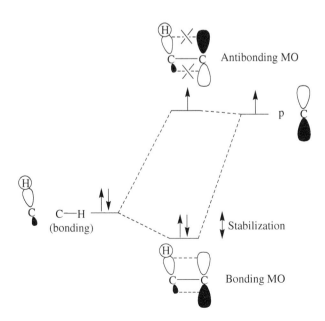

FIGURE 2.2   Molecular orbital model for the hyperconjugation effect in the ethyl radical.

is shown in Figure 2.2 using a molecular orbital (MO) model [2]. The interactions (linear combinations) of a low-energy C—H bond and a high-energy p orbital lead to the formation of a bonding MO (with the energy level lower than that of the C—H bonding orbital) and an antibonding MO (with the energy level essentially the same as that of the p orbital). The difference in energy between the C—H bond and the bonding MO represents stabilization of the radical by a methyl group. The stabilization can also be accounted for using the resonance structures of ethyl radical (Fig. 2.3a) showing the electron delocalization, which gives rise to a **resonance stabilization** (Valence Bond model) [2].

In $(CH_3)_2CH^{\cdot}$, two C—H bonds (each from one methyl group) can overlap simultaneously with a lobe of the unhybridized p orbital in the secondary CH carbon. In $(CH_3)_3C^{\cdot}$, three C—H bonds (each from one methyl group) can overlap simultaneously with a lobe of the unhybridized p orbital in the tertiary carbon. As a result, the increase in number of the C—H bonds overlapping with the unhybridized p orbital (hyperconjugation effects) makes the unpaired electron delocalize to greater domains and further lowers the energies of the radicals, resulting in further decreases in BDEs. Correspondingly, more possible resonance structures can be written for the radicals $[(CH_3)_2CH^{\cdot}$ and $(CH_3)_3C^{\cdot}]$ with increasing in number of the methyl groups in the central carbon (Fig. 2.3b and c), giving rise to greater resonance stabilization.

When unsaturated groups, such as vinyl and phenyl, are attached to radicals on the carbon atom that holds an unpaired electron, the radicals are greatly stabilized. The stabilization is indicated by relatively low BDEs associated to these radicals as shown below [2]:

$$CH_2 = CHCH_2 - H \rightarrow CH_2 = CHCH_2^{\cdot} + H^{\cdot} \quad \Delta H^{\circ}(BDE) = 89 \, kcal/mol$$
$$PhCH_2 - H \rightarrow PhCH_2^{\cdot} + H^{\cdot} \quad\quad\quad\quad \Delta H^{\circ}(BDE) = 89 \, kcal/mol$$

As a result, both the allyl $CH_2=CHCH_2^{\cdot}$ and benzyl $PhCH_2^{\cdot}$ radicals are about 3 kcal more stable than the tertiary butyl $(CH_3)_3C^{\cdot}$ radical. Such a remarkable stabilization is due to large conjugation effects of the unsaturated groups.

In each of the allyl and benzyl radicals, the singly occupied unhybridized p orbital overlaps in sideways with the π bond of the unsaturated group (**conjugation effect**), which delocalizes the unpaired electron to the vinyl or phenyl group and lowers energy of the radical. The conjugation effects which result in decrease in energy in the radicals are demonstrated in Figures 2.4 and 2.5 using MO models. In addition, the electron delocalizations for the radicals are also accounted for by the resonance stabilization (Fig. 2.6) in allyl and benzyl radicals.

Radical species formed by homolytic cleavage of the C_{sp^2}—H and C_{sp}—H bonds possess relatively high energies and are very unstable, which is indicated by relatively large BDEs associated to the radicals (Fig. 2.7).

In all the radicals of Figure 2.7, the unpaired electron is localized in a sp^2 or sp orbital in a multiply-bonded carbon atom. Their stability is lower than that of all types of alkyl radicals presumably due to lack of any delocalization of the single electron. In addition, due to greater percentage of s-characters in the sp (50%)

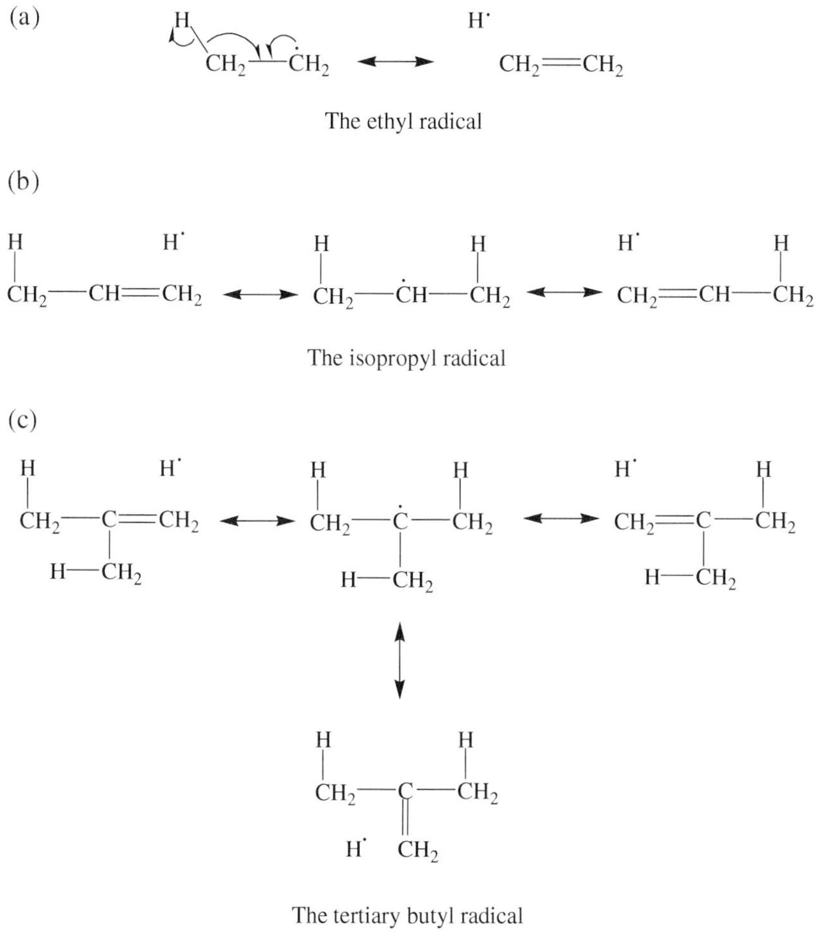

(a)

The ethyl radical

(b)

The isopropyl radical

(c)

The tertiary butyl radical

FIGURE 2.3   Resonance stabilization for (a) the ethyl radical, (b) the isopropyl radical, and (c) the tertiary butyl radical. (a) Based on Bruckner [2].

and sp^2 (33.3%) hybridized orbitals than that for a sp^3 orbital (25%), the sp and sp^2 orbitals are more electronegative than that of a sp^3 orbital. As a result, each of sp and sp^2 orbitals holds the corresponding C–H electrons more strongly than does the sp^3 orbital. It would take more energy to separate the bonding electrons in each of the C_{sp}–H and C_{sp^2}–H bonds than those in the C_{sp^3}–H bond. This leads to substantially higher BDEs for the C–H bonds in multiply-bonded carbon atoms than those for the C_{sp^3}–H bonds. Here is a general principle: *A single electron formally residing on a multiply-bonded carbon atom is highly energetic, and the corresponding radical*

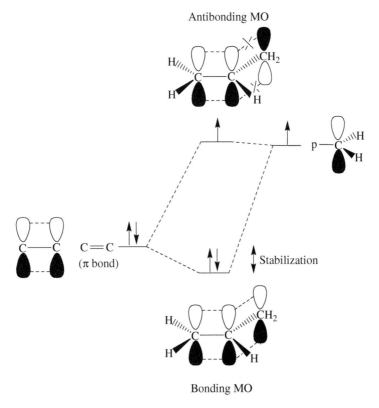

FIGURE 2.4   Molecular orbitals in the allyl radical.

species is extremely unstable. The relative energy levels of different types of radicals represented by their BDEs are illustrated in Figure 2.8.

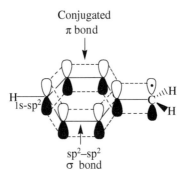

FIGURE 2.5   Conjugation effect in the benzyl radical.

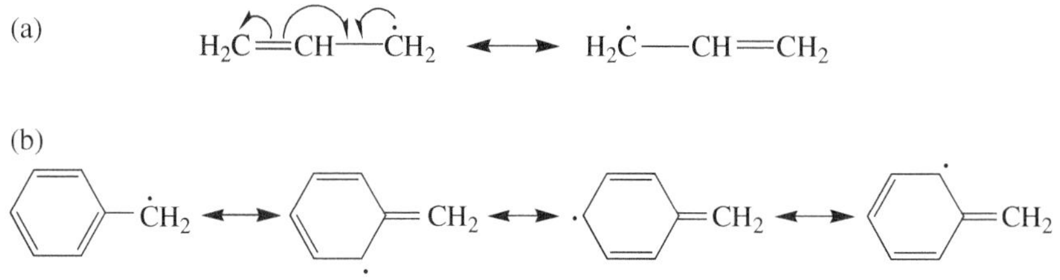

(a)

$$H_2C = CH - \dot{C}H_2 \quad \longleftrightarrow \quad H_2\dot{C} - CH = CH_2$$

(b)

FIGURE 2.6   Resonance stabilization for allyl (a) and benzyl radicals (b).

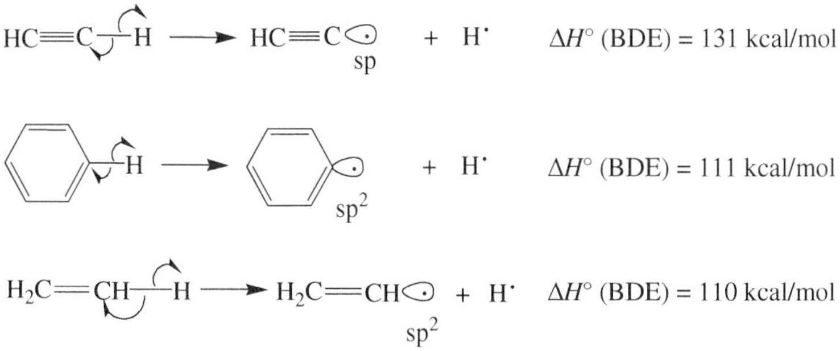

$$HC \equiv C - H \longrightarrow HC \equiv C \cdot \quad + \quad H^{\bullet} \qquad \Delta H° \text{ (BDE)} = 131 \text{ kcal/mol}$$
$$\underset{sp}{}$$

$$\longrightarrow \quad + \quad H^{\bullet} \qquad \Delta H° \text{ (BDE)} = 111 \text{ kcal/mol}$$
$$\underset{sp^2}{}$$

$$H_2C = CH - H \longrightarrow H_2C = CH \cdot \quad + \quad H^{\bullet} \qquad \Delta H° \text{ (BDE)} = 110 \text{ kcal/mol}$$
$$\underset{sp^2}{}$$

FIGURE 2.7   Bond dissociation energies (BDEs) for $HC \equiv C^{\bullet}$, $C_6H_5{}^{\bullet}$, and $H_2C = CH^{\bullet}$ radicals.

2.2   RADICAL HALOGENATIONS OF THE C–H BONDS ON SP³-HYBRIDIZED CARBONS: MECHANISM AND NATURE OF THE TRANSITION STATES

Mechanism   Alkanes (R–H) react with elemental chlorine (Cl_2) and bromine (Br_2) photochemically in ultraviolet light or thermally at high temperatures (~400 °C) giving chloroalkane (alkyl chloride, R–Cl) and bromoalkane (alkyl bromide, R–Br), respectively [2, 3].

$$R-H \ + \ Cl_2 \,(Br_2) \ \xrightarrow[\text{or } h\nu]{\Delta} \ R-Cl \,(Br) \ + \ HCl(Br)$$

When the alkane substrate is in excess, monosubstitution takes place as the major reaction. If the molar quantities of the alkane and halogen are comparable, multiple substitutions happen. The halogenations follow a radical mechanism. We will demonstrate the reaction mechanism using monochlorination of methane (Fig. 2.9) as follows.

In the first mechanistic step (initiation), Cl_2 undergoes photochemical or thermal homolytic cleavage (reversible) producing chlorine $Cl^{\bullet}$ radicals. Then, $Cl^{\bullet}$ strikes a methane molecule along its C–H bond and breaks the bond homolytically (activation of the C–H bond). As a result, a reactive methyl $CH_3{}^{\bullet}$ radical and a HCl molecule are formed. $CH_3{}^{\bullet}$ in turn strikes an unreacted Cl_2 molecule along its Cl–Cl

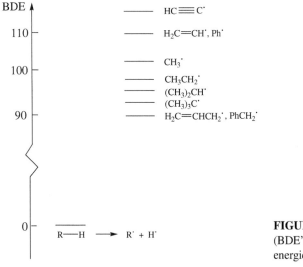

FIGURE 2.8   Energy levels (BDE's—bond dissociation energies) for various radicals.

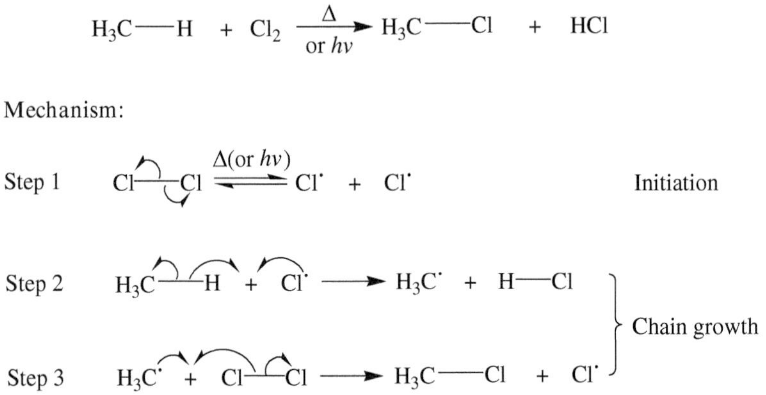

$$H_3C\text{—}H \; + \; Cl_2 \; \xrightarrow[\text{or } h\nu]{\Delta} \; H_3C\text{—}Cl \; + \; HCl$$

Mechanism:

Step 1   Cl—Cl $\overset{\Delta(\text{or } h\nu)}{\rightleftharpoons}$ Cl$^{\cdot}$ + Cl$^{\cdot}$ Initiation

Step 2   H$_3$C—H + Cl$^{\cdot}$ ⟶ H$_3$C$^{\cdot}$ + H—Cl ⎫

⎬ Chain growth

Step 3   H$_3$C$^{\cdot}$ + Cl—Cl ⟶ H$_3$C—Cl + Cl$^{\cdot}$ ⎭

FIGURE 2.9   Mechanism for radical chlorination of methane.

bond and breaks the bond homolytically. As a result, a product CH$_3$Cl molecule is formed (functionalization process) and a chlorine Cl$^{\cdot}$ radical is regenerated. Then, the Cl$^{\cdot}$ radical starts a second cycle of Steps 2 and 3 (chain-growth steps). These two steps keep repeating until all the Cl$_2$ molecules (in deficit) are converted to the CH$_3$Cl product.

In the course of the reaction, radical species may possibly combine as shown below to form diamagnetic molecules which do not contain any single electrons.

$$CH_3^{\cdot} + Cl^{\cdot} \rightarrow CH_3Cl$$
$$CH_3^{\cdot} + CH_3^{\cdot} \rightarrow CH_3CH_3$$

These termination steps would destroy the radicals essential for the chlorination reaction, and as a result, the reaction could be terminated. However, the termination steps only consist of very minor processes because the concentrations of radicals are extremely low in the course of the reaction, and their interactions are considered disfavored and essentially ineffective. Instead, the Cl˙ and CH₃˙ radicals, once generated, will predominantly interact with relatively more highly concentrated CH_4 and Cl_2, respectively, as shown in Figure 2.9 (chain-growth). In the presence of ultraviolet light or upon heating, the termination steps can only take place to a very small extent. By large, the overall chlorination will be a continuous process and essentially, all the limiting reactant molecules will be converted to the CH_3Cl product.

The transition states   Now let us have a closer look at the pathways for the two chain-growth steps in Figure 2.9 so that we can achieve a better understanding of the nature of the transition states for those steps. In Step 2, as a Cl˙ radical approaches the hydrogen linearly in a C—H bond of methane, the single electron in Cl˙ and a bonding electron in C–H with the same spin render Pauli repulsion (Fig. 2.10) [4]. Such a repulsion force expels the bonding electron toward carbon and simultaneously, the other bonding electron in C–H with the different spin starts moving toward the chlorine atom and begins pairing up with the single electron of Cl˙ in the mid-region between the hydrogen and chlorine atoms. As a result, a linear transition state (TS₁ in Figure 2.10) is formed in which the old C—H bond is being partially broken homolytically and a new bond between hydrogen and chlorine is being partially formed. Eventually, as the progress continues, the two bonding electrons in C–H will be completely separate, and a methyl CH₃˙ radical bearing a single electron

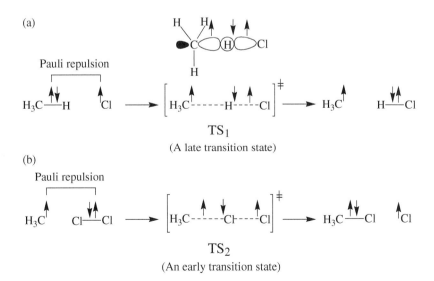

FIGURE 2.10   Nature of the transition states for the chain-growth steps of chlorination of methane. (a) Cleavage of the CH bond and (b) Formation of the CCl bond. Based on Sun [4].

on carbon will be formed, together with simultaneous formation of a H–Cl molecule. In Step 3, $CH_3{}^{\cdot}$ approaches a Cl_2 molecule along its Cl—Cl bond linearly. The methyl single electron and a bonding electron in Cl–Cl with the same spin render Pauli repulsion resulting in an analogous linear transition state (TS_2 in Figure 2.10) to that in Step 2 [4]. Eventually, as the progress continues, the Cl—Cl bond will break homolytically giving rise to the formations of a $Cl^{\cdot}$ radical and a new C—Cl bond in the CH_3Cl product molecule. The overall process and the corresponding energy profiles are shown in Figures 2.10 and 2.11, respectively. Step 2, the formation of the methyl $CH_3{}^{\cdot}$ radical effected by abstraction of a hydrogen atom from methane by a $Cl^{\cdot}$ radical, has much greater activation energy (E_a) than that (E_a') for Step 3, the product formation step. Therefore, Step 2 (abstraction of hydrogen from a C—H bond by chlorine radical) is the rate-determining step for the overall chlorination of methane.

Multiple halogenations   When molar quantities of methane and chlorine are comparable, multiple substitutions happen giving CH_2Cl_2, $CHCl_3$, and CCl_4 in addition to CH_3Cl..

$$CH_4 + Cl_2 \xrightarrow[\text{or } h\nu]{\Delta} CH_3Cl \xrightarrow{Cl_2} CH_2Cl_2 \xrightarrow{Cl_2} CHCl_3 \xrightarrow{Cl_2} CCl_4$$

Comparable molar quantities ⏟   A mixture of several chlorination products

In the other extreme when Cl_2 is in large excess, all the methane molecules will be fully chlorinated to tetrachloromethane CCl_4. Bromination of alkanes follows the same radical mechanism as that for chlorination. The reaction will take place via bromine $Br^{\cdot}$ and alkyl radicals.

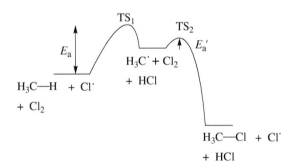

FIGURE 2.11   Reaction profiles for the chain-growth steps of chlorination of methane.

2.3 ENERGETICS OF THE RADICAL HALOGENATIONS OF ALKANES AND THEIR REGIOSELECTIVITY

2.3.1 Energy Profiles for Radical Halogenation Reactions of Alkanes

Figure 2.11 shows energy profile for chlorination of methane. The chlorination and bromination of all the alkanes have the similar features in their overall energy profiles. For halogenations of all types of alkanes, **abstraction of the hydrogen atom from a C—H bond by a halogen radical giving an intermediate alkyl radical (the TS$_1$ step of Fig. 2.11 and below) is the rate-determining step** [5]. This is also supported by the kinetic deuterium isotope effect for the radical bromination of toluene ($k_H/k_D \sim 5$) (refer to Section 1.10).

$$R\text{—}H \ + \ X^{\cdot} \ \xrightarrow[E_a]{k_H} \ R^{\cdot} \ + \ HX \qquad \Delta H_r{}^{\circ} = BDE(C\text{–}H) - BDE(H\text{–}X)$$

Its activation energy (E_a) is directly determined by the C–H BDE. Qualitatively, the stronger the C—H bond, the higher is the E_a for the hydrogen abstraction.

For a given halogen and at a certain temperature, E_a increases linearly with the reaction enthalpy ($\Delta H_r{}^{\circ}$) as expected by the Bell–Evans–Polanyi relationship (refer Section 1.6) [2].

$$E_a = c_1 \Delta H_r{}^{\circ} + c_2 \tag{2.1}$$

$\Delta H_r{}^{\circ} = BDE(C\text{–}H) - BDE(H\text{–}X)$. c_1 and c_2 are positive constants. Therefore, we have

$$E_a = c_1\, BDE(C - H) + c_2 - c_1\, BDE(H - X) \tag{2.2}$$

By employing Arrhenius equation [$k_H = A \exp(-E_a/RT)$] [5, 6], we have

$$\ln(k_H) = -E_a/RT + \ln A$$
$$= -(c_1\Delta H_r{}^{\circ} + c_2)/RT + \ln A$$
$$= -(c_1/RT)\Delta H_r{}^{\circ} - c_2/RT + \ln A$$

By using the common logarithm (log), we have

$$\log(k_H) = -2.303(c_1/RT)\Delta H_r{}^{\circ} + 2.303(\ln A - c_2/RT)$$
$$= -2.303(c_1/RT)\Delta H_r{}^{\circ} + \text{constant}$$

Let $\alpha = 2.303(c_1/RT)$ (a positive constant). We have

$$\log(k_H) = -\alpha\Delta H_r{}^{\circ} + \text{constant}$$
$$\text{or} \quad \log(k_H) = -\alpha[BDE(C - H) - BDE(H - X)] + \text{constant} \tag{2.3}$$

k_H is the rate constant of the abstraction of hydrogen from a C—H bond, and α is a constant. Research has shown [5] that for the highly reactive Cl· radical, the α value of the reaction is about 0.3. For the less reactive Br· radical, the α value is about 0.7.

Equation 2.3 shows that $\log(k_H)$ decreases linearly as a function of BDE(C–H). In general, the BDEs for C—H bonds in different types of carbons follow the order of $BDE(C_{tert}–H) < BDE(C_{sec}–H) < BDE(C_{prim}–H)$ (Fig. 2.8) due to the hyperconjugation effect as discussed in Section 2.1. Therefore, for a given halogen [with a fixed BDE(H–X)], the order of values for rate constants k_H is that $k_H(C_{tert}–H) > k_H(C_{sec}–H) > k_H(C_{prim}–H)$. The rate constant for the overall radical halogenations is determined by k_H, the rate constant for the rate-determining step.

2.3.2   Regioselectivity for Radical Halogenation Reactions

Simple alkanes   Regioselectivity of alkane halogenations is subject to kinetic control. The product distribution (relative yields) is determined by combination of the relative rate constants for halogenations on specific carbons (a primary, secondary, and tertiary carbon atom) and the number of available hydrogen atoms on each of the carbons. Tables 2.1 and 2.2 show the regioselectivity of chlorination and bromination, respectively, for 2-methylbutane, a simple alkane which contains primary, secondary, and tertiary carbon atoms [2]. For each of the halogenations (chlorination and bromination), four monosubstituted haloalkane isomers are produced in different yields. They are 2-halo-2-methylbutane (a tertiary haloalkane), 2-halo-3-methylbutane (a secondary haloalkane), 1-halo-2-methylbutane (a primary haloalkane), and 1-halo-3-methylbutane (a primary haloalkane). On the per-H-atom basis, the yield of tertiary haloalkane is greater than that of secondary haloalkane, and the yield of secondary haloalkane is greater than that of primary haloalkane. These yields are directly determined by the relative rate constants for halogenations on primary, secondary, and tertiary carbons. Figure 2.12 shows the energy profiles for rate-determining steps (hydrogen abstraction

TABLE 2.1   Regioselectivity of Radical Chlorination of 2-Methylbutane

The relative yield of the monochlorination products	22%	33%	30%	15%
Number of available hydrogen atoms for the reaction	1	2	6	3
Yields on the per-H-atom basis	22%	16.6%	5%	5%
Relative rate constant k_{rel} for monochlorination	4.4	3.3	1	1
Type of reacting CH bond	C_{tert}—H	C_{sec}—H	C_{prim}—H	C_{prim}—H

Reprinted from Ref. [2], Page 23, with permission from Elsevier.

TABLE 2.2   Regioselectivity of Radical Bromination of 2-Methylbutane

2-Methylbutane $\xrightarrow[\Delta]{Br_2}$ (products shown with Br substituents) + + +

The relative yield of the monobromination products	92.2%	7.38%	0.28%	0.14%
Number of available hydrogen atoms for the reaction	1	2	6	3
Yields on the per-H-atom basis	92.2%	3.69%	0.047%	0.047%
Relative rate constant k_{rel} for monochlorination	2000	79	1	1
Type of reacting CH bond	C_{tert}–H	C_{sec}–H	C_{prim}–H	C_{prim}–H

Reprinted from Ref. [2], Page 25, with permission from Elsevier.

from a C—H bond by a halogen radical) for alkane chlorination and bromination on primary, secondary, and tertiary carbons. The relative values of activation energy E_a for the rate-determining steps are proportional to the energy levels (stability, represented by BDEs) of the intermediate alkyl radicals (Eq. 2.2). As established in Section 2.3.1, for a given halogen, $k_H(C_{tert}–H) > k_H(C_{sec}–H) > k_H(C_{prim}–H)$. This accounts for the relative k_{rel} values for halogenations of different types of alkanes in Tables 2.1 and 2.2. The difference in the k_{rel} values gives rise to the observed regioselectivity.

Although the formation of $CH_3{}^{\cdot}$ from hydrogen abstraction of CH_4 by $Cl^{\cdot}$ is endothermic [$\Delta H_r^\circ = BDE(H_3C–H) – BDE(H–Cl) = 104–103 = 1\,kcal/mol$)], the formations of primary, secondary, and tertiary radicals from hydrogen abstractions of the corresponding C—H bonds by $Cl^{\cdot}$ are exothermic, and their enthalpies (ΔH_r°) are –5, –8, and –11 kcal, respectively. This is because the BDEs for these C—H bonds are relatively lower than that of CH_4. As a result, the reaction enthalpy values [$\Delta H_r^\circ = BDE(C–H) – BDE(H–Cl)$] become negative. Therefore, all these rate-determining steps for chlorination have early TS_1 transition states, closer to the starting C—H bonds (Fig. 2.12). As a result, the differences in the E_a values for the formation of different types of radicals are small, giving rise to small differences in the relative rate constants k_{rel}. This accounts for the relatively small regioselectivity for chlorinations. However, in bromination reactions, the formations of primary, secondary, and tertiary radicals from hydrogen abstractions of the corresponding C—H bonds by $Br^{\cdot}$ are endothermic, and their enthalpies (ΔH_r°) are +10, +7, and +4 kcal/mol, respectively. This is because the BDE of H–Br (88 kcal/mol) is much smaller than that of H–Cl (103 kcal/mol). The difference in the BDEs enhances the reaction enthalpy [$\Delta H_r^\circ = BDE(C–H) – BDE(H–Br)$] by 15 kcal/mol for the formations of different types of radicals in bromination relative to those for chlorination, making the rate-determining steps in bromination endothermic. As a result, all these rate-determining steps in bromination have late TS_1

(a)

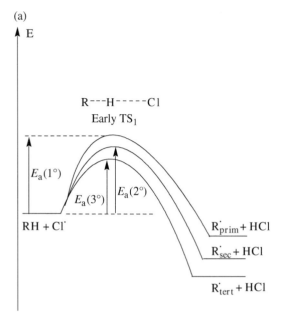

(b)

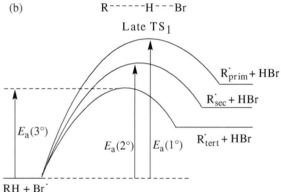

FIGURE 2.12 Reaction profiles, transition states, and activation energies for rate-determining steps for radical chlorination (a) and bromination (b) of primary (1°), secondary (2°), and tertiary (3°) C–H bonds: R–H + X· → R· + HX (X = Cl or Br).

transition states, closer to the corresponding alkyl radicals (Fig. 2.12). It makes relatively large differences in the E_a values for the formation of different types of radicals in brominations. Correspondingly, the relative rate constants k_{rel} vary a lot, the differences being much greater than those for chlorination reactions. Therefore, the alkane bromination reaction exhibits remarkable regioselectivity, much more appreciable than that for chlorination. Equation 2.3 shows that for a given alkane substrate, $\log(k_H)$ increases linearly as a function of BDE(H–X). Since BDE(H–Cl) is 15 kcal/mol greater than BDE(H–Br), k_H for Cl_2 is greater than k_H for Br_2. Cl_2 is more reactive toward an alkane than Br_2.

Hydrocarbons containing an unsaturated group   Unsaturated groups (such as phenyl and vinyl) stabilize radicals via conjugation effect. Therefore, an alkyl which is attached to an unsaturated group undergoes radical halogenations readily with the unsaturated entity being intact in the overall reaction. For example, at a temperature of 100 °C, toluene reacts with elemental chlorine giving benzyl chloride [2]. The phenyl ring is intact.

(Excess)   Benzyl chloride

This is the industrial method for synthesis of benzyl chloride. At 100 °C, Cl_2 dissociates reversibly giving a small amount of $Cl^.$ radical. $Cl^.$ can readily attack the methyl group in toluene producing the benzyl $PhCH_2^.$ radical. It is even more stable than a regular tertiary radical due to the conjugation effect of phenyl. As a result, a radical chlorination on $-CH_3$ takes place giving benzyl chloride. No hydrogen atom in the phenyl ring is abstracted by $Cl^.$ because the BDE of $C_{sp^2}-H$ in benzene (111 kcal/mol) is high and the corresponding $Ph^.$ radical is highly unstable. The formation of $Ph^.$ would require a large activation energy and be unfavorable. As a result, no reaction takes place in the aromatic ring.

With increase in quantity of chlorine, the monosubstituted benzyl chloride undergoes further radical chlorinations, but at lower rates:

Benzyl chloride

Studies have shown [2] that the second chlorination is about 10 times slower than the first one, and the third chlorination is about 10 times slower than the second one.

Similar to halogenations of toluene, reaction of ethylbenzene with elemental bromine at a high temperature (or in the light) results in bromination predominantly in the secondary benzylic carbon.

(Excess)   (Major)

The reaction takes place via a secondary benzyl $PhCH^.CH_3$ radical. It is even more stable than the primary benzyl $PhCH_2^.$ radical. This stabilization favors bromination overwhelmingly on the benzylic carbon. Bromination on the methyl group would proceed via a regular primary $PhCH_2CH_2^.$ radical whose BDE (~98 kcal/mol) is about 9 kcal/mol higher than that of $PhCH^.CH_3$ (~89 kcal/mol). Therefore, the reaction on the methyl group would be energetically unfavorable.

At a high temperature of 500 °C, propene reacts with elemental chlorine giving 3-chloropropene (allyl chloride). The C=C double bond is intact.

$$H_2C\!\!=\!\!CH\!\!-\!\!CH_3 \ + \ Cl_2 \ \xrightarrow{\ 500\ °C\ } \ H_2C\!\!=\!\!CH\!\!-\!\!CH_2Cl \ + \ HCl$$

(Excess)   3-Chloropropene

This is the industrial method for synthesis of allyl chloride [2]. At 500 °C, Cl_2 dissociates producing a considerable amount of $Cl^{\cdot}$ radical. $Cl^{\cdot}$ can readily attack the methyl group in propene producing the allyl $H_2C=CHCH_2^{\cdot}$ radical. Its stability is comparable to that for $PhCH_2^{\cdot}$ due to the conjugation effect of vinyl. As a result, a radical chlorination on $-CH_3$ takes place almost exclusively at this temperature giving allyl chloride as essentially a sole product. Addition of Cl_2 to the C=C bond does not take place at this temperature. In the following section, we will further study the factors which affect the regioselectivity (radical substitution vs. addition) for propene halogenation.

2.4   KINETICS OF RADICAL HALOGENATIONS OF ALKANES

2.4.1   Alkanes

The following equation shows the overall reaction of radical halogenation (X = Cl or Br) of a general alkane (RH) which takes place thermally at a high temperature:

$$RH + X_2 \rightarrow RX + HX$$

According to the mechanism studied in the previous sections, the reaction consists of the following three microscopic elementary steps:

$$X_2 \ \underset{k_{-1}}{\overset{k_1}{\rightleftharpoons}} \ 2\,X^{\cdot}$$

$$RH \ + \ X^{\cdot} \ \xrightarrow{k_2} \ R^{\cdot} \ + \ HX$$

$$R^{\cdot} \ + \ X_2 \ \xrightarrow{k_3} \ RX \ + \ X^{\cdot}$$

As discussed in Section 2.2, termination steps are very minor upon heating. They will not have appreciable effect on the reaction kinetics and are neglected.

In the overall reaction of thermally induced alkane halogenation, the first step is the reversible dissociation of a halogen X_2 molecule to two halogen $X^{\cdot}$ radicals. k_1 and k_{-1} are the rate constants for the forward and backward processes, respectively. At equilibrium, the forward reaction rate $(k_1[X_2])$ is equal to the backward reaction rate $(k_{-1}[X^{\cdot}]^2)$, namely, that

$$k_1[X_2] = k_{-1}[X]^2$$

Its equilibrium constant (K) can be expressed as

$$K = [X^\cdot]^2/[X_2] = k_1/k_{-1}$$

Therefore,

$$[X^\cdot] = K^{1/2}[X_2]^{1/2} = (k_1/k_{-1})^{1/2}[X_2]^{1/2} \tag{2.4}$$

k_2 and k_3 are the rate constants for the two chain-growth steps. The intermediate alkyl radical $R^\cdot$ is highly reactive, and it is remained at approximately a constant low concentration throughout the overall reaction. Therefore, we can apply the steady-state assumption to $R^\cdot$ [6], formulated mathematically as:

$$d[R^\cdot]/dt = 0 = k_2[RH][X^\cdot] - k_3[R^\cdot][X_2] \tag{2.5}$$

The first term $k_2[RH][X^\cdot]$ in Equation 2.5 is the rate of the formation for $R^\cdot$, while the second term $k_3[R^\cdot][X_2]$ is the rate of the disappearance for $R^\cdot$. Therefore, the net change in the $R^\cdot$ molar concentration per unit time ($d[R^\cdot]/dt$) is equal to the rate of formation ($k_2[RH][X^\cdot]$) minus the rate of disappearance ($k_3[R^\cdot][X_2]$).

Equation 2.5 can be rewritten as

$$k_3[R^\cdot][X_2] = k_2[RH][X^\cdot] \tag{2.6}$$

The rate for the overall halogenation reaction can be defined as $d[RX]/dt$ (increase in molar concentration of the product RX per unit time). It also represents the rate of the k_3 step (the second step of the chain-growth) because RX is only produced in this step. The rate-law for the k_3 step can be written as

$$d[RX]/dt = k_3[R^\cdot][X_2] \tag{2.7}$$

Combining Equations 2.6 and 2.7 leads to

$$d[RX]/dt = k_2[RH][X^\cdot] \tag{2.8}$$

Substitute Equation 2.4 for Equation 2.8

$$d[RX]/dt = k_2 K^{1/2} [RH][X_2]^{1/2} = k_2 (k_1/k_{-1})^{1/2} [RH][X_2]^{1/2} \tag{2.9}$$

Equation 2.9 is the rate-law for the overall radical halogenation reaction. The reaction is half-order in X_2, indicative of the fact that it is initiated by $X^\cdot$ which is produced from dissociation of X_2. Equation 2.9 shows that the rate of the overall reaction is proportional to k_2, but independent of k_3. The rate law has confirmed that the k_2 step (abstraction of hydrogen from the R—H bond by a halogen radical) is the rate-determining step, and it is consistent with the mechanistic studies.

Comparison of the energy profile of chlorination with that of bromination (Fig. 2.12) shows that for a given alkane substrate, the activation energy for chlorination is smaller than that for bromination. Therefore, k_2 (rate constant of the rate-determining step) for chlorination is greater than k_2 for bromination. In other words, Cl_2 is more reactive than Br_2 toward a given alkane. We have seen above that radical bromination of alkanes is more regioselective than chlorination. This implies a general principle: *A highly reactive reagent is usually less regioselective than a reagent with lower reactivity.*

2.4.2 Hydrocarbons Containing an Unsaturated Group

Now let us consider the factors that affect regioselectivity for reaction of propene ($CH_2=CHCH_3$) with an elemental halogen X_2 (X = Cl or Br) at a high temperature.

$$XCH_2CH(X)CH_3 \leftarrow CH_2=CHCH_3 + X_2 \rightarrow CH_2=CHCH_2X + HX$$

Reaction on the right side is a substitution giving 3-halopropene ($CH_2=CHCH_2X$), while reaction on the left side is an addition giving 1,2-dihalopropane [$XCH_2CH(X)CH_3$]. The two competing reactions can take place concurrently, resulting in different product distributions (different yields) depending on reaction conditions. The mechanism for the competing reactions is presented in Figure 2.13a.

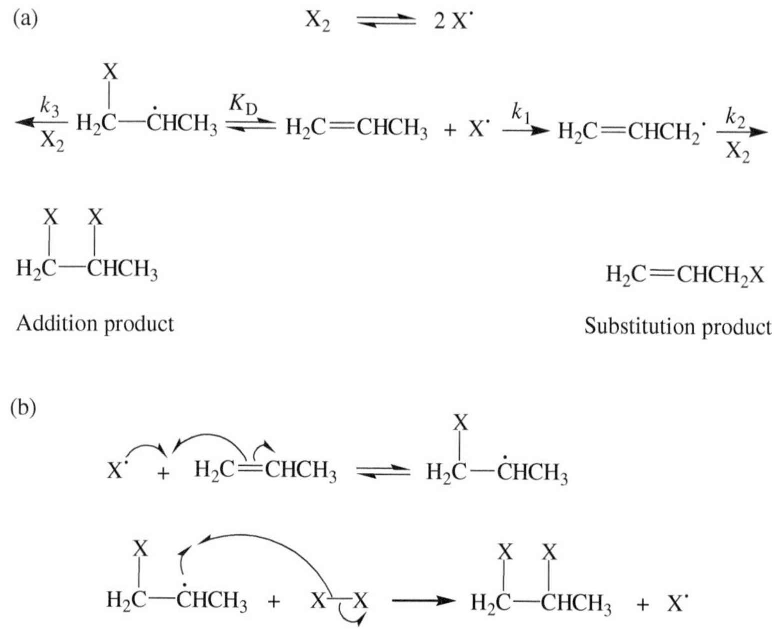

FIGURE 2.13   (a) Overall mechanism for radical halogenation (substitution and addition) of propene; and (b) further mechanistic details for the radical addition.

Both the substitution and addition are initiated by an $X^\cdot$ radical produced from reversible thermal dissociation of X_2. The substitution follows a general alkane radical halogenation mechanism. The addition is stepwise. First, $X^\cdot$ is added reversibly to the doubly bonded CH_2 carbon in propene giving an intermediate $XCH_2CH^\cdot CH_3$ radical, while the reversal dissociation of $XCH_2CH^\cdot CH_3$ would regenerate propene and $X^\cdot$. Both the addition and dissociation are fast, reaching an equilibrium rapidly. K_D represents the dissociation constant for $XCH_2CH^\cdot CH_3$. After $XCH_2CH^\cdot CH_3$ is formed, it subsequently attacks a halogen X_2 molecule to break the $X{-}X$ bond homolytically and produce an $X^\cdot$ radical and a $XCH_2CH(X)CH_3$ molecule, the addition product. The mechanism for the radical addition is further detailed in Figure 2.13b.

The substitution product $CH_2{=}CHCH_2X$ is formed in the k_2 step. The rate-law can be formulated as

$$d[CH_2{=}CHCH_2X]/dt = k_2[CH_2{=}CHCH_2^\cdot][X_2] \tag{2.10}$$

Applying the steady-state assumption to $CH_2{=}CHCH_2^\cdot$, we have

$$d[CH_2{=}CHCH_2^\cdot]/dt = 0 = k_1[propene][X^\cdot] - k_2[CH_2{=}CHCH_2^\cdot][X_2] \tag{2.11}$$

Equation 2.11 can be rewritten as

$$k_2[CH_2{=}CHCH_2^\cdot][X_2] = k_1[propene][X^\cdot] \tag{2.12}$$

Combining Equations 2.10 and 2.12, we have

$$d[CH_2{=}CHCH_2X]/dt = k_1[propene][X^\cdot] \tag{2.13}$$

The addition product $XCH_2CH(X)CH_3$ is formed in the k_3 step. The rate-law can be formulated as:

$$d[XCH_2CH(X)CH_3]/dt = k_3[XCH_2CH^\cdot CH_3][X_2] \tag{2.14}$$

The dissociation constant K_D for $XCH_2CH^\cdot CH_3$ can be expressed as

$$K_D = [propene][X^\cdot]/[XCH_2CH^\cdot CH_3]$$

Therefore,

$$[XCH_2CH^\cdot CH_3] = (1/K_D)[propene][X^\cdot] \tag{2.15}$$

Substituting Equation 2.15 for Equation 2.14 leads to

$$d[XCH_2CH(X)CH_3]/dt = (k_3/K_D)[propene][X^\cdot][X_2] \tag{2.16}$$

Referring to Equations 2.13 and 2.16, ratio of the rate for the formation of $CH_2=CHCH_2X$ $\{d[CH_2=CHCH_2X]/dt\}$ to the rate for the formation of XCH_2CH $(X)CH_3$ $\{d[XCH_2CH(X)CH_3]/dt\}$ can be written as

$$
\begin{aligned}
&\{d[CH_2 = CHCH_2X]/dt\}/\{d[XCH_2CH(X)CH_3]/dt\} \\
&= k_1[\text{propene}][X^{\cdot}]/(k_3/K_D)[\text{propene}][X^{\cdot}][X_2] = K_D(k_1/k_3)(1/[X_2])
\end{aligned}
\tag{2.17}
$$

This would be equal to ratio of the yield of $CH_2=CHCH_2X$ (substitution product) to the yield of $XCH_2CH(X)CH_3$ (addition product).

Equation 2.17 shows that at a given temperature, the product distribution (yield ratio of $CH_2=CHCH_2X$ to $XCH_2CH(X)CH_3$) for reaction of a given halogen with propene is inversely proportional to the molar concentration of the starting X_2. High concentrations of X_2 favors the formation of the addition product $XCH_2CH(X)CH_3$ because its rate-determining step involves X_2, while X_2 is not involved in the rate-determining step for $CH_2=CHCH_2X$. When concentrations of X_2 are low, the rate for the formation of $XCH_2CH(X)CH_3$ decreases. By default, the substitution reaction becomes more competitive, giving rise to higher yield of $CH_2=CHCH_2X$. When the total amount of a halogen (bromine or chlorine) is fixed, variation in reaction temperature can alter the molar concentration of the undissociated halogen ($[X_2]$). As a result, a very high temperature (such as ~500 °C) leads to a high percentage of dissociation of X_2, giving rise to a relatively low $[X_2]$ concentration. This will favor the substitution reaction giving $CH_2=CHCH_2X$. When temperature is lowered, percentage of dissociation of X_2 is getting smaller, and $[X_2]$ becomes relatively greater. It will lead to the formation of more addition product $XCH_2CH(X)CH_3$.

Halogenations of cyclohexene follow analogous regioselectivity (Eq. 2.18).

At low X_2 concentrations (or high temperatures), the reaction is dominated by the substitution giving 3-halocyclohexene as the major product. The substitution takes place highly selectively in the secondary allyl C_3 carbon because the corresponding allyl radical is more stable than a regular secondary radical due to conjugation effect.

An allyl radical   A regular
                          secondary radical

Increase in the X_2 concentration (or lowering reaction temperature) will facilitate the addition giving 1,2-dihalocyclohexane.

2.5 RADICAL INITIATORS

Some substances are thermally unstable. Upon moderate heating (usually less than 100 °C), they decompose to radical species which can interact with certain reagents to produce another radical which initiates a C–H functionalization. Such substances are called **radical initiators**. When a radical initiator is employed in an organic functionalization, usually, only catalytic amount is required.

2.5.1 Azobisisobutyonitrile (AIBN)

A common radical initiator used for halogenations of alkanes is azobisisobutyonitrile (AIBN). At 80 °C, it undergoes thermal decomposition to the $NCMe_2C^{\cdot}$ radical and nitrogen gas [2].

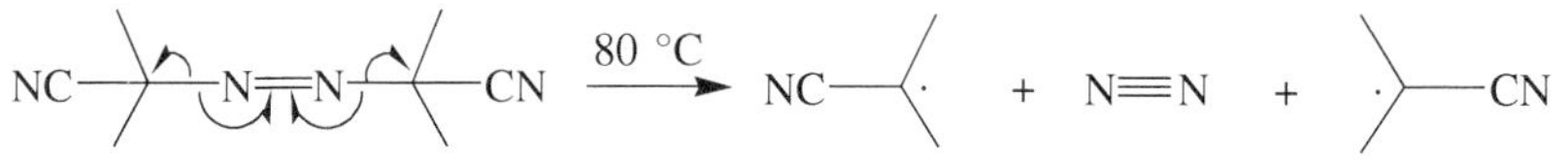

Azobisisobutyonitrile $(NCMe_2C^{\cdot})$ $(NCMe_2C^{\cdot})$
(AIBN)

In the presence of a catalytic amount of AIBN, halogenations (such as bromination) of alkanes (such as 2-methylpropane) can take place under moderate heating (~80 °C).

$$(CH_3)_3C-H \; + \; Br_2 \quad \xrightarrow[\text{catalytic amount}]{\text{AIBN}} \quad (CH_3)_3C-Br$$

First, AIBN undergoes a thermal decomposition to produce the $NCMe_2C^{\cdot}$ radical. Then, $NCMe_2C^{\cdot}$ attacks a Br_2 molecule to produce a $Br^{\cdot}$ radical. $Br^{\cdot}$ attacks $(CH_3)_3C-H$ to initiate the chain-growth steps for the radical bromination reaction. The mechanism is shown below:

$$AIBN \rightarrow 2\,NCMe_2C^{\cdot} + N_2$$
$$NCMe_2C^{\cdot} + Br_2 \rightarrow NCMe_2CBr + Br^{\cdot} \,(\text{initiation})$$
$$(CH_3)_3C - H + Br^{\cdot} \rightarrow (CH_3)_3C^{\cdot} + HBr$$
$$(CH_3)_3C^{\cdot} + Br_2 \rightarrow (CH_3)_3C - Br + Br^{\cdot}$$

Normally, light or high temperature is required for bromination of alkanes (such as 2-methylpropane). In the presence of a radical initiator (such as AIBN), light is not necessary, and reaction can take place at relatively mild conditions.

AIBN can catalyze bromination of o-xylene. The reaction takes place only on the two methyl groups, and the aromatic ring is intact.

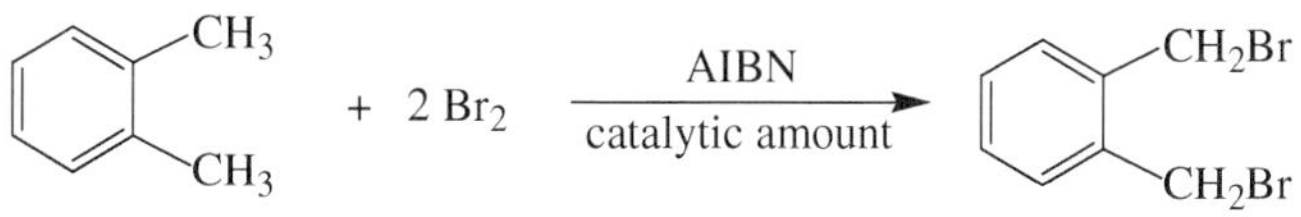

An aromatic C—H bond has a large BDE (~111 kcal/mol). The formation of a Ph˙–like radical would be highly energetically unfavorable. Therefore, an aromatic ring does not undergo a radical halogenation. The reaction of *o*-xylene with bromine may also be conducted photochemically in the ultraviolet light or thermally at high temperature with the same regioselectivity.

AIBN catalyzed bromination reaction of propene gives 1,2-dibromopropane (an addition product) as the major product.

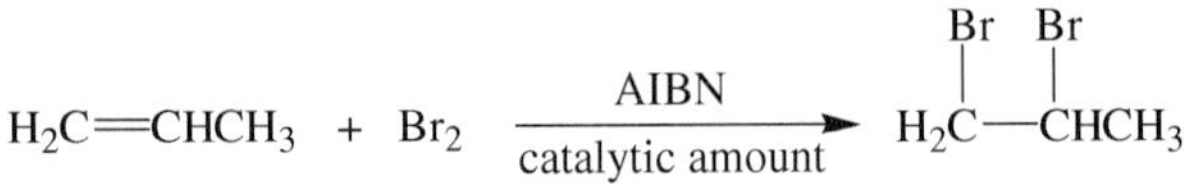

$$\text{H}_2\text{C}\!\!=\!\!\text{CHCH}_3 \ + \ \text{Br}_2 \ \xrightarrow[\text{catalytic amount}]{\text{AIBN}} \ \overset{\overset{\displaystyle \text{Br}}{|}}{\text{H}_2\text{C}}\!\!-\!\!\overset{\overset{\displaystyle \text{Br}}{|}}{\text{CHCH}_3}$$

In the reaction conditions, NCMe$_2$C˙, the radical initiated by AIBN, can only induce dissociation of a small percentage of Br$_2$ to Br˙. As a result, the Br$_2$ concentration in the reaction system will be relatively high. According to Equation 2.17, it will facilitate a radical addition reaction.

2.5.2   Dibenzol Peroxide

Another important radical initiator is dibenzol peroxide (PhCOO)$_2$. Upon moderate heating (95 °C), it decomposes to produce very reactive phenyl Ph˙ radical (Fig. 2.14). Ph˙ (BDE = 111 kcal/mol) is even less stable or more reactive than the methyl CH$_3$˙ radical (BDE = 104 kcal/mol).

Ph˙ can initiate autoxidation of a tertiary, benzylic or allylic C—H bond of a hydrocarbon. Selective oxidation of an organic compound (typically on a C–H or a C=C bond) by molecular oxygen is called **autoxidation**. Autoxidation is a major functionalization process for hydrocarbons. Commonly, it is initiated by a radical species, although nonradical mechanisms have also been found for some of this type of reactions.

Autoxidation of cumene that is facilitated by a catalytic amount of (PhCOO)$_2$ (radical initiator) is an industrially important reaction (Fig. 2.15) [2]. First,

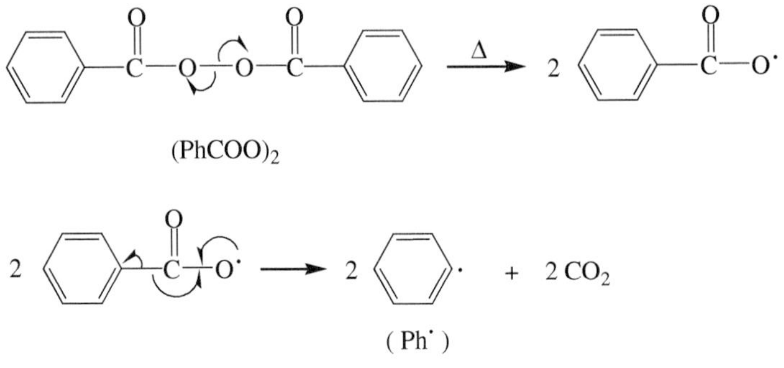

(PhCOO)$_2$

(Ph˙)

FIGURE 2.14   Thermal dissociation of (PhCOO)$_2$ giving the Ph˙ radical.

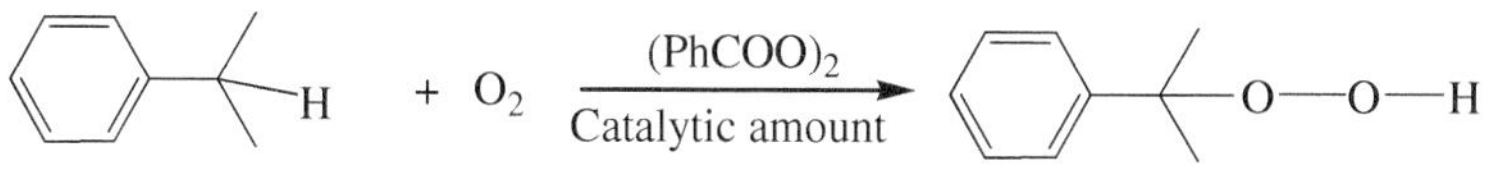

Mechanism:

$$(PhCOO)_2 \xrightarrow{\Delta} 2\ PhCOO^{\cdot} \longrightarrow 2\ Ph^{\cdot} + 2\ CO_2$$

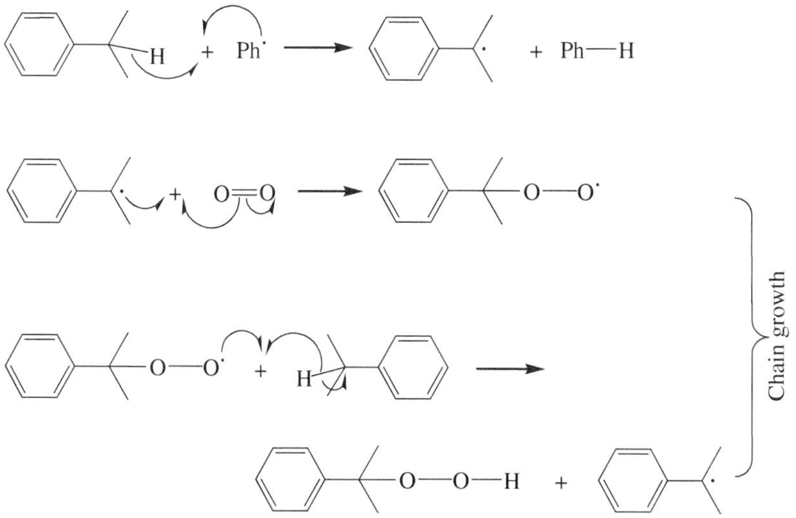

FIGURE 2.15 Mechanism for $(PhCOO)_2$ initiated autoxidation of cumene (isopropylbenzene). Based on Bruckner [2].

$(PhCOO)_2$ undergoes a thermal decomposition to ultimately initiate phenyl $Ph^{\cdot}$ radical. Then, the highly reactive $Ph^{\cdot}$ attacks the C_{tert}—H bond in cumene, abstracting the hydrogen atom and producing a cumyl radical plus a benzene Ph–H molecule. The large BDE(Ph–H) makes this step exothermic with an early transition state and a relatively low activation energy. Its reaction enthalpy ΔH_r° is estimated as follows:

$$\begin{aligned}\Delta H_r^\circ &= \mathrm{BDE}(PhCH_2^{\cdot}) - \mathrm{BDE}(Ph - H)\\ &= 89\ \mathrm{kcal/mol} - 111\ \mathrm{kcal/mol} = -22\ \mathrm{kcal/mol}\end{aligned}$$

The cumyl radical subsequently attacks an O_2 molecule to initiate the autoxidation. The product cumyl peroxide can undergo rearrangement to phenol and acetone (refer to this rearrangement in Chapter 10). In industry, this approach has been used as a major method to make phenol and acetone, two important fundamental chemicals.

2.6 TRANSITION-METAL-COMPOUNDS CATALYZED ALKANE C—H BOND ACTIVATION AND FUNCTIONALIZATION

2.6.1 The C—H Bond Activation via Agostic Bond

In general, an overall functionalization process that occurs to a C—H bond in organic compounds consists of two steps: First, the strong C—H bond is cleaved. It is called **activation** and usually, is the rate-determining step for the overall reaction. The C—H bond activation is followed by reaction of the carbon with an inorganic functionalizing agent to make a functional group. The second step is referred to as **functionalization**. In the previous sections, we have learned a general method of the C—H bond activation by reactive radicals (such as chlorine and bromine radicals), namely, that a radical species attacks the C—H bond breaking it homolytically to give an alkyl radical. In this section, we will study a second general method, which is the C—H bond activation by transition metal compounds, typically organometallic or coordination compounds of the late transition metals.

Coordination compounds (ML_n, L represents a ligand) of some late transition metals, such as Pt, Ir, Rh, and Hg, contain vacant orbitals (typically d orbitals) in the valence shell of the metal center of ML_n. These vacant orbitals are highly electrophilic (strong electron-pair acceptors). They can even accept the very inactive (of low reactivity) bonding electron-pair of an alkane C—H bond via an **agostic** interaction [the interaction of a σ-bonding orbital (filled) in one molecule with a vacant atomic (or hybridized) orbital in another molecule] (Fig. 2.16). As the σ bonding electron-pair is transferred into the vacant orbital in the metal center, the C—H bond is cleaved, resulting in an **oxidative addition** to the metal center by forming relatively weak C—M and H—M bonds. The formation of the oxidative $L_nM(H)R$ adduct takes place via a three-center agostic transition state in which the C—H bond is being partially broken coincident with partial formation of the C—M and H—M bonds. The nature of the transition state is evidenced by large kinetic deuterium isotope effect ($k_H/k_D = 3–6$) for the activation step of many transition metal complexes, indicating

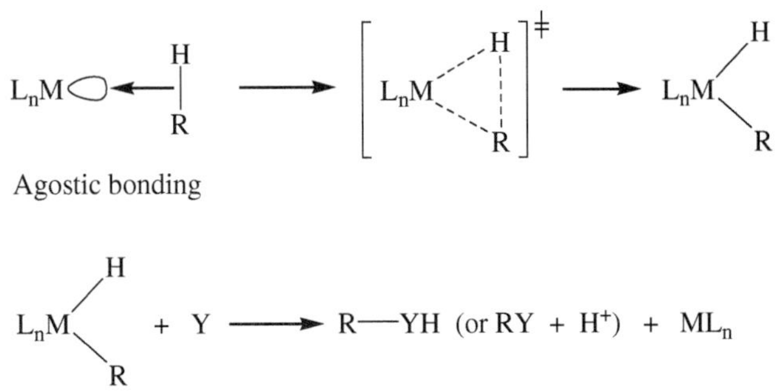

Agostic bonding

FIGURE 2.16 The alkane C—H bond activation by transition metal complex (ML_n) via agostic bonding.

significant C—H bond cleavage in the transition state [1]. The $L_nM(H)R$ adduct, upon being treated with certain inorganic reagents (functionalizing reagents Y), undergoes a reductive elimination to lead to functionalization (the formation of a C—Y bond) and regeneration of ML_n. The general activation and functionalization of a C—H bond via an agostic transition state are demonstrated in Figure 2.16.

Figure 2.17 shows examples of activation of alkane C—H bonds by several transition metal organometallic compounds [1, 7]. It has been found that treatment of $Cp*(PMe_3)M(H)R$, an activated alkane (RH) adduct of the $Cp*(PMe_3)M$ (M = Ir or Rh) complex (the first product in Fig. 2.17), by Br_2 in bromoform affords bromoalkane.

$$Cp*(PMe_3)M(H)R + Br_2 \rightarrow RBr + HBr + Cp*(PMe_3)M$$

For the third organoiridium complex, the –OTf (triflate) ligand undergoes a rearrangement in the process of the oxidative C—H bond addition. The iridium metal center selectively activates a tertiary C_{tert}—H bond leading to transformation of the H—CR_3 bond to Ir—CR_3 and Ir—H bonds [7].

Much investigation has revealed that Hg(II) and Pt(II) exhibit highly strong electrophilicity toward inactive C—H bonds. The Hg(II) and Pt(II) compounds facilitate oxidative functionalization of hydrocarbon C—H bonds by molecular oxygen. This type of functionalization is of a great synthetic significance in chemical industry. We will give it a brief introduction below.

2.6.2   Mechanisms for the C—H Bond Oxidative Functionalization

Among the most important C–H functionalizations is selective oxidation of methane to methanol by molecular oxygen. Methane, the major component (95%) of natural gas, is an abundant resource (comparable to petroleum) that could serve as an efficient fuel source and chemical feedstock. However, its utilization is greatly hindered by transportation of the gas. An effective solution is to convert it to methanol (referred to as "gas-to-liquid" conversion in industry) which is much more readily transported to utilization sites. An ideal method for this conversion would be to selectively oxidize a C—H bond in methane by O_2 (a very cheap oxidant) to make methanol. For many years, this has been an extremely difficult task due to a very high BDE(C–H) = 104 kcal/mol of methane. In addition, the oxidation renders side reactions the major of which is an overoxidation of CH_4 to CO_2. Many efforts have been made in development of the oxidative C—H bond functionalization for methane.

A remarkable way to activate the C—H bond in methane is to use highly electrophilic Hg(II) compounds. It has been found that stoichiometric reaction of methane with mercury(II) triflate in triflic acid at 180 °C results in almost a quantitative formation of methyl triflate as shown below [8]:

$$CH_4 + 2Hg(CF_3SO_3)_2 \rightarrow CF_3SO_3CH_3 + Hg_2(CF_3SO_3)_2 + CF_3SO_3H$$

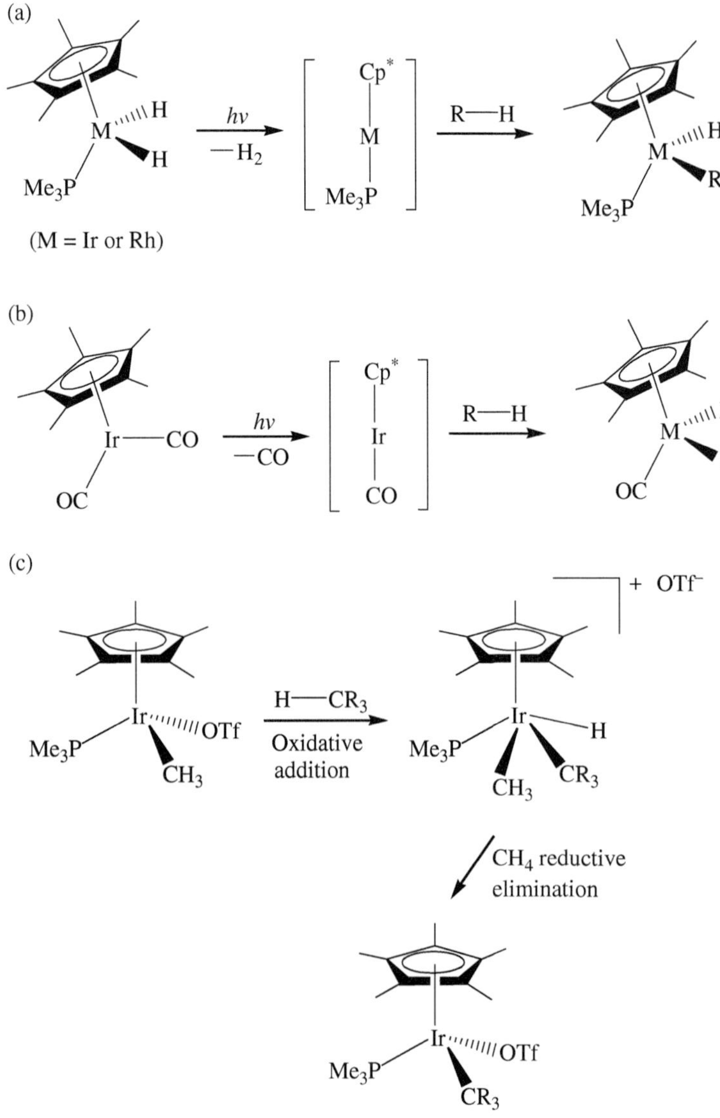

(a)

(M = Ir or Rh)

(b)

(c)

FIGURE 2.17 The alkane C—H bond activation by various transition metal complexes. (a) Metal hydrides, (b) Metal carbonyl, and (c) Metal triflate. Arndtsen et al. [1] and Klei et al. [7].

In the course of the reaction, methane is oxidized by Hg(II) in Hg(CF$_3$SO$_3$)$_2$. Hydrolysis of CF$_3$SO$_3$CH$_3$ gives CH$_3$OH readily.

Further studies have revealed that in 100% sulfuric acid, the reaction of methane can proceed at the same temperature using only a catalytic amount of Hg(II) [in the form of mercury(II) bisulfate Hg(OSO$_3$H)$_2$] [8]. The reaction gives methyl bisulfate (yield: 85%) and sulfur dioxide as shown below:

$$CH_4 + 2H_2SO_4 \xrightarrow[\text{catalytic amount}]{Hg(OSO_3H)_2} CH_3OSO_3H + SO_2 + H_2O$$

During the course of the reaction, an organometallic intermediate CH_3HgOSO_3H (containing a C—Hg bond) has been identified experimentally by ^{13}C NMR spectroscopy to form at a low steady-state concentration. Hydrolysis of CH_3OSO_3H affords CH_3OH quantitatively and readily. In addition, reoxidation of SO_2 to H_2SO_4 (via SO_3) by air (20% O_2) is performed readily.

The mechanism of the overall oxidative functionalization of methane is shown as follows [8]:

$$CH_4 + Hg(OSO_3H)_2 \rightarrow CH_3HgOSO_3H + H_2SO_4$$

$$CH_3HgOSO_3H + 3H_2SO_4 \rightarrow CH_3OSO_3H + Hg(OSO_3H)_2 + SO_2 + 2H_2O$$

$$CH_3OSO_3H + H_2O \rightarrow CH_3OH + H_2SO_4$$

$$\tfrac{1}{2}O_2 + SO_2 + H_2O \rightarrow H_2SO_4$$

Net : $CH_4 + \tfrac{1}{2}O_2 \rightarrow CH_3OH$

In the first step of the reaction, a C—H bond in methane is activated by Hg(II) resulting in the formation of a C—Hg bond. Then, C–Hg in the intermediate CH_3HgOSO_3H is oxidized by sulfuric acid H_2SO_4 (the functionalizing agent) giving the methyl ester CH_3OSO_3H. Sulfuric acid is reduced to SO_2 and 2 mol of H_2O. $Hg(OSO_3H)_2$ (the catalyst) is regenerated. In the following step, hydrolysis of CH_3OSO_3H is performed to afford CH_3OH, the targeted product. Finally, SO_2 is reoxidized to H_2SO_4. The net result is an oxidation of methane to methanol by molecular oxygen (a very cheap oxidant). This approach represents a practical method which can be potentially useful for industry.

We believe that activation of the very inactive C—H bond in methane (the rate-determining step of and the key to the overall oxidative functionalization) is achieved by an agostic interaction of the C—H bond with electrophilic Hg(II) in $Hg(OSO_3H)_2$ as shown in Figure 2.18. The agostic interaction can possibly result in the formation of a postulated four-center transition state in which the C—H bond is substantially cleaved coincident with significant formation of a C—Hg bond. Finally, the collapse of the transition state leads to the formation of CH_3HgOSO_3H and H_2SO_4. Overall, the activation of the H_3C—H bond by $Hg(OSO_3H)_2$ giving CH_3HgOSO_3H and H_2SO_4 is a concerted electrophilic substitution.

Similar to mercury(II), some planar platinum(II) ($5d^8$ configuration) complexes can also catalyze the selective oxidation of the C—H bond in methane to a methanol derivative, which in turn can be readily hydrolyzed to methanol. A remarkable Pt(II) complex used in the catalysis is dichloro(η-2-{2,2′-bipyrimidyl})platinum(II), denoted as **(bpym)PtCl₂** (Fig. 2.19). The central platinum in the complex is dsp^2–hybridized. Research has shown [9] that in the presence of (bpym)PtCl₂ in a catalytic amount, CH_4 can be oxidized by 102% H_2SO_4 to CH_3OSO_3H.

$$CH_4 + 2H_2SO_4 \xrightarrow[\text{Sulfuric acid}]{(bpym)PtCl_2} CH_3OSO_3H + SO_2 + 2H_2O$$

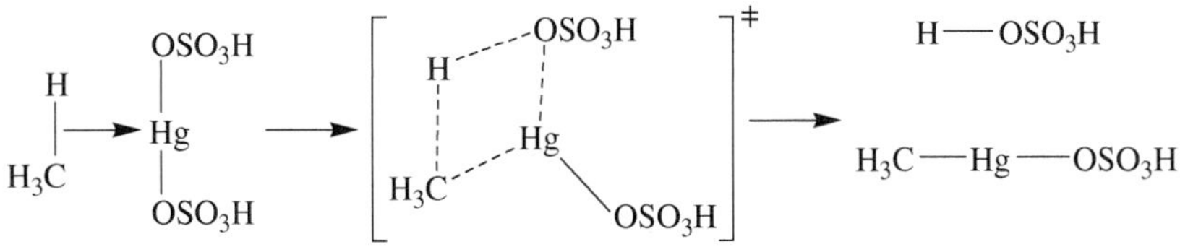

FIGURE 2.18   Mechanism for the methane C—H bond activation by mercury(II) bisulfate.

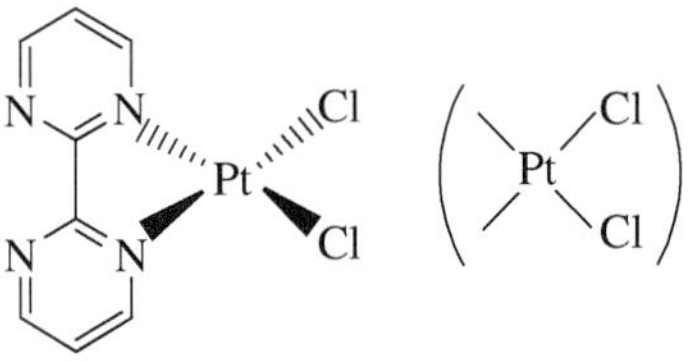

FIGURE 2.19   Structure of dichloro(η-2-{2,2′-bipyrimidyl})platinum(II), denoted as (bpym)PtCl$_2$.

The reaction is conducted in the sulfuric acid medium at 220 °C with the yield of CH$_3$OSO$_3$H being 81%. The NMR study shows that the only observed carbon product in the liquid phase during the reaction is CH$_3$OSO$_3$H [9].

The mechanism for the reaction has been studied [9] and is presented in Figure 2.20. At the reaction temperature, (bpym)PtCl$_2$, first, undergoes a thermal dissociation (reversible) to give the [(bpym)PtCl]$^+$ cation. The vacant dsp^2 orbital in the Pt(II) center of the cation is believed highly electrophilic. It can strongly attract the bonding electron pair in a C—H bond of CH$_4$ to possibly effect an oxidative addition and cleave the C—H bond (activation) via an agostic transition state. As a result, a Pt—CH$_3$ σ bond is formed. Then, the intermediate platinum(II)–methyl complex is oxidized by H$_2$SO$_4$ to CH$_3$OSO$_3$H (oxidative functionalization), and concurrently, H$_2$SO$_4$ is reduced to SO$_2$ and H$_2$O. The catalyst (bpym)PtCl$_2$ is regenerated. In principle, SO$_2$ can be readily reoxidized by O$_2$ back to H$_2$SO$_4$, while CH$_3$OSO$_3$H can be easily hydrolyzed to the targeted CH$_3$OH. Therefore, analogous to the Hg(II) compounds, **the (bpym)PtCl$_2$ catalyst can make possible an overall net selective oxidation of CH$_4$ to CH$_3$OH by O$_2$.**

Different from the Hg(II) catalyzed reaction performed in the neat H$_2$SO$_4$ in which a mercury(II)–methyl complex (CH$_3$HgOSO$_3$H) is observed to form at a low concentration by NMR spectroscopy, the Pt—CH$_3$ σ bond is very unstable and no platinum(II)–methyl complex is observed experimentally in the course of the Pt(II) complex catalyzed reaction.

2.7   SUPERACIDS CATALYZED ALKANE C—H BOND ACTIVATION AND FUNCTIONALIZATION

A **superacid** is any protic acid whose strength (dissociation constant) is greater than that of sulfuric acid, or any Lewis acid which is stronger than aluminum chloride. Common examples of superacids are fluoro sulfuric acid (FSO$_3$H) and triflic acid

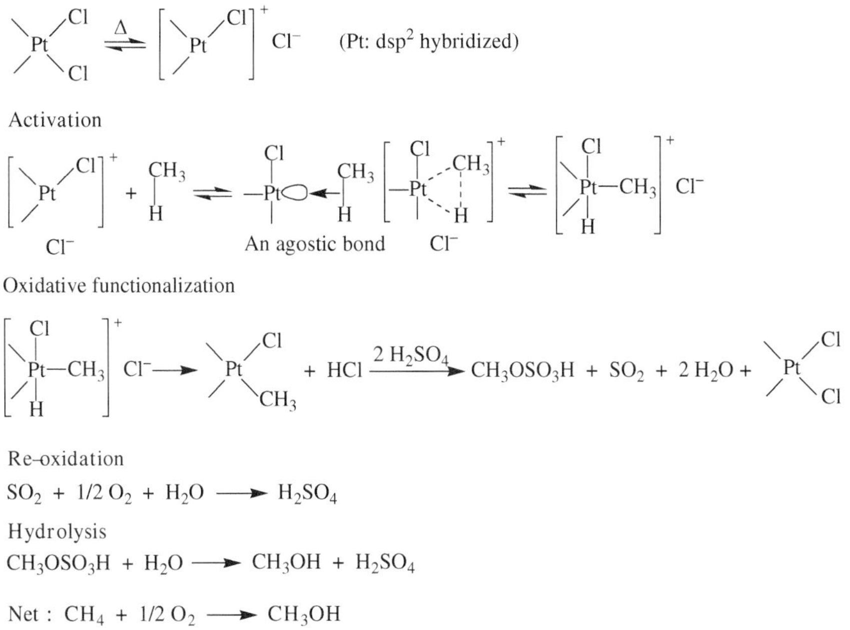

FIGURE 2.20   Mechanism for the (bpym)PtCl$_2$ catalyzed functionalization of methane by sulfuric acid.

(CF$_3$SO$_3$H). Both of them are derivatives of sulfuric acid. The attachment of a highly electronegative group (–F or –CF$_3$) to the central sulfur atom of the acids substantially enhances their acidity. In addition, a strong Lewis acid can also enhance the acidity of a protic acid, and as a result, a mixture of a protic acid and a strong Lewis acid becomes a superacid. For example, HF–SbF$_5$ and FSO$_3$H–SbF$_5$ (magic acid) are very effective superacids. Due to its highly strong ability to donate protons, a superacid may possibly protonate a C—H bond (a very poor electron-pair donor) in alkanes to facilitate its heterogeneous cleavage (activation) forming a reactive carbocation as illustrated below ([H$^+$] symbolizes a protic superacid) [10–13]:

$$R - H + [H^+] \rightarrow R^+ + H_2$$

R$^+$ in turn reacts with an available functionalizing agent completing the overall functionalization process.

For example, the tertiary hydrogen in 2-methyl-propane (CH$_3$)$_3$CH can be abstracted by FSO$_3$H–SbF$_5$ in liquid SO$_2$ClF to produce a tertiary carbocation (CH$_3$)$_3$C$^+$, which is observed in the reaction media by NMR spectroscopy (Reaction 2.19) [11, 13].

$$(CH_3)_3C - H + FSO_3H - SbF_5 \rightarrow (CH_3)_3C^+ + H_2 + FSO_3^- - SbF_5 \quad (2.19)$$

Some secondary carbocations have also been produced this way by hydrogen abstraction from secondary carbons (such as those in propane and cyclopentane) by superacids. In general, the order of reactivity toward the C–H hydrogen abstraction by a superacid is that $C_{tertiary}$–H > $C_{secondary}$–H > $C_{primary}$–H [10].

The exact nature of the transition state for the alkane C–H hydrogen abstraction by superacids is not certain. Both linear and triangular transition states (three-center, two-electron bonds) have been proposed (Fig. 2.21) [10, 11, 13].

A superacid can also protonate an inorganic functionalizing agent (Y) (e.g., elemental sulfur, as shown below) making it highly electrophilic presumably due to the positive charge introduced onto the agent.

$$Y + [H^+] \rightarrow HY^+$$

As a result, the highly electrophilic HY^+ may activate (cleave) and functionalize a C–H bond subsequently.

Triflic acid (CF_3SO_3H, $pK_a = -15$) is a common superacid. In the medium of CF_3SO_3H, various alkanes such as 2-methylpropane, cyclopentane, and propane have been found to react with elemental sulfur at 125–150 °C giving dialkyl sulfides (yields: 30–50%) and hydrogen sulfide as a by-product (Fig. 2.22) [14].

Reaction of propane takes place only on the secondary carbon, but not on any primary carbon. For $(CH_3)_3CH$, the reaction takes place on a methyl carbon (primary carbon). It does not take place on the tertiary carbon due to steric hindrance.

Two alternative mechanisms have been proposed (Fig. 2.23) to account for the reactions [14].

(A) For each of the reactions, the R—H bond to be functionalized could be first protonated by CF_3SO_3H resulting in ionization of R–H (heterogeneous cleavage of the C—H bond) to a reactive carbocation R^+ (activation process). Then, R^+ reacts with S_8 giving an electrophilic intermediate $R–S–S–S_5–S^+$ (functionalization process), which activates a second R–H with its positively charged sulfur by abstracting the hydrogen atom heterogeneously from the C—H bond (mechanism comparable to that of the hydrogen abstraction by superacids). The resulting $R–S–S–S_5–SH$ is then further protonated by CF_3SO_3H yielding another electrophilic intermediate $R–S–^+SH–S_5–SH$.

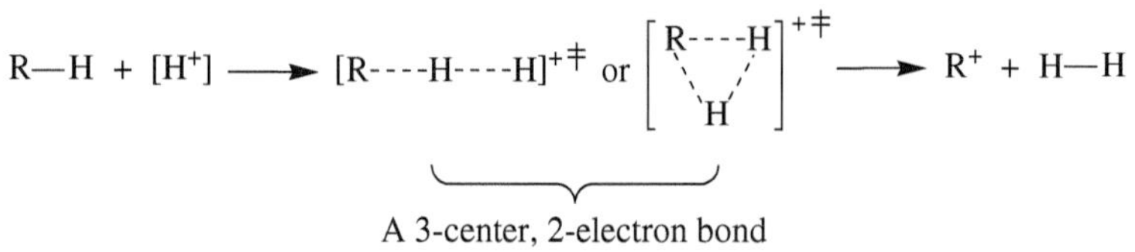

$$R{-}H + [H^+] \longrightarrow [R{-}{-}{-}H{-}{-}{-}H]^{+\ddagger} \text{ or } \left[\begin{matrix} R{-}{-}{-}H \\ \diagdown \; \diagup \\ H \end{matrix} \right]^{+\ddagger} \longrightarrow R^+ + H{-}H$$

A 3-center, 2-electron bond

FIGURE 2.21 The proposed transition states for the hydrogen abstraction of the alkane C—H bond by a superacid ([H⁺]). Olah et al. [10], Olah [11], and Olah [13].

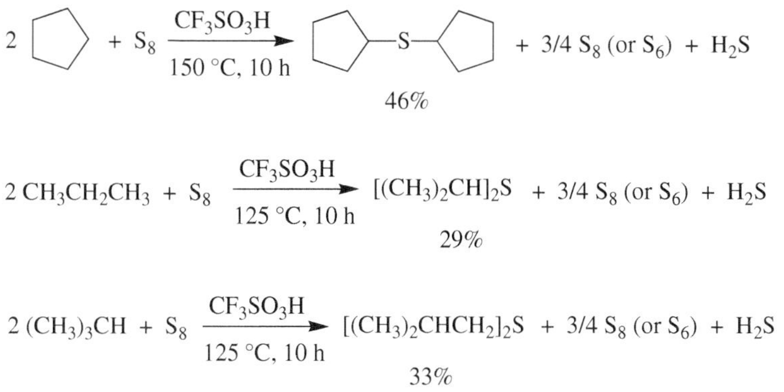

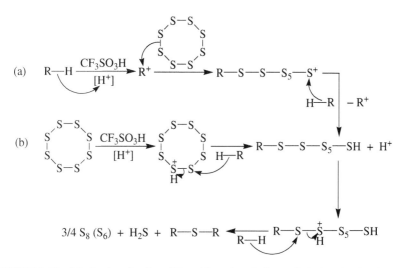

2 + S8
CF3SO3H
150 °C, 10 h
S
+ 3/4 S8 (or S6) + H2S
46%
2 CH3CH2CH3 + S8
CF3SO3H
125 °C, 10 h
[(CH3)2CH]2S + 3/4 S8 (or S6) + H2S
29%
2 (CH3)3CH + S8
CF3SO3H
125 °C, 10 h
[(CH3)2CHCH2]2S + 3/4 S8 (or S6) + H2S
33%

FIGURE 2.22 Reactions of alkanes with elemental sulfur in triflic acid. Based on Olah et al. [14].

(a)
R—H
CF3SO3H
[H+]
R+
R—S—S—S5—S+
H—R
– R+
(b)
CF3SO3H
[H+]
H—R
R—S—S—S5—SH + H+
3/4 S8 (S6) + H2S + R—S—R
R—H
R—S—S—S5—SH

FIGURE 2.23 Mechanism for the triflic acid catalyzed alkane C—H bond activation and function. Based on Olah et al. [14].

Finally, it activates and functionalizes an R–H via an electrophilic substitution leading to the formation of R–S–R.

(B) Alternatively, it is also possible that an S_8 molecule is protonated by CF_3SO_3H first. Then, the highly electrophilic protonated sulfur molecule attacks an R—H bond (electrophilic substitution) resulting in cleavage (activation) of the bond and simultaneous formation of an R—S bond (functionalization). In the next step, the intermediate electrophile R–S–$^+$SH–S_5–SH attacks a second R–H bond giving rise to the formation of the final R–S–R product.

Since the reaction of 2-methylpropane with sulfur occurs only on the primary methyl carbon (Fig. 2.22), it unlikely follows the mechanism (A), as the formation of a primary carbocation would be unfavorable.

2.8   NITRATION OF THE ALIPHATIC C—H BONDS VIA THE NITRONIUM NO_2^+ ION

Unlike aromatic nitration which can be effected by a mixture of nitric acid and sulfuric acid and some other nitrating agents, nitration of an alkane C—H bond is much more difficult and cannot be achieved by the aforementioned nitrating agents. However, some alkanes can be nitrated by the highly electrophilic nitronium NO_2^+ ion under certain conditions.

Research has shown [15] that methane reacts with nitronium hexafluorophosphate at 25 °C to give nitromethane (Reaction 2.20).

$$H_3C—H + NO_2^+ PF_6^- \rightleftharpoons \left[H_3C \cdots \begin{matrix} H \\ \\ NO_2 \end{matrix} \right]^+ \left(or\ \begin{matrix} H_3C \cdots H \\ \oplus \\ NO_2 \end{matrix} \right) \longrightarrow H_3C—NO_2 + H^+$$

A 3-center, 2-electron bond

(2.20)

The yield is, however, only 0.1%. Reaction 2.20 proceeds via a three-center, two-electron bond carbocationic intermediate (transition state) in which the C—H bond is partially broken and the nitrogen of NO_2^+ is partially bonded to both carbon and hydrogen in the activated C—H bond. The bonding is analogous to the previously discussed agostic bond in Section 2.6. The linear nitronium $[O=N=O]^+$ ion has no vacant orbital on nitrogen (similar to the ammonium NH_4^+ ion). Its electrophilicity originates from the polarizability of the N=O bonds, namely that partial shift of the π electrons from nitrogen to the more electronegative oxygen. In contrast to π-donor aromatics, the C—H σ-bond donors in alkanes are weak and do not bring about ready polarization on NO_2^+. If $[O=N=O]^+$ becomes bent, an empty p orbital will be developed on the central nitrogen atom, and its electrophilicity will be enhanced. This can be achieved by protonation on the cation.

Research has indicated [15] that an oxygen atom of nitronium can be protonated by superacids, such as a mixture of HF and HSO_3F, to give the angular NO_2H^{2+} dication (Fig. 2.24). An empty p orbital is formed in NO_2H^{2+} holding the +2 positive charge. NO_2H^{2+} is more electrophilic than NO_2^+ because of the vulnerable positively charged empty p orbital. As a result, when reaction of methane with nitronium is performed in HF-HSO_3F, the yield of nitromethane is enhanced up to 10%. The reaction takes place via the more electrophilic NO_2H^{2+} (Fig. 2.24).

Nitration of higher alkanes by NO_2^+ affords much greater yields. For example, in the purified nitromethane medium reaction of adamantane with $NO_2^+BF_4^-$ at ambient temperature gives 1-nitroadmantane in 65% yield (Fig. 2.25) [15]. Kinetic

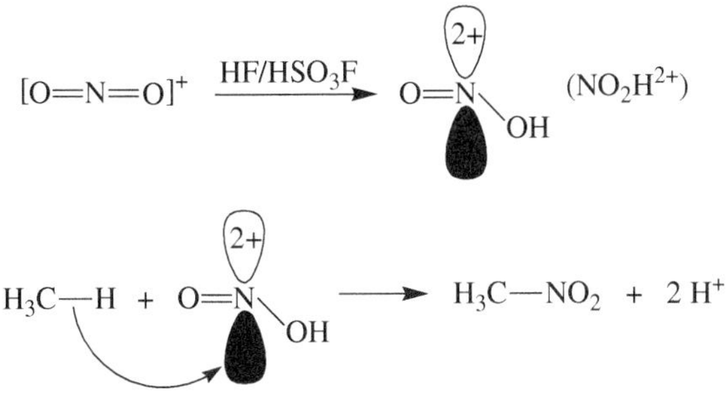

$$[O{=}N{=}O]^+ \xrightarrow{\text{HF/HSO}_3\text{F}} O{=}N\underset{OH}{\overset{2+}{\diagup}} \quad (NO_2H^{2+})$$

FIGURE 2.24 The protonated nitronium NO_2H^{2+} dication and its reaction with methane.

$$H_3C{-}H \; + \; O{=}N\underset{OH}{\overset{2+}{\diagup}} \; \longrightarrow \; H_3C{-}NO_2 \; + \; 2\,H^+$$

hydrogen isotope effect study of the rate of nitration for the deuterated 1,3,5,7-tetradeuterioadmantane compared with that for the light (^{1}H) adamantane shows no primary kinetic hydrogen isotope effect ($k_H/k_D = 1.06$). It supports that the reaction proceeds via a carbocationic intermediate (Fig. 2.25) in which the C—H bond is only partially broken and an N—C bond is partially formed. Because adamantane assumes a cage structure, geometric constraints make the electrophilic attack on the C—H σ-bond by NO_2^+ occur in the front side of the adamantane molecule. The reaction proceeds via an S_E2-like electrophilic substitution. Similarly, reaction of cyclohexane with NO_2^+ at ambient temperature is shown to give nitrocyclohexane in 30% yield.

Different from nitrations of adamantane and cyclohexane, reaction of ethane with nitronium hexafluorophosphate results in cleavage of the C—C bond in addition to the C—H bond nitration (Fig. 2.26a) [15]. Nitration of propane by nitronium hexafluorophosphate gives both 1- and 2-nitropropanes. In addition, nitromethane and nitroethane are formed as a result of the concurrent C—C bond cleavage by NO_2^+ (Fig. 2.26b). In both cases, the yields of the products due to the C—C bond cleavage are greater than those of the C—H bond nitration products. On the other hand, nitration of 2-methylpropane [$(CH_3)_3CH$] by nitronium hexafluorophosphate mainly occurs on the tertiary C—H bond giving $(CH_3)_3CNO_2$ (~70%). CH_3NO_2 is formed as a minor side product (~30%) due to the C—C bond cleavage by NO_2^+ (Fig. 2.26c). The C—H bond nitrations of 2-methylpropane and adamantane indicate that the steric hindrance to the electrophilic attack by NO_2^+ is insignificant. In general, the reactivity order of the alkane C—H bonds toward nitration is that $C_{tertiary}$–H $>$ $C_{secondary}$–H $>$ $C_{primary}$–H.

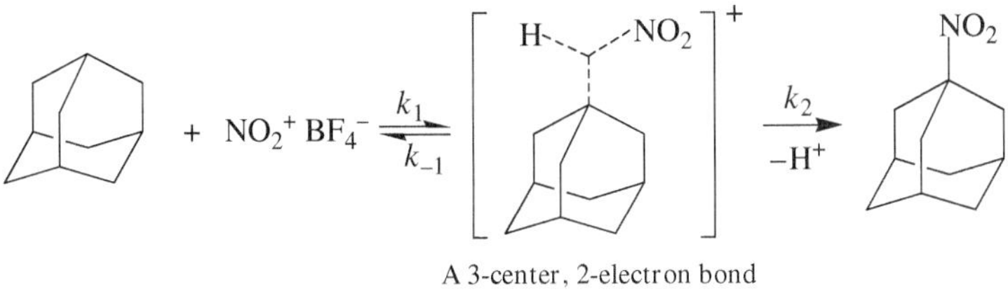

$$\text{adamantane} \; + \; NO_2^+ \, BF_4^- \; \underset{k_{-1}}{\overset{k_1}{\rightleftharpoons}} \; \left[\begin{array}{c} H\cdots \cdots NO_2 \\ \text{adamantane} \end{array} \right]^+ \; \xrightarrow[-H^+]{k_2} \; \text{adamantane-}NO_2$$

A 3-center, 2-electron bond

k_1 (slow) $< k_2$ (fast) and k_1 (slow) $< k_{-1}$ (fast)

FIGURE 2.25 Nitration of adamantane via a three-center, two-electron bond carbocationic intermediate.

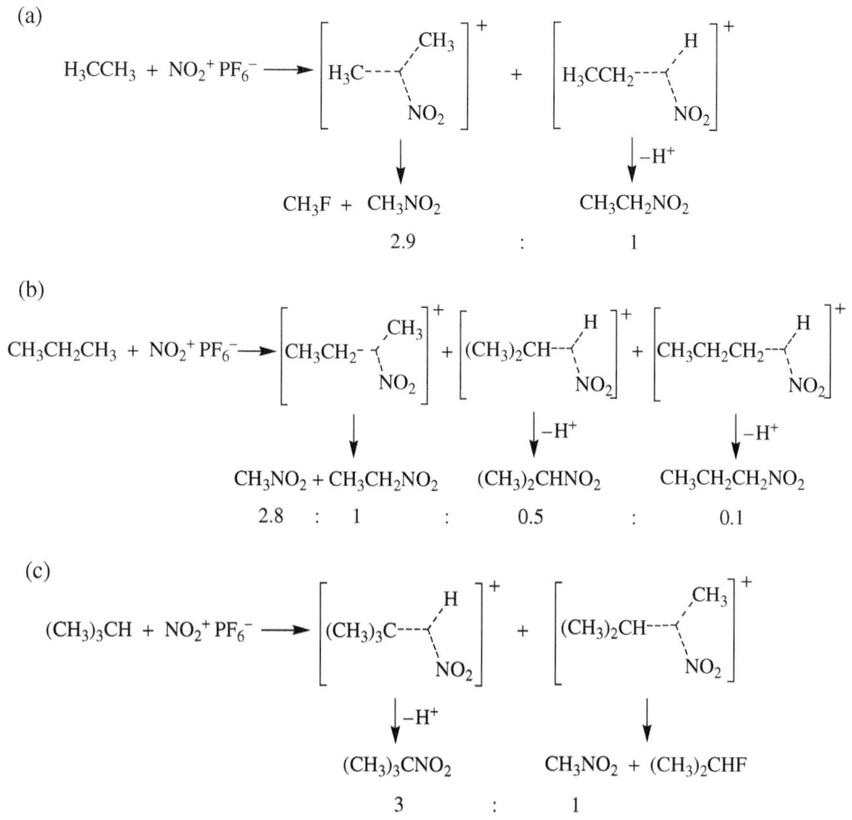

(a)

$$H_3CCH_3 + NO_2{}^+ PF_6{}^- \longrightarrow \left[H_3C\overset{CH_3}{\underset{NO_2}{-\!-\!-\!\!\diagup\!\!\!\diagdown}} \right]^+ + \left[H_3CCH_2\overset{H}{\underset{NO_2}{-\!-\!-\!\!\diagup\!\!\!\diagdown}} \right]^+$$

$$\downarrow \qquad\qquad\qquad\qquad \downarrow -H^+$$

$$CH_3F + CH_3NO_2 \qquad\qquad CH_3CH_2NO_2$$

$$2.9 \qquad\qquad : \qquad\qquad 1$$

(b)

$$CH_3CH_2CH_3 + NO_2{}^+ PF_6{}^- \longrightarrow \left[CH_3CH_2\overset{CH_3}{\underset{NO_2}{-\!\!\diagup\!\!\!\diagdown}} \right]^+ + \left[(CH_3)_2CH\overset{H}{\underset{NO_2}{-\!-\!\!\diagup\!\!\!\diagdown}} \right]^+ + \left[CH_3CH_2CH_2\overset{H}{\underset{NO_2}{-\!-\!\!\diagup\!\!\!\diagdown}} \right]^+$$

$$\downarrow \qquad\qquad\qquad\qquad \downarrow -H^+ \qquad\qquad \downarrow -H^+$$

$$CH_3NO_2 + CH_3CH_2NO_2 \qquad (CH_3)_2CHNO_2 \qquad CH_3CH_2CH_2NO_2$$

$$2.8 \quad : \quad 1 \qquad : \qquad 0.5 \qquad : \qquad 0.1$$

(c)

$$(CH_3)_3CH + NO_2{}^+ PF_6{}^- \longrightarrow \left[(CH_3)_3C\overset{H}{\underset{NO_2}{-\!-\!-\!\!\diagup\!\!\!\diagdown}} \right]^+ + \left[(CH_3)_2CH\overset{CH_3}{\underset{NO_2}{-\!-\!-\!\!\diagup\!\!\!\diagdown}} \right]^+$$

$$\downarrow -H^+ \qquad\qquad\qquad\qquad \downarrow$$

$$(CH_3)_3CNO_2 \qquad\qquad CH_3NO_2 + (CH_3)_2CHF$$

$$3 \qquad : \qquad 1$$

FIGURE 2.26   Reactions of nitronium hexafluorophosphate with (a) ethane, (b) propane, and (c) 2-methylpropane resulting in the C—H bond nitration and concurrent C—C bond cleavage.

2.9   PHOTOCHEMICAL AND THERMAL C—H BOND ACTIVATION BY THE OXIDATIVE URANYL UO$_2{}^{2+}$(VI) CATION

The linear uranyl UO$_2{}^{2+}$(VI) ion ([O=U=O]$^{2+}$, D$_\infty$ symmetry) is a major form of uranium(VI) and exhibits diverse chemistry [16, 17]. Particularly interesting is its high reduction potential in both excited and ground states and thus, UO$_2{}^{2+}$ possesses an extraordinary capability to oxidize many substances, including the very inactive C$_{sp^3}$—H bonds in some organic compounds [16]. Such C—H bond oxidation by UO$_2{}^{2+}$ takes place photochemically (via the excited *UO$_2{}^{2+}$) as well as thermally in some cases, and the reactions are effected by a single-electron transfer (charge-transfer [CT]) from the C—H bond to the valence shell of the uranium center of UO$_2{}^{2+}$ [18, 19]. As a result, the C—H bond is cleaved homolytically (activation) to lead to the formation of a carbon radical, and UO_2^{2+} is in turn reduced to a less oxidative $UO_2^+(V)$ ion.

Research has shown that the UO_2^{2+}–doped polyvinyl alcohol [PVA, $-CH_2-CH(OH)-CH_2-$] film is irradiated by γ-ray, and the C–H bond in the α–carbon is cleaved homolytically by the oxidative UO_2^{2+} to give an α–hydroxyalkyl radical (Reaction 2.21) [20].

$$\overset{\overset{\displaystyle e}{\frown}}{U}O_2^{2+} + -(CH_2-\underset{\underset{\displaystyle OH}{|}}{C}H-CH_2)- \xrightarrow[CT]{\gamma\text{-ray}} UO_2^+ + -(CH_2-\underset{\underset{\displaystyle OH}{|}}{\dot{C}}-CH_2)- + H^+$$

$$\text{(PVA)} \qquad\qquad\qquad\qquad \text{(PVA·)}$$

In the reaction, simultaneous CT and deprotonation occur to the α—CH bond in PVA to lead to the formation of a PVA· radical (with the unpaired electron residing in the OH carbon). UO_2^{2+} in turn accepts an electron from PVA to transform into UO_2^+.

The thermal oxidation of liquid dimethyl sulfoxide (DMSO) by UO_2^{2+} (as the nitrate salt) in the presence of sulfuric acid has been performed in the dark at ambient temperature [18]. The reaction leads to homolytic cleavage of a C—H bond giving the DMSO· $(CH_3SOCH_2·)$ radical (Reaction 2.22).

(DMSO)   (DMSO· radical)

The reaction occurs via a CT from a CH bond in DMSO to the uranium valence shell in UO_2^{2+} [18]. Sulfuric acid enhances solubility of UO_2^{2+} in DMSO and also makes UO_2^{2+} more oxidative. All this facilitates the reaction to proceed. The DMSO· radical produced in the reaction is stabilized by the DMSO host (solvent) molecules so that it is observable by EPR spectroscopy [18, 21].

It has also been found that UO_2^{2+} can oxidize the relatively inert C—H bond in methanol and the α—CH bonds in primary and secondary alcohols (such as ethanol, 2-propanol, and cyclohexanol) to lead to the formations of hydroxymethyl and α-hydroxyalkyl radicals (Reactions 2.23–2.26).

$$UO_2^{2+} + CH_3OH \rightarrow UO_2^+ + \cdot CH_2OH + H^+ \qquad (2.23)$$

$$UO_2^{2+} + CH_3CH_2OH \rightarrow UO_2^+ + CH_3\dot{C}HOH + H^+ \qquad (2.24)$$

$$UO_2^{2+} + (CH_3)_2CHOH \rightarrow UO_2^+ + (CH_3)_2\dot{C}OH + H^+ \qquad (2.25)$$

$$UO_2^{2+} + \langle\text{hexagon}\rangle-OH \longrightarrow UO_2^+ + \langle\text{hexagon}\rangle\cdot-OH + H^+ \qquad (2.26)$$

While Reactions 2.23 and 2.24 (methanol and ethanol reactions) have been shown to take place rapidly on laser photolysis or irradiation of the UV light (photochemical reactions via the excited $*UO_2^{2+}$) [19, 22, 23], all the reactions (Reactions 2.23–2.26) are also found to occur slowly in the dark at ambient temperature (thermal reactions of UO_2^{2+} in the ground state) [19]. In all the reactions, the radical products are possibly stabilized by the alcohol hosts (solvents) and thus, they are observable by EPR spectroscopy [19].

Reaction 2.23 (the methanol reaction) has been shown to proceed via an intermediate electron–donor–acceptor (EDA) [UO_2^{2+}, $HOCH_3$] complex [19]. The EDA complex is formed reversibly via a weak coordination bond between the alcohol OH oxygen and the uranium center in UO_2^{2+}. A subsequent irreversible CT possibly occurs between a methanol CH bond and the uranium center of UO_2^{2+} leading to the formation of an ion–radical [UO_2^+, $HOCH_2^·$] adduct weakly linked by the same oxygen–uranium coordination bond. Dissociation of the ion–radical adduct gives separate UO_2^+ and $HOCH_2^·$ (the hydroxymethyl radical). The overall mechanism is illustrated in Equation 2.27 [19].

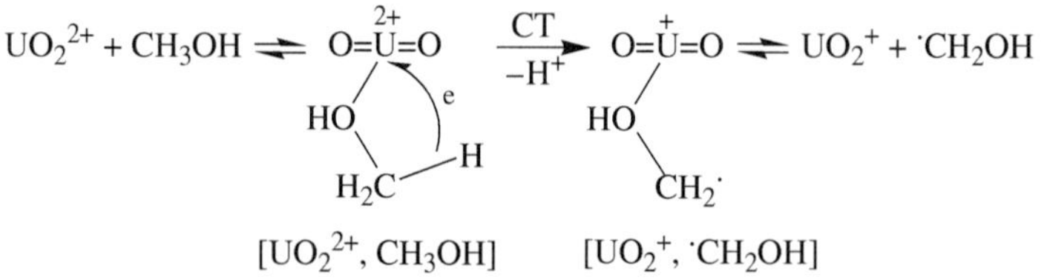

$$UO_2^{2+} + CH_3OH \rightleftharpoons O=\overset{2+}{\underset{\underset{\overset{|}{H_2C}}{\overset{|}{HO}}}{U}}=O \xrightarrow[-H^+]{CT} O=\overset{+}{\underset{\underset{CH_2^·}{\overset{|}{HO}}}{U}}=O \rightleftharpoons UO_2^+ + {}^·CH_2OH$$

$$[UO_2^{2+}, CH_3OH] \qquad [UO_2^+, {}^·CH_2OH]$$

(2.27)

The oxidation of primary and secondary alcohols to α-hydroxyalkyl radicals by UO_2^{2+} follows the analogous charge–transfer mechanism.

2.10   ENZYME CATALYZED ALKANE C–H BOND ACTIVATION AND FUNCTIONALIZATION: BIOCHEMICAL METHODS

Living nature performs alkane oxidative functionalization continually at ambient temperature, and sometimes with great selectivity through the use of **oxygenase enzymes**. The most important oxygenase enzymes include monooxygenase cytochrome P-450 and methane monooxygenase (MMO) [1]. The enzymes catalyze incorporation of molecular oxygen into alkane C–H bonds with concurrent loss of water, and the process is accompanied by oxidation of NADH or NADPH as shown below:

$$R - H + 2e + 2H^+ + O_2 \rightarrow R - OH + H_2O$$

$$NAD(P)H \rightarrow NAD(P)^+ + 2e + H^+$$

NADH or NADPH, a reduced coenzyme found in living organisms, functions as an electron pool for the oxidative functionalization of an alkane to an alcohol.

Cytochrome P-450 is a heme protein. Its active site contains an iron(II)–porphyrin complex (heme), which is attached to a protein chain through a Fe–S

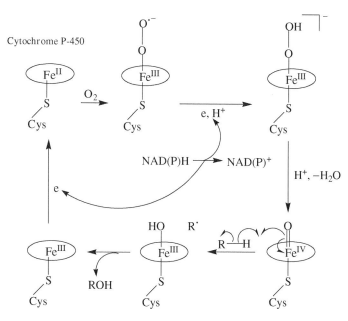

FIGURE 2.27 Cytochrome P-450 catalyzed alkane C–H bond oxidative functionalization. Shilov and Shteinman [24] and Newcomb and Toy [25].

linkage on a cysteine residue of the protein chain (Fig. 2.27) [24, 25]. The Fe–S linkage facilitates cleavage of the O_2 molecule. Figure 2.27 shows that the first step in the mechanism is binding of an O_2 molecule to the heme Fe(II) in cytochrome P-450. The Fe(II) is subsequently oxidized by O_2 to F(III). Then, the Fe^{III} –O–O$^{-\cdot}$ linkage accepts an electron from NAD(P)H to form a peroxide [Fe^{III} –O–OH]$^-$ intermediate, which upon losing a water molecule leads to the formation of a Fe^{IV}=O bond. The NAD(P)H, upon losing electrons, is oxidized to NAD(P)$^+$ simultaneously. The iron(IV) atom in the Fe^{IV}=O moiety is highly oxidative. It is now capable to abstract a hydrogen atom from an R–H bond through the doubly bonded Fe^{IV}=O oxygen, and to break the R–H bond homolytically giving an intermediate alkyl R˙ radical. Concurrently, Fe(IV) is reduced to Fe(III). Then, R˙ cleaves the Fe^{III} –OH bond homolytically to form ROH, and finally Fe(III) is reduced to Fe(II) to regenerate a free enzyme and complete a catalytic cycle.

The biochemical mechanism for cytochrome P-450 catalyzed R–H bond oxidation to R–OH by molecular oxygen is demonstrated in Figure 2.27. Kinetic studies for oxidative functionalizations of cyclohexane (C_6H_{12}) and deuterium cyclohexane (C_6D_{12}) catalyzed by cytochrome P-450 have been performed. A very large kinetic isotope effect ($k_H/k_D = 21.9$ at 20 °C) is observed [24]. The data supports a possible proton tunneling pathway (a low E_a process) associated to the alkyl R˙ radical formation step (rate-determining step).

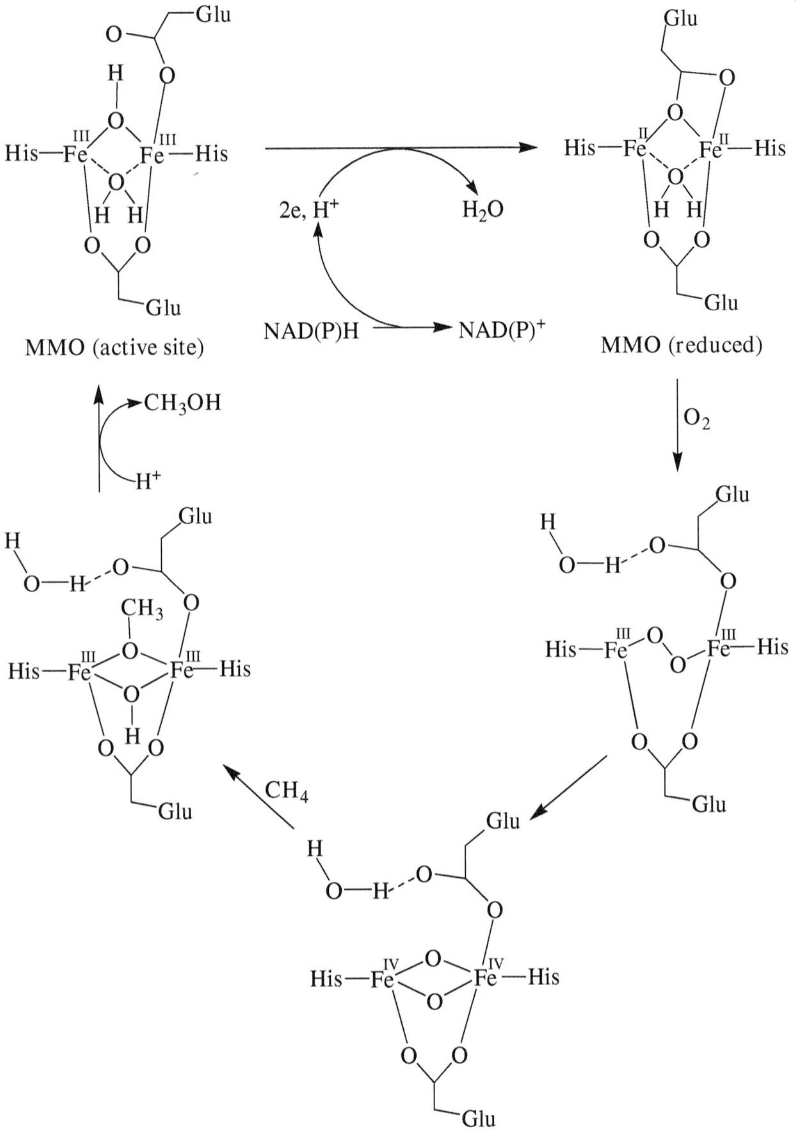

FIGURE 2.28 Structure of active site of methane monooxygenase (MMO) and mechanism for MMO catalyzed oxidative functionalization of methane to methanol. Adapted from Shilov and Shteinman [24].

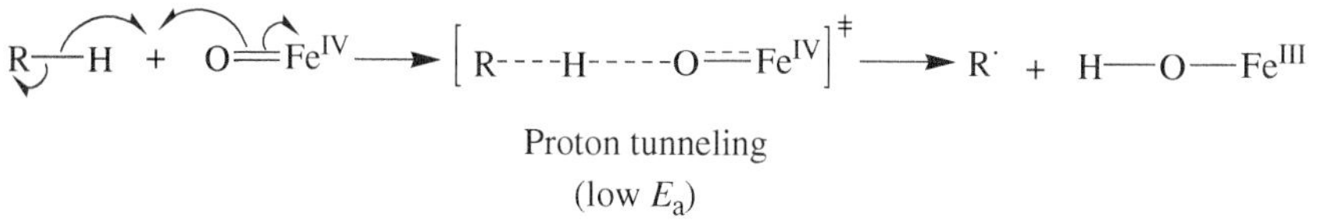

$$R \overset{\frown}{} H \;+\; O \overset{\frown}{=\!=} Fe^{IV} \longrightarrow \left[R\text{-}\text{-}\text{-}H\text{-}\text{-}\text{-}\text{-}\text{-}O \overset{=\!=\!=}{} Fe^{IV} \right]^{\ddagger} \longrightarrow R^{\cdot} \;+\; H\!-\!O\!-\!Fe^{III}$$

Proton tunneling

(low E_a)

Since the oxidative functionalization takes place via an intermediate alkyl radical, a C_{tert}—H bond which has relatively low BDE undergoes the reaction more easily than other types of C—H bonds.

MMO is found in some bacteria. At ambient temperature, it catalyzes oxidation of methane to methanol by molecular oxygen in the air [1]. The active site of MMO contains an iron–oxygen cluster that consists of two Fe atoms connected to a carboxylate group of a glutamate residue in the protein chain (Fig. 2.28) [24]. In addition, the two Fe atoms are bridged by a hydroxide group, and each of them is also attached to a histidine residue in the protein chain. The structure of the MMO active site and the mechanistic steps are illustrated in Figure 2.28. The oxidative cleavage and functionalization of a C—H bond in methane are mediated by Fe atoms of the iron–oxygen cluster in MMO. In the first step, both Fe^{III} atoms are reduced to Fe^{II} by NAD(P)H. Then, an oxygen molecule O_2 binds to the reduced MMO, oxidizing Fe^{II} back to Fe^{III} and resulting in the formation of a peroxide –O–O– entity. The peroxide O—O bond, once formed, will be cleaved spontaneously, and the Fe^{III} atoms are further oxidized to Fe^{IV}. Finally, a C—H bond in methane is cleaved by the highly oxidative Fe^{IV} atoms, and a methanol CH_3OH molecule is produced. The hydroxylation of methane to methanol by MMO is quite specific, but oxidations of more complicated alkanes by MMO typically give product mixtures.

PROBLEMS

2.1 Give the major organic product for the following reaction

$$+ \quad Br_2 \quad \xrightarrow{\text{Heat}}$$

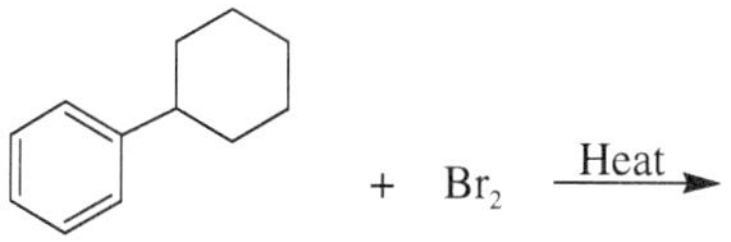

Show the reaction mechanism. Sketch the reaction profile in enthalpy for the chain-growth (propagation) steps. Calculate the enthalpy ($\Delta H°$) for each of the chain-growth steps. Known that BDE(C_{tert}–H) = 93 kcal/mol, BDE(C_{sec}–H) = 96 kcal/mol, BDE(C_{prim}–H) = 100 kcal/mol, BDE(Br–Br) = 46 kcal/mol, and BDE(H–Br) = 87 kcal/mol.

2.2 Deduct the rate law for the overall reaction in Problem 2.1 on the basis of the mechanism. What species in the reaction mechanism is (are) in the steady-state?

2.3 Consider the following reactions.

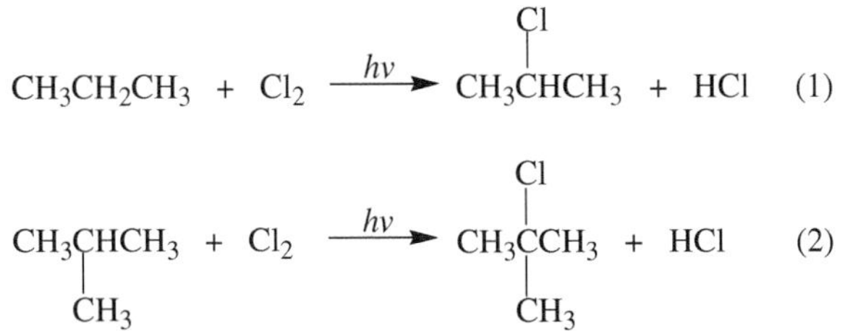

$$CH_3CH_2CH_3 + Cl_2 \xrightarrow{\textit{hv}} CH_3\overset{\displaystyle Cl}{\overset{|}{C}}HCH_3 + HCl \quad (1)$$

$$CH_3\overset{\displaystyle}{\underset{\overset{|}{CH_3}}{C}}HCH_3 + Cl_2 \xrightarrow{\textit{hv}} CH_3\overset{\displaystyle Cl}{\underset{\overset{|}{CH_3}}{\overset{|}{C}}}CH_3 + HCl \quad (2)$$

Draw reaction profile for the rate-determining step for each reaction. Which reaction goes faster under the same conditions? Explain in terms of relative stability of the intermediate alkyl radicals and activation energies (enthalpies) of the reactions. Estimate the relative activation enthalpies for the reactions (rate-determining steps) using the Evans–Polanyi relationship (Eq. 2.1).

2.4 Using molecular orbital diagram, show stabilization of an intermediate alkyl radical formed in the reactions of Problem 2.3 by a C—H bond.

2.5 Fluorination of 2-methylbutane follows the radical chain mechanism analogous to the radical chlorination and bromination and gives a mixture of four products as shown below:

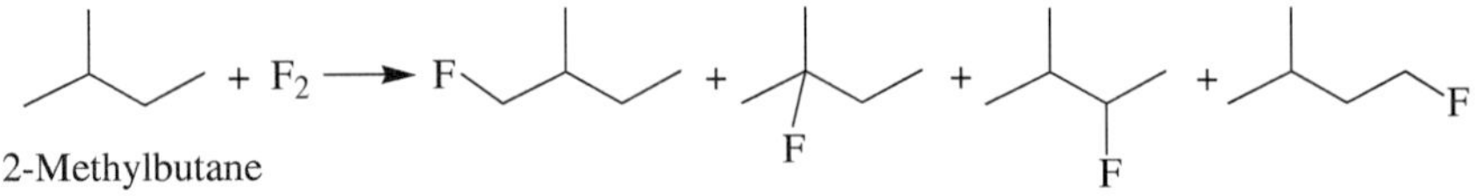

2-Methylbutane

Propose a pair of chain-growth (propagation) steps for formation of each product and calculate $\Delta H°$ for each step. Known that $BDE(C_{tert}-H) = 93$ kcal/mol, $BDE(C_{sec}-H) = 96$ kcal/mol, $BDE(C_{prim}-H) = 100$ kcal/mol, BDE (F–F) = 38 kcal/mol, and $BDE(H-F) = 136$ kcal/mol.

2.6 For Problem 2.5, what is the rate-determining step for formation of each product? Draw energy profile in enthalpy for each of the rate-determining steps on the basis of the Hammond postulate. Predict regioselectivity of the radical fluorination (percentage of each product) relative to that of the radical chlorination and bromination. Account for the difference.

2.7 Write out the detailed mechanisms for radical halogenation (substitution and addition) of cyclohexene as shown below (also see Eq. 2.18).

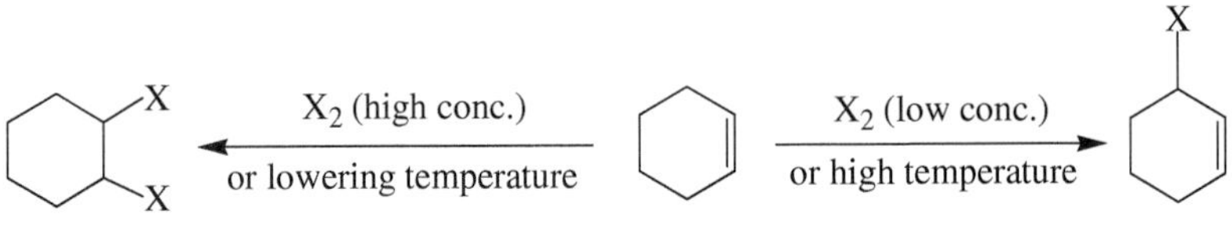

Account for effect of the temperature on the type of reactions on the basis of the reaction rate laws.

2.8 What is an agostic bond? Account for the roles agostic bonds play in the alkane C—H bond functionalization.

2.9 Both mercury(II) and the nitronium NO_2^+ ion can activate the C—H bond in methane. Mechanistically, what is the difference between the two species in activation of the C—H bond?

2.10 Account for the general roles superacids play in the alkane C—H bond functionalization.

2.11 Cytochrome P-450 can catalyze selective oxidation of cyclohexane (C_6H_{12}) to cyclohexanol ($C_6H_{11}OH$) by molecular oxygen (O_2). Write out the mechanism and indicate the rate-determining step.

REFERENCES

1. Arndtsen, B. A.; Bergman, R. G.; Mobley, T. A.; Peterson, T. H. Selective Intermolecular Carbon-Hydrogen Bond Activation by Synthetic Metal Complexes in Homogeneous Solution. *Acc. Chem. Res.* 1995, *28*, 154–162.

2. Bruckner, R. *Advanced Organic Chemistry: Reaction Mechanisms*, Chapter 1, pp. 1–42, Harcourt/Academic Press, Orlando, FL, USA (2002).

3. Fox, M. A.; Whitesell, J. K. *Core Organic Chemistry*, 2nd ed., Jones and Bartlett, Sudbury, MA, USA (1997).

4. Sun, X. An Integrated Approach to the Lewis Model, Valence Bond Theory, and Molecular Orbital Theory: A New Model for Simple Molecular Orbitals and a Quicker Way of Learning Covalent Bonding in General Chemistry. *Chem. Educator*, 2007, *12*, 331–334.

5. Finn, M.; Friedline, R.; Suleman, N. K.; Wohl, C. J.; Tanko, J. M. Chemistry of the *t*-Butoxyl Radical: Evidence that Most Hydrogen Abstractions from Carbon are Entropy-Controlled. *J. Am. Chem. Soc.* 2004, *126*, 7578–7584.

6. Silbey, R. J.; Alterty, R. A.; Bawendi, M. G. *Physical Chemistry*, 4th ed., Wiley, Danvers, MA, USA (2005).

7. Klei, S. R.; Tilley, T. D.; Bergman, R. G. The Mechnaism of Silicon–Hydrogen and Carbon–Hydrogen Bond Activation by Iridium(III): Production of a Silylene Complex and the First Direct Observation of Ir(III)/Ir(V) C–H Bond Oxidative Addition and Reductive Elimination. *J. Am. Chem. Soc.* 2000, *122*, 1816–1817.

8. Periana, R. A.; Taube, D. J.; Evitt, E. R.; Loffler, D. G.; Wentrcek, P. R.; Voss, G.; Masuda, T. A Mercury-Catalyzed, High-Yield System for the Oxidation of Methane to Methanol. *Science*, 1993, *259*, 340–343.

9. Periana, R. A.; Taube, D. J.; Gamble, S.; Taube, H.; Satoh, T.; Fujii, H. Platinum Catalysts for the High-Yield Oxidation of Methane to a Methanol Derivative. *Science*, 1998, *280*, 560–564.

10. Olah, G. A.; Prakash, G. K. S.; Wade, K.; Molnar, A.; Williams, R. E. *Hypercarbon Chemistry*, 2nd ed., John Wiley & Sons, Inc., Hoboken, NJ, USA (2011).

11. Olah, G. A. Crossing Conventional Boundaries in Half a Century of Research. *J. Org. Chem.* 2005, *70*, 2413–2429.

12. Olah, G. A. 100 Years of Carbocations and Their Significance in Chemistry. *J. Org. Chem.* 2001, *66*, 5843–5957.

13. Olah, G. A. Carbocations and Electrophilic Reactions. *Angew. Chem. Int. Ed.* 1973, *12*, 173–212.

14. Olah, G. A.; Wang, Q.; Prakash, G. K. S. Trifluoromethanesulfonic Acid Catalyzed Electrophilic Sulfuration of Alkanes (Cycloalkanes) with Elemental Sulfur to Dialkyl (Dicycloalkyl) Sulfides. *J. Am. Chem. Soc.* 1990, *112*, 3697–3698.

15. Olah, G. A.; Malhotra, R.; Narang, S. C. *Nitration: Methods and Mechanisms*, VCH, New York, USA (1989).

16. Kats, J. J.; Seaborg, G. T.; Morss, L. R. *The Chemistry of the Actinide Elements*, Chapmand and Hall, London, UK (1986).

17. Sun, X.; Samples, C.; Daia, K.; Meyers, M.; Bumgarner, M. Behaviour of the Uranyl (UO_2^{2+}) Ion in Different Strongly Acidic Media: Characterisation of UO_2^{2+} in Common Acids by Electronic Absorption Spectroscopy. *J. Chem. Res.* 2009, *33*, 351–355.

18. Sun, X.; Kolling, D. R. J.; Deskins, S.; Adkins, E. The Thermal Charge-Transfer Reduction of Uranyl UO_2^{2+}(VI) to UO_2^+(V) by Various Functionalized Organic Compounds, and Evidence for Possible Spin–Spin Interactions Between UO_2^+(V) and Hydroxymethyl (CH_2OH) Radical and Between UO_2^+(V) and Diphenyl Sulfide Radical Cation (Ph_2S^+). *Inorg. Chim. Acta*, 2018, *438*, 12–20.

19. Sun, X.; Kolling, D. R. J.; Mazagri, H.; Karawan, B.; Pierron, C. Investigation of Charge-Transfer Absorptions in Uranyl UO_2^{2+}(VI) Ion and Related Chemical Reduction of UO_2^{2+}(VI) to UO_2^+(VI) by UV–Vis and Electron Paramagnetic Resonance Spectroscopies. *Inorg. Chim. Acta*, 2015, *435*, 117–124.

20. Kumar, M.; Kadam, R. M.; Dhobale, A. R.; Sastry, M. D. Radiation Effects on the Property of ^{238}U(VI) or ^{239}Pu(IV) Doped Polyvinyl Alcohol Films Investigated by EPR and PAS. *J. Nucl. Radiochem. Sci.* 2000, *1*, 77–80.

21. Opstad, C. L.; Melo, T.-B.; Sliwka, H.-R.; Partali, V. Formation of DMSO and DMF Radicals with Minute Amounts of Base. *Tetrahedron*, 2009, *65*, 7616–7619.

22. Azenha, M. E. D. G.; Burrows, H. D.; Formosinho, S. J.; Miguel, M. de G. M. Photophysics of the Excited Uranyl Ion in Aqueous Solutions. *J. Chem. Soc., Faraday Trans. 1*, 1989, 85, 2625–2634.

23. Tsushima, S. Photochemical Reduction of UO_2^{2+} in the Presence of Alcohol Studied by Density Functional Theory Calculations. *Inorg. Chem.* 2009, *48*, 4856–4862.

24. Shilov, A. E.; Shteinman, A. A. Oxygen Atom Transfer into C–H Bond in Biological and Model Chemical Systems. Mechanistic Aspects. *Acc. Chem. Res.* 1999, *32*, 763–771.

25. Newcomb, M.; Toy, P. H. Hypersensitive Radical Probes and the Mechanisms of Cytochrome P450-Catalyzed Hydroxylation Reactions. *Acc. Chem. Res.* 2000, *33*, 449–455.

3

FUNCTIONALIZATION OF THE ALKENE C=C BOND BY ELECTROPHILIC ADDITIONS

Alkenes are characterized by electrophilic addition reactions essentially due to special nucleophilicity associated to the π bond in their C=C double bond domain. The general pattern of the reactions can be written as

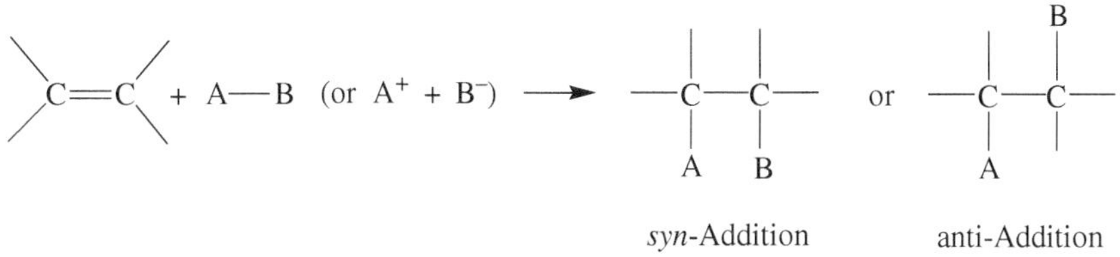

syn-Addition   anti-Addition

$$\Delta H^\circ = [\text{BDE}(\pi \, \text{bond}) + \text{BDE}(A - B)] - [\text{BDE}(C - A) + \text{BDE}(C - B)] < 0$$

By the end, an unsaturated alkene is transformed into a saturated functionalized hydrocarbon. In the course of the reaction, one π bond (C=C) and one σ bond (A–B) are broken, and two new σ bonds (C–A and C–B) are formed. Since the bond dissociation energy (BDE) of a π bond is in general smaller than a σ bond, the enthalpy (ΔH°) of the overall reaction is negative. At ambient temperature, the addition reactions are thermodynamically favorable as the entropy effect is less significant.

Organic Mechanisms: Reactions, Methodology, and Biological Applications,
Second Edition. Xiaoping Sun.
© 2021 John Wiley & Sons, Inc. Published 2021 by John Wiley & Sons, Inc.
Companion website: www.wiley.com/go/Sun/OrgMech_2e

Kinetically, the π bond in an alkene C=C domain is vulnerable due to diffusive p_π orbital lobes (electron cloud) distributed above and below the C–C linkage. They make a C=C π bond nucleophilic, readily donating a pair of the π-electrons to an available electrophilic reagent (an electron-deficient species readily accepting a pair of electrons). As a result, the aforementioned bond-breaking and bond-formation take place fast in relatively mild conditions, leading to an overall addition reaction. In addition, the vulnerable C=C π bond can also be effectively attacked by some radical species giving rise to homolytic cleavage of the π bond. As a result, in the presence of certain radical initiators the alkene additions can follow radical mechanisms.

Mechanistically, the alkene addition reactions can be divided into two categories, stepwise reactions (via an intermediate carbocation, radical, or other cationic species) and concerted reactions (via a single transition state). The reactions can also be categorized as *syn*-additions (*cis*-additions), with two groups (A and B) added to the alkene in the same side of the C=C bond, and anti-additions (*trans*-additions), with two groups (A and B) added in different sides of the C=C bond. These mechanistic features will determine the overall regiochemistry and stereochemistry of the addition reactions. In this chapter, major types of the alkene addition reactions and their mechanisms are presented. Their regiochemistry, stereochemistry, and synthetic and biological applications will be discussed.

The alkene C=C bond undergoes additions to some unsaturated entities forming cyclic molecules. This type of additions is called **cycloaddition reactions.** They will be discussed in Chapter 4.

3.1 MARKOVNIKOV ADDITIONS VIA INTERMEDIATE CARBOCATIONS

3.1.1 Protonation of the Alkene C=C π Bond by Strong Acids to form Carbocations

It has been well established [1, 2] that the C=C π bond in various alkenes can be protonated by strong acids (including superacids) to form carbocations, generalized as follows:

p_π

A strong acid $H^+(1s)$ A 3-center, 2-electron TS

The protonation has been shown to occur via a transient π complex (a 3-center, 2-electron transition state) in which a hydrogen (proton) is partially bonded to two carbon atoms by the interaction of the empty H^+ 1s orbital (electrophilic)

with the filled p_π orbital (nucleophilic), and the C=C π bond is partially broken [1]. If the alkene is unsymmetrical (the two C=C carbons contain different numbers of hydrogens), the protonation occurs preferably on such a carbon that generates the most stable carbocation (molecular and mechanistic basis of the Markovnikov's rule). The following are specific examples for protonation of alkenes [1, 2].

$$RCH=CH_2 \xrightarrow[\text{HF-SbF}_5]{\text{FSO}_3\text{H-SbF}_5} R\overset{\oplus}{C}HCH_3$$

$$\text{(cyclopentane ring)}=CH_2 \xrightarrow{\text{FSO}_3\text{H-SbF}_5} \text{(cyclopentane ring)}\overset{\oplus}{-}CH_3$$

$$\text{(isobutylene)} + H^+ \longrightarrow \text{(tert-butyl cation)}$$

In the first two examples, the alkenes are protonated by superacids and the resulting carbocations are stabilized and observed experimentally. The conjugate base of a superacid (e.g., FSO_3^-) is an extremely weak base and nucleophile. Therefore, it does not interact with the electrophilic carbocation. This accounts for the carbocation stabilization by a superacid. The third example shows the protonation of 2-methylpropene to make *tert*-butyl cation, which is used to initiate alkylation reactions. We will later study the alkylation reactions by carbocations in this chapter. In addition, in the first example, the terminal CH_2 carbon is selectively protonated to make a more stable 2° carbocation. Protonation of the CH carbon would lead to the formation of a less stable 1° carbocation. Thus, it does not happen. In the second and third examples, protonation selectively occurs to the terminal carbons to make more stable 3° carbocations. Protonation of the other doubly bonded carbon in each case would lead to the formation of a much less stable 1° carbocation. Thus, it does not happen.

When the conjugate base of a strong acid is nucleophilic (e.g., HCl, HBr, or HI), it can be added to the carbocation, initially produced by protonation of an alkene with the acid, to complete an electrophilic addition reaction.

$$-\overset{|}{C}=\overset{|}{C}- + H-A \rightleftharpoons -\overset{|}{C}-\overset{|}{\underset{H}{C}}- + A^- \longrightarrow -\overset{|}{\underset{H}{C}}-\overset{|}{\underset{A}{C}}-$$

A strong acid   H   A nucleophilic   H   A
                              base

The protonation of the C=C bond gives the most stable carbocation. As a result, the addition reaction is regioselective following the Markovnikov's rule.

3.1.2   Additions of Hydrogen Halides (HCl, HBr, and HI) to Alkenes: Mechanism, Regiochemistry, and Stereochemistry

Hydrogen halides HX (X = Cl, Br, or I) are strong protic acids. They are strong electrophiles with the reactive center on the proton. Therefore, each of them can be added to alkenes (nucleophiles) via carbocation intermediates. The rate of the addition is determined by the relative stability of the intermediate carbocation. If the reaction produces a primary carbocation which is highly unstable, it will be slower than that which produces a secondary carbocation. The reaction via a secondary carbocation will be slower than that which proceeds through a tertiary carbocation. If the alkene is unsymmetrical with the two doubly bonded carbons attached to different numbers of hydrogen atoms, the regiochemistry of the addition will be dictated by the relative stability of the intermediate carbocation (kinetically controlled), giving the **Markovnikov product** as the major product.

Mechanism and regiochemistry   First, we will demonstrate the regiochemistry using addition of hydrogen chloride (HCl) to 1-methylcyclohexene in dichloromethane, an **unreactive solvent** (Fig. 3.1). The first step in the reaction mechanism (the rate-determining step) is that HCl interacts with the π bond in the alkene C=C bond domain. In principle, the electrophilic attack (protonation of the C=C bond) can possibly lead to the formation of either a secondary (2°) carbocation (less stable with a higher energy level) or a tertiary (3°) carbocation (more stable with a lower energy level).

According to Equation 1.55 (the Bell–Evans–Polanyi principle, Section 1.6.3) and on the basis of Hammond postulation (Section 1.6.2), the activation energy (E_a) for the formation of a carbocation (endothermic) in the first step of the electrophilic addition is expected to increase linearly as a function of enthalpy (ΔH) for the formation of the carbocation from the initial reactants (1-methylcyclohexene and HCl). The relationship is shown in Equation 3.1 (with c_1 and c_2 being positive constants) [3].

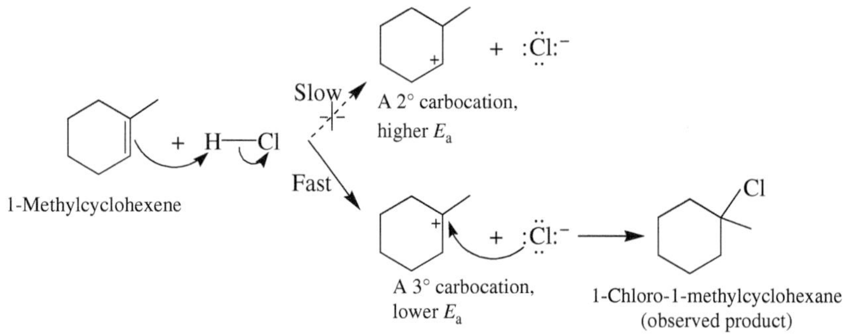

FIGURE 3.1   Mechanism and regioselectivity for electrophilic addition of hydrogen chloride to 1-methylcyclohexene in dichloromethane, an unreactive solvent.

$$E_a = c_1 \Delta H + c_2 \tag{3.1}$$

ΔH is equal to the difference between energy levels of the carbocation ($E_{carbocation}$) and the initial reactants ($E_{reactants}$) in enthalpy as formulated in Equation 3.2,

$$\Delta H = E_{carbocation} - E_{reactants} \tag{3.2}$$

Substituting Equation 3.2 for Equation 3.1, we have

$$E_a = c_1 \left(E_{carbocation} - E_{reactants} \right) + c_2 = c_1 E_{carbocation} - c_1 E_{reactants} + c_2$$

Since $(-c_1 E_{reactants} + c_2) = $ constant, we have

$$E_a = c_1 E_{carbocation} + \text{constant} \tag{3.3}$$

Equation 3.3 shows that E_a in enthalpy ($\Delta H^{\neq}$) for the formation of an intermediate carbocation (rate-determining step) from a given alkene and HCl is directly proportional to energy level of the intermediate carbocation. Since a 2° carbocation possesses a higher energy level than does a 3° carbocation, the activation energy for the formation of a 2° carbocation (E_{a2}) will be higher than the activation energy for the formation of a 3° carbocation (E_{a3}) (Fig. 3.2). As a result, the formation of the 2° carbocation is slow and uncompetitive, while the formation of the 3° carbocation is fast and a dominant pathway (Fig. 3.1). Preferably, in the reaction of 1-methylcyclohexene with HCl almost all the reactant molecules follow the fast, competitive pathway (with a lower activation energy E_{a3}) giving rise to the formation of an intermediate 3° carbocation, which is subsequently attacked by chloride

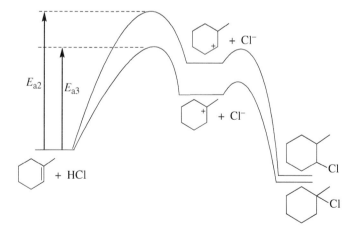

FIGURE 3.2 Reaction profiles for electrophilic addition of hydrogen chloride to 1-methylcyclohexene. The activation energy (E_a) for the formation of an intermediate carbocation (rate-determining step) increases linearly as a function of the energy level of the carbocation ($E_{carbocation}$) as formulated by Eq. 3.3.

Cl⁻ (nucleophile) leading to 1-chloro-1-methylcyclohexane, the observed product. In addition, the observed tertiary chloroalkane product is slightly thermodynamically more stable than the alternative secondary chloroalkane [4]. Therefore, the formation of the tertiary chloroalkane is both kinetically and thermodynamically favorable.

In general, when HX reacts with an unsymmetrical alkene, the hydrogen from HX is preferably added to the doubly bonded carbon which contains more hydrogen atoms, while the –X group is added to another carbon. This is the well-known **Markovnikov's rule.** Clearly, this rule is dictated by the relative stability of the intermediate carbocations possibly formed in the reaction: When the hydrogen from HX is added (protonate) to the doubly bonded carbon which contains more hydrogen atoms (Markovnikov's addition), the reaction gives a more stable intermediate carbocation possessing lower energy level. This makes the reaction faster, and the corresponding product will be the major product. The alternative pathway of addition (non-Markovnikov's addition) would proceeds via a less stable carbocation giving rise to a slower reaction. As a result, the yield of the product will be low (minor or not formed).

The regioselectivity dictated by the carbocation stability is further demonstrated by the reaction in Figure 3.3. Of the two C=C bonds in the alkene substrate, the C=C π bond in the cyclohexene ring is connected to an electron-withdrawing (deactivating) carbonyl (C=O) group. The C=O group lowers electron density in the ring C=C bond domain through a conjugation effect, making the π bond relatively inactive (unreactive). As a result, the addition reaction with HBr only takes place on the acylic C=C π bond [3]. As HBr interacts with the acylic C=C π bond, the hydrogen will be selectively added (protonate) to a methylene (=CH₂) carbon giving a more stable intermediate 3° carbocation. The 3° carbocation is subsequently attacked by

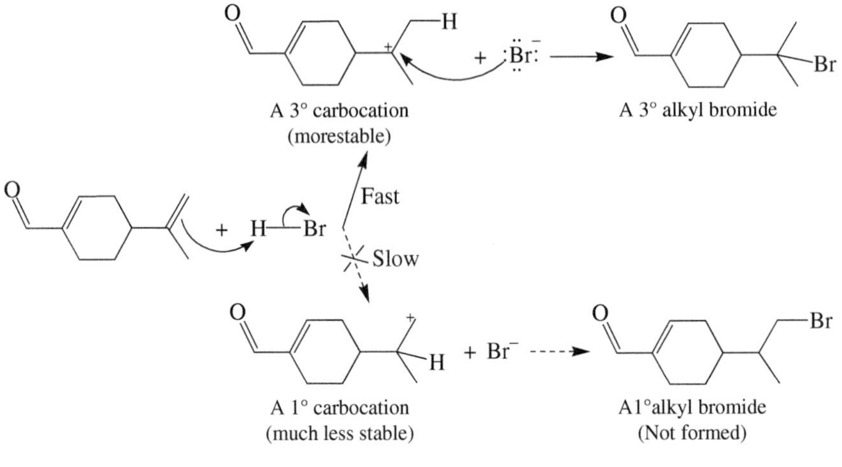

FIGURE 3.3 Regioselectivity for electrophilic addition of hydrogen bromide to a diene.

bromide leading to a 3° alkyl bromide, the observed product. An alternative pathway of addition through a 1° carbocation (very unstable) would be very slow due to a high activation energy associated to the formation of the carbocation. The process would be uncompetitive, and the corresponding 1° alkyl bromide product is not formed.

Stereochemistry   Next, we will address the feature of stereochemistry associated to the Markovnikov addition reactions which take place via carbocation intermediates. While the regiochemistry of the HX additions to alkenes is determined by the **relative stability** of the possibly formed intermediate carbocations, the stereochemistry of the reactions is dictated by the **structure** of the intermediate carbocations. Figure 3.4 illustrates reaction of *cis*-2-butene with HCl. The reaction is regiochemically inactive. As HCl interacts with the alkene π bond, only one carbocation can be generated. Then the chloride Cl⁻ anion (nucleophile) attacks equally either of the p orbital lobes in the carbocation giving a racemic mixture. Such stereochemistry feature determined by the structure of the intermediate carbocation is further demonstrated by reaction of styrene with HBr in Figure 3.5. The reaction follows Markovnikov's rule and takes place via a relatively more stable secondary benzylic cation. Then the bromide Br⁻ anion (nucleophile) attacks equally either of the p orbital lobes in the carbocation giving a racemic mixture, the same as the outcome of the *cis*-2-butene reaction (Fig. 3.4). In both reactions, the intermediate carbocations are **prochiral**, namely, that the positively charged trivalent carbon in each cation is connected to three different groups. Although each carbocation is achiral, the central carbon will become asymmetric once attached to a nucleophile. In general, a prochiral carbocation leads to the formation of a racemic mixture if two faces of the planar carbocation have the same environment. This feature will be seen again in S_N1 reactions (Chapter 6). In the case when two faces of a planar carbocation have

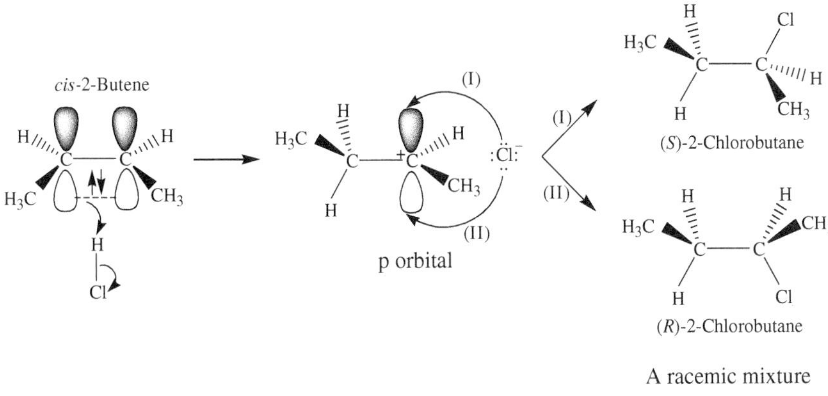

FIGURE 3.4   Mechanism and stereochemistry for electrophilic addition of hydrogen chloride to *cis*-2-butene.

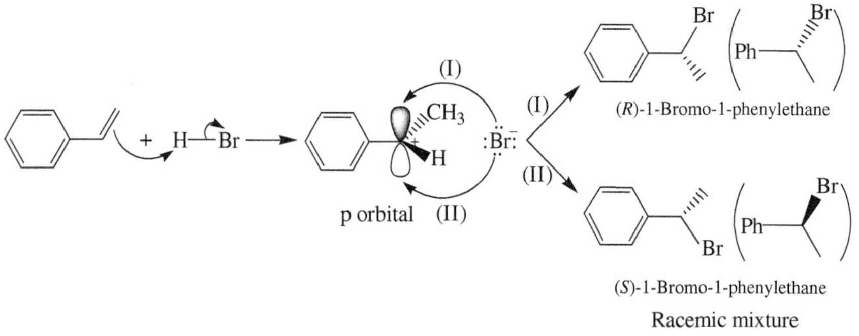

FIGURE 3.5   Mechanism, regiochemistry, and stereochemistry for electrophilic addition of hydrogen bromide to styrene.

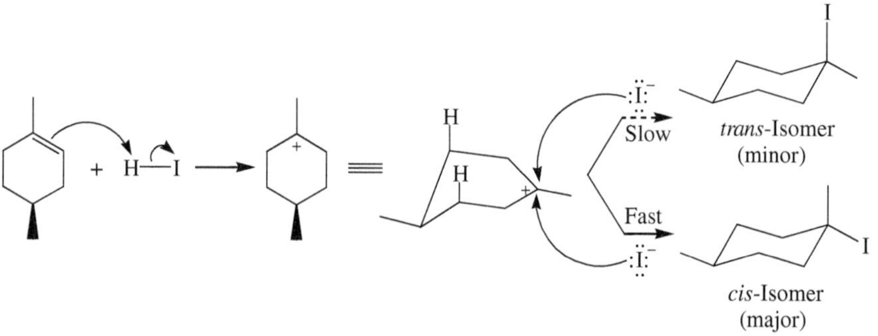

FIGURE 3.6   Stereoselectivity of electrophilic addition of hydrogen iodide to (S)-1,4-dimethylcyclohexene. Adapted from Fox and Whitesell [4].

different environments, the reaction may be stereoselective. This type of stereose-lectivity is demonstrated by electrophilic addition of HI to (S)-1,4-dimethylcyclohexene (Fig. 3.6) [4].

Figure 3.6 shows that similar to electrophilic addition of HX to 1-methylcyclo-hexene, reaction of (S)-1,4-dimethylcyclohexene with HI proceeds via a tertiary intermediate carbocation. In its favorable conformation, the C_4-methyl stays in an equatorial position and both C_3-hydrogen and C_5-hydrogen stay in axial positions. The two axial hydrogen atoms hinder the iodide from approaching the carbocation from the top. As a result, the corresponding *trans*-isomer will be produced in rela-tively low yield as a minor product. Instead, the attack of the iodide from the bottom will be essentially free from steric hindrance leading to a fast reaction. The corre-sponding *cis*-isomer will be formed in high yield as a major product [4]. In addition, the *cis*-isomer is slightly thermodynamically more stable than the *trans*-isomer. This makes the formation of the *cis*-isomer thermodynamically more favorable as well.

3.1.3   Acid and Transition-Metal Catalyzed Hydration of Alkenes and Its Applications

Acid catalyzed hydration   While strong protic acids can be added (protonate) to an alkene π bond readily because of high electrophilicity of the proton, a weak protic acid does not undergo appreciable addition (protonation) to alkenes primarily due to its low electrophilicity making the reaction very slow. However, **addition of a weak protic acid to alkenes can be generally catalyzed by a strong acid**. Acid-catalyzed hydration (addition of water) is a good example to demonstrate such reactivity.

H_2O is considered a very weak protic acid with pK_a (pK_w) being 14. In the presence of a strong acid (typically sulfuric acid), addition of H_2O (hydration) to alkenes can take place via the hydronium (H_3O^+) ion (Fig. 3.7). H_3O^+ is a strong proton donor, being highly electrophilic. In all the reactions illustrated in Figure 3.7, the π bonds in alkenes are effectively protonated by H_3O^+ giving carbocation intermediates. The reactions follow Markovnikov's rule: Protonation by H_3O^+ takes place on the doubly bonded carbon bearing more hydrogen atoms. As a result, a more stable carbocation is formed. The carbocations are subsequently attacked by a H_2O molecule (nucleophile) leading to the formation of Markovnikov alcohols. Since the reactions are regioselective, they can be employed in industry to make some useful alcohols. On the other hand, the non-Markovnikov alcohols can be made by hydroboration of the alkenes followed by oxidation with H_2O_2. For reactions of 1-butene and styrene, the intermediate carbocations are prochiral (with the trivalent carbon connecting to three different groups). The prochiral carbocations lead to the formation of a racemic mixture in each of the reactions. Acid-catalyzed hydration of the (*S*)-1,4-dimethylcyclohexene is analogous to the addition of HI (Fig. 3.6). It is both regioselective and stereoselective.

The presence of a strong electron-donating group (activating group) on an alkene C=C bond increases electron density in the π bond and accelerates acid-catalyzed hydration of the alkene. In addition, the electron donating group dictates the regiochemistry of the hydration such that it directs the positive charge of the carbocation intermediate formed on the carbon which is attached to the activating group. This effect is demonstrated by acid-catalyzed hydration of 1-methoxycyclohexene (Fig. 3.8) [5]. In the first step of the reaction mechanism, protonation of the π bond favorably occurs on the doubly bonded carbon bearing one hydrogen atom. As a result, in the carbocation intermediate (methoxy carbocation) the carbon atom bearing a methoxy $-OCH_3$ group (a strong electron-donating group) becomes trivalent, holding a positive charge in its unhybridized p orbital. The hyperconjugation effect of an oxygen lone-pair of electrons in methoxy stabilizes the carbocation by delocalizing the positive charge (resonance stabilization). An alternative pathway of protonation on the methoxy carbon would lead to the formation of a regular secondary carbocation, which is much less stable making the reaction slow and uncompetitive. In the second step, the methoxy carbocation intermediate is attacked by nucleophilic water eventually leading to introduction of a hydroxyl $-OH$ group onto the methoxy

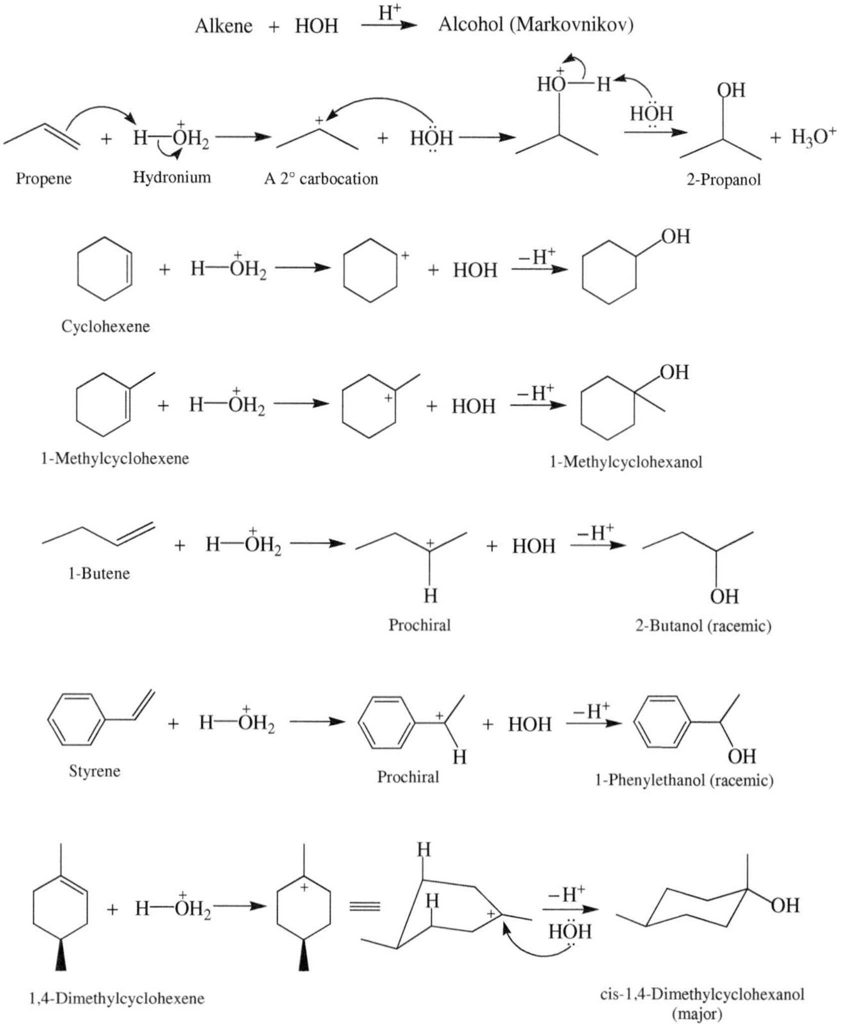

Alkene + HOH $\xrightarrow{H^+}$ Alcohol (Markovnikov)

Propene + Hydronium → A 2° carbocation + HÖH → ... → 2-Propanol + H_3O^+

Cyclohexene + H—ÖH$_2$ → + HOH $\xrightarrow{-H^+}$

1-Methylcyclohexene + H—ÖH$_2$ → + HOH $\xrightarrow{-H^+}$ 1-Methylcyclohexanol

1-Butene + H—ÖH$_2$ → Prochiral + HOH $\xrightarrow{-H^+}$ 2-Butanol (racemic)

Styrene + H—ÖH$_2$ → Prochiral + HOH $\xrightarrow{-H^+}$ 1-Phenylethanol (racemic)

1,4-Dimethylcyclohexene + H—ÖH$_2$ → ≡ $\xrightarrow{-H^+}$ cis-1,4-Dimethylcyclohexanol (major)

FIGURE 3.7   The acid-catalyzed hydration of various alkenes.

carbon and subsequent protonation of the –OCH$_3$ group. The protonated methoxy group [–$^+$O(H)CH$_3$, a good leaving group] is readily eliminated. As it leaves, the positive charge initially formed on the protonated methoxy oxygen starts being transferred to the carbon which is attached to a hydroxyl –OH group. Simultaneously, an active lone-pair of electrons in –OH is moving down to the domain of the carbon–oxygen linkage, resulting in the formation of a C=O π bond via a carbocation-like transition state. In the end, cyclohexanone is formed together with the release of methanol as a by-product.

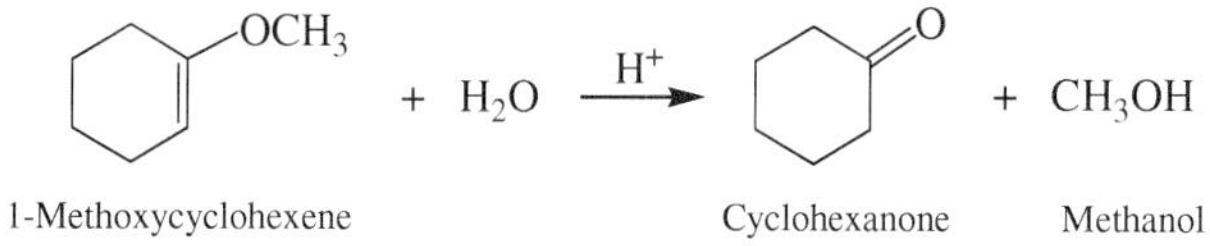

1-Methoxycyclohexene   + H_2O $\xrightarrow{H^+}$ Cyclohexanone + CH_3OH Methanol

Mechanism:

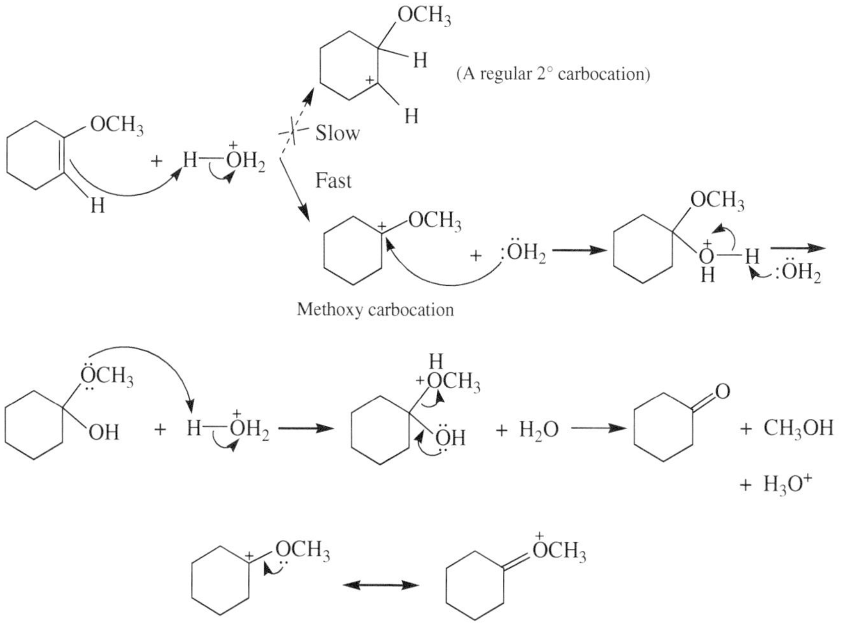

OCH₃ / H (A regular 2° carbocation)

OCH₃ + H—$\overset{+}{O}H_2$ Slow Fast

Methoxy carbocation   + $:\overset{..}{O}H_2$

OCH₃ / $\overset{+}{O}$—H $:\overset{..}{O}H_2$

$\overset{..}{O}CH_3$ / OH + H—$\overset{+}{O}H_2$ ⟶ H+OCH₃ / $\overset{..}{O}H$ + H_2O ⟶ + CH_3OH

+ H_3O^+

OCH₃ ⟷ $\overset{+}{O}CH_3$

Resonance stabilization due to charge delocalization

FIGURE 3.8   Mechanism for the acid-catalyzed hydration of 1-methoxycyclohexene.

While reaction of an alkene such as $RCH=CH_2$ with hydrogen halide (HCl, HBr, or HI) in an **unreactive solvent** (e.g., CH_2Cl_2) gives a 2-haloalkane [$RCH(X)CH_3$], a reaction performed in an aqueous medium will result in the formation of a mixture of 2-haloalkane [$RCH(X)CH_3$] and 2-alcohol [$RCH(OH)CH_3$]. It is equivalent to a strong-acid (HX) catalyzed hydration in addition to electrophilic addition of HX to an alkene.

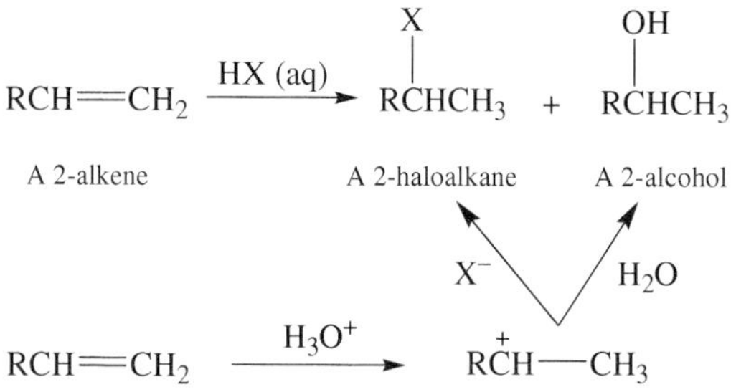

$$RCH=CH_2 \xrightarrow{HX\ (aq)} \underset{\substack{| \\ RCHCH_3}}{X} + \underset{\substack{| \\ RCHCH_3}}{OH}$$

A 2-alkene   A 2-haloalkane   A 2-alcohol

X^- H_2O

$$RCH=CH_2 \xrightarrow{H_3O^+} R\overset{+}{C}H-CH_3$$

The first step in the mechanism is the protonation of the alkene π bond by hydronium giving a secondary carbocation intermediate, which is subsequently attacked by halide and water, leading to the formation of a haloalkane and an alcohol, respectively.

Mercury(II) catalyzed hydration   The alkene hydration reactions can also be catalyzed by Hg(II), typically in the form of mercury(II) acetate [Hg(OAc)$_2$] [3]. The hydration catalyzed by Hg(OAc)$_2$ is usually more efficient than that which is catalyzed by a simple strong acid in terms of product yield, regioselectivity, and milder reaction conditions. Therefore, the Hg(OAc)$_2$ catalyzed hydration of alkenes possesses better synthetic utility for making various Markovnikov alcohols.

Mercury(II) acetate is a covalent compound with two acetate ligands bonded to the central mercury through empty sp-hybrid orbitals in Hg(II). Structure of the Hg(OAc)$_2$ molecule is shown in Figure 3.9 with the O–Hg–O linkage in the molecule being linear [6]. The valence shell of the mercury center in Hg(OAc)$_2$ contains empty 6p orbitals (but the d-subshell, 5d^{10}, filled) making Hg(II) in the molecule highly electrophilic. Such high electrophilicity associated to Hg(II) has been utilized to activate the very inactive alkane C—H bond as demonstrated in Section 2.6.2. The electrophilic Hg(OAc)$_2$ can readily attack an nucleophilic alkene C=C bond leading to the formation of a reactive mercury(II) π complex and bringing about the subsequent hydration reaction.

Figure 3.10 shows the Hg(OAc)$_2$ catalyzed hydration of alkenes, which is followed by reduction with NaBH$_4$ giving Markovnikov alcohols. The counterpart of non-Markovnikov alcohol for each alkene can be made by hydroboration coupled with a subsequent oxidation by H$_2$O$_2$. In the first step of mechanism for the Hg(OAc)$_2$ catalyzed hydration, Hg(OAc)$_2$ approaches an alkene C=C bond perpendicularly toward the alkene molecule plane, which leads to the formation of an intermediate π complex. The π complex is then attacked by a nucleophilic water molecule leading to an addition product with a hydroxyl –OH group and a C—Hg bond formed. The positive charge in the π complex is delocalized, mainly distributed to the secondary carbon in the 1-alkene (C$_{sec}$–Hg is longer than C$_{prim}$–Hg), and to the tertiary carbon in 1-methylcyclohexene (C$_{tert}$–Hg is longer than C$_{sec}$–Hg), so that the complex is better stabilized [3]. As a result, the water molecule will

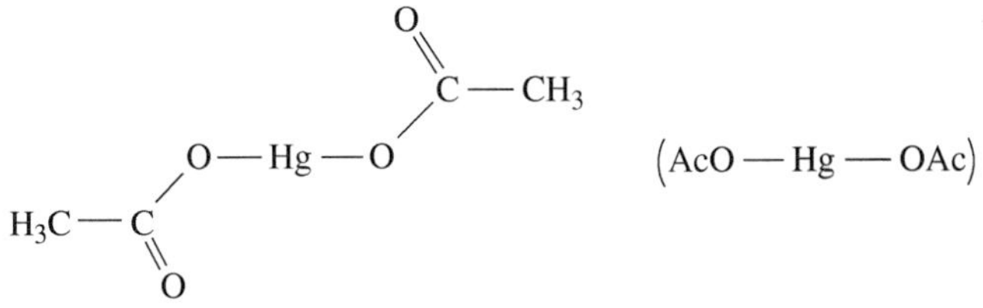

FIGURE 3.9   Structure of mercury(II) acetate.

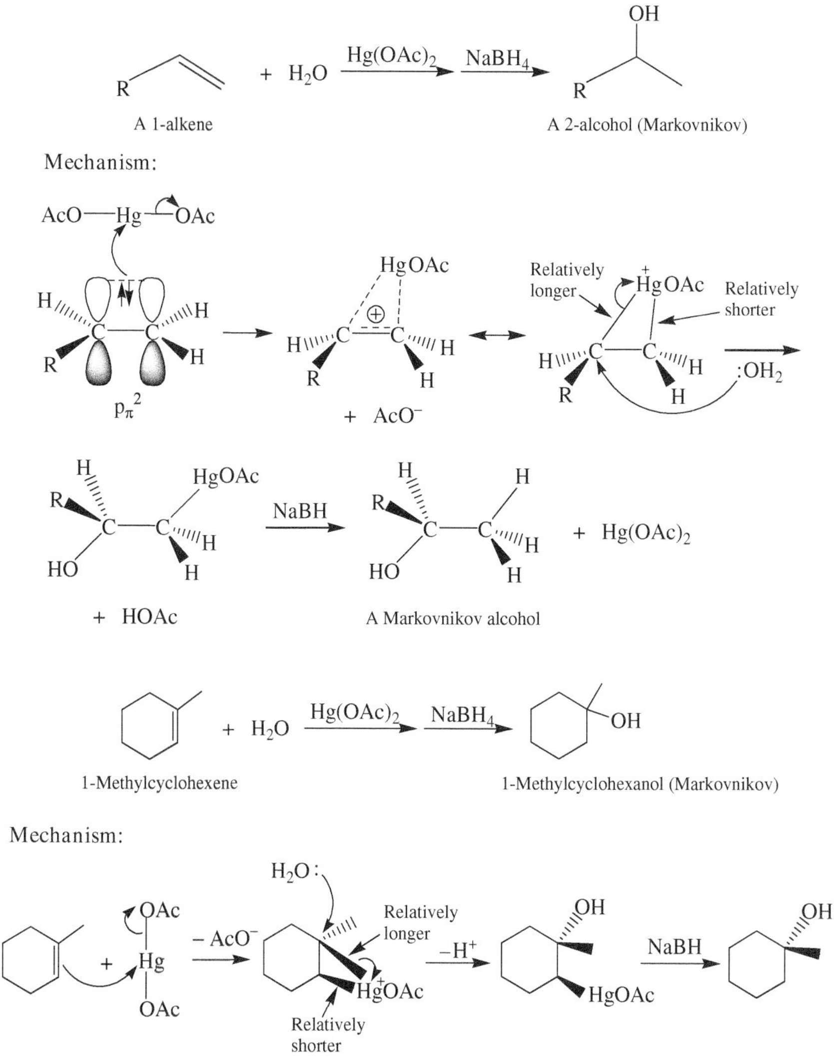

A 1-alkene + H_2O $\xrightarrow{Hg(OAc)_2}$ $\xrightarrow{NaBH_4}$ A 2-alcohol (Markovnikov)

Mechanism:

1-Methylcyclohexene + H_2O $\xrightarrow{Hg(OAc)_2}$ $\xrightarrow{NaBH_4}$ 1-Methylcyclohexanol (Markovnikov)

Mechanism:

FIGURE 3.10   Mechanism and regioselectivity for mercury(II) acetate catalyzed hydration of alkenes.

preferably attack the secondary carbon and tertiary carbon in the 1-alkene and 1-methylcyclohexene reactions, respectively. In each reaction, the nucleophilic attack by water occurs from the different side of mercury giving rise to an anti-addition. Upon treated by $NaBH_4$, the C–Hg bond is reduced and a C—H bond is formed. The overall hydration is completed.

3.1.4   Acid Catalyzed Additions of Alcohols to Alkenes

Analogous to water, alcohols also undergo acid (typically H_2SO_4 and TsOH) catalyzed additions to alkenes via carbocation intermediates. The reactions give ethers and follow Markovnikov's rule (Figure 3.11). In the first step of the reaction mechanism, an alkene C=C bond is protonated by the acid catalyst to generate a carbocation intermediate. Then a nucleophilic alcohol attacks the carbocation to lead to the formation of an ether product (via an oxonium intermediate). The reactions can also be catalyzed by $Hg(OAc)_2$, the mechanism of which is analogous to that of hydration.

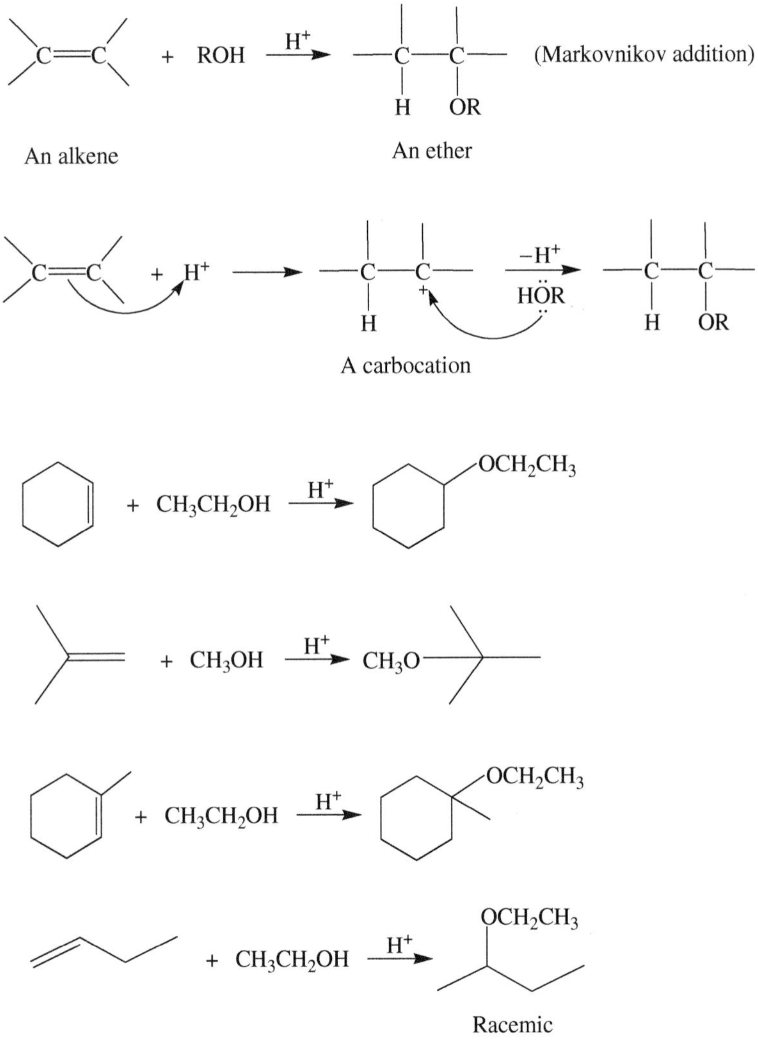

An alkene   An ether

A carbocation

Racemic

FIGURE 3.11   The acid-catalyzed electrophilic addition of alcohols to various alkenes.

3.1.5 Special Electrophilic Additions of the Alkene C=C Bond: Mechanistic and Synthetic Aspects

Reactions with carboxylic acids   As mentioned in Section 3.1.3, addition of a weak protic acid to alkenes can take place in the presence of a catalytic amount of a strong acid, and the reaction follows Markovnikov's rule. Carboxylic acids (RCO_2H) are among the most common weak organic acids (with pK_a ~4). The additions of RCO_2H to alkenes can be catalyzed by *p*-toluenesulfonic acid (TsOH) [1]. The reactions give esters, an example of which is shown in Figure 3.12. This type of reactions is utilized to synthesize tertiary butyl esters, while the corresponding Fischer esterification (acid-catalyzed reaction of a carboxylic acid with tertiary butyl alcohol) would render steric hindrance and does not take place.

The reaction described in Figure 3.12 occurs via a tertiary carbocation which is effectively generated by protonation of an alkene C=C π bond with TsOH (a strong acid). Then the carbocation is attacked by the nucleophilic carbonyl oxygen in the carboxylic acid resulting in the formation of a tertiary butyl ester.

The following specialty tertiary butyl ester is made by the acid-catalyzed addition of a carboxylic acid to 2-methylpropene [3]:

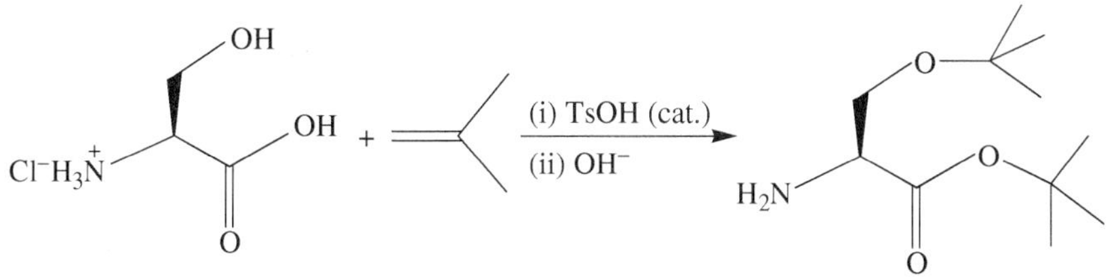

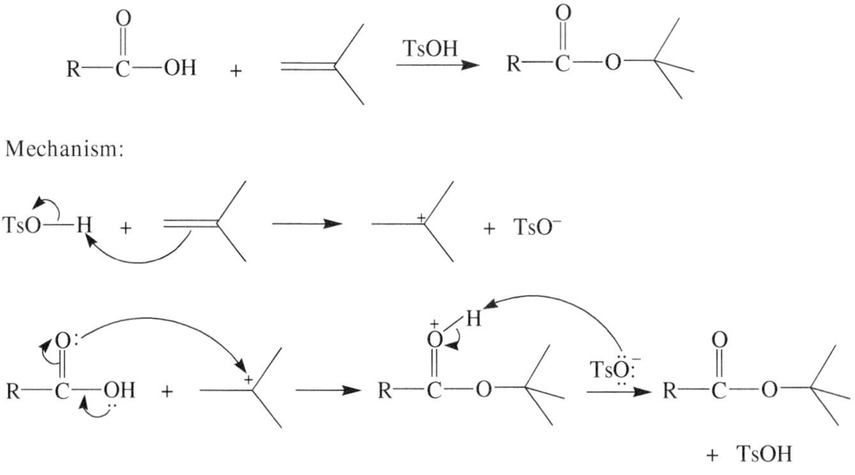

Mechanism:

FIGURE 3.12   Mechanism for toluenesulfonic acid catalyzed electrophilic addition of a carboxylic acid to 2-methylpropene.

The starting compound is a β-hydroxycarboxilic acid. In the presence of a catalytic amount of TsOH, the carboxyl group ($-CO_2H$) is added to the alkene giving a tertiary butyl ester. In addition, the hydroxyl –OH is also added to the alkene giving a tertiary butyl ether. Because neither the carboxyl nor the hydroxyl group is sufficiently acidic to protonate the alkene, a strong acid, such as TsOH, is required to catalyze the reaction by protonating the C=C bond to make a carbocation intermediate. For both groups, the addition takes place via a tertiary butyl cation generated by TsOH.

Lewis-acid catalyzed electrophilic addition to alkenes   Sometimes, alkene addition reactions can also be catalyzed by a Lewis acid. An industrially useful example is the reaction of benzene with cyclohexene which gives cyclohexylbenzene in the presence of aluminum chloride in a catalytic amount. The mechanism is shown in Figure 3.13. This reaction takes place via a secondary carbocation which is generated by facile electrophilic addition of $AlCl_3$ to the cyclohexene π bond. It takes place via a Zwitterionic transient π-complex (transition state) in which the aluminum

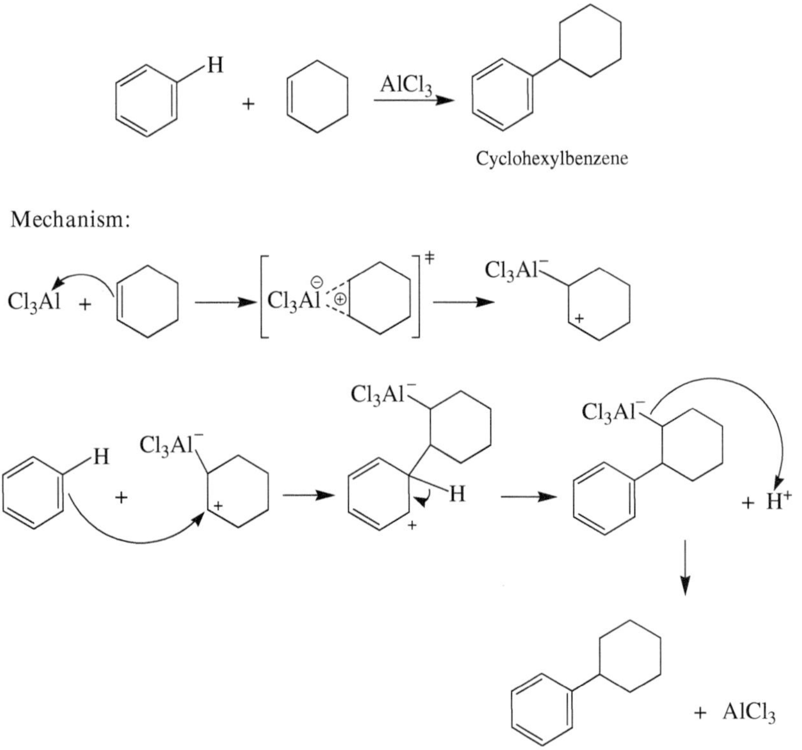

Cyclohexylbenzene

Mechanism:

FIGURE 3.13   Aluminum chloride catalyzed addition of benzene to cyclohexene.

atom in $AlCl_3$ is weakly (partially) bonded to two carbon atoms and the C=C π bond is partially broken. Then the intermediate carbocation is attacked by nucleophilic benzene effecting the overall reaction. The second step, addition of benzene to the carbocation, is also known as an electrophilic aromatic substitution reaction. This type of reactions will be further studied in Chapter 5. Typically, alkylbenzenes are made by $AlCl_3$ catalyzed electrophilic aromatic substitution reactions with alkyl halides (Friedel–Crafts reactions). Because alkenes are in general much cheaper than alkyl halides, the reaction described in Figure 3.13 is of industrial significance.

Additions of carbocations   Due to high electrophilicity of carbocations, they can be effectively added to alkenes as shown below.

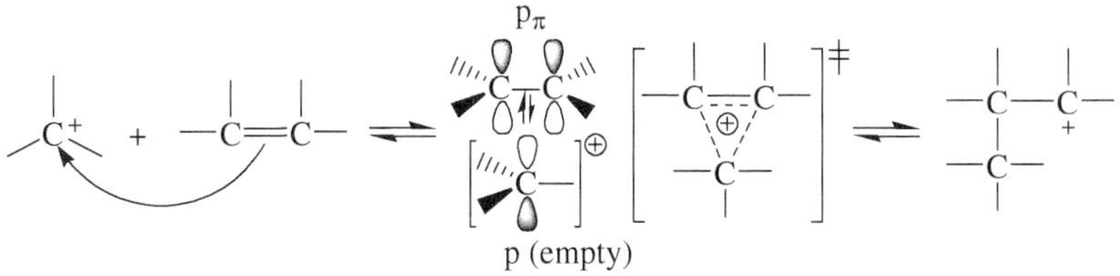

The empty p orbital in a carbocation interacts with the p_π bond in the C=C domain of an alkene to lead to the formation of a 3-center, 2-electron π-complex (transition state) in which the original carbocation carbon is partially bonded to the two carbons in the alkene C=C, and the C=C π bond is partially broken. As the transition state collapses, the original carbocation is converted to another in the end of the reaction. The regiochemistry follows Markovnikov's rule.

Particularly effective are additions of tertiary carbocations to the C=C double bonds. They possess synthetic perspectives. The following shows a remarkable example of alkene addition to carbocations. This addition is found as a mechanistic step in the *in vivo* **biosynthesis** of cholesterol and has been used for **biomimetic synthesis** of the steroid structure (see the following scheme, reprinted from Ref. [3], Page 125, with permission from Elsevier).

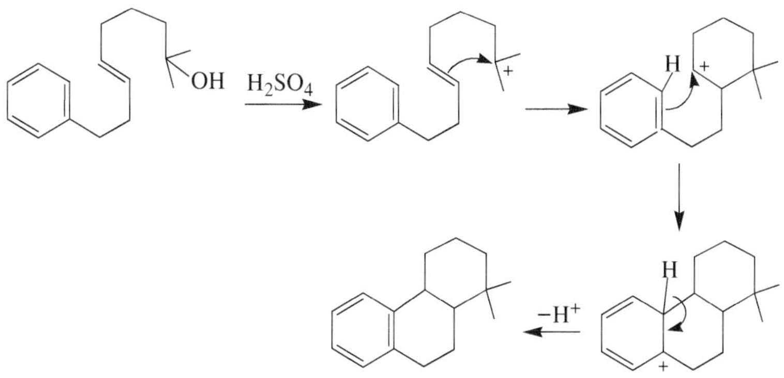

The overall reaction involves a thermodynamically favorable intramolecular addition of a tertiary carbocation to a C=C double bond, leading to the formation of a six-membered ring with the positive charge transferred to a cyclic carbon (Step 2). Then a second ring-closure reaction takes place via an electrophilic aromatic substitution (or addition) of a carbocation to a substituted benzene (Steps 3 and 4), analogues to the reaction described in Figure 3.13.

A tertiary-carbocation effected addition of an alkene has found an application in synthesis of useful 2,2,4-trimethylpentane (isooctane) from 2-methylpropene and 2-methylpropane (isobutane) in the presence of a catalytic amount of hydrogen fluoride (Fig. 3.14) [3]. Isooctane is used as the standard for the gasoline octane ratings. The overall reaction is initiated by a tertiary butyl cation and follows a chain-reaction mechanism. In the first step, HF protonates 2-methylpropene to give an intermediate tertiary butyl cation. Then the carbocation is added to a second 2-methylpropene molecule (Markovnikov addition) generating a second carbocation intermediate, which in turn abstracts the tertiary hydrogen from an isobutane molecule giving an isooctane (product) molecule and regenerating a tertiary butyl cation. The second and third steps are chain-growth steps. They keep repeating until all the reactant molecules are converted to the isooctane product. The feature of this

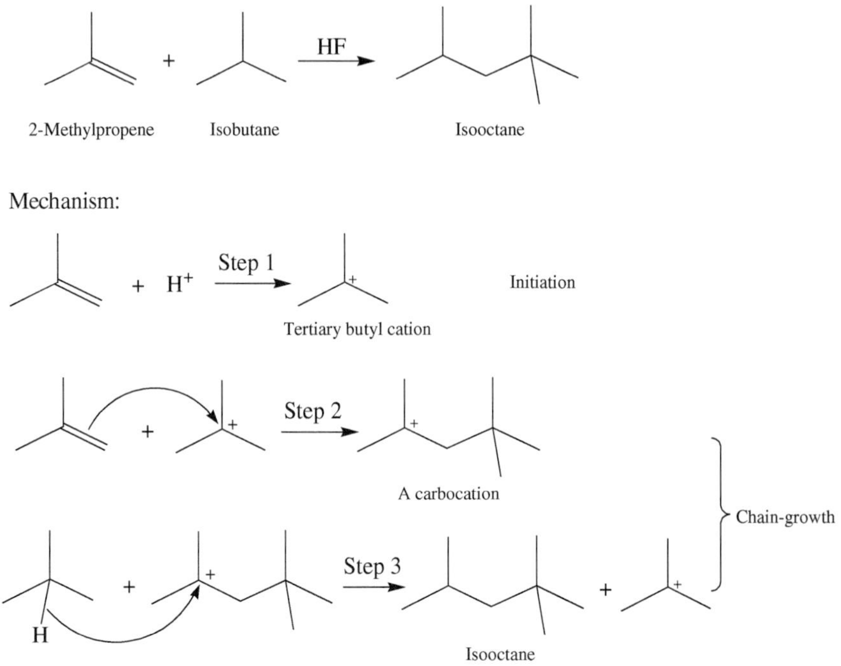

FIGURE 3.14   Acid catalyzed reaction of 2-methylpropene with isobutane effected by tertiary butyl cation.

carbocation chain reaction resembles the pattern of the radical chain-mechanism for alkane halogenations as described in Chapter 2.

The hydrogen abstraction from isobutene by a carbocation intermediate (Step 3 in Fig. 3.14) takes place via a triangular transition state as shown below.

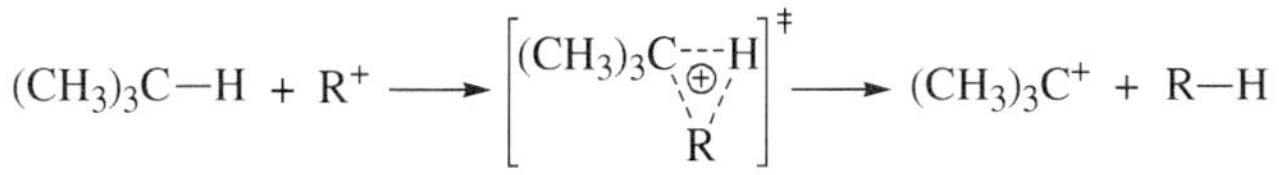

$$(CH_3)_3C-H \ + \ R^+ \ \longrightarrow \ \left[\begin{array}{c} (CH_3)_3C\text{---}H \\ \oplus \\ R \end{array} \right]^{\ddagger} \ \longrightarrow \ (CH_3)_3C^+ \ + \ R-H$$

3.1.6   Electrophilic Addition to the C≡C Triple Bond via a Vinyl Cation Intermediate

Analogous to the alkene addition, research has shown that a carbocation can be added to a π bond in an alkyne C≡C triple bond domain giving a vinyl cation intermediate, and the addition follows Markovnikov regiochemistry [7].

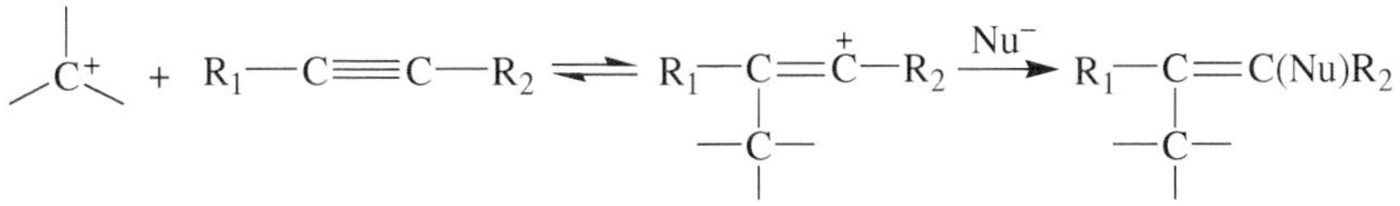

$$\begin{array}{c} | \\ C^+ \\ | \end{array} \ + \ R_1-C\equiv C-R_2 \ \rightleftharpoons \ R_1-\underset{\underset{|}{\overset{|}{\underset{C}{|}}}{C}}{=}\overset{+}{C}-R_2 \ \xrightarrow{Nu^-} \ R_1-\underset{\underset{|}{\overset{|}{\underset{C}{|}}}{C}}{=}C(Nu)R_2$$

A vinyl cation

One way to generate a carbocation intermediate is dissociation of a haloalkane using Friedel–Crafts method. The above mechanism is supported by the observed stereochemistry for $AlCl_3$-catalyzed reaction of deuterated $PhCD_2Cl$ with $PhC\equiv CCH_2Ph$ (Fig. 3.15).

The net reaction in Figure 3.15 is an addition of $PhCD_2Cl$ to an alkyne C≡C bond to give (E)-alkene and (Z)-alkene products in equal amount [7]. The overall reaction is effected by addition of the deuterated benzylic $PhCD_2^+$ cation to the $PhC\equiv CCH_2Ph$ triple bond. This addition affords a Markovnikov vinyl cation intermediate $Ph^+C=C(CD_2Ph)CH_2Ph$ (stabilized by –Ph via conjugation effect), which is subsequently attacked by the nucleophilic $AlCl_4^-$ (chloride donor) from top to bottom of the vinyl cation equally to give (E)-alkene and (Z)-alkene, respectively, in equal amount. An alternative vinyl cation $Ph(PhCD_2)C=C^+CH_2Ph$ would be less stable because there is no aromatic ring attached to the positively charged carbon, and it is not formed. Both stereochemistry and regiochemistry of the reaction are consistent with the formation of a vinyl cation intermediate during the reaction. The mechanism has been shown to be fully valid for the phenyl-substituted acetylenes (R_1 or R_2 is an aromatic ring). However, the reactions of alkyl-substituted acetylenes (both R_1 and R_2 are alkyl groups) deviate the stereochemistry expected for the vinyl cation intermediate. In addition, studies on the proton addition (acid-catalyzed hydration and addition of carboxylic acids) to alkynes indicate that the reaction follows the analogous vinyl cation mechanism as shown below [7].

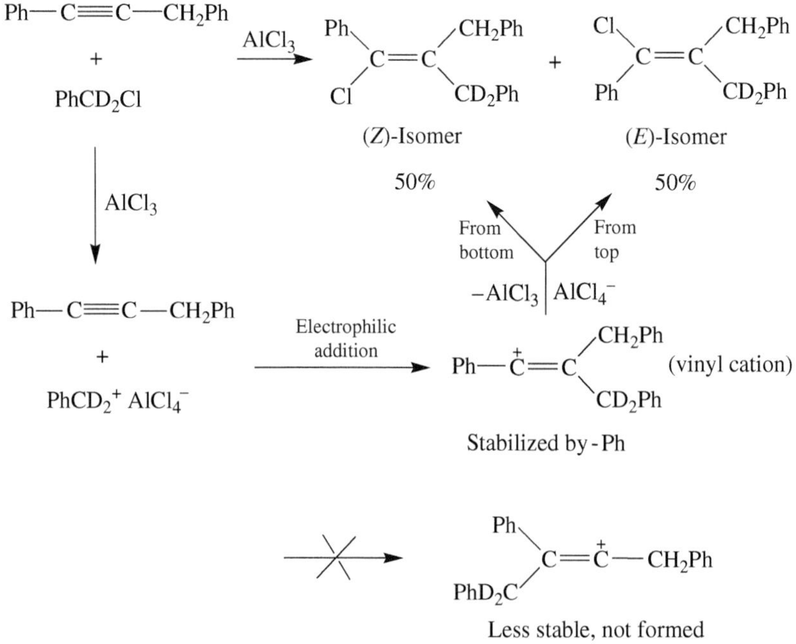

FIGURE 3.15 The AlCl$_3$-catalyzed electrophilic addition of deuterated PhCD$_2$Cl to PhC≡CH$_2$Ph via a vinyl cation intermediate.

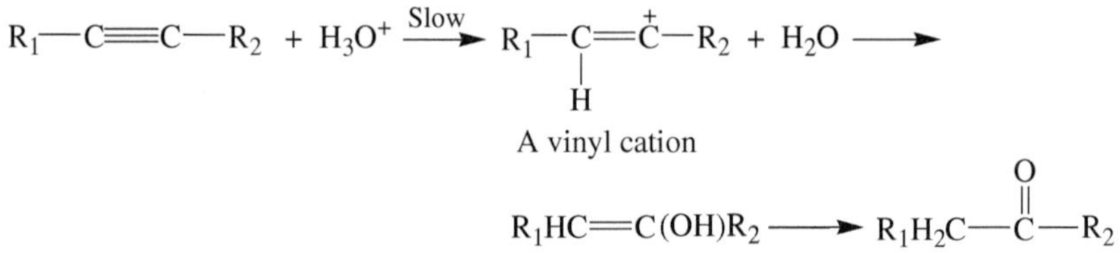

3.2 ELECTROPHILIC ADDITION OF HYDROGEN HALIDES TO CONJUGATED DIENES

Hydrogen halides (such as HCl) can be added to conjugated dienes (such as 1,3-butadiene) giving a 1,2-adduct or a 1,4-adduct (Fig. 3.16) depending on reaction temperatures [2]. The first step in reaction of HCl with 1,3-butadiene is protonation of a π bond by HCl on C$_1$-carbon giving a secondary (2°) allylic cation (Markovnikov addition). Then the carbocation is attacked by chloride possibly on different carbons to lead to the formations of two isomeric products. While the first step (the formation of an intermediate carbocation) is the rate-determining step for the overall reaction, the identity of the final product is determined by the second step on the basis of its activation energy and the product stability. The activation energy

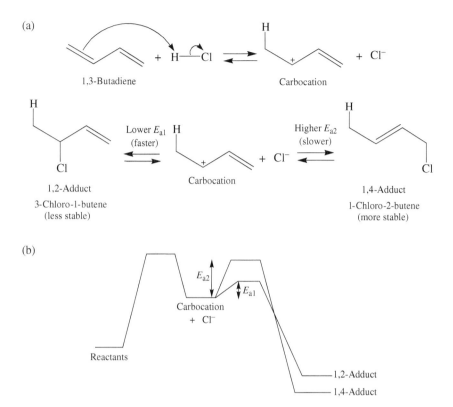

FIGURE 3.16 Electrophilic addition of hydrogen chloride to 1,3-butadiene. (a) The reaction mechanism; and (b) the reaction profile.

for the formation of the 1,2-adduct (E_{a1}) from the carbocation is lower than that for the formation of the 1,4-adduct (E_{a2}). Therefore, the formation of the 1,2-adduct is fast and the formation of the 1,4-adduct is relatively slow. At low temperatures, the overall reaction is essentially irreversible and under kinetic control, with the 1,2-adduct (less stable, but fast formation) being the major product.

As temperature increases, backward reactions (elimination of HCl) for both products become significant. At normal ambient temperature, the reaction is readily reversible. As a result, the overall reaction is equilibrated to the more stable 1,4-adduct as the major product. The reaction is subject to thermodynamic control.

1,2-Adduct
(less stable)
Minor

1,3-Butadiene   + H—Cl

1,4-Adduct
(more stable)
Major

3.3   NON-MARKOVNIKOV RADICAL ADDITION

In the media of the unreactive dichloromethane which favors a carbocation intermediate, additions of HBr to alkenes (such as 1-methylcyclohexene) follow Markovnikov's rule. However, the reactions in diethyl ether which contains a small
amount of peroxides (such as ROOH, a radical initiator) are non-Markovnikov as
shown below [4].

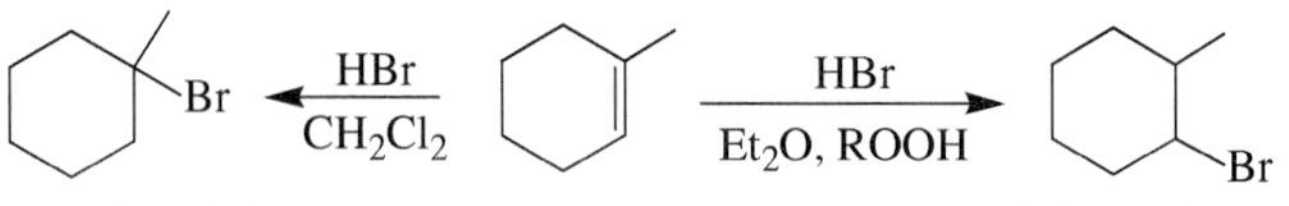

1-Bromo-1-methyl-
cyclohexane
(Markovnikov)

1-Bromo-2-methyl-
cyclohexane (unstereospecific)
(Non-Markovnikov)

Since the reaction of 1-methylcyclohexene with HBr possesses a different regiochemistry in the presence of a radical initiator, its mechanism must be different from
that of a Markovnikov reaction. A radical addition takes place as shown in
Figure 3.17.

The first step in the reaction of 1-methylcyclohexene with HBr in the presence of
a peroxide (ROOH) is the dissociation of ROOH in the light or by gentle warming. It
gives the alkoxy (RO˙) radical, which subsequently attacks HBr initiating a bromine
(Br˙) radical. Then Br˙ interacts with the π bond in 1-methylcyclohexene and results
in the formation of a C—Br bond in the doubly bonded carbon which contains one

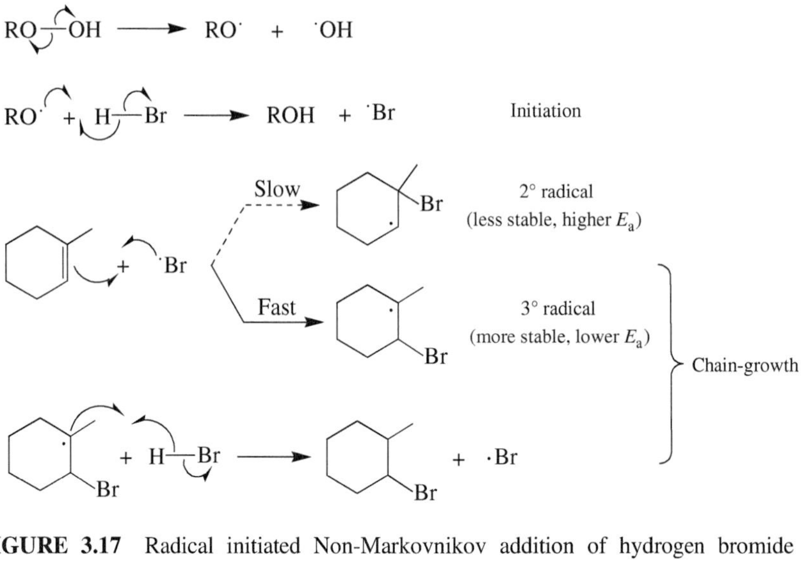

FIGURE 3.17   Radical initiated Non-Markovnikov addition of hydrogen bromide to
1-methylcyclohexene.

hydrogen atom (non-Markovnikov addition), giving an intermediate tertiary (3°) alkyl radical. An alternative pathway of addition of Br˙ to the doubly bonded carbon which does not contain any hydrogen (Markovnikov addition) would generate a less stable secondary (2°) alkyl radical with a higher energy level. Therefore, the Markovnikov addition would render higher activation energy, and be slow and uncompetitive. In the following step, the 3° alkyl radical will attack an HBr molecule resulting in the formation of a C—H bond and regenerating a Br˙ radical, which makes the chain-growth steps continue. The C—H bond can be formed equally in both faces of the planar 3° alkyl radical, giving a mixture of anti-addition and *syn*-addition products. As a result, this addition reaction is not stereochemically controlled. The radical initiated non-Markovnikov addition only works for HBr and fails for other hydrogen halides.

3.4 HYDROBORATION: CONCERTED, NON-MARKOVNIKOV *syn*-ADDITION

3.4.1 Diborane (B$_2$H$_6$): Structure and Properties

Diborane (B$_2$H$_6$) is the central reagent for hydroboration of alkenes. Before we study this type of reactions, let us first go over structure and properties of diborane.

Figure 3.18 shows the structure of diborane. It is dimeric, consisting of two BH$_3$ (borane) units [3, 8]. Each of the boron atoms in B$_2$H$_6$ is nearly tetrahedral. The two BH$_3$ units are linked by two equivalent B–H–B three-center, two-electron bridge bonds. Each of them is formed by overlap of an H 1s orbital with two sp^3 orbitals from two B atoms. Such bridge bonds are relatively weak. As a result, diborane

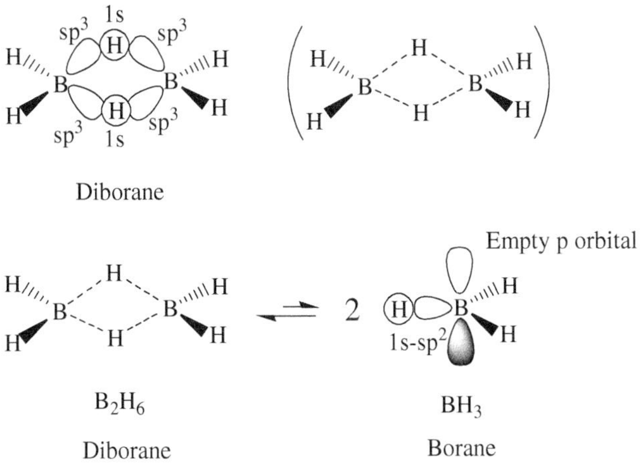

Diborane

B$_2$H$_6$

Diborane

BH$_3$

Borane

FIGURE 3.18   Structure of and bonding in diborane and borane. The three-center, two-electron B—H—B bridge bond. Bruckner [3] and Huheey et al. [8].

undergoes dissociation partly at ambient temperature. Dissociation of each B_2H_6 molecule gives two BH_3 molecules.

BH_3 is trigonal planar with the boron atom sp^2 hybridized. In the valence shell of boron, there is an unhybridized p orbital perpendicular to the BH_3 molecule plane, and this p orbital remains unoccupied (empty). The empty p orbital in BH_3 has a strong ability to accept a pair of electrons making BH_3 highly electrophilic. As a result, BH_3 can be added to the nucleophilic π bond of alkenes (hydroboration).

3.4.2   Concerted, Non-Markovnikov *syn*-Addition of Borane (BH_3) to the Alkene C=C Bond: Mechanism, Regiochemistry, and Stereochemistry

An alkene C=C double bond undergoes hydroboration reaction with B_2H_6 in THF or ether (Fig. 3.19). The reaction takes place via a borane BH_3 molecule, which is generated from dissociation of B_2H_6 to a small extent at ambient temperature. If the alkene is unsymmetrical, the addition is non-Markovnikov: The boron from BH_3 is added to the doubly bonded carbon containing more hydrogen atoms, while a hydrogen from BH_3 is added to the other carbon containing less hydrogen. It is a *syn*-addition with the born and a hydrogen from borane added to the same side of the C=C bond. The second and third B—H bonds in BH_3 undergo the same type of non-Markovnikov *syn*-additions. Subsequent treatment of the hydroboration product (trialkyl borane) by hydrogen peroxide (H_2O_2) in a strong alkaline medium (such as NaOH) affords a non-Markovnikov alcohol. On the other hand, synthesis of Markovnikov alcohols can be achieved by acid or mercury(II) catalyzed hydration reactions of alkenes.

Mechanism and regiochemistry   First, let us demonstrate mechanism and regiochemistry of hydroboration reactions using the reaction of a 1-alkene ($RCH=CH_2$) (Fig. 3.19). The alkene undergoes a **concerted, non-Markovnikov, *syn*-addition** with borane ($H–BH_2$). In the course of the reaction, the electrophilic borane molecule approaches the alkene π bond such that the boron empty p orbital attacks the p_π orbital of the CH_2 (a doubly bonded carbon containing more hydrogens) [9]. Simultaneously, the hydrogen from a borane B—H bond attacks the CH carbon (the other doubly bonded carbon containing less hydrogen). As the pair of π electrons is being transferred to the electrophilic boron, the B—H bonding electrons will be transferred to the CH carbon simultaneously. A four-center cyclic transition state is formed in which the C=C π bond and a B—H bond are partially broken, coincident with the partial formation of C—B and C—H bonds [9]. Dissociation of the transition state affords the formation of the addition product of an alkylborane ($R'BH_2$) in which the H and BH_2 from borane are added in the same side of the C—C bond (*syn*-addition). An alternative pathway of addition, namely that approach of the BH_2 group to the CH carbon, will render steric hindrance (interaction of BH_2 with the R– group in the alkene molecule), leading to a slow, uncompetitive reaction. Subsequently, the

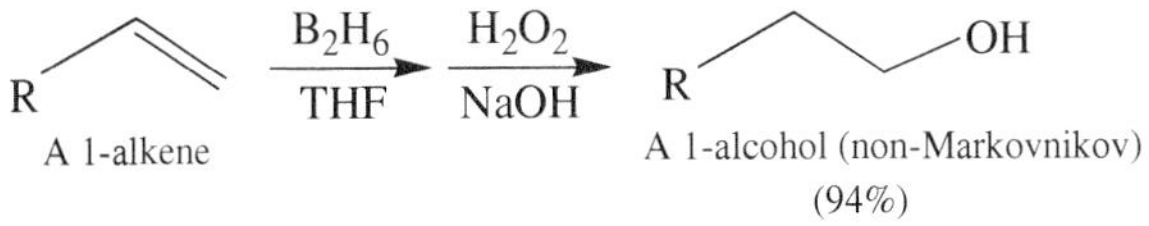

Mechanism:

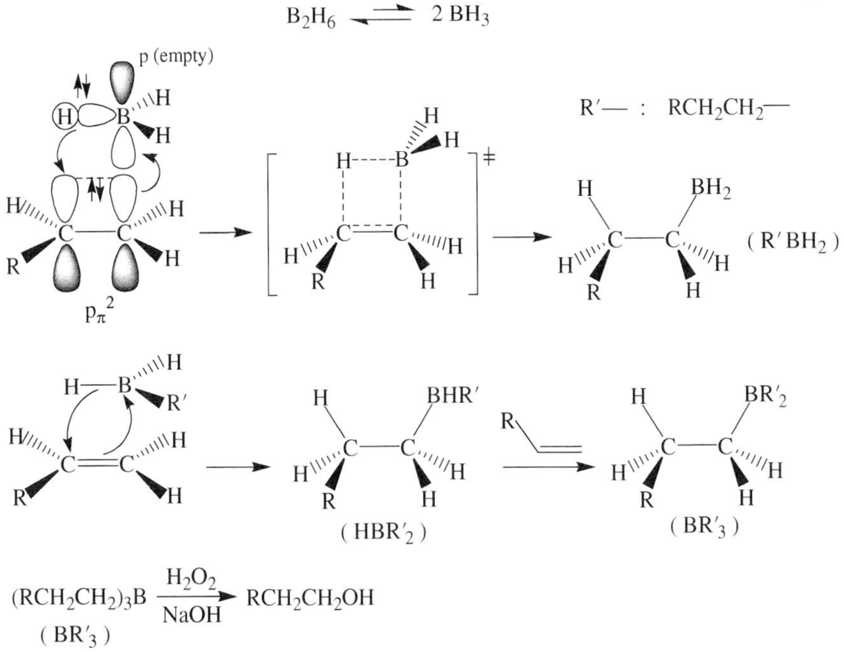

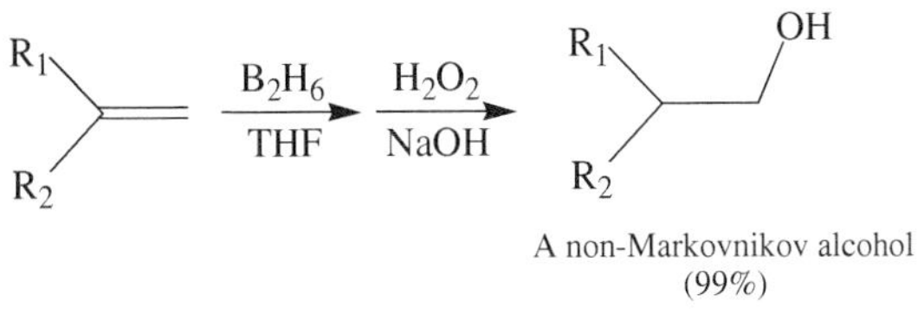

$$B_2H_6 \rightleftharpoons 2\ BH_3$$

$R' —\ :\ RCH_2CH_2—$

$$(RCH_2CH_2)_3B \xrightarrow[NaOH]{H_2O_2} RCH_2CH_2OH$$
(BR'_3)

A non-Markovnikov alcohol
(99%)

Mechanism:

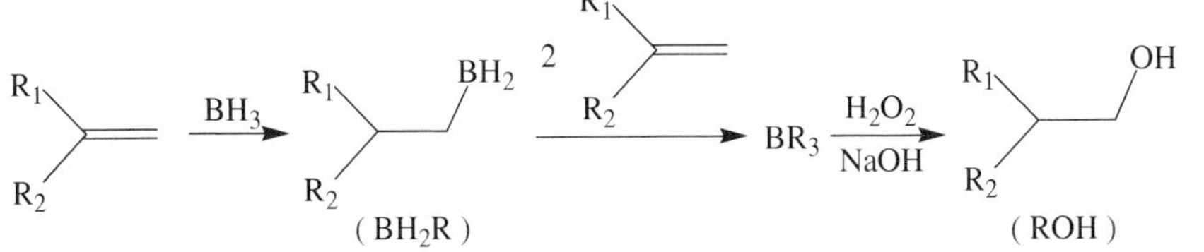

FIGURE 3.19   Mechanism for hydroboration of alkenes: Concerted, non-Markovnikov *syn*-addition.

alkylborane $R'BH_2$ formed in the first step of addition will undergo a similar concerted, non-Markovnikov *syn*-addition to $RCH=CH_2$ giving a dialkyl borane HBR'_2, which in turn undergoes one more addition to a third $RCH=CH_2$ eventually giving a trialkyl borane BR'_3. Treatment of BR'_3 with H_2O_2 in the alkaline medium leads to the formation of RCH_2CH_2OH (a 1-alcohol) in 94% yield. Hydroboration of the second 1-alkene in Figure 3.18 which contains two alkyl groups on the C_2-carbon exhibits higher regioselectivity, which, after oxidation by H_2O_2, gives an anti-Markovnikov 1-alcohol in 99% yield.

Stereochemistry   The reactions of the two 1-alkenes in Figure 3.19 have demonstrated characteristic regioselectivity associated to alkene hydroboration. However, its stereochemistry (the *syn*-addition feature) cannot be proved by either of the reactions. The *syn*-addition feature of alkene hydroboration will be proved by examining the product distribution for the hydroboration of 1-methylcyclopentene followed by oxidation with H_2O_2 as shown below.

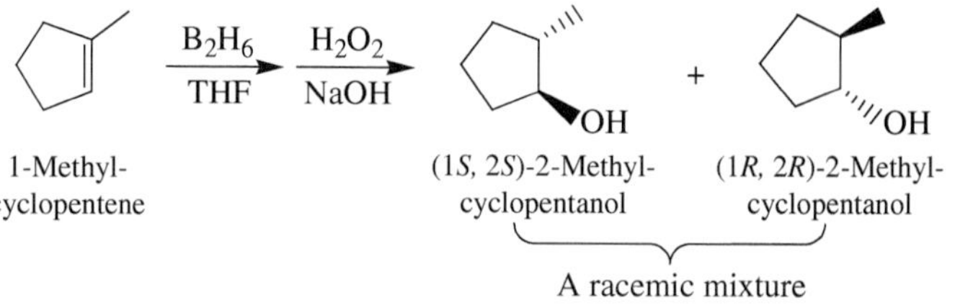

| 1-Methyl-cyclopentene | (1*S*, 2*S*)-2-Methyl-cyclopentanol | (1*R*, 2*R*)-2-Methyl-cyclopentanol |

A racemic mixture

The overall reaction gives a racemic mixture of two non-Markovnikov 2-methylcyclopentanols both of which are in *trans*-configurations. It shows that in the initial non-Markovnikov hydroboration the boron and hydrogen are added in the same side of the C=C bond in 1-methylcyclopentene. As a result, the methyl group will be positioned to the different side of boron in the addition product. The subsequent oxidation by H_2O_2 will lead to a net replacement of boron with a hydroxyl –OH group, with original configuration of the carbon retained (i.e., the orientation of the –OH group remains consistent with that of the boron). Therefore, only *trans*-alcohols are formed in the overall reaction. The detailed mechanism is illustrated in Figure 3.20.

In the first step of the hydroboration of 1-methylcyclopentene described in Figure 3.20, a BH_3 molecule can be added equally to both faces (front and back) of the alkene C=C bond, generating two enantiomeric alkylboranes $R'BH_2$ (addition from front) and $R''BH_2$ (addition from back) in an equal amount (a racemic mixture). Then both $R'BH_2$ and $R''BH_2$ undergo additions to a second 1-methylcyclopentene molecule, respectively, in both faces of the alkene giving dialkylboranes. Since addition from front of the alkene double bond makes R' and addition from back makes R'' (enantiomeric to R'), the second hydroboration of $R'BH_2$ gives R'_2BH (front addition) and $R'R''BH$ (back addition), while the second hydroboration of $R''BH_2$ gives $R'R''BH$ (front addition) and R''_2BH (back addition). Then each of the

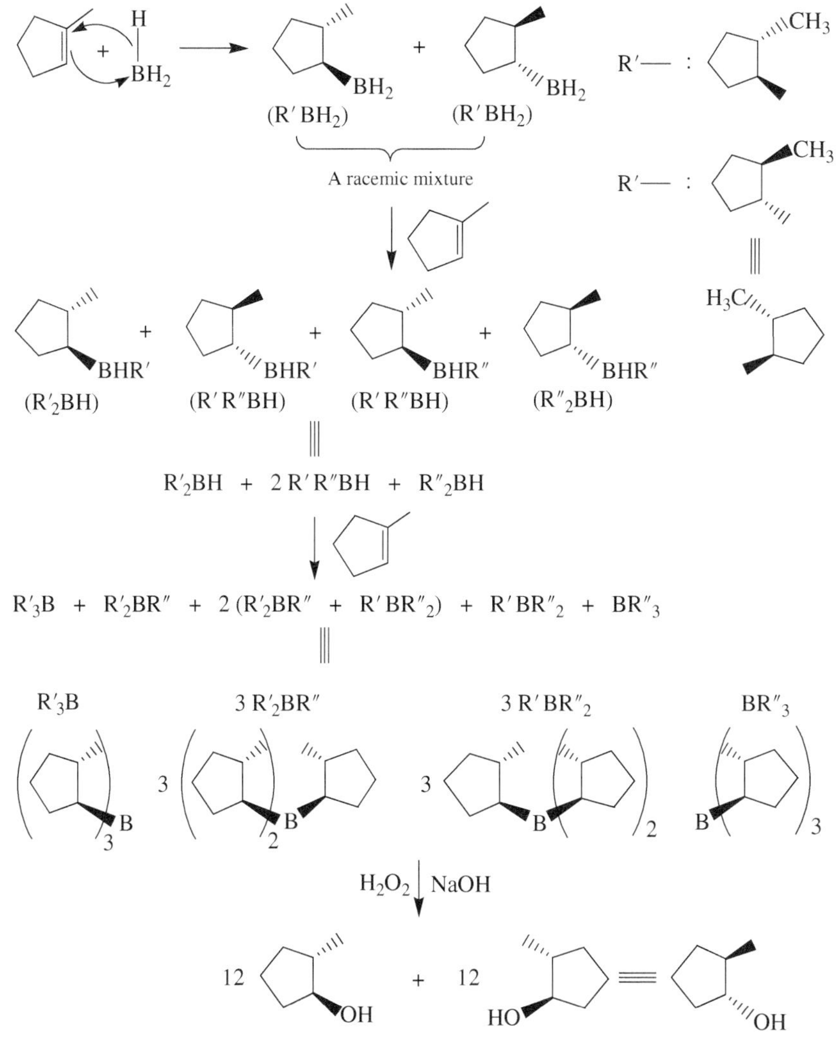

FIGURE 3.20 Mechanism, regiochemistry, and stereochemistry for hydroboration of 1-methylcyclopentene.

dialkylboranes (R'_2BH, 2 $R'R''BH$, and R''_2BH) undergoes addition to a third 1-methylcyclopentene molecule from the front face and the back face, respectively, giving four types of trialkylboranes (R'_3B, 3 R'_2BR'', 3 $R'BR''_2$, and BR''_3). The resulting trialkylboranes contain an equal number (12) of R' and R'' groups, which, after oxidation by H_2O_2, are converted to $R'OH$ (*S*-configuration) and $R''OH$ (*R*-configuration), respectively.

3.4.3   Synthesis of Special Hydroborating Reagents

Reaction of diborane (B_2H_6) with 1,5-cyclooctadiene affords 9-borabicyclo[3.3.1] nonane (9-BBN) [3]. It is a valuable hydroborating agent with significant synthetic applications. The reaction takes place via one BH_3 molecule with its two B—H bonds added to the two C=C bonds in the diene consecutively as shown in Figure 3.21. Due to steric hindrance of the very bulky bicyclic alkyl group attached to the boron in 9-BBN, its third B—H bond cannot be added to another 1,5-cyclooctadiene.

Steric hindrance on hydroboration of the second and third B—H bonds of borane to alkenes is also seen in the reaction of 2,3-dimethyl-2-butene with B_2H_6 [3].

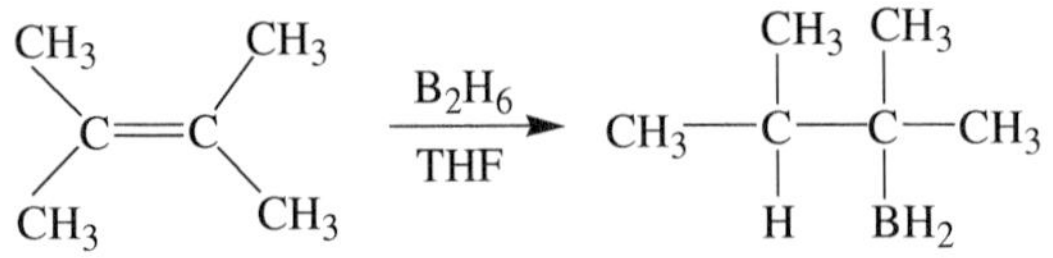

2,3-Dimethyl-2-butene

Only a monoalkylborane is formed, and the second and third B—H bonds of borane do not undergo further hydroborations with the alkene.

Figure 3.22 shows that α-pinene (a sterically very active alkene) reacts with diborane (via BH_3) in THF giving a stereospecific dialkylborane (Ipc_2BH) with the methyl and –BH groups *trans* to each other in the ring [10]. The overall reaction consists of two consecutive hydroborations. In the first hydroboration, a B—H bond in BH_3 is selectively added to the alkene π bond from back face of the ring giving a *trans*-monoalkylborane ($IpcBH_2$). The front face of the ring is greatly sterically hindered by the –CMe_2 group, and addition in the front would render a large activation energy. For the same reason, addition of the B—H bond in $IpcBH_2$ to α-pinene (the second hydroboration) also occurs selectively on the back face of the α-pinene ring giving the stereospecific Ipc_2BH. The very bulky cycloalkyl rings in Ipc_2BH

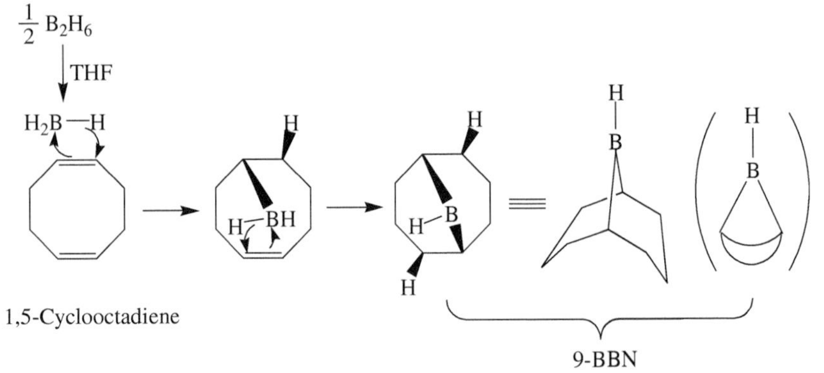

1,5-Cyclooctadiene

9-BBN

FIGURE 3.21   Synthesis of 9-borabicyclo[3.3.1]nonane (9-BBN).

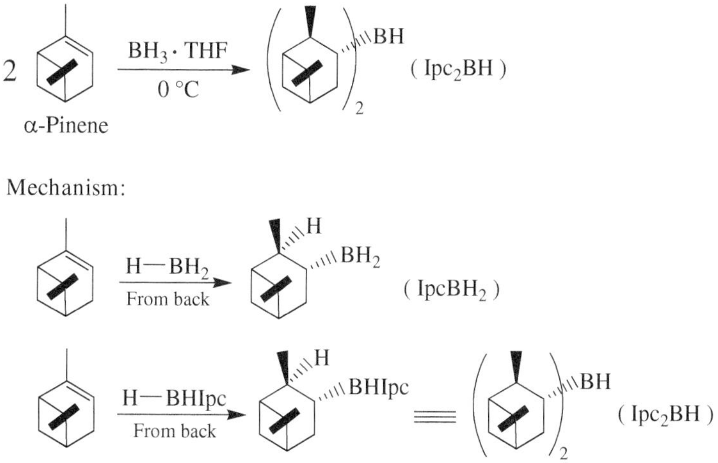

2 α-Pinene $\xrightarrow[0\,°C]{BH_3 \cdot THF}$ (Ipc$_2$BH)

Mechanism:

$\xrightarrow[\text{From back}]{H-BH_2}$ (IpcBH$_2$)

$\xrightarrow[\text{From back}]{H-BHIpc}$ $\equiv$ (Ipc$_2$BH)

FIGURE 3.22   Hydroboration of a-pinene by diborane (via BH$_3$) in THF.

prohibit the further hydroboration of the B—H bond in Ipc$_2$BH to a third α-pinene molecule. α-Pinene is characterized by facile rearrangement in many reactions (see Chapter 10 for more details). However, it does not rearrange in the course of hydroboration. This has provided further evidence for concertedness of hydroboration reactions.

3.4.4   Reactions of Alkenes with Special Hydroborating Reagents: Regiochemistry, Stereochemistry, and Their Applications in Chemical Synthesis

As a hydroborating reagent, a remarkable advantage of 9-BBN lies in its great susceptibility to any subtle difference in steric environments of the two doubly bonded carbons in an alkene molecule. This susceptibility originates from the bulkiness of the large bicyclic alkyl group attached to the boron atom. As a result, hydroboration of an alkene by 9-BBN exhibits much higher regioselectivity than that by diborane. For example, reaction of the 1-alkene RCH=CH$_2$ with 9-BBN followed by oxidation with H$_2$O$_2$ affords 99% 1-alcohol RCHCH$_2$OH, while the hydroboration of the alkene by B$_2$H$_6$ followed by oxidation gives 94% RCHCH$_2$OH (Fig. 3.19). Figure 3.23 shows comparison of hydroborations of 2-methyl-(*E*)-3-hexene by 9-BBN and B$_2$H$_6$, respectively [3]. Each of the C=C carbons in the alkene contains one hydrogen. The two alkyl groups (ethyl and isopropyl) attached to the two C=C carbons are very similar with isopropyl being only slightly more sterically hindered. This subtle difference in steric hindrance of ethyl and isopropyl is even greatly perceived by 9-BBN so that ~98% of the final alcohol product (the alcohol on the left) is derived from addition of boron in 9-BBN to the slightly less sterically hindered C$_4$-carbon of the alkene which is attached to an ethyl. Only ~2% of the

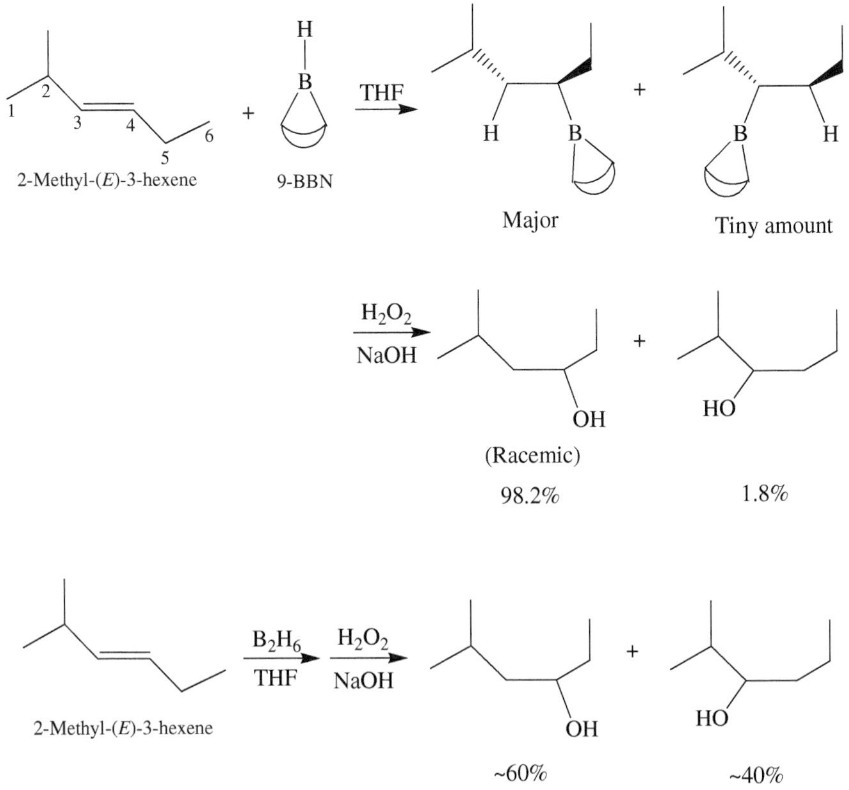

FIGURE 3.23   Comparison of regioselectivity for alkene hydroborations effected by 9-BBN and diborane.

alcohol product (the alcohol on the right) originates from addition of boron in 9-BBN to the slightly more sterically hindered C_3-carbon of the alkene which is attached to an isopropyl. When hydroboration of 2-methyl-(E)-3-hexene is performed by B_2H_6, the reaction is much less regioselective, with the yield of the alcohol on the left (~60%) being only slightly greater than that for the alcohol on the right (~40%).

Figure 3.24 has further demonstrated stereochemistry associated to hydroboration by 9-BBN [3]. The first alkene substrate is the stereospecific (R)-3-ethyl-1-methylcyclohexene. Its addition to 9-BBN only takes place in the back side of the C=C bond resulting in the formation of (1R,2R,6S)-2-ethyl-6-methylcyclohexanol (R,R,S-alcohol) after oxidation. The addition of 9-BBN from the front side of the C=C bond would render a big steric hindrance due to strong interactions of the bulky bicyclic alkyl group from 9-BBN with the ethyl group present in the front side of the alkene. The steric hindrance slows down the reaction and makes it uncompetitive. As a result, the corresponding (1S,2R,6R)-2-ethyl-6-methylcyclohexanol is not formed. Analogously, hydroboration of the stereospecific

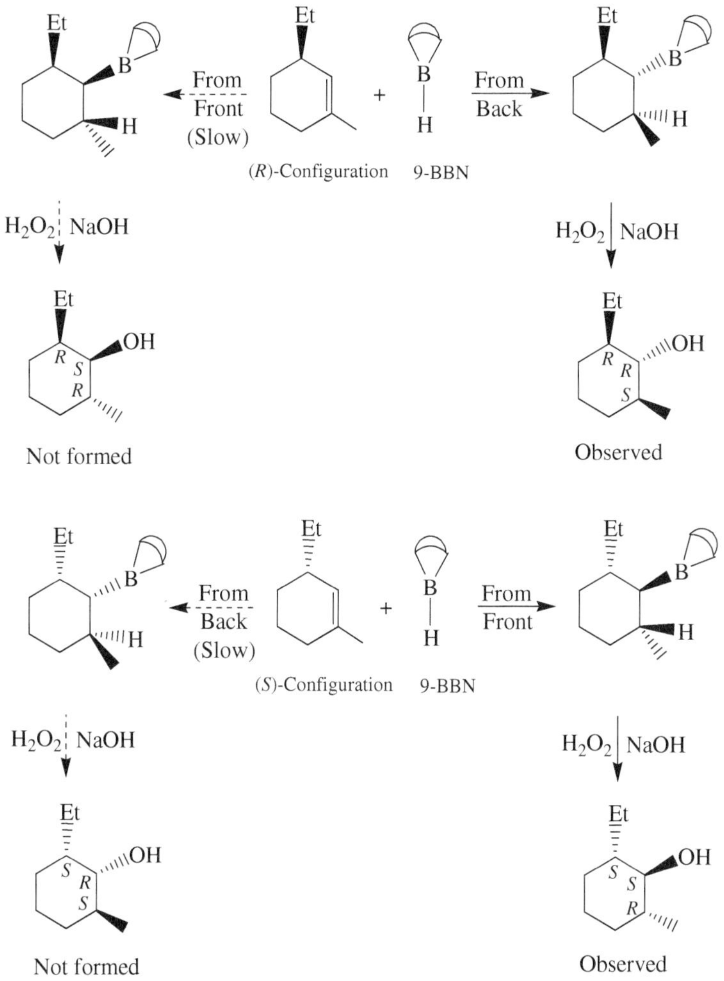

FIGURE 3.24 Stereoselectivity for hydroboration reactions of (*R*)- and (*S*)-3-ethyl-1-methylcylohexene by 9-BBN.

(*S*)-3-ethyl-1-methylcyclohexene by 9-BBN only takes place in the front of the C=C bond giving (1*S*,2*S*,6*R*)-2-ethyl-6-methylcyclohexanol (*S*,*S*,*R*-alcohol) after oxidation. The addition from the back side renders strong steric hindrance due to interactions of the bicyclic alkyl from 9-BBN and the ethyl present at the back side of the alkene. The reaction would be slow, and the corresponding (1*R*,2*S*,6*S*)-2-ethyl-6-methylcyclohexanol is not formed. The two observed alcohol products, *R*,*R*,*S*-alcohol and *S*,*S*,*R*-alcohol, are enantiomers, produced from the (*R*)-alkene and (*S*)-alkene, respectively. Therefore, when a racemic (unstereospecific) 3-ethyl-1-methylcyclohexene reagent is employed, its reaction with 9-BBN, after oxidation, gives a racemic mixture of *R*,*R*,*S*-alcohol and *S*,*S*,*R*-alcohol.

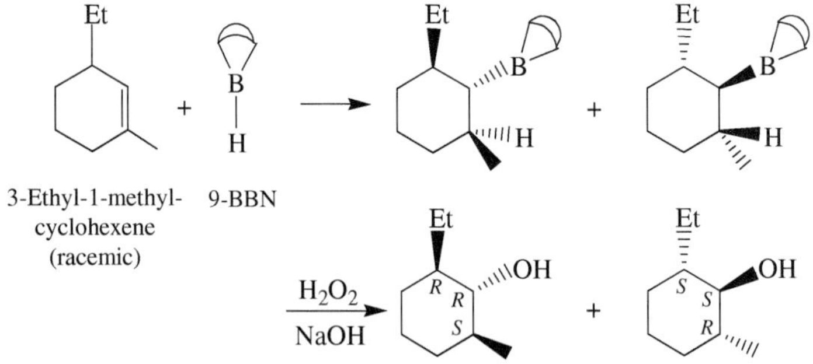

3-Ethyl-1-methyl-   9-BBN
cyclohexene
(racemic)

Ipc$_2$BH, the product from hydroboration of α-pinene by BH$_3$ (Fig. 3.22), is a useful hydroborating reagent for asymmetric synthesis. Research has shown that reaction of Ipc$_2$BH with a *cis*-alkene (such as *cis*-2-butene) followed by oxidation with H$_2$O$_2$ in alkaline medium gives a (*R*)-alcohol selectively in ~90% optical purity. On the other hand, a (*S*)-alcohol with high optical purity can be synthesized from a *trans*-alkene (such as *trans*-2-butene) [10].

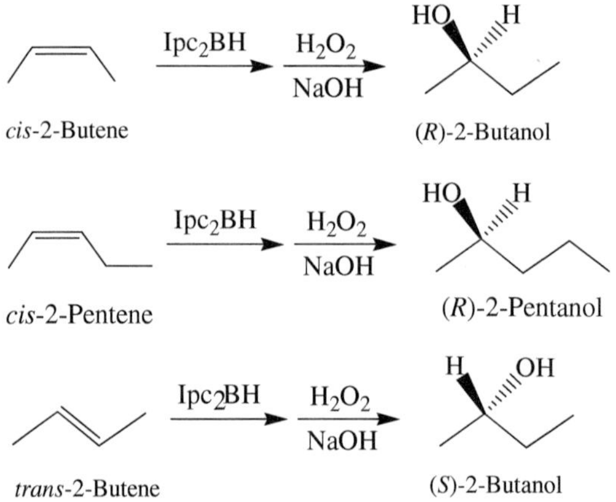

cis-2-Butene   (*R*)-2-Butanol

cis-2-Pentene   (*R*)-2-Pentanol

trans-2-Butene   (*S*)-2-Butanol

3.5 TRANSITION-METAL CATALYZED HYDROGENATION OF THE ALKENE C=C BOND (*syn*-ADDITION)

Addition of molecular hydrogen (hydrogenation) to alkenes giving saturated hydrocarbons is in general thermodynamically favorable. For example, hydrogenation of ethylene to ethane has negative enthalpy and free energy at standard conditions [8].

$$H_2C = CH_2 + H_2 \rightarrow H_3C - CH_3 \quad \Delta H° = -32.5 \, \text{kcal/mol}, \Delta G° = -24.2 \, \text{kcal/mol}$$

However, the reaction would be extremely slow and actually does not take place in any conditions [8]. The low rate for hydrogenation of alkenes such as ethylene is primarily attributable to the strong H—H bond (BDE = 103 kcal/mol). Some transition metals, such as Pd, Pt, and Ni, and their compounds can effectively interact with H_2 through their d orbitals and weaken or break the H—H bond. In addition, they can also effectively interact with the alkene π bond to weaken it. As a result, the activation energy for hydrogenation of alkenes can be lowered by these metals or their compounds and the reaction will be able to take place in feasible conditions with reasonable rates. When metallic Pd, Pt, or Ni is employed as a catalyst, the catalysis is **heterogeneous**, namely that the reaction takes place between gaseous (or liquid) and solid phases. When a transition metal compound is employed as a catalyst (Wilkinson catalyst), the catalysis is **homogeneous**. The reaction takes place in a homogenous solution.

Heterogeneous catalysis by metallic transition metals is more common and more effective for hydrogenation of alkenes, and has better synthetic utility than homogeneous catalysis. In heterogeneous catalysis, a powdered metal (such as Pd and Pt, or their oxides) is dispersed over an inorganic supporter (such as aluminum oxide and/ or carbon). A mixture of alkene and hydrogen gases is passed through the catalyst. Alternatively, the alkene substrate (if the molecule is large) can be dissolved in an organic solvent. The solution is then in contact with the solid catalyst dispersed in a supporter. Then the hydrogen gas is passed through the liquid–solid system to bring about the heterogeneous hydrogenation. The heterogeneous catalysis and the overall reaction take place on the **surface** of a transition metal. In this section, emphasis of the discussion will be placed on the mechanism and synthetic applications for the heterogeneous hydrogenation.

3.5.1   Mechanism and Stereochemistry

Figure 3.25 shows the general mechanism for hydrogenation of an alkene catalyzed by a transition metal (Pd, Pt, or Ni) (heterogeneous catalysis). When a mixture of the alkene and hydrogen gases is passed through the catalyst surface, a H_2 molecule is adsorbed on the metal. The H_2 antibonding orbital ($\sigma_{1s}*$, empty) is of symmetry-match to a d_π^2 orbital (d_{xy}, d_{yz}, or d_{xz}, filled) in the metal center. The two orbitals overlap effectively, which allows the d_π^2-electrons to flow into the $\sigma_{1s}*$ orbital. Population of the H_2 antibonding $\sigma_{1s}*$ orbital with electrons results in cleavage of the H—H bond. Consequently, the initial overlap of the transition metal d_π orbital lobes with those of the H_2 $\sigma_{1s}*$ orbital will lead to dissociation of H_2 and the simultaneous formation of relatively weak H—M—H bonds on the surface of the transition metal. Concurrently, an alkene molecule is adsorbed to the surface of the transition metal as well, and this results in the formation of a weak π-complex with the metal. Then the two H–M–H hydrogen atoms which are weakly, but covalently bonded to a metal center are added consecutively through a σ-complex to the alkene C=C bond which is initially weakly bonded to a nearby metal atom (alkene-metal π-complex). The consecutive addition of the two hydrogen atoms to the alkene takes place in the same

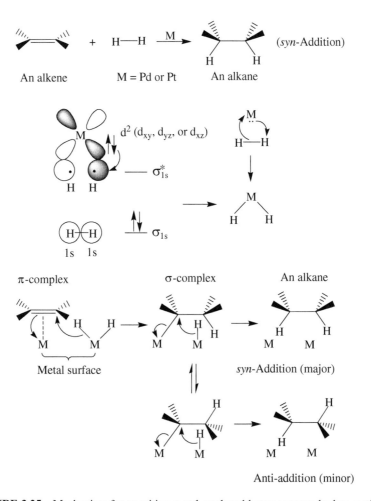

FIGURE 3.25 Mechanism for transition-metal catalyzed heterogeneous hydrogenation of alkene.

side of the C=C bond which is facing toward the metal surface, giving rise to an overall *syn*-addition (major). In the course of the reaction, a very small percentage of the σ-complex intermediate may possibly undergo free rotation about its C—C bond prior to addition of the second M–H hydrogen. This will give rise to an anti-addition (minor) to a very small extent. Since the second step of addition to the σ-complex is fast, the majority of the σ-complex molecules will not have sufficient time to undergo rotation before the addition occurs although the rotation about the C—C bond from the initial eclipsed conformation to a staggered conformation would be slightly thermodynamically favorable. Overall, the *syn*-addition is the major reaction.

The key to the catalytic hydrogenation of alkenes is to break the strong H—H bond by a transition metal catalyst. Figure 3.26 shows the MO diagram for the formation of the H—M—H bonds from the empty H–H antibonding MO ($\sigma_{1s}*$) and a filled metal d_π orbital. The linear combinations of $\sigma_{1s}*$ and d_π lead to the formations of a bonding MO (constructive combination) and an antibonding MO (destructive combination) for H–M–H.

$$\text{Bonding MO (HMH)} = d_\pi + \sigma_{1s}*$$

$$\text{Antibonding MO (HMH)} = d_\pi - \sigma_{1s}*$$

The bonding MO is filled and responsible for the dissociation of H_2 and the formation of the H—M—H bonds. The H_2 bonding MO (σ_{1s}) does not participate in bonding for H–M–H.

The stereochemistry of the catalytic hydrogenation determined by the *syn*-addition pathway can be demonstrated by the platinum-catalyzed hydrogenation of 1,2-dimethylcyclohexene giving a *cis*-product (from *syn*-addition) as the major product along with a small amount of *trans*-product (from anti-addition).

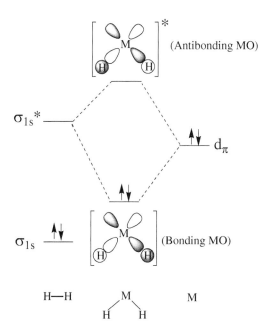

FIGURE 3.26   The MO diagram for the formation of the H—M—H bonds from H_2 and a transition metal (M) atom.

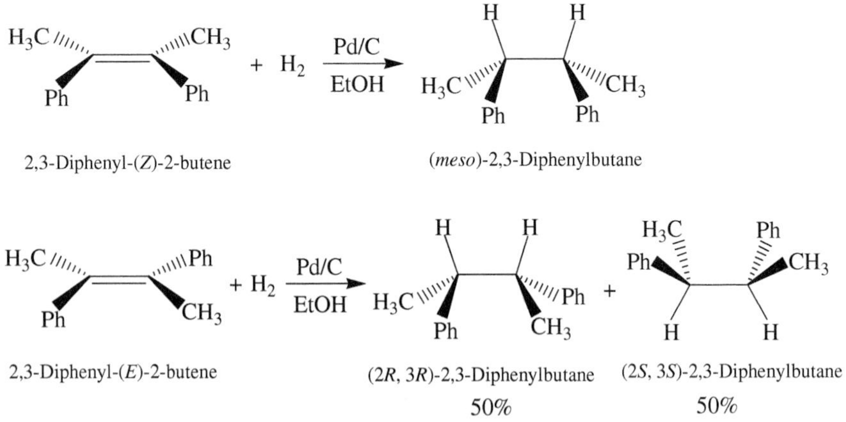

2,3-Diphenyl-(Z)-2-butene   (meso)-2,3-Diphenylbutane

2,3-Diphenyl-(E)-2-butene   (2R, 3R)-2,3-Diphenylbutane   (2S, 3S)-2,3-Diphenylbutane
                                              50%   50%

FIGURE 3.27   Palladium catalyzed hydrogenation of 2,3-diphenyl-2-butenes in (Z)- and (E)-configurations.

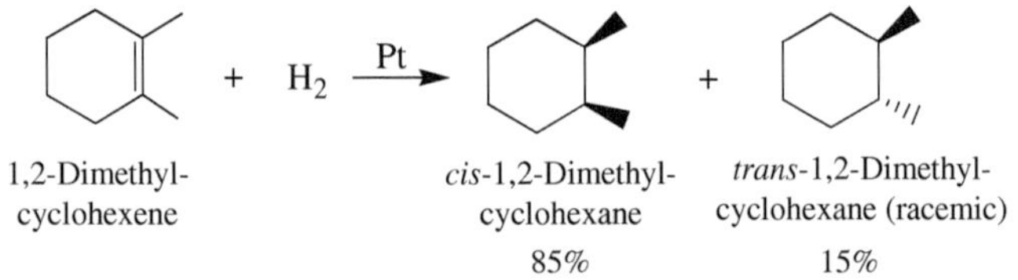

1,2-Dimethyl-   cis-1,2-Dimethyl-   trans-1,2-Dimethyl-
cyclohexene   cyclohexane   cyclohexane (racemic)
                              85%   15%

The syn-addition feature of alkene hydrogenation is further unambiguously demonstrated by the palladium-catalyzed hydrogenation of two ethylene derivatives in cis- and trans-configurations in Figure 3.27. The hydrogenation of the alkene in cis-configuration gives a meso-compound, while the hydrogenation of the trans-alkene results in the formation of a racemic mixture. The outcomes show that in each of the reactions, the two hydrogen atoms are added to the same side of the alkene C=C bond, either from the top or from the bottom.

3.5.2   Synthetic Applications

Hydrogenation of unsaturated carbonyl compounds   Transition-metal catalyzed hydrogenation of a π bond in a carbon–carbon linkage (including the C=C double bond and C=C triple bond) is an effective method to reduce a carbon–carbon multiple bond. The method has been successfully used in organic synthesis. In contrast, hydrogenation of the C=O bond in ketones, aldehydes, carboxylic acids, and/or esters to hydroxyl (–OH) under the same conditions will proceed much more slowly. Therefore, the following unsaturated ketones can be converted to the saturated ones by Pt-catalyzed hydrogenation under mild conditions, with the carbonyl group in each compound being intact in the reaction.

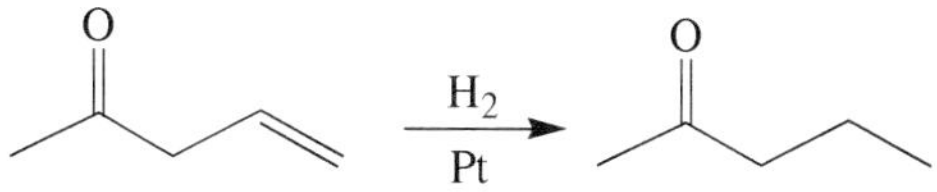

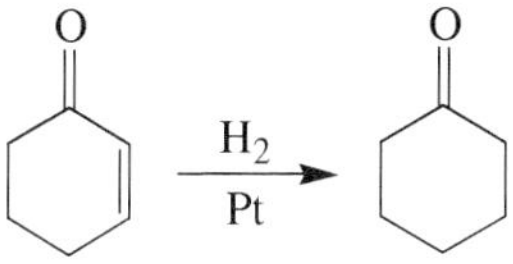

Hydrogenation of alkynes   The C≡C triple bond contains two π bonds. Under the same conditions, catalytic hydrogenation of the first π bond making a C=C double bond is substantially faster than the hydrogenation of the second π bond making a C—C single bond. This is because the two π bonds in a C≡C triple bond are individually weaker (more reactive) than the π bond in a C=C double bond. Therefore, with appropriate care, an alkyne can be hydrogenated to an alkene and then to an alkane consecutively, with the intermediate alkene product being isolable when needed [4].

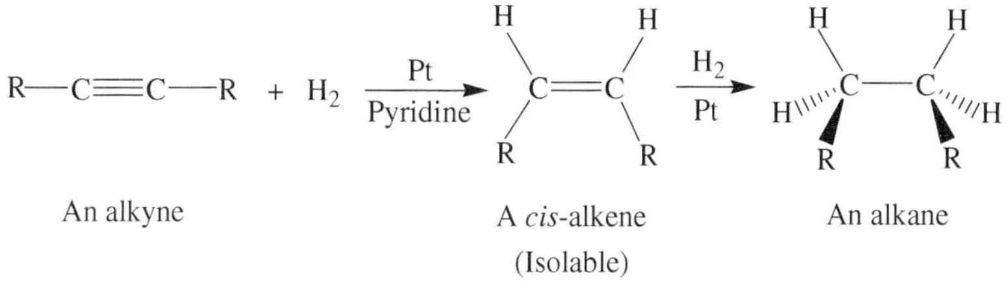

An alkyne   A *cis*-alkene   An alkane
                             (Isolable)

The same as for the catalytic hydrogenation of an alkene, the hydrogenation of an alkyne also follows a *syn*-addition mechanism, namely that the two hydrogen atoms in the H–M–H transition metal hydride on the surface of the metal are added consecutively to the same side of a π bond which faces to the metal surface. The isolable intermediate alkene possesses a *cis*-configuration.

Hydrogenation of dienes   Another remarkable feature of the catalytic hydrogenation of alkenes is that the reaction rate is greatly affected by the crowdedness of the C=C bond. A highly substituted alkene on the C=C bond will hinder the adsorption of the alkene to the catalyst surface, reducing the overall reaction rate. Such an effect can be used to selectively hydrogenate a particular C=C bond in a diene as shown below on the basis of difference in steric environments of the two C=C bonds [3].

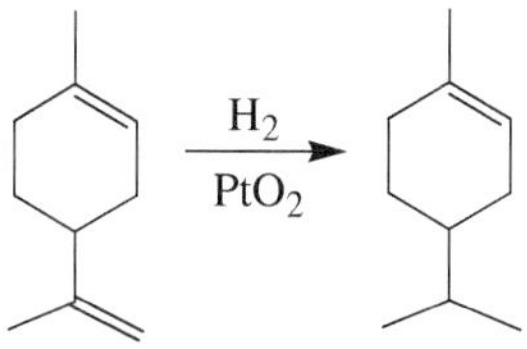

The C=C bond in the cyclohexene ring contains only one hydrogen, and is obviously more heavily substituted than the acyclic C=C bond which is attached to two hydrogen atoms in the terminal carbon. Therefore, the catalytic hydrogenation on the acyclic C=C bond (less crowded) is faster than that on the more crowded cyclic C=C bond. With appropriate care, the catalytic hydrogenation will only take place in the acyclic C=C bond, and the cyclic C=C bond remains intact.

3.5.3   Biochemically Related Applications: Hydrogenated Fats (Oils)

The transition-metal catalyzed hydrogenation has found important applications in food industry for manufacturing **hydrogenated fats (oils)** from natural vegetable oils. Hydrogenated fats are solids or semi-solids, depending on the extent of saturation. They are added to processed foods to provide a desirable firmness along with a moist texture and pleasant taste. Both fats (saturated and being solid) and oils (unsaturated and being liquid) are **lipids**, which are in general fatty acid esters of glycerol (triacylglycerols) (Fig. 3.28). A natural vegetable oil (unsaturated lipid) contains unsaturated fatty acid residues with all the C=C bonds in *cis*-configuration. Figure 3.28 illustrates that upon catalytic hydrogenation which are analogous to the reactions for the alkenes discussed above, some or all the C=C bonds in the fatty acid residues in an unsaturated lipid (such as a vegetable oil) can be reduced (saturated). Reaction conditions can be modified so that the percentage of the hydrogenated C=C bonds will be controllable resulting in different degrees of saturation, which would depend on the needs. In general, the higher percentage of the saturation, the higher the melting point for the hydrogenated fat. Under the same conditions, the carbonyl C=O groups in a lipid molecule have low reactivity toward hydrogen. Therefore, the entity of the ester groups will be retained in the course of the catalytic hydrogenation.

A major concern about hydrogenated fats is that in the course of catalytic hydrogenation, a small amount of harmful ***trans*-fats** is produced due to stereoisomerization of some *cis*-(C=C) bonds to *trans*-(C=C) in the fatty acid residues. Recent studies have shown that consuming a significant amount of *trans*-fats can lead to serious health problems related to serum cholesterol levels. In principle, the transition-metal catalyzed hydrogenation of alkenes described in Figure 3.24 is reversible. As free rotation about the C—C bond in the intermediate σ-complex (eclipsed conformation) occurs to make a staggered conformation, the backward dissociation of the C—M and C—H bonds in the staggered conformation will result in the formation of a thermodynamically more stable *trans*-(C=C) bond. In addition, the π-complex intermediate may also undergo a free rotation due to weakening of the π bond by the metal. Such a rotation will directly lead to isomerization of *cis*-(C=C) to *trans*-(C=C) in a fatty acid residue. Overall, the forward hydrogenation process is thermodynamically favorable, and it is a dominating process. The backward dissociation is thermodynamically unfavorable and only occurs to a small extent, giving a very small amount of *trans*-fats. The Food and Drug Administration announced recently

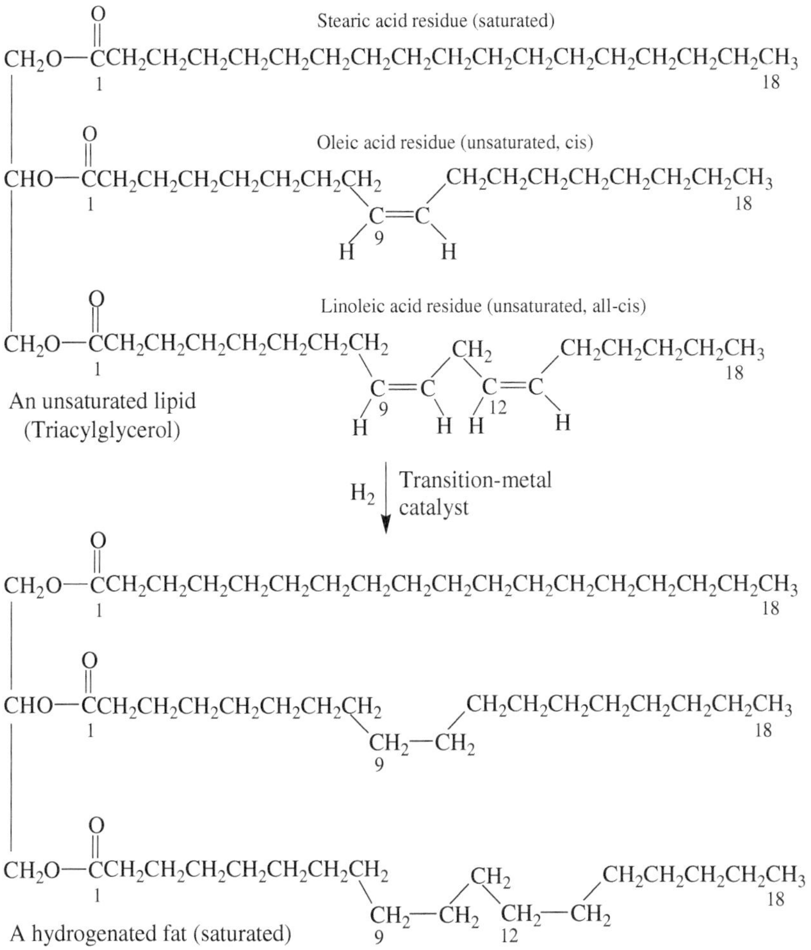

FIGURE 3.28   A hypothesized hydrogenated fat (saturated) made by transition-metal catalyzed addition of the molecular hydrogen to an unsaturated lipid.

that processed foods with hydrogenated oils will soon be required to list the content of *trans* fatty acids [5].

3.6 HALOGENATION OF THE ALKENE C=C BOND (ANTI-ADDITION): MECHANISM AND ITS STEREOCHEMISTRY

In an unreactive solvent such as dichloromethane, elemental chlorine (Cl_2) and bromine (Br_2) undergo facile electrophilic addition to alkenes giving vicinal dihaloalkanes. The reactions follow an anti-addition mechanism as the major outcome [11].

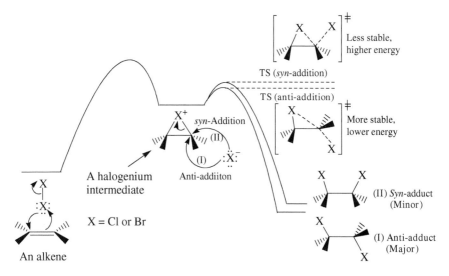

FIGURE 3.29 Mechanism and transition states for halogenation of alkene in dichloromethane, an unreactive solvent.

Mechanism   The mechanism of alkene halogenations is illustrated in Figure 3.29. In the first step of the reaction, an X_2 (X = Cl or Br) molecule approaches the C=C π bond perpendicularly to the alkene molecule plane. This electrophilic attack leads to the formation of a cationic halogenium (chlorinium or brominium) intermediate (rate-determining step) which is characterized by a positively charged C–X–C three-membered ring [12]. The halogenium intermediate is analogous to the mercury(II)–alkene π complex as described in Section 3.1.3 and Figure 3.10. The positive charge in the intermediate is delocalized to the two carbons. As a result, either of the carbon atoms can be subsequently attacked by a halide generated in the first step giving the final addition product (product-development step). The identity of the product is determined by the relative stability of the transition state (TS) of this product-development step. The nucleophilic attack by X^- (Cl^- or Br^-) occurs preferably from the opposite side of the halogen atom in the halogenium intermediate giving rise to an overall major anti-addition (two halogen atoms from X_2 are added to the different sides of the C=C bond). The anti-addition proceeds via a transition state (TS) in a staggered conformation. It is relatively more stable (than an eclipsed conformation) and possesses lower energy level, leading to a faster reaction for the product-development step. The alternative *syn*-addition would proceed via a relatively less stable transition state in an eclipsed conformation, with the two bromine atoms strongly interacting resulting in enhancement of the energy level and slowing down the reaction for the product-development step. Therefore, the *syn*-addition only takes place to a small extent as a minor reaction.

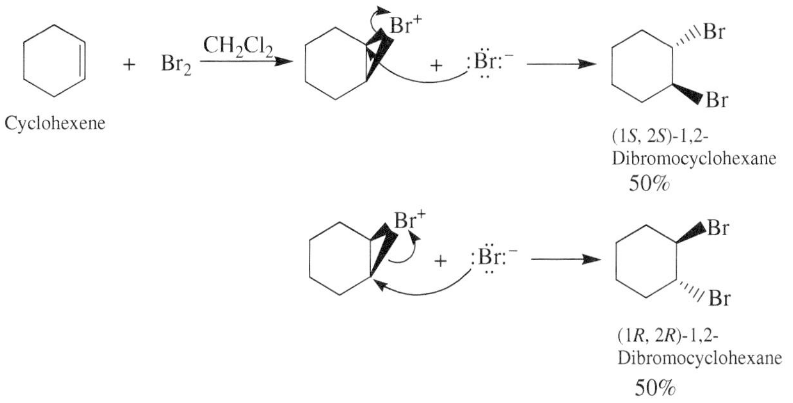

Cyclohexene

(1S, 2S)-1,2-
Dibromocyclohexane
50%

(1R, 2R)-1,2-
Dibromocyclohexane
50%

FIGURE 3.30 Mechanism and stereochemistry for bromination of cyclohexene in dichloromethane.

Stereochemistry Reaction of cyclohexene with bromine in CH_2Cl_2 gives a racemic mixture of *trans*-1,2-dibromocyclohexanes, showing that the reaction follows an anti-addition mechanism (Fig. 3.30). The first step in the reaction is the formation of a brominium intermediate in the substrate ring. Then both carbons can be attacked by bromide from the different side of the brominium giving (1R, 2R)-1,2-dibromocyclohexane and (1S, 2S)-1,2-dibromocyclohexane, respectively.

Figure 3.31 shows reaction of (Z)-1-phenylpropene with bromine in tetrachloromethane giving 78% anti-addition products and 22% *syn*-addition products [3]. In the first step of the reaction, Br_2 approaches the alkene substrate equally from the top and bottom of the C=C bond giving two enantiomeric brominium intermediates. In each of the brominium intermediates, the positive charge is mainly distributed to the carbon that is attached to an electron-donating phenyl group. As a result, the intermediate is better stabilized. Therefore, the Br—C_1 (attached to Ph) bond is relatively longer than the Br—C_2 (attached to CH_3) bond [c.f. the mercury(II)–alkene π complex in Figure 3.10]. Preferably, in the second step Br^- will attack the C_1-carbon in each of the brominium intermediates from the different side or from the same side giving different percentage of anti-addition (major) or *syn*-addition (minor) product, respectively.

Bromination of chiral (R)-4-*tert*-butylcyclohexene has been shown to give a stereospecific (1S,2S,4R)-1,2-dibromo-4-*tert*-butylcyclohexane (Fig. 3.32) [5]. The (1R,2R,4R)-isomer is not formed. In the thermodynamically most favorable chair conformation of (R)-4-*tert*-butylcyclohexene, the bulky *tert*-butyl group stays in an equatorial position. The C=C π bond in the cyclohexene ring can be approached by Br_2 from both bottom and top faces to give intermediate Brominium (A) and Brominium (B), respectively. In the following step, Br^- attacks each of the brominium intermediates only from an axial direction (but not from an equatorial direction).

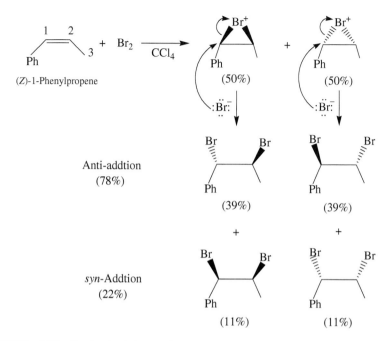

FIGURE 3.31 Regioselectivity and stereoselectivity for bromination of (Z)-1-phenylpropene. Based on Bruckner [3].

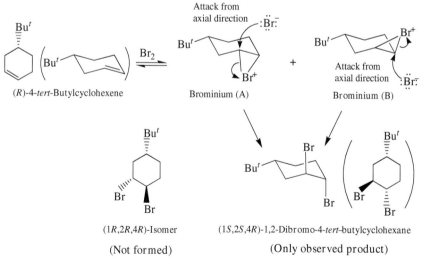

FIGURE 3.32 Bromination of (R)-4-*tert*-butylcyclohexene giving stereospecific (1S,2S,4R)-1,2-dibromo-4-*tert*-butylcyclohexane. Adapted from Brown et al. [5].

As a result, both brominium intermediates lead to the formation of the same (1S,2S,4R)-1,2-dibromo-4-*tert*-butylcyclohexane isomer. The attacks of brominiums by Br$^-$ from equatorial directions would give the (1R,2R,4R)-isomer. However, they would be kinetically unfavorable. Thus, (1R,2R,4R)-isomer is not formed. The key to the stereochemical control of this reaction lies in the bulkiness of the *tert*-butyl group, which makes the group always occupy an equatorial position in a chair conformation of the cyclohexene ring. The overall stereoselectivity for the bromination reaction is directed by the bulky *tert*-butyl group.

Halogenations in nucleophilic solvents   When the alkene halogenation reactions are performed in nucleophilic solvents (such as water and alcohols), the solvent molecules will participate in the reactions. The net result would be that as if a HOX or "XOR" molecule was added to the C=C bond as shown in Figure 3.33.

Figure 3.33 shows the reactions of cyclohexene with bromine performed in water and methanol, respectively. Similar to the reaction performed in CH_2Cl_2, the first step in the reaction performed in water is that the C=C bond in cyclohexene is attacked by Br_2 giving a cationic brominium intermediate, which will be

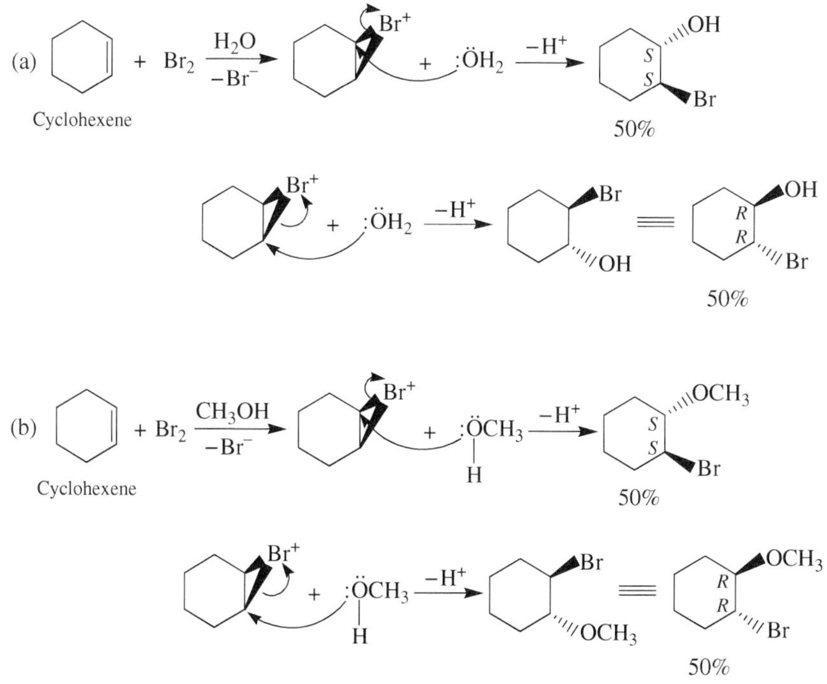

FIGURE 3.33   Reactions of cyclohexene with elemental bromine in water (a) and methanol (b), nucleophilic solvents.

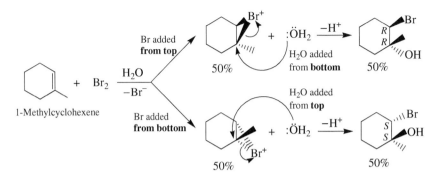

FIGURE 3.34   Regiochemistry and stereochemistry for reactions of 1-methylcyclohexene with elemental bromine in water.

subsequently surrounded heavily by the solvent molecules. The bromide generated in the first step will be separated from the brominium intermediate by the overwhelmingly concentrated water molecules. In the second step, the brominium intermediate is attacked by a surrounding nucleophilic water molecule, preferably, from the opposite side of bromine giving an anti-addition product in a *trans*-configuration. The nucleophilic attacks of water on the two different carbons in the intermediate brominium result in different enantiomers. The reaction performed in methanol follows an analogous anti-addition mechanism giving two enantiomers in an equal amount (a racemic mixture).

Figure 3.34 shows reaction of bromine with 1-methylcyclohexene, an unsymmetrical alkene, in water. The reaction is regioselective giving 2-bomo-1-methylcyclohexanols as the major products. In the first step of the reaction, a bromine molecule can attack the alkene from either side of the C=C bond equally, giving two stereospecific brominium intermediates. In each of the brominium intermediates, the positive charge is mainly distributed to the carbon that is attached to an electron-donating methyl group. As a result, the intermediate is better stabilized. Therefore, the Br—C$_{tert}$ bond is relatively longer than the Br—C$_{sec}$ bond [c.f. the mercury(II)–alkene π complex in Figure 3.10]. Preferably, a water molecule will attack the tertiary carbon of each of the brominium intermediates from different side of bromine giving a both regio- and stereo-specific product.

PROBLEMS

3.1 Consider the following reactions.

$$\text{C}_6\text{H}_5\text{—CH}=\text{CH}_2 \ + \ \text{HBr} \ \longrightarrow \qquad (1)$$

$$\text{C}_6\text{H}_{11}\text{—CH}=\text{CH}_2 \ + \ \text{HBr} \ \longrightarrow \qquad (2)$$

Write out the mechanism and give the major product for each of the reactions. Draw reaction profiles for both reactions. Estimate the relative activation energies for rate-determining steps of the two reactions using the Evans–Polanyi principle (Eq. 3.1). Which reaction goes faster?

3.2 Treatment of a β,γ-unsaturated acid with catalytic amount of *p*-toluenesulfonic acid (TsOH) affords a lactone as shown below.

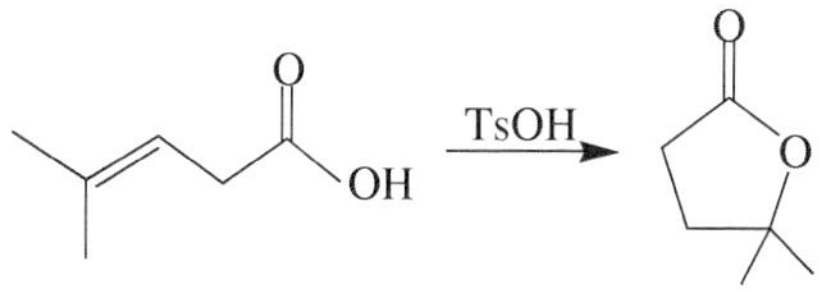

An β, γ -unsaturated acid   A lactone

Show mechanism for the reaction.

3.3 Write out mechanism and draw energy profile for the following reaction.

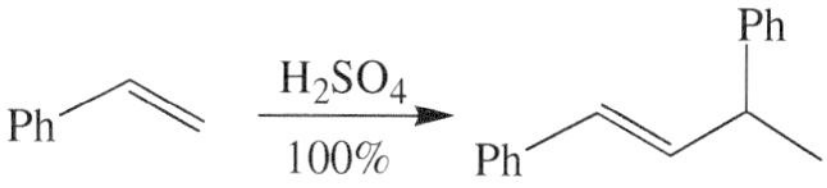

3.4 Reaction of the following bicycloalkene with bromine in CH_2Cl_2 gives a *trans*-dibromide. In both (a) and (b), the bromine atoms are *trans* to each other. However, only one of the products is formed.

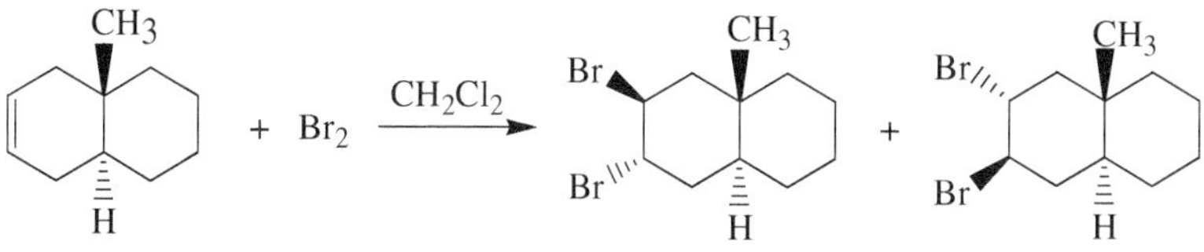

(a)   (b)

Which *trans*-dibromide is formed? Write out mechanism and draw energy profile for the reaction. Account for formation of a single *trans*-dibromide product with the exclusion of the other *trans*-isomer.

3.5 When 2-pentene is treated with Cl_2 in methanol, three products are formed as shown below.

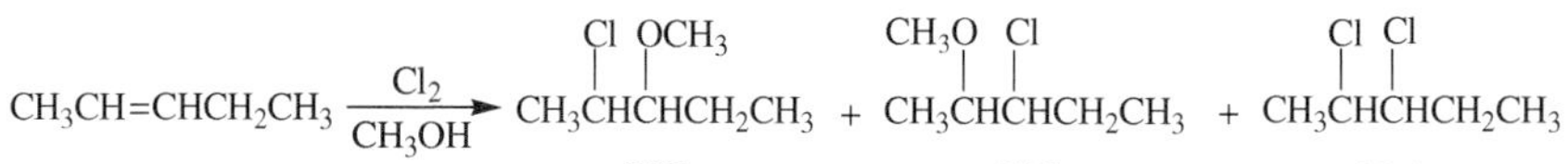

Write the reaction mechanism and account for formation of each product. Explain the relative percentages of the three products.

3.6 Reaction of 4-penten-1-ol with bromine in CH_2Cl_2 gives a cyclic bromoether.

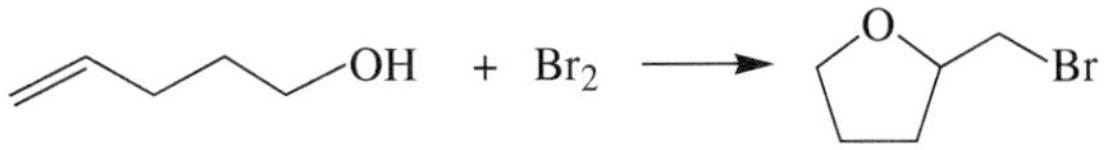

Propose the reaction mechanism. Account for formation of this product rather than a *trans*-dibromide.

3.7 Propose a mechanism to account for the following reaction.

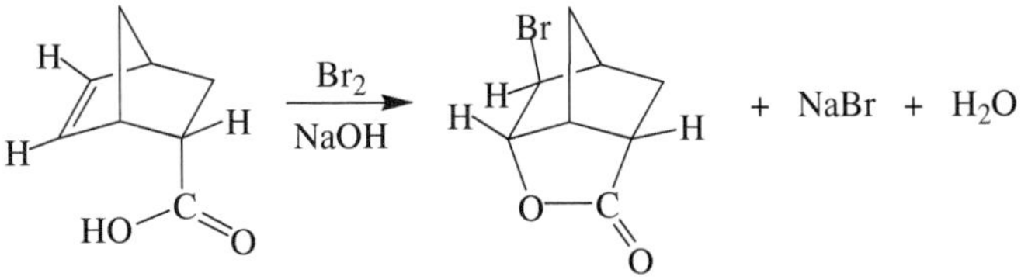

3.8 Reaction of α-pinene with borane followed by treatment of the resulting alkylborane intermediate with alkaline hydrogen peroxide affords one stereoisomer of the following alcohol in 85% and another stereoisomer in less than 15%.

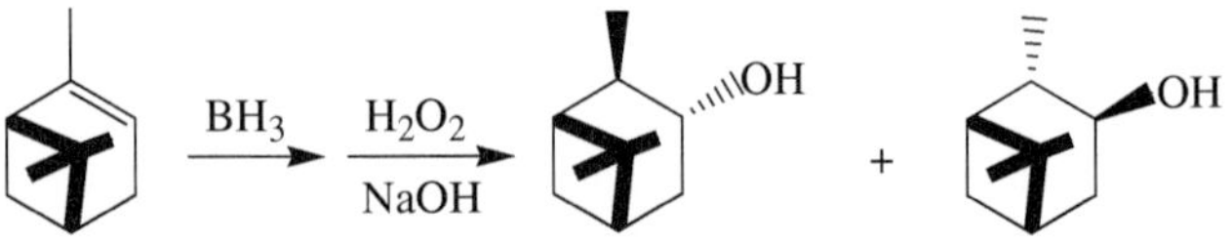

α-Pinene   85%   <15%

Provide a mechanism to account for formations of the corresponding alkylborane intermediates which lead to formation of both stereoisomers of the alcohol. Explain the observed stereoselectivity associated to the hydroboration reaction.

3.9 Show possible mechanism and account for stereochemistry for the following transition metal catalyzed hydrogenation reaction.

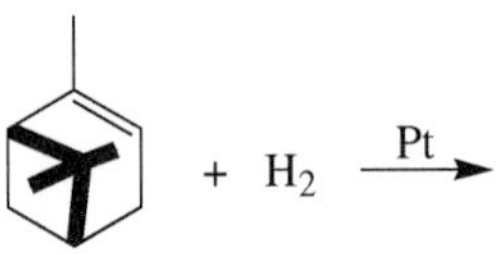

α-Pinene

3.10 Reaction of the following alkene with HBr in diethyl ether containing a small amount of diethyl peroxide (Et_2O_2, a radical initiator) follows a radical addition mechanism.

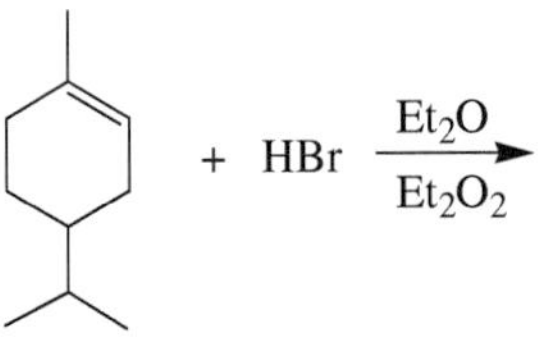

Write out the mechanism and account for the regiochemistry and stereochemistry for the reaction.

3.11 Provide a mechanism to account for the following reaction.

REFERENCES

1. Olah, G. A. Carbocations and Electrophilic Reactions. *Angew. Chem. Int. Ed.* 1973, *12*, 173–254.

2. Olah, G. A. 100 Years of Carbocations and Their Significance in Chemistry. *J. Org. Chem.* 2001, *66*, 5943–5957.

3. Bruckner, R. *Advanced Organic Chemistry: Reaction Mechanisms*, Chapter 3, pp. 85–128, Harcourt/Academic Press, Orlando, FL, USA (2002).

4. Fox, M. A.; Whitesell, J. K. *Core Organic Chemistry*, 2nd ed., Jones and Bartlett, Sudbury, MA, USA (1997).

5. Brown, W. H.; Foote, C. S.; Iverson, B. L.; Anslyn, E. V. *Organic Chemistry*, 6th ed., Brooks/Cole, Belmont, CA, USA (2012)

6. Allmann, R. The Crystal Structure of Mercury(II) Acetate. *Z. Kristallogr., Kristallgeom., Kristallphys., Kristallchem.* 1973, *138*, 366–373.

7. Melloni, G.; Modena, G.; Tonellato, U. Relative Reactivities of Carbon–Carbon Double and Triple Bonds Toward Electrophiles. *Acc. Chem. Res.* 1981, *14*, 227–233.

8. Huheey, J. E.; Keiter, E. A.; Keiter, R. L. *Inorganic Chemistry: Principles of Structure and Reactivity*, 4th ed., Harper Collins, New York, NY, USA (1993).

9. Jackson, R. A. *Mechanisms in Organic Reactions*, The Royal Society of Chemistry, Cambridge, UK (2004).

10. Brown, H. C.; Singaram, B. Development of a Simple General Procedure for Synthesis of Pure Enantiomers via Chiral Organoboranes. *Acc. Chem. Res.* 1988, *21*, 287–293.

11. Hoffman, R. V. *Organic Chemistry: An Intermediate Text*, 2nd ed., Wiley, Hoboken, NJ, USA (2004).

12. Ruasse, M.-F. Brominium Ions or β-Bromocarbocations in Olefin Bromination. A Kinetic Approach to Product Selectivities. *Acc. Chem. Res.* 1990, *23*, 87–93.

4

FUNCTIONALIZATION OF THE ALKENE C=C BOND BY CYCLOADDITION REACTIONS

Cyclic and heterocyclic organic compounds have many fundamental and practical applications in chemistry and biochemistry. Very often, these compounds can be effectively and efficiently synthesized by addition of an unsaturated reagent to an alkene C=C bond. As a result, a ring molecule is formed, with a net transformation of two π bonds (one from each reactant) to two σ bonds (forming linkage of the ring structure). Such an addition of two acyclic unsaturated molecules to form a ring is called **cycloaddition.** Because a σ bond is stronger [with higher bond dissociation energy (BDE)] than a π bond, a cycloaddition reaction is very often thermodynamically favorable. Particularly, this is true, in general, for the reactions leading to the formations of five-membered and six-membered rings that essentially do not have molecular strain.

Usually, cycloaddition reactions of various unsaturated reagents to alkenes can lead to the formations of three-membered rings, four-membered rings, five-membered rings, and six-membered rings. Most of the reactions are concerted [one microscopic step involving a single transition state (TS)], and they take place via the interaction between the highest occupied molecular orbital (HOMO) of one reactant and the lowest unoccupied molecular orbital (LUMO) of the other. The HOMO and LUMO must have symmetry match for the cycloaddition to take place effectively (to be kinetically favorable) via their overlap. Therefore, depending on the relative symmetries of the frontier molecular orbitals (FMOs, including HOMO and LUMO) of the two unsaturated reactants, the reactions can be either thermally

Organic Mechanisms: Reactions, Methodology, and Biological Applications,
Second Edition. Xiaoping Sun.
© 2021 John Wiley & Sons, Inc. Published 2021 by John Wiley & Sons, Inc.
Companion website: www.wiley.com/go/Sun/OrgMech_2e

allowed or photochemically allowed (thermally forbidden). Certain compounds which contain a triple bond such as alkynes (C≡C) and nitriles (C≡N) can also undergo cycloaddition reactions with unsaturated reagents via analogous HOMO–LUMO interactions to those for the alkene reactions.

In this chapter, detailed mechanisms for various types of cycloaddition reactions and some related pericyclic reactions (cyclizations) will be discussed.

4.1 CYCLOADDITION OF THE ALKENE C=C BOND TO FORM THREE-MEMBERED RINGS

In some molecules, an active electron pair (a bonding or nonbonding pair of electrons) is contained in the valence shell of an electrophilic center (E–:) which is electron-deficient and readily accept a pair of electrons. Such an electrophilic center can be cycloadded to a nucleophilic alkene C=C bond leading to the formation of a three-membered ring (C–E–C) (Fig. 4.1). As the π electrons from the C=C bond flow to the electrophilic center (E–:), its active electron pair will be transferred to a carbon in C=C simultaneously effecting a concerted cycloaddition reaction. In this section, we will discuss this type of cycloaddition reactions, including epoxidation of alkenes forming epoxides (characterized by a C–O–C ring) and synthesis of cyclopropane derivatives from the alkene C=C bond.

4.1.1 Epoxidation

In general, an alkene C=C bond can be oxidized to an epoxide by a relatively stable **percarboxylic acid** (RCO_3H), such as m-chloroperbenzoic acid (MCPBA) or peracetic acid, as shown in Figure 4.2. The reaction is concerted, with transfers of several electron pairs taking place simultaneously [1]. The initial driving force for this reaction can be attributed to the electronegative carbonyl (C=O) group in the percarboxylic acid. The peroxy O—O bond can overlap with the C=O π bond (hyperconjugation). As a result, the electronegative C=O strongly withdraws electrons from the O—O bond toward the doubly bonded oxygen, making the O–H oxygen

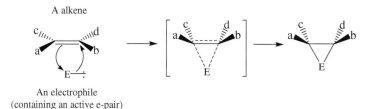

FIGURE 4.1 Prototype for electrophilic cycloaddition of an alkene forming a three-membered ring.

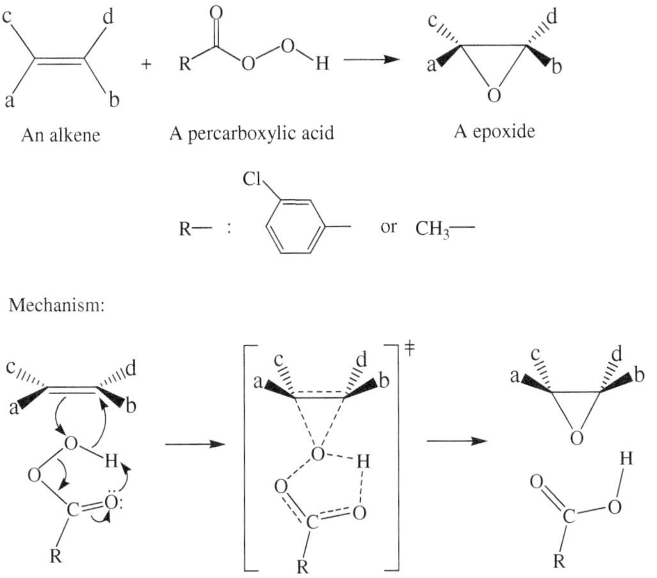

FIGURE 4.2 Epoxidation of an alkene by a percarboxylic acid following a concerted cycloaddition mechanism.

electron-deficient (electrophilic) so that it can attack the alkene π bond readily along a pathway perpendicular to the alkene molecule plane. As the electrophilic attack proceeds, a pair of π electrons is being transferred into the electrophilic oxygen leading to the formation of a carbon–oxygen bond. Simultaneously, the O–H group protonates the electronegative oxygen in the carbonyl C=O intramolecularly. This results in breaking of the O—H bond, and concurrently, its bonding electron pair is being transferred to an alkene C=C carbon to form a second carbon–oxygen bond.

Figure 4.2 shows that the formation of a five-membered ring structure in the transition state from the reactant molecule (peracid) in the concerted epoxidation only involves minimum atomic movement. This gives rise to a small activation entropy ($\Delta S^{\ddagger}$), resulting in stabilization of the transition state. It makes the overall concerted electron-transfer steps kinetically favorable, and the reaction possesses a low activation energy, eventually giving the epoxide product and a carboxylic acid by-product. The concerted mechanism gives rise to such a stereochemistry feature that all the four groups initially attached to the alkene C=C bond will retain their orientations in the epoxide product.

Figure 4.3 shows several examples of concerted epoxidation of alkenes by m-chloroperbenzoic acid. Reaction of cis-2-butene with the percarboxylic acid gives a cis-peroxide (a $meso$ compound), while the reaction of $trans$-2-butene leads to the formation of a racemic mixture of two $trans$-epoxides. (Z)-1-Phenylpropene has both the Ph and CH_3 groups in the same side of the C=C bond. It reacts with the percarboxylic acid giving a racemic mixture of two cis-epoxides. In each of the

cis-2-Butene   m-Chloroperbenzoic acid   cis-Configuration

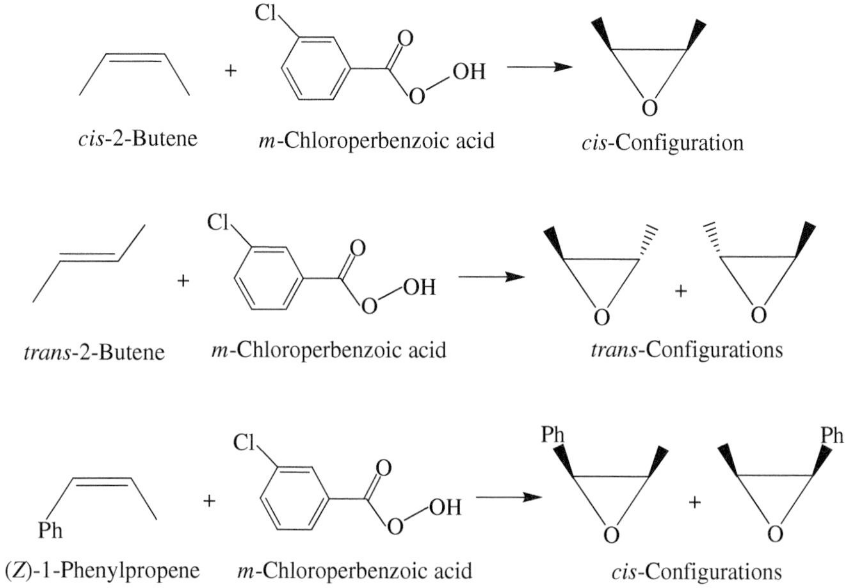

trans-2-Butene   m-Chloroperbenzoic acid   trans-Configurations

(Z)-1-Phenylpropene   m-Chloroperbenzoic acid   cis-Configurations

FIGURE 4.3   Examples of alkene epxidation reactions.

reactions, any two groups attached to a C=C bond in a certain side remain in the same side of the C—C bond in the epoxide product.

4.1.2   Cycloadditions via Carbenes and Related Species

Dichlorocarbene (CCl₂)   Upon heating, an alkene reacts with a mixture of chloroform ($CHCl_3$) and sodium hydroxide (NaOH) giving a dichlorocyclopropane derivative (Fig. 4.4). The reaction takes place via an intermediate dichlorocarbene (CCl_2) that is generated by an α-elimination of $CHCl_3$ [1]. The carbon atom in CCl_2 is sp^2-hybridized with a lone-pair of electrons held in a sp^2 orbital and an unhybridized p orbital being empty. Therefore, the molecule is singlet (with the total electron-spin $S = 0$), and the carbon atom is electrophilic due to the empty p orbital. It attacks the alkene π bond with the empty p orbital effecting a concerted cycloaddition [2]. As shown in Figure 4.4, as the alkene π electrons flow into the empty p orbital of CCl_2, the lone pair electrons in the CCl_2 sp^2 orbital are being transferred to a C=C carbon simultaneously forming a three-membered carbon ring. The concertedness of the reaction gives rise to a similar stereochemistry feature to that of the alkene epoxidation: The product of the dichlorocyclopropane derivative has the same configuration as that of the original alkene molecule. This feature can be further demonstrated by the reactions in Figure 4.5.

Figure 4.5 shows that cycloaddition of dichlorocarbene to *cis*-2-butene gives a *cis*-dichlorocyclopropane derivative, while the cycloaddition to *trans*-2-butene

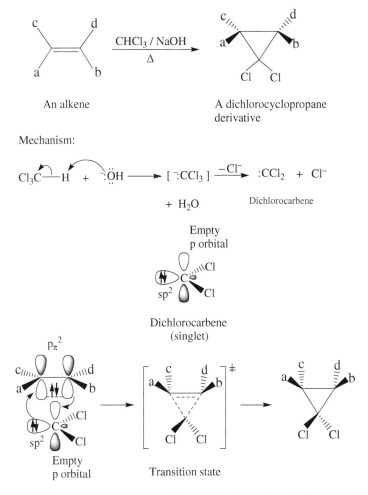

FIGURE 4.4 Mechanism for cycloaddition of dichlorocarbene (CCl$_2$) to an alkene.

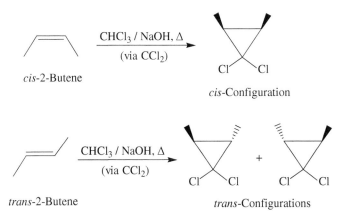

FIGURE 4.5 Stereochemistry for cycloaddition of dichlorocarbene (CCl$_2$) to alkenes.

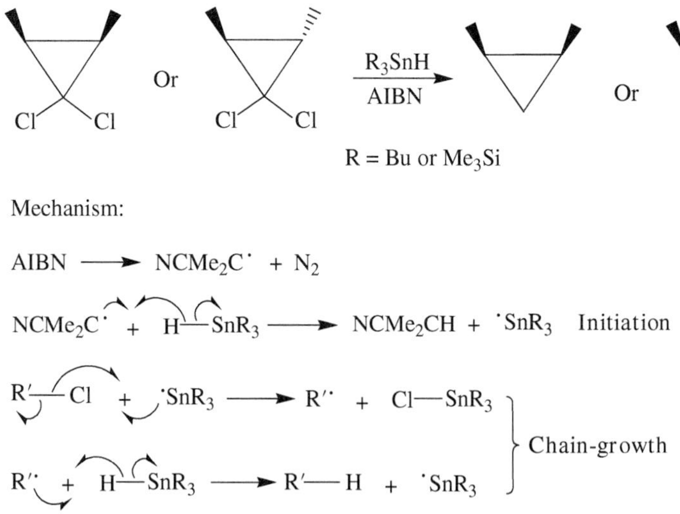

R = Bu or Me$_3$Si

Mechanism:

AIBN $\longrightarrow$ NCMe$_2$C$^{\cdot}$ + N$_2$

NCMe$_2$C$^{\cdot}$ + H$-$SnR$_3$ $\longrightarrow$ NCMe$_2$CH + $^{\cdot}$SnR$_3$ Initiation

R$^{\prime}-$Cl + $^{\cdot}$SnR$_3$ $\longrightarrow$ R$^{\prime\prime}$ + Cl$-$SnR$_3$ ⎫
⎬ Chain-growth
R$^{\prime\prime}$ + H$-$SnR$_3$ $\longrightarrow$ R$^{\prime}-$H + $^{\cdot}$SnR$_3$ ⎭

FIGURE 4.6   Removal of chloro groups (defunctionalization) from 1,1-dicloropropane.

gives a racemic mixture of two *trans*-dichlorocyclopropane derivatives. The chloro groups in a dichlorocyclopropane derivative can be removed (defunctionalized) readily by treatment with R$_3$SnH in the presence of catalytic amount of azobisiso-butyronitrile (AIBN) as shown in Figure 4.6. The reaction has provided a general method for preparation of substituted cyclopropanes. It makes the cycloaddition of dichlorocarbene to alkenes synthetically attractive.

Diphenylcarbene (CPh$_2$)   In contrast to CCl$_2$, CPh$_2$ is a triplet molecule (a dira-dical) with each of a sp^2 and the unhybridized p orbitals in the divalent carbon hold-ing a single electron and the total electron-spin $S = 1$ (Fig. 4.7) [2]. The formation of a triplet state (diradical) in CPh$_2$ is facilitated by the conjugation effects of the two aromatic rings stabilizing the single electron in the unhybridized p orbital. Cycload-dition of CPh$_2$ to alkenes (such as *cis*-2-butene) is stepwise via the formation of an intermediate diradical (Fig. 4.7). It has sufficient time to undergo free rotation about the C$-$C bond prior to the second step of the radical addition. As a result, the reaction gives a mixture of *cis*- and *trans*-configurations of cyclopropane derivatives.

Carbenoids   A carbenoid is a carbene complex with a metal such as zinc. It undergoes cycloaddition to an alkene C=C bond giving a substituted cyclo-propane [2]. Figure 4.8 shows reaction of *trans*-2-butene with CH$_2$I$_2$/Zn giving a racemic mixture of two *trans*-1,2-dimethylcyclopropanes. The first step in the reaction is the formation of a carbenoid I$-$CH$_2-$Zn$-$I via a radical coupling mechanism. The carbenoid is then cycloadded to the alkene

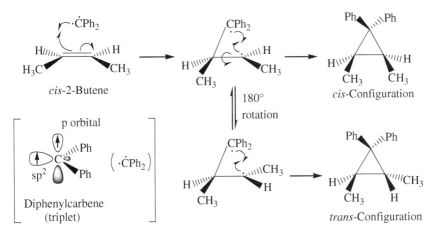

FIGURE 4.7 Stepwise radical mechanism for cycloaddition of the triplet diphenylcarbene (CPh$_2$) to an alkene. Adapted from Fox and Whitesell [2].

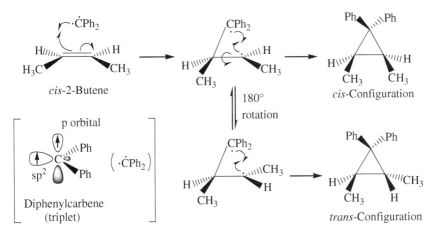

FIGURE 4.8 Mechanism and stereochemistry for cycloadditions of a carbenoid to alkenes giving cyclopropane derivatives.

concertedly from either side of the alkene π bond (top or bottom) giving two *trans*-products (racemic). An iodo (–I) group (a very good leaving group) makes the carbenoid carbon electrophilic, being readily attacked by the nucleophilic alkene π bond with a simultaneous cleavage of the C—I bond (an S_N2-type process). As the π electrons in the alkene C=C bond are being transferred to the carbenoid carbon and a new C—C bond is being formed on one C=C carbon, the other carbon in C=C becomes electron-deficient (electrophilic). It is concurrently attacked by the nucleophilic C—Zn bond (bonding electron-pair) forming a ring structure. The concertedness of the cycloaddition dictates the overall stereochemistry in the same manner as that for the cycloadditions of percarboxylic acid and dichlorocarbene to alkenes as discussed above.

Reaction of CH_2I_2/Zn with cyclohexene affords a bicyclic system following the same mechanism.

Cyclohexene

Clearly, these types of reactions have significant synthetic utility for making cyclopropane rings.

4.2   CYCLOADDITIONS TO FORM FOUR-MEMBERED RINGS

Although cycloaddition of two ethylene molecules giving cyclobutane is thermodynamically favorable, it does not take place at any temperatures due to a kinetic barrier [3].

Ethylene   Ethylene   Cyclobutane

This situation can be well understood by examining the symmetry and the possible interactions of FMOs (HOMO and LUMO) of ethylene molecules.

Figure 4.9 shows that the linear combinations of p orbitals (p_1 and p_2) from two carbon atoms of ethylene lead to the formation of two molecular orbitals, π orbital ($p_1 + p_2$) that is the bonding orbital and occupied (HOMO) (symmetric), and π* orbital ($p_1 - p_2$) that is the antibonding orbital and unoccupied (LUMO) (antisymmetric).

The possible FMO interactions between two ethylene molecules (Fig. 4.10) are the interaction of the HOMO of one molecule with the HOMO of the other (HOMO–HOMO interaction), the interaction of the LUMO of one molecule with the LUMO of the other (LUMO–LUMO interaction), and the interaction of the HOMO of one molecule with the LUMO of the other (HOMO–LUMO interaction). We can see

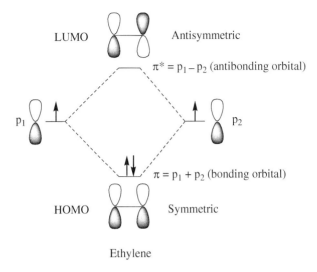

FIGURE 4.9 π molecular orbital diagram of ethylene.

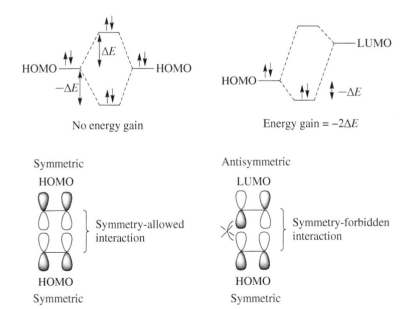

FIGURE 4.10 Possible frontier molecular orbital (FMO) interactions between two ethylene molecules in the ground state.

from Figure 4.10 that the HOMO–HOMO interaction is symmetry-allowed, but it does not have an energy gain. Such an interaction will not lead to the formation of any bonds. The LUMO does not contain any electrons. Clearly, the interaction of two empty orbitals does not have any energy gain and cannot lead to the formation of any bonds. The only interaction that could have an energy gain would be the HOMO (symmetric)–LUMO (antisymmetric) interaction. However, this interaction is symmetry-forbidden. The overlap of two p orbital lobes with different phases would make the wavefunction zero in the midregion of the two carbons. Therefore, the electron density in the midregion of the two carbons is zero and no bond is formed. As a result, this symmetry-forbidden interaction cannot lead to a cycloaddition. Because of all this, the cycloaddition of two ethylene molecules giving cyclobutane is **thermally symmetry-forbidden** [4].

The HOMO–LUMO energy gap in ethylene and many other alkenes matches the photon energy of UV-light. Therefore, an electron in the HOMO of an alkene (such as ethylene or 2-butane) can be excited photochemically to its LUMO (Fig. 4.11). As a result, both FMOs become **singly occupied molecular orbitals** (SOMOs), and the excited alkene molecule is triplet (with the total spin $S = 1$).

Figure 4.12 shows that one alkene molecule (such as ethylene or 2-butene) in the ground state (singlet with the total spin $S = 0$) can interact with another excited molecule (triplet) leading to a **photochemical cycloaddition.** Both HOMO (symmetric)–SOMO$_{(1)}$ (symmetric) and LUMO (antisymmetric)–SOMO$_{(2)}$ (antisymmetric) interactions have energy gains and are symmetry-allowed. In principle, the two interactions will work together resulting in the cycloaddition reaction. Such a cycloaddition is a **photochemically symmetry-allowed, but thermally symmetry-forbidden**.

In the UV-light, two 2-butene molecules (*cis* or *trans*) undergo cycloaddition giving two stereoisomers of 1,3-dimethylcyclobutanes [4]. Figures 4.13 and 4.14 illustrate detailed mechanisms that account for the stereochemistry of photochemical

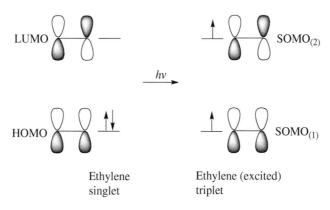

Ethylene
singlet

Ethylene (excited)
triplet

FIGURE 4.11   HOMO–LUMO electronic transition in ethylene.

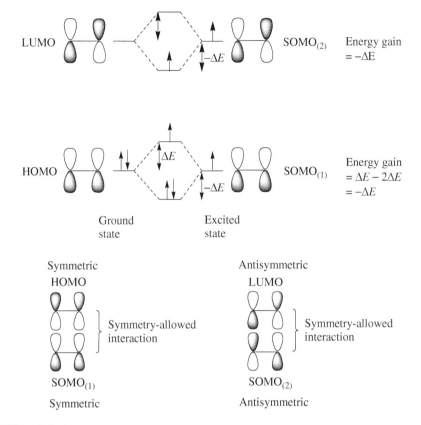

FIGURE 4.12 Possible frontier molecular orbital (FMO) interactions between an ethylene molecule in the ground state and an ethylene molecule in an excited state.

cycloadditions of *cis*-2-butene and *trans*-2-butene. For both alkenes, the reactions take place via the symmetry allowed HOMO–SOMO$_{(1)}$ and LUMO–SOMO$_{(2)}$ inter-actions. For each of the reactions, the two stereospecific alkene molecules (*cis* or *trans*) can approach each other in two different orientations giving rise to the forma-tions of two different stereoisomers of 1,3-dimethylcyclobutanes.

4.3 DIELS–ALDER CYCLOADDITIONS OF THE ALKENE C=C BOND TO FORM SIX-MEMBERED RINGS

In general, a conjugated diene (C=C—C=C) can undergo a facile concerted thermal cycloaddition reaction to an alkene C=C bond leading to the formation of a six-membered ring (a cyclohexene derivative). This type of cycloaddition reactions is called **Diels–Alder reactions.** The simplest Diels–Alder reaction is the cycloaddition of 1,3-butadiene to ethylene giving cyclohexene upon heating [1, 2, 4, 5].

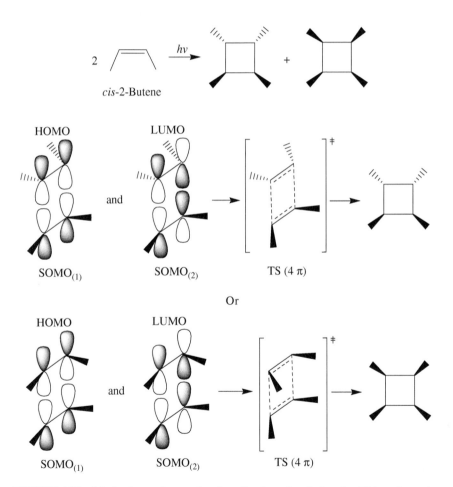

1,3-Butadiene   Ethylene   Cyclohexene
                   (a dienophile)

The reaction serves as a prototype for all the Diels–Alder reactions. First of all, we will present a detailed discussion on the mechanism of this reaction.

FIGURE 4.13   Mechanism and stereochemistry for photochemical cycloaddition of two *cis*-2-butene molecules.

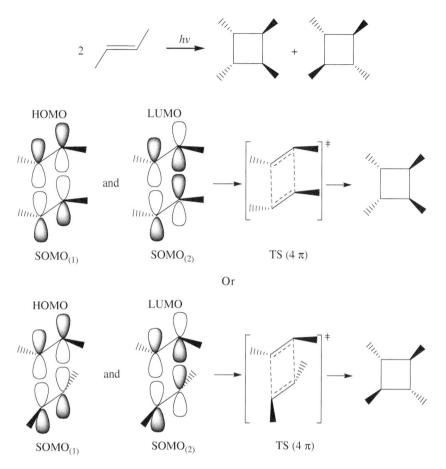

FIGURE 4.14 Mechanism and stereochemistry for photochemical cycloaddition of two *trans*-2-butene molecules.

4.3.1 Frontier Molecular Orbital Interactions

We have conducted molecular orbital analysis for ethylene (Section 4.2). Now let us do an analysis of molecular orbitals (π orbitals) for 1,3-butadiene. The sideway overlap of p orbitals in 1,3-butadiene results in a conjugated π bond. The linear combinations of four p orbitals (p_1, p_2, p_3, and p_4) on the carbon atoms of the conjugated diene molecule give rise to the formations of four molecular orbitals (Ψ's) as shown below:

$$\Psi_1 = p_1 + p_2 + p_3 + p_4$$
$$\Psi_2 = p_1 + p_2 - p_3 - p_4 \quad \text{(HOMO)}$$
$$\Psi_3 = p_1 - p_2 - p_3 + p_4 \quad \text{(LUMO)}$$
$$\Psi_4 = p_1 - p_2 + p_3 - p_4$$

Figure 4.15 shows the FMOs of 1,3-butadiene and ethylene (a **dienophile**). Due to the delocalization of the p orbitals, the HOMO–LUMO energy difference in the conjugated diene is smaller than the energy difference for the HOMO–LUMO in ethylene. As a result, the energy level of the ethylene HOMO is below that of the diene HOMO. The energy level of the ethylene LUMO is located between the energy levels of HOMO and LUMO of the diene. As illustrated above, for thermal reactions, only a HOMO–LUMO interaction between reactant molecules has an energy gain and can possibly lead to a cycloaddition reaction. The HOMO (diene) and LUMO

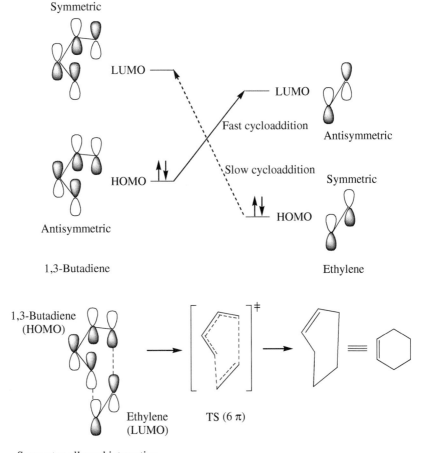

FIGURE 4.15 Frontier molecular orbital (FMO) interactions involved in thermal cycloaddition of 1,3-butadiene with ethylene.

(dienophile) are antisymmetric, while the HOMO (dienophile) and LUMO (diene) are symmetric. Both HOMO (diene)–LUMO (dienophile) and HOMO (dienophile)–LUMO (diene) interactions are symmetry-allowed. However, the energy difference of HOMO (diene)–LUMO (dienophile) is relatively smaller (312 kcal/mol) than that (317 kcal/mol) of HOMO (dienophile)–LUMO (diene). The activation energy (E_a) of a reaction is directly proportional to the energy gap between the reacting HOMO and LUMO whose interaction leads to the reaction. Therefore, the HOMO (diene)–LUMO (dienophile) interaction will lead to a relatively faster cycloaddition (with a smaller E_a), while the HOMO (dienophile)–LUMO (diene) interaction would lead to a relatively slower cycloaddition (with a greater E_a). The overall concerted cycloaddition of 1,3-butadiene to ethylene is effected by the interaction of HOMO of the diene with the LUMO of ethylene (dienophile) as illustrated in Figure 4.15. The reaction is **thermally symmetry-allowed.**

Figure 4.15 also shows that the symmetry-allowed overlap of the 1,3-butadiene HOMO with the ethylene LUMO results in the formation of a cyclic transition state (TS) that contains 6 π electrons (each carbon in the TS contributes one π electron). This is because all the 6 π electrons, 4 π electrons from 1,3-butadiene and 2 π electrons from ethylene, participate in the cycloaddition and are responsible for the formation of the transition state. For the cycloadditions of two ethylene derivatives (Section 4.2), each of the reactions involves a total of 4 π electrons. The cyclic transition state contains 4 π electrons. According to the Huckel's $(4n + 2)$ rule, a conjugated cyclic molecule that contains $(4n + 2)$ π electrons (n is a whole number) possesses special stability (aromaticity), while a conjugated cyclic molecule that contains $(4n)$ π electrons is relatively unstable (antiaromatic). This rule is also applicable to cyclic transition states. Therefore, it can be used to predict reactivity of cycloaddition reactions. We have the following general principle [4]:

If a cycloaddition reaction proceeds via a cyclic transition state containing $(4n + 2)$ π electrons (n is a whole number, and typically, $n = 1$ or 2), the transition state will experience stabilization with a relatively low energy level, and the cycloaddition is thermally allowed. If a cycloaddition reaction proceeds via a cyclic transition state containing $(4n)$ π electrons (n is a whole number, and typically, $n = 1$ or 2), the transition state will be relatively unstable possessing a high energy level, and the cycloaddition is photochemically allowed, but thermally forbidden.

We have seen such cases from the above discussions on the cycloaddition reactions giving four-membered and six-membered rings. This general principle is also applicable to cycloadditions that lead to the formation of five-membered rings. We will discuss those reactions in Section 4.4.

4.3.2   Substituent Effects

Reaction rates   When an electron-withdrawing group (such as –CO$_2$CH$_3$) is attached to the dienophile (an ethylene derivative), the Diels–Alder reaction is accelerated [1].

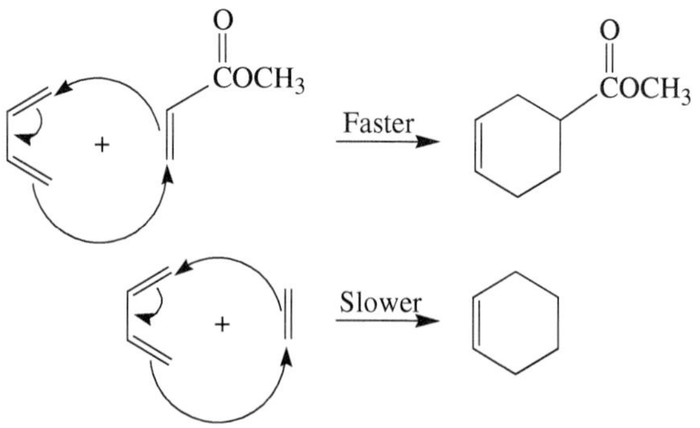

This substituent effect owes to an electronic control. The electron-withdrawing group attracts π electrons from the C=C double bond and lowers the energy level of LUMO of the dienophile (ΔE). As a result, the HOMO (diene)–LUMO (dienophile) energy gap is reduced (Fig. 4.16). Since the reaction is effected by the HOMO (diene)–LUMO (dienophile) interaction, the decrease in the HOMO–LUMO energy gap makes the reaction faster relative to the ethylene reaction.

On the other hand, the reaction can also be accelerated by an electron-donating group which is attached to the conjugated diene. The electron-donating group enhances the HOMO energy level of the diene reducing the HOMO (diene)–LUMO (dienophile) energy gap and making the reaction faster. The following example shows the case:

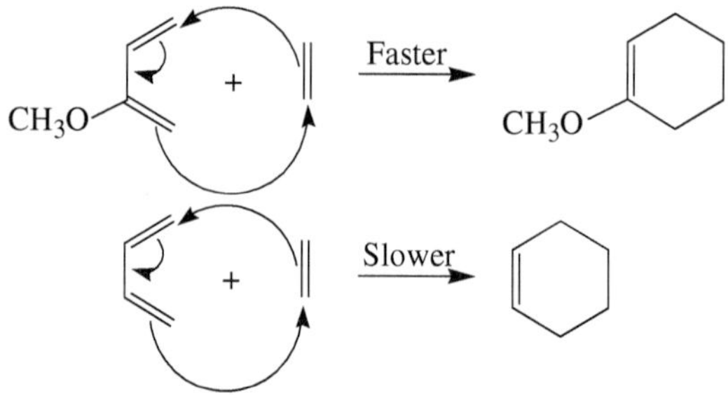

More examples of Diels–Alder reactions are shown in Figure 4.17. For each of the reactions, both the electron-withdrawing group(s) (–CN or –SO$_2$Ph) attached to the dienophile (ethylene derivative) and the electron-donating group(s) (–Ph or – CH$_3$) attached to the conjugated diene accelerate the reaction. By a comparison of the reactions (1) and (2) in Figure 4.17, the electron-donating phenyl (Ph–) group attached to a diene enhances the rate constant for ~200 times [1].

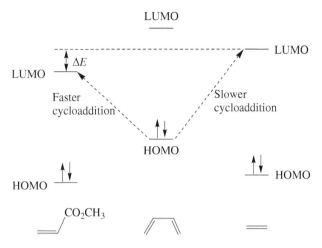

FIGURE 4.16   Effect of an electron-withdrawing group on the energy level of LUMO of ethylene.

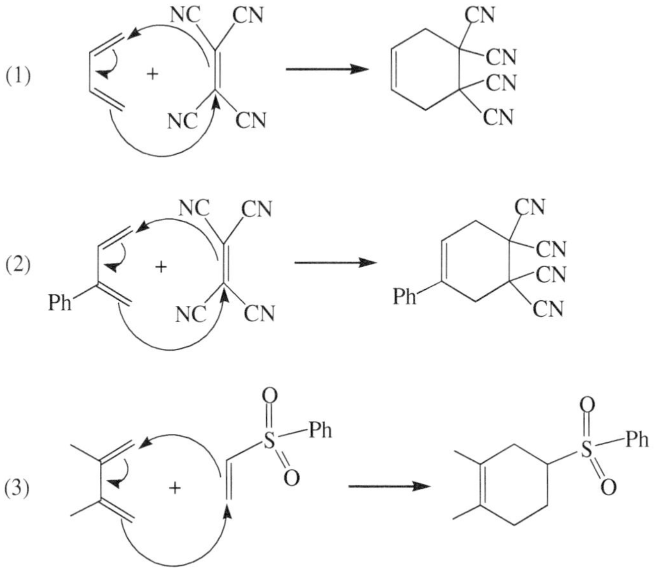

FIGURE 4.17   Examples of Diels–Alder reactions of 1,3-butadiene derivatives with ethylene derivatives.

Regiochemistry   If either the conjugated diene or dienophile (ethylene derivative) is symmetrical, the Diels–Alder reaction only gives one structural isomer (regioisomer). However, if both of the reactants are unsymmetrical, a mixture of two

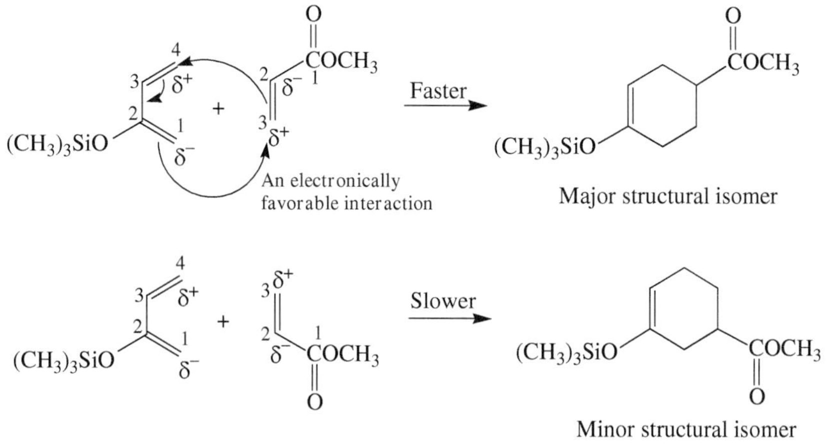

FIGURE 4.18   Regiochemistry for Diels–Alder reaction.

structural isomers will be formed with one isomer sometimes being more favored (Fig. 4.18). The regioselectivity is primarily due to an electronic control.

Figure 4.18 shows that the electron-withdrawing group ($-CO_2CH_3$) makes the C3-carbon of the dienophile (ethylene derivative) partially positively charged (electron-deficient) and the adjacent C2-carbon partially negatively charged (electron-rich), while the electron-donating group ($(CH_3)_3SiO-$) makes the C1-carbon of the conjugate diene partially negatively charged (electron-rich) and the C4-carbon partially positively charged (electron-deficient) [5]. The interactions between electron-rich C1 of the diene and electron-deficient C3 of the dienophile and between electron-deficient C4 of the diene and electron-rich C2 of the dienophile are energetically more favorable, giving rise to a faster reaction and the corresponding product is the observed major structural isomer. The alternative interactions are electronically unfavorable giving rise to a slower reaction and the corresponding product is the minor isomer.

Although Diels–Alder cycloadditions of unsymmetrical alkene reactants are sometimes regioselective, the regioselectivity is in general very poor. Therefore, for most of Diels–Alder reactions, at least one symmetrical reactant (diene or dienophile) is employed in order to obtain a single product.

4.3.3   Other Diels–Alder Reactions

Formation of bicyclic molecules   Diels–Alder reactions can be utilized to make various bicyclic molecules. One of the simplest reactions of this category is the cycloaddition of 1,3-cyclopentadiene with ethylene as shown in Figure 4.19. Analogous to the reaction of 1,3-butadiene, the cycloaddition of 1,3-cyclopentadiene takes place on its C1 and C4 carbons via the symmetry-allowed interaction of

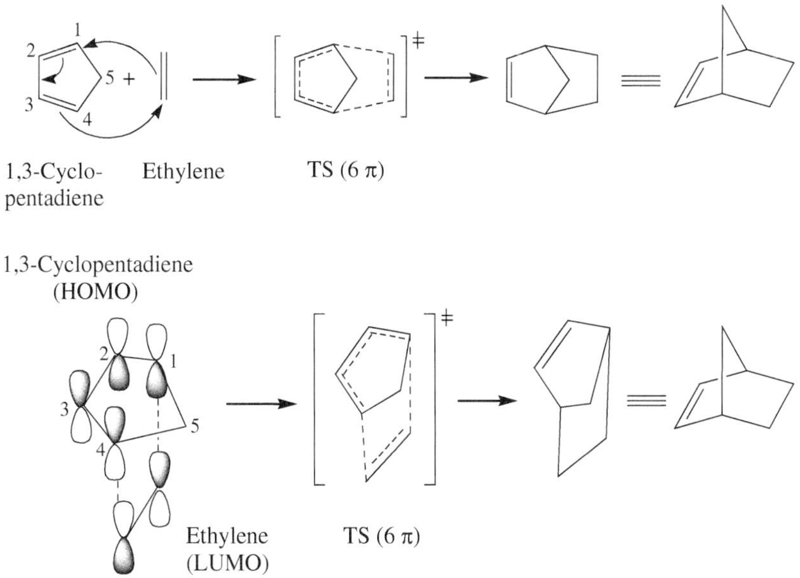

1,3-Cyclo-   Ethylene   TS (6 π)
pentadiene

1,3-Cyclopentadiene
(HOMO)

Ethylene   TS (6 π)
(LUMO)

FIGURE 4.19   Diels–Alder reaction of 1,3-cyclopentdiene with ethylene giving a bicyclic molecule.

HOMO (diene)–LUMO (ethylene). Since the conjugated diene is cyclic, a bicyclic molecule is formed after the cycloaddition, with the CH_2 carbon from 1,3-cyclopentadiene bridging across the carbon ring.

A similar Diels–Alder reaction of 1,3-cyclohexadiene with methyl acrylate (an ethylene derivative) takes place giving a bicyclic compound as shown below [4]:

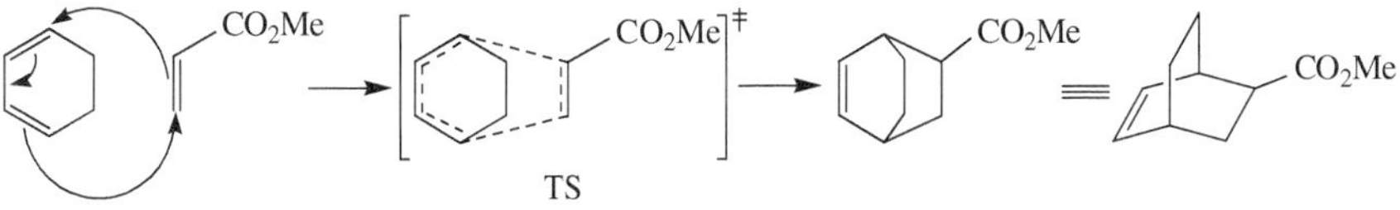

An alkyne contains two π bonds in its C≡C triple bond that are perpendicular to each other. It can also undergo a Diels–Alder reaction with its one π bond to a conjugated diene, while the second π bond is retained in the reaction. As a result, the C≡C triple bond is transformed into a C=C double bond after the cycloaddition. An example of an alkyne Diels–Alder reaction to a conjugated diene is shown below [5]:

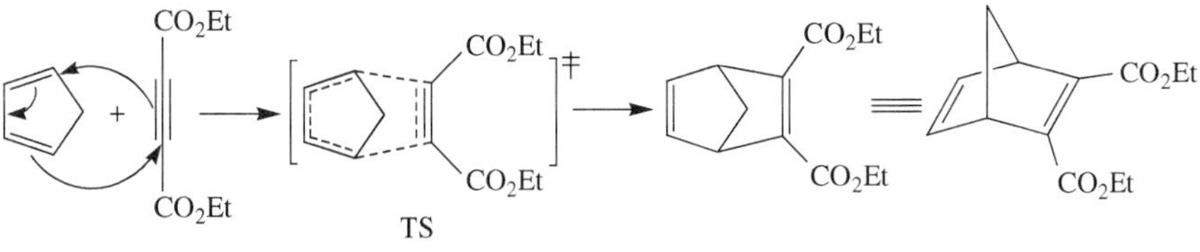

Since the conjugated diene is cyclic, the cycloaddition reaction gives a bicyclic compound. The reaction is accelerated by the two electron-withdrawing $-CO_2Et$ groups attached to the C≡C bond (dienophile).

A cyclic alkene (with a C=C bond contained in a carbon ring) can be cycloadded to a 1,3-butadiene derivative giving a fused bicyclic molecule [4].

TS

The reaction is accelerated by electron-withdrawing carbonyls (C=O) attached to the C=C double bond (refer to Fig. 4.16 for the effect of an electron-withdrawing group on LUMO of an alkene).

Cycloaddition of the C=N bond to conjugated dienes   Similar to the C=C bond, the C=N π bond can interact with the conjugated π bond in a diene (C=C—C=C) to effect a Diels–Alder cycloaddition with a 6 π transition state. A particularly interesting C=N containing reagent is *t*-butyl 2-azidoacrylate (Fig. 4.20). It undergoes Diels–Alder cycloadditions with 1,3-pentadiene, 1-methoxybutadiene, 1-methoxy-3-trimethylsilyloxybutadiene, and 1,3-pentadiene, respectively (Fig. 4.20) [6]. All the reactions occur via the similar interaction of the diene HOMO and the C=N LUMO. The C=N bond contains an electronegative nitrogen atom and is attached to an electronegative carbonyl group. Both factors work for lowering the energy level of the LUMO of the C=N bond. In addition, the electron-donating groups [$-CH_3$, $-OCH_3$, and $-OSi(CH_3)_3$] enhance the energy level of the diene HOMO. All this substantially decreases the HOMO–LUMO energy gap between the diene and dienophile to make the reactions occur at ambient temperature. The reactions are stereoselective as shown in Figure 4.20.

Cycloaddition of 2,3-dicyano-p-benzoquinone to 1,3-cyclopentadiene   2,3-Dicyano-*p*-benzoquinone contains two separated C=C π bonds in the molecule. Each of them can interact with the conjugated π bond in 1,3-cyclopentadiene at certain conditions to effect a selective Diels–Alder reaction (Fig. 4.21) [7]. Both reactions in Figure 4.21 occur via the interaction of the diene HOMO with the LUMO of a separated C=C bond. The other C=C bond is intact in the course of the reaction. The first reaction takes place at the dicyano C=C bond. The electronegative –CN groups lower the energy level of the C=C LUMO. This decreases the HOMO–LUMO energy gap between the two reactants, lowering the activation energy for the reaction. Therefore, the first reaction proceeds fast and is subject to a kinetic control. It takes place at room temperature. On the other hand, the HOMO–LUMO energy gap for the second reaction is relatively greater because

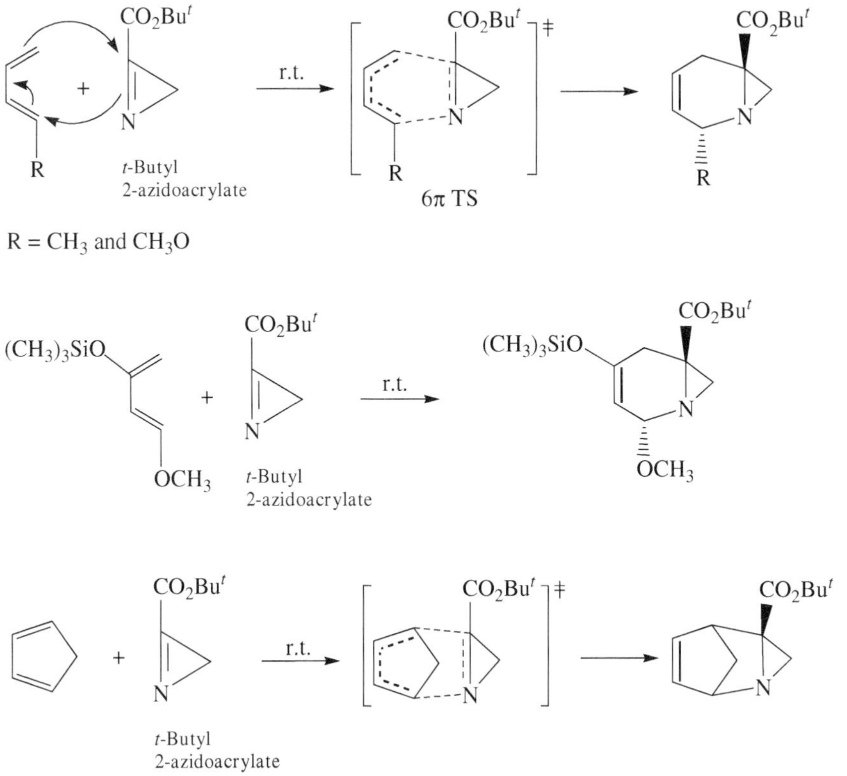

R = CH$_3$ and CH$_3$O

FIGURE 4.20   Diels–Alder reactions of the C=N containing t-butyl 2-azidoacrylate with 1,3-diene derivatives. Based on Alves and Gilchrist [6].

the reacting C=C bond in the separated diene does not have electronegative groups and its LUMO energy level is relatively higher than that for the dicyano C=C. The greater HOMO–LUMO energy gap gives rise to a higher activation energy for the second reaction and makes it relatively slow. Thus, it does not occur at room temperature. The product from the second reaction is more stable than the product from the first reaction. At elevated temperatures upon heating, both products undergo backward dissociations and the reactions become reversible. Therefore, the cycloaddition reactions are subject to a thermodynamic control, and the less stable product from the first reaction is equilibrated to the more stable product in the second reaction. Overall, at elevated temperatures the second reaction occurs and is subject to a thermodynamic control.

Cycloaddition of larger alkenes   Figure 4.22 shows the cycloaddition of a conjugated cyclic triene (6 π) to 1,3-butadiene (4 π) [4]. The reaction takes place readily. All the C=C π electrons from the reactants participate in the reaction, giving a bicyclic product via a 10 π [(4n + 2) with n = 2, aromatic] transition state. The carbonyl

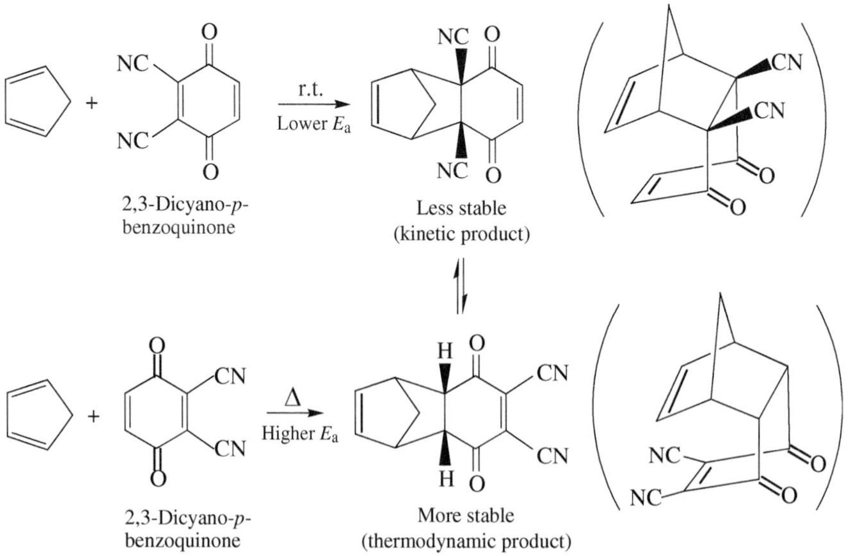

2,3-Dicyano-*p*-benzoquinone

Less stable
(kinetic product)

2,3-Dicyano-*p*-benzoquinone

More stable
(thermodynamic product)

FIGURE 4.21 Diels–Alder reactions of 1,3-cyclopentadiene with 2,3-dicyano-*p*-benzoquinone (a separated diene): kinetic and thermodynamic controls. Based on Marchand et al. [7].

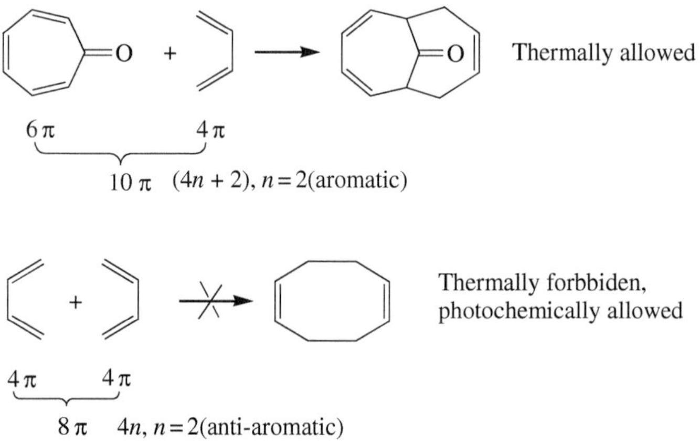

Thermally allowed

$6\,\pi$ $4\,\pi$

$10\,\pi$ $(4n + 2), n = 2$(aromatic)

Thermally forbbiden, photochemically allowed

$4\,\pi$ $4\,\pi$

$8\,\pi$ $4n, n = 2$(anti-aromatic)

FIGURE 4.22 Diels–Alder reactions for larger cyclic systems.

group in the cyclic triene does not participate in the reaction. According to the Huckel's rule, the cycloaddition is thermally allowed. However, the cycloaddition of two 1,3-butadiene ($4\,\pi$) molecules to 1,5-octadiene does not take place thermally (Fig. 4.22). The reaction would proceed via an $8\,\pi$ [$(4n)$ with $n = 2$, antiaromatic] transition state. According to the Huckel's rule, the cycloaddition is thermally forbidden, but photochemically allowed.

Cyclohexene derivatives   Regiospecific and stereospecific cyclohexene derivatives can be effectively synthesized by Diels–Alder reactions of 1,3-butadiene
derivatives with ethylene derivatives (Fig. 4.23). In general, a preferred condition for the reactions is that the diene is electron-rich (containing an electron-
donating group) enhancing the energy level of its HOMO, while the dienophile
(ethylene derivative) is electron-deficient (containing an electron-withdrawing
group) lowering the energy level of its LUMO [5]. As a result, the HOMO
(diene)–LUMO (dienophile) energy gap will be relatively small leading to a
fast reaction. Both reactions in Figure 4.23 satisfy the condition.

Figure 4.23a shows reaction of (*E,E*)-2,4-hexadiene (a symmetrical conjugated
diene) with an unsymmetrical ethylene derivative in a *trans*-configuration. The reaction only gives one structural isomer. The stereochemistry feature of the reaction is
determined by the favorable configuration of the transition state in which the substituent groups on two adjacent carbons are staggered. This configuration effectively
reduces the steric interactions of the substituent groups and stabilizes the TS, leading
to the formation of the corresponding observed stereoisomer. The formations of
alternative stereoisomers with different configurations would proceed via less stable
transition states and be unfavorable. Figure 4.23b shows reaction of an unsymmetrical conjugate diene with an unsymmetrical ethylene derivative. The reaction is
regioselective as well as stereoselective. Preferably, the cycloaddition takes place
via the interaction of the partially negatively charged, relatively electron-rich carbon
in the diene (created by the electron-donating methoxy group) with the partially positively charged, electron-deficient carbon in the dienophile (created by the electron-
withdrawing carbonyl group); and the interaction of the partially positively charged,
relatively electron-deficient carbon in the diene with the partially negatively
charged, electron-rich carbon in the dienophile. The stereochemistry feature of
the reaction is determined by the favorable configuration of the transition state in
the same way as that for the reaction in Figure 4.23a.

Synthetic application by intramolecular Diels–Alder cycloaddition   *Tofogliflozin*
functions as a sodium glucose cotransporter two inhibitor for the treatment of dia-
betes. Traditionally, its chemical synthesis involves halogen–metal exchange reactions catalyzed by transition metals such as Rh and Ru. The synthetic methodology,
among other things, has suffered with troublesome removal of residual transition
metals. Recently, a novel synthetic approach to making tofogliflozin has been developed by employing an intramolecular Diels–Alder cycloaddition between a conjugated diene (C=C—C=C) and a C≡C triple bond in a diene-yne compound
(Fig. 4.24) [8]. The diene-yne compound in Figure 4.24 can be prepared by a series
of synthetic reactions starting with commercially available precursors [8]. Then the
compound undergoes an intramolecular Diels–Alder cycloaddition in the air upon
mild heating, and the reaction proceeds via the interaction of the diene HOMO
and the C≡C LUMO (a 6π-cycloaddition). It is a thermally symmetry-allowed
cycloaddition. The cycloaddition, followed by a facile aromatization (–2H), leads

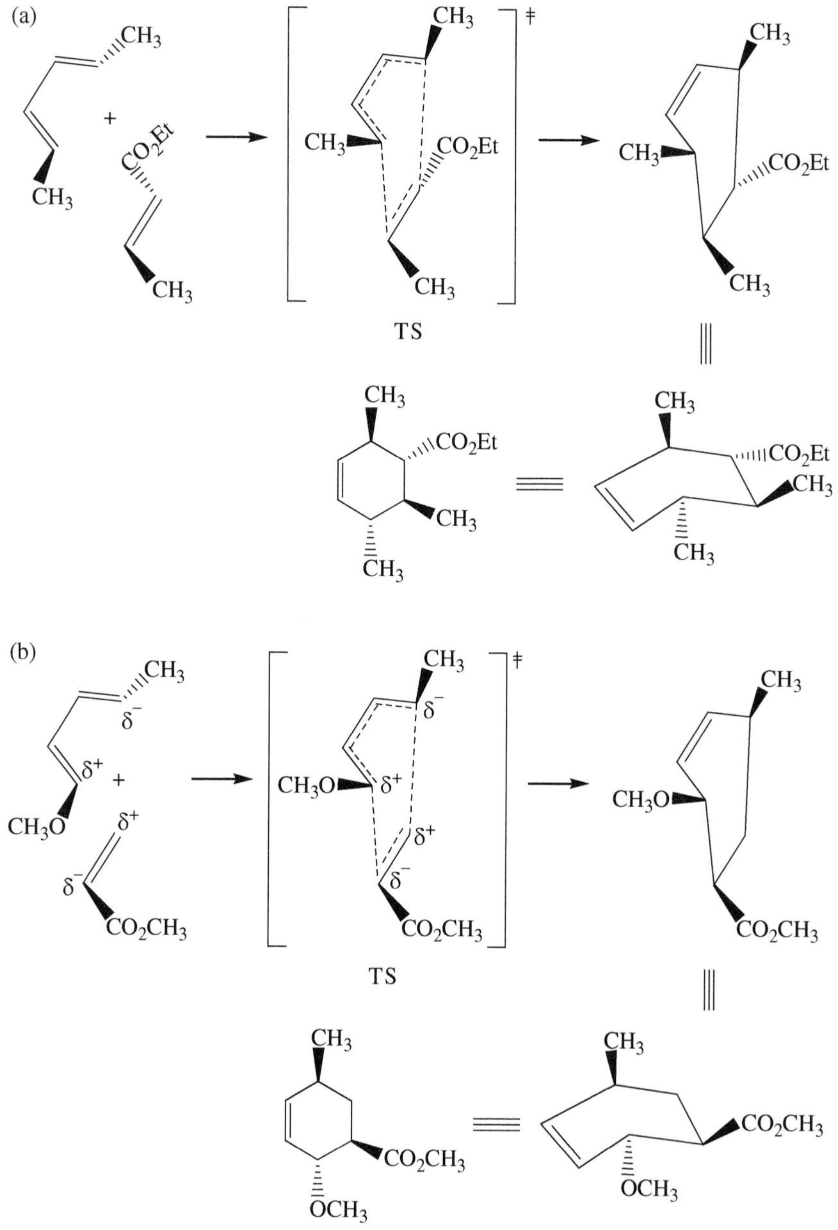

FIGURE 4.23   Regioselectivity and stereoselectivity for Diels–Alder reactions.

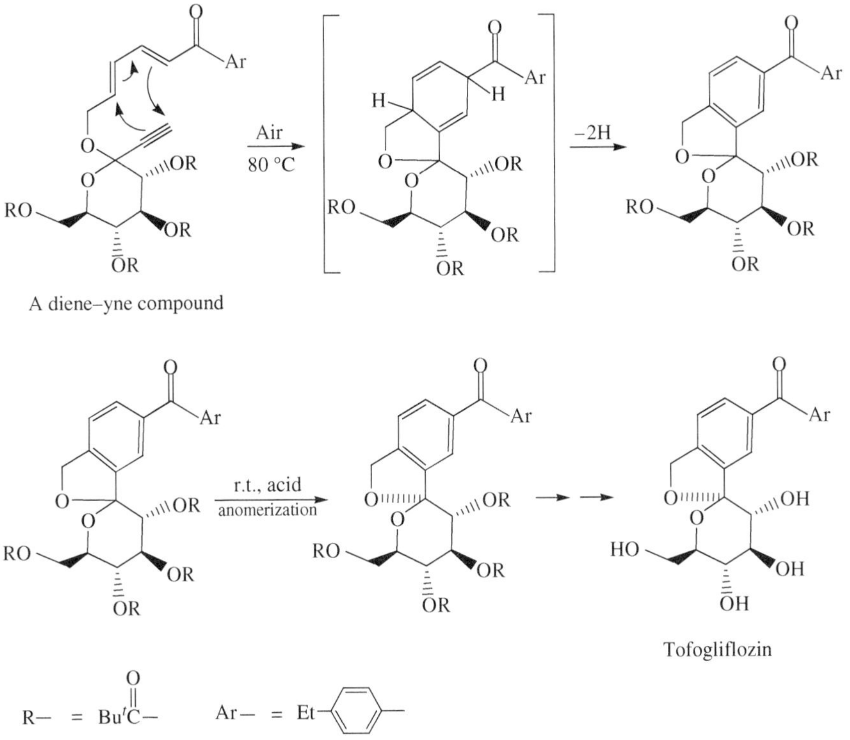

A diene–yne compound

FIGURE 4.24   Synthesis of tofogliflozin by a 6π intramolecular diene-yne Diels–Alder cycloaddition.

to the formation of a six-membered aromatic ring. The initial reaction product is a mixture of two anomers. Treatment of the product with an acid results in an anomerization. Further chemical treatment, including hydrolysis and a redox reaction, leads to the formation of the targeted tofogliflozin product.

4.4   1,3-DIPOLAR CYCLOADDITIONS OF THE C=C AND OTHER MULTIPLE BONDS TO FORM FIVE-MEMBERED RINGS

4.4.1   Oxidation of Alkenes by Ozone (O₃) and Osmium Tetraoxide (OsO₄) via Cycloadditions

Ozone (O₃) is a 1,3-dipolar-like angular molecule with its central oxygen atom sp^2 hybridized. It can be characterized by two resonance structures as shown below:

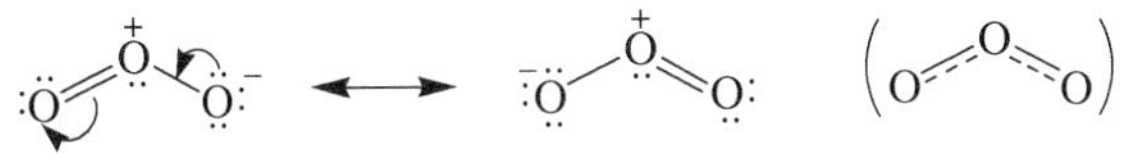

There are a total of 4 π electrons in O_3. The O=O bond formally contributes 2 π electrons. The other 2 π electrons are formally contributed from a lone pair of electrons in a terminal oxygen atom. A delocalized (conjugated) π bond (three-center, four-electron bond) is formed in the molecule (Fig. 4.25) [4].

An alkene C=C bond contains 2 π electrons. Therefore, it undergoes a facile concerted cycloaddition to O_3 via a 6 π [(4n + 2) with n = 1, aromatic] cyclic transition state giving a molozonide, and the cycloaddition is thermally allowed (refer to Section 4.3.1).

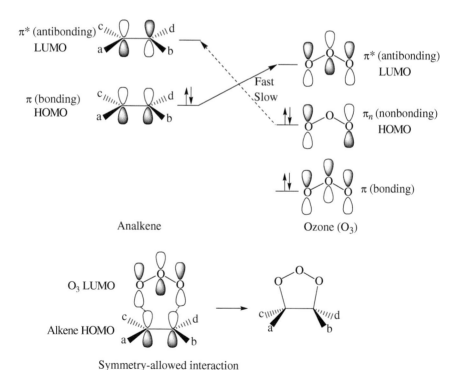

FIGURE 4.25 Frontier molecular orbital (FMO) interactions involved in thermal cycloaddition of ozone (O_3) to an alkene. Adapted from Jackson [4].

Figure 4.25 shows the molecular orbital diagram for ozone. A delocalized, three-center, four-electron π bond is formed by sideway overlap of three p orbitals in the oxygen atoms. Linear combinations of the three p orbitals (p_1, p_2, and p_3) result in three molecular orbitals, π (bonding), π_n (nonbonding, HOMO), and $\pi*$ (antibonding, LUMO) as shown below:

$$\pi = p_1 + p_2 + p_3$$
$$\pi_n = p_1 - p_3 \quad \text{(HOMO)}$$
$$\pi* = p_1 - p_2 + p_3 \quad \text{(LUMO)}$$

The electronegative oxygens in O_3 greatly lower the energies of the O_3 FMOs. As a result, the O_3 HOMO is below the alkene HOMO (π). The energy level of the alkene HOMO (π) is situated between the energy levels of O_3 FMOs, while the energy level of the alkene LUMO ($\pi*$) is above that of the O_3 LUMO. Therefore, the HOMO (alkene)–LUMO (O_3) interaction (with a smaller energy gap) would be energetically more favorable than the alternative HOMO (O_3)–LUMO (alkene) interaction (with a greater energy gap), although both interactions are symmetry-allowed. The cycloaddition is effected by the energetically more favorable HOMO (alkene)–LUMO (O_3) interaction giving rise to a faster reaction than the alternative interaction. This situation is opposite from the orbital interaction for the Diels–Alder reaction of 1,3-butadiene and ethylene (Section 4.3.1, Fig. 4.15). The cycloaddition of O_3 to an alkene is **reverse electron demand**. We will see that cycloadditions of all the 1,3-dipolar-like molecules to alkenes presented in this section are reverse electron demand, effected by the interaction of the alkene HOMO with the 1,3-dipole LUMO, because all the 1,3-dipolar-like molecules contain electronegative atoms.

The initial cycloaddition product (molozonide) is unstable. It dissociates spontaneously once formed, and the two fragments combine giving an ozonide. Usually, the reaction of alkene with ozone is followed by treatment with a weakly reducing agent (such as dimethyl sulfide Me_2S) [9]. The intermediate ozonide will be transformed into two carbonyl groups, formed in the original C=C carbon atoms, and the C=C bond is eventually cleaved. The overall reaction is shown in Figure 4.26.

This approach can be a useful synthetic method for making specialty ketones and aldehydes from readily available alkenes. An example is shown below:

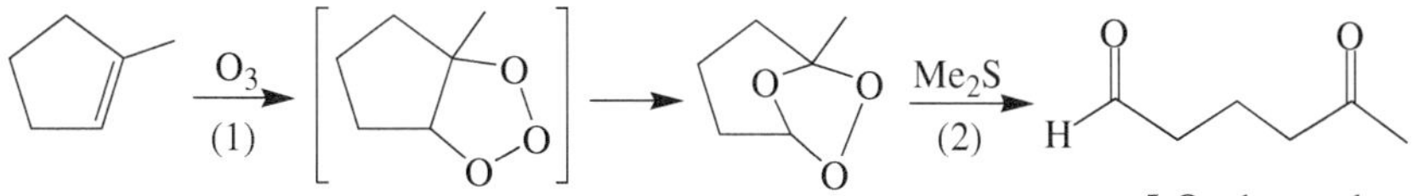

1-Methyl-
cyclopentene

5-Oxohexanal
(a ketoaldehyde)

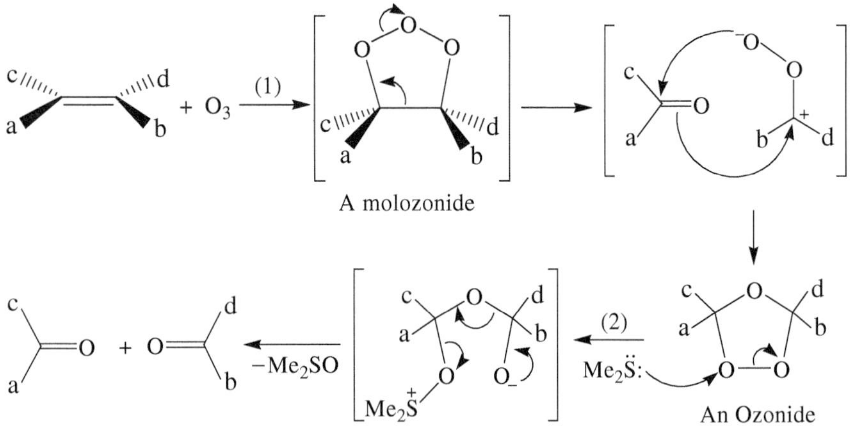

FIGURE 4.26   Overall mechanism for oxidation of an alkene to two carbonyl groups by ozone (O_3).

Osmium tetraoxide (OsO_4) undergoes a similar concerted thermal cycloaddition to an alkene C=C bond as shown below:

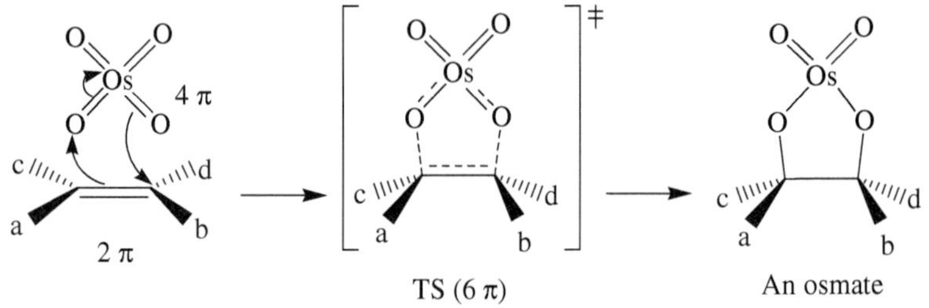

Each of the two O=Os bonds involved in the cycloaddition formally contributes 2π electrons. This gives rise to the formation of a 6π transition state in the cycloaddition, making the reaction kinetically favorable. The five-membered osmium-containing ring is bonded to the original alkene in a *cis*-configuration. The cycloadduct (osmate) can be isolated and characterized. Usually, the intermediate osmate is treated *in situ* by a reducing agent, such as $NaHSO_3$, which cleaves the O–Os bonds and leads to the formation of a *cis* glycol.

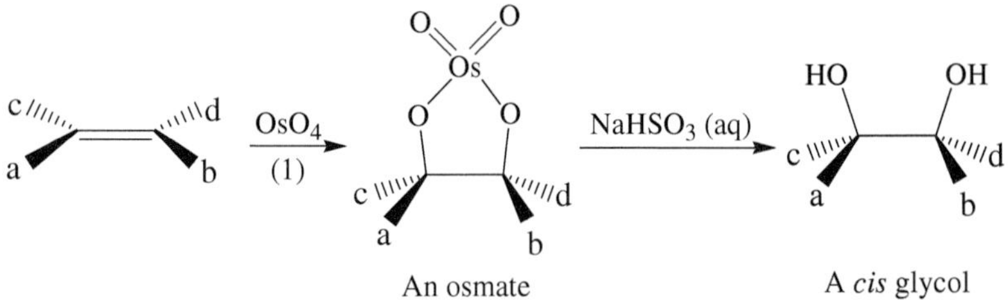

4.4.2   Cycloadditions of Nitrogen-Containing 1,3-Dipoles to Alkenes

Diazoalkanes This type of molecules are linear 1,3-dipoles. The simplest diazoalkane is diazomethane ($^-N=N^+=CH_2$). It can be characterized by two resonance structures as shown below:

$$:\overset{-}{N}=\overset{+}{N}=CH_2 \longleftrightarrow :N≡\overset{+}{N}-\overset{..}{C}H_2$$

Similar to ozone, diazomethane has a delocalized three-center, four-electron π bond, with each of the N=N and N=C bonds formally contributing 2 π electrons. It is a conjugated 4 π system. Diazomethane has a similar molecular orbital structure to that of ozone (Fig. 4.25), namely, that linear combinations of three p orbitals from the nitrogen and carbon atoms in diazomethane result in bonding, nonbonding (HOMO), and antibonding (LUMO) π orbitals. It can undergo a concerted thermal cycloaddition to an alkene (Fig. 4.27) via a 6 π cyclic transition state. Similar to the ozone cycloaddition, the cycloaddition of diazomethane to an alkene C=C bond is accomplished by a symmetry-allowed interaction of HOMO (alkene) with LUMO (diazomethane) (reverse electron demand) as shown in Figure 4.27.

When CH_2N_2 reacts with a cyclic alkene, the cycloaddition gives a bicyclic molecule as shown below [10]:

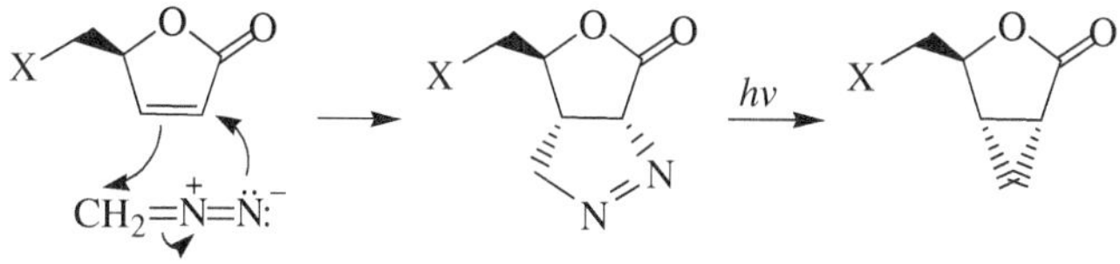

X = OH, OSTBDP, OTs, and I

If the C=C bond is attached to an electron-withdrawing group, its cycloaddition product with CH_2N_2 undergoes a facile photochemical dissociation to give a cyclopropane derivative [10]. This has provided a general method to make functionalized cyclopropanes.

An alkyne C≡C bond can also undergo cycloaddition with its one π bond to a diazoalkane as shown below [5]:

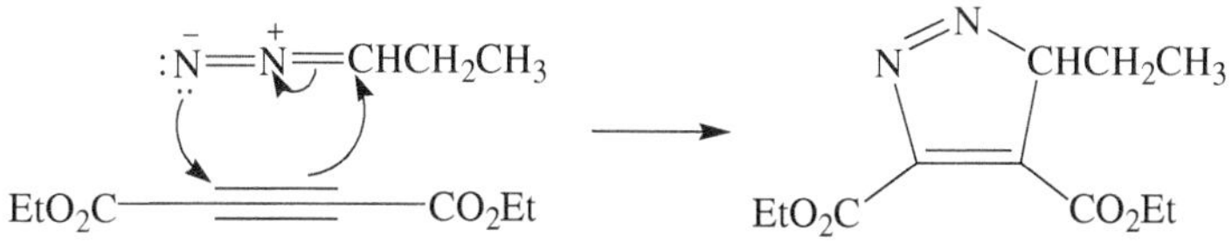

Azides ($R-N=N^+=N^-$) are another type of nitrogen-containing 1,3-dipoles, characterized by the following resonance structures:

$$R-N=\overset{+}{N}=\overset{-}{N}: \longleftrightarrow R-\overset{-}{N}-\overset{+}{N}=N:$$

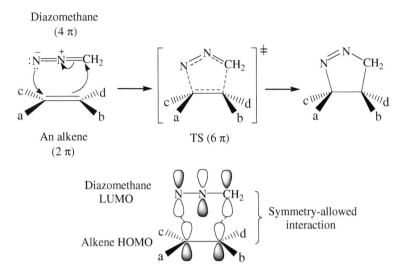

FIGURE 4.27 Cycloaddition reaction of an alkene to diazomethane.

Analogous to other 1,3-dipoles, an azide also has a delocalized three-center, four-electron π bond, with each of the N=N bonds formally contributing 2 π electrons forming a conjugated 4 π system. It undergoes concerted thermal cycloaddition to an alkene as shown below:

Other nitrogen-containing 1,3-dipoalr molecules include **Nitrile oxides** as shown below:

A nitrile oxide

The molecule contains 4 π electrons that are contributed formally from the C=N bond (2 π) and a lone pair of electrons in oxygen (2 π).

Similar to other 1,3-dipoles, a nitrile oxide ($R—C\equiv N^+—O^-$) (4 π) undergoes a concerted thermal cycloaddition to an alkene C=C bond (2 π) via a 6 π cyclic transition state.

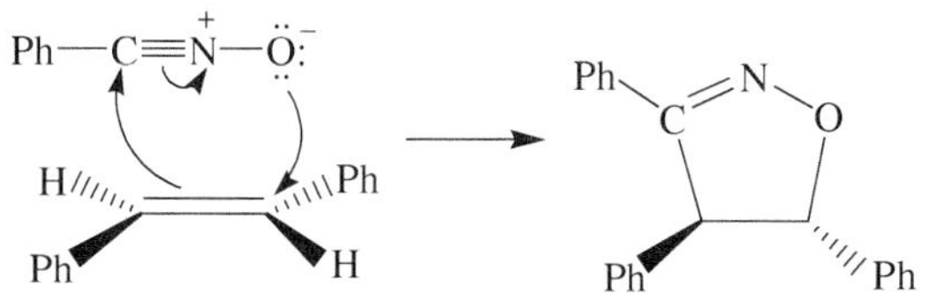

Since the cycloaddition is concerted, it is stereospecific. The geometry of the alkene is maintained in the cyclic product. This feature is found in all the above 1,3-dipolar cycloadditions to alkenes.

Similar to Diels–Alder reactions, the regioselectivity for cycloadditions of 1,3-dipolar molecules to alkenes can be understood on the basis of electron distributions of the 1,3-dipole and the ethylene derivative. The following example shows the case [5].

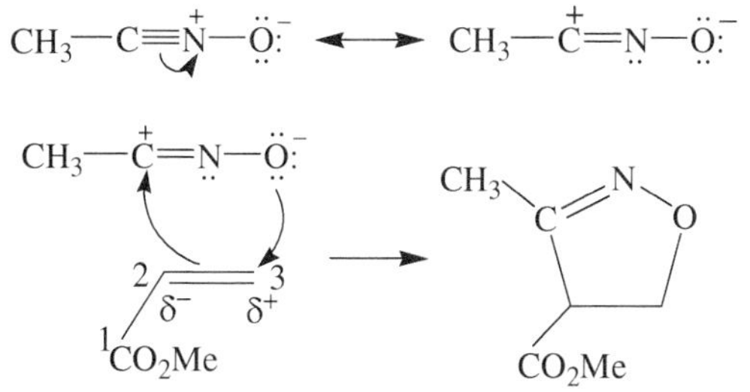

The nitrile oxide has a relatively electron-deficient carbon (positively charged) and an electron-rich oxygen (negatively charged). The electron-withdrawing group ($-CO_2Me$) makes C3-carbon of the ethylene derivative (methyl acrylate) relatively electron-deficient (positively charged) and C2-carbon negatively charged. As a result, the interaction of the electron-rich oxygen in the nitrile oxide with the electron-deficient C3-carbon in methyl acrylate and the interaction of the positive carbon in the nitrile oxide with the negative C2-carbon in methyl acrylate are energetically favorable, leading to an observed major cycloaddition product.

4.4.3   Cycloadditions of the Dithionitronium (NS_2^+) Ion to Alkenes, Alkynes, and Nitriles: Making CNS-Containing Aromatic Heterocycles

Dithionitronium (NS_2^+ or $[S=N=S]^+$) ion is a linear 1,3-dipolar-like cationic molecule [11]. The species contains two sets of perpendicular three-center, four-electron π bonds, each of which is formed by the in-plane overlap of p orbitals in sideways (Fig. 4.28a). Similar to other 1,3-dipolar molecules, the linear combinations of each set of coplanar p orbitals result in bonding (π), nonbonding (π_n, HOMO), and antibonding ($\pi*$, LUMO) orbitals (Fig. 4.28b). Due to the difference in electronegativities of nitrogen and sulfur, the LUMO in NS_2^+ is sulfur-based, namely, that the eigenvector coefficient for the p orbital on each of the sulfur atoms (+0.51) is greater

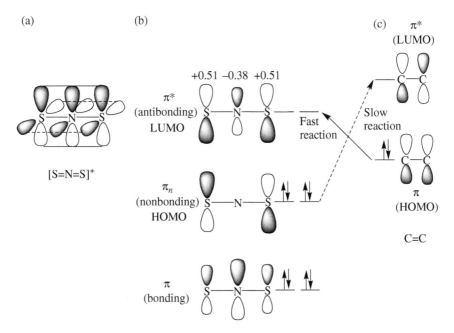

FIGURE 4.28 (a) The in-plane sideway overlaps of p orbitals in two perpendicular planes for the dithionitronium $[S=N=S]^+$ ion giving rise to the formation of two sets of perpendicular three-center, four-electron π bonds; (b) Bonding (π), nonbonding (π_n), and antibonding ($\pi*$) orbitals for each three-center, four-electron π bond in $[S=N=S]^+$; and (c) HOMO and LUMO of the alkene (C=C).

than that (absolute value) for the p orbital on nitrogen (-0.38) [12]. The remarkable features in the electronic structure of NS_2^+ allow it to undergo double cycloaddition reactions to two alkene molecules. Each cycloaddition takes place on one three-center, four-electron π bond in NS_2^+ [11, 12]. In addition, NS_2^+ undergoes cycloaddition reactions with alkynes (C≡C) and nitriles (C≡N) [11, 13]. One set of the three-center, four-electron π bond in NS_2^+ is cycloadded to a π bond in C≡C or C≡N. When a ring is formed, the other three-center, four-electron π bond in NS_2^+ overlaps in sideways with the second, unreacted π bond in C≡C or C≡N forming a 6 π aromatic system. In this section, we will give a brief account of the NS_2^+ cycloaddition reactions.

Cycloadditions with alkenes With its one set of three-center, four-electron π bond, NS_2^+ (typically, in the form of $NS_2^+AsF_6^-$) can be cycloadded to an alkene (ethylene, *cis-* and *trans-*2-butene, 2-methylpropene, 2,3-dimethyl-2-butene, or nor-bornene) giving a planar five-membered cationic ring (1,3,2-dithiazolidine) (Fig. 4.29) [12]. Similar to cycloadditions of other 1,3-dipolar molecules [14], the reaction takes place via a 6 π [(4n + 2), n = 1] transition state (NS_2^+ contributes 4 π

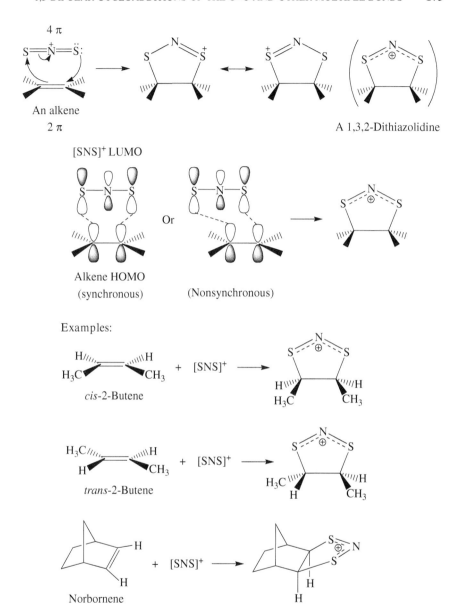

FIGURE 4.29 Thermally symmetry-allowed cycloadditions of the dithionitronium [S=N=S]$^+$ ion to alkenes: mechanism and examples. Based on Brooks et al. [12].

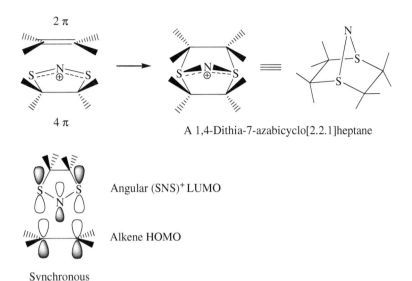

2 π

4 π

A 1,4-Dithia-7-azabicyclo[2.2.1]heptane

Angular (SNS)⁺ LUMO

Alkene HOMO

Synchronous

FIGURE 4.30 Mechanism for cycloaddition of an alkene to an alkene-NS₂⁺ cycloadduct. Based on Brooks et al. [12].

electrons and the alkene contributes $2\,\pi$ electrons) and, therefore, it is thermally allowed. Since the cycloaddition is concerted, the product maintains the configuration of the starting alkene. For example, the reactions of *cis*- and *trans*-2-butene give cycloadducts in *cis*- and *trans*-configurations, respectively. The reaction of norbornene only gives one stereoisomer. The energy level of NS_2^+ LUMO is situated between the energy levels of HOMO and LUMO of an alkene (Fig. 4.28c). Therefore, the interaction of the alkene (C=C) HOMO with the NS_2^+ LUMO (reverse electron demand) gives rise to a faster cycloaddition than does the interaction of NS_2^+ HOMO with the alkene LUMO. As a result, the cycloadditions of NS_2^+ to alkenes take place via a symmetry-allowed, synchronous or nonsynchronous interaction of the alkene HOMO with the NS_2^+ LUMO (Fig. 4.29) [12].

In the alkene-NS_2^+ cycloadduct (1,3,2-dithiazolidine), the C–C linkage becomes saturated, while the angular (SNS)⁺ moiety maintains an unreacted, three-center, four-electron π bond, analogous to the π bond in the original linear [S=N=S]⁺ cation. Therefore, the angular (SNS)⁺ moiety in the ring can undergo a similar cycloaddition to a second alkene molecule (a molecule of either the same or a different alkene) giving a bicyclic cationic product (1,4-dithia-7-azabicyclo[2.2.1]heptanes) (Fig. 4.30) [12]. The cycloaddition to a second alkene takes place via a symmetry-allowed, synchronous interaction of the alkene HOMO with LUMO of the angular (SNS)⁺ moiety in the first cycloadduct. The alkene p_π orbitals approach the angular (SNS)⁺ moiety perpendicularly to the ring plane, bringing about the reaction.

The cycloadditions of the first alkene-NS_2^+ cycloadducts to a second alkene molecule are subject to steric hindrance of the substituents in the alkene C=C bond, and the reactions are stereoselective. The reactions in Figure 4.31 demonstrate the case. In each of the reactions shown in Figure 4.31, the cycloaddition of the (*cis*-2-butene)-NS_2^+ cycloadduct to a second *cis*-2-butene molecule and the cycloaddition

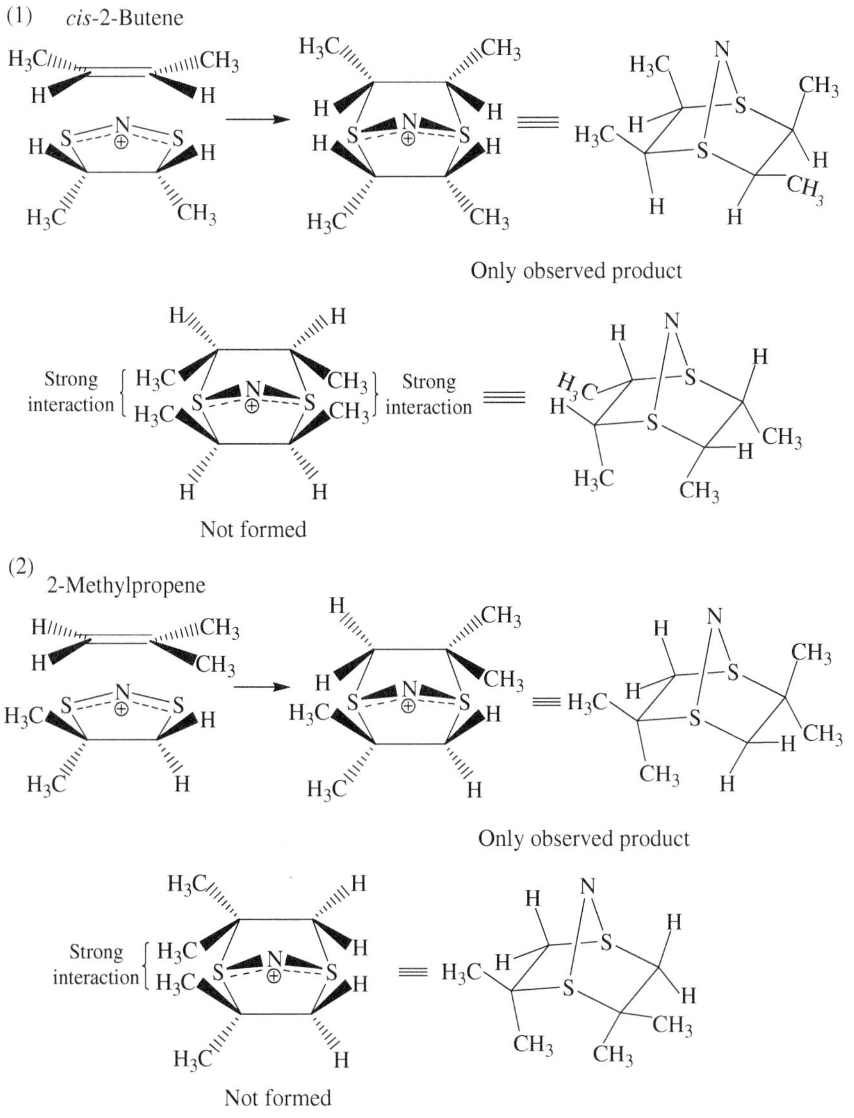

(1)   *cis*-2-Butene

Only observed product

Strong interaction / Strong interaction

Not formed

(2) 2-Methylpropene

Only observed product

Strong interaction

Not formed

FIGURE 4.31   Stereochemistry of cycloadditions of alkenes to alkene-NS_2^+ cycloadducts.

of the (2-methylpropene)-NS_2^+ cycloadduct to a second 2-methylpropene molecule, only one stereoisomer is formed. The formation of an alternative isomer would render strong steric interactions of two methyl groups, which would be kinetically unfavorable. For the same reason, reaction of NS_2^+ with 2,3-dimethyl-2-butene only gives the alkene-NS_2^+ cycloadduct (Fig. 4.32) [12]. It does not undergo cycloaddition with a second 2,3-dimethyl-2-butene molecule even though the alkene is in large excess. The cycloaddition with a second alkene molecule would render strong steric hindrance due to the methyl–methyl interactions, which prohibits the second cycloaddition from happening.

Cycloadditions with alkynes and nitriles   Analogous to the alkene cycloadditions, one set of three-center, four-electron π bond in NS_2^+ can react with a π bond in $C\equiv C$ of an alkyne or in $C\equiv N$ of a nitrile forming a five-membered cationic heterocycle (1,3,2-dithiazolium or 1,3,2,4-dithiadiazolim) [13]. The reaction is concerted and thermally allowed. In each of the cycloaddition product molecule, the unreacted three-center, four-electron π bond in the angular $(SNS)^+$ moiety (perpendicular to the ring) will overlap in sideways with the unreacted π bond (2 electrons) in the C–C or C–N domain (also perpendicular to the ring) forming a conjugated 6 π aromatic system. The cycloadditions of NS_2^+ (in $NS_2^+AsF_6^-$) with alkynes ($RC\equiv CR'$) (such as R,R' =H,H; CH_3,H; CF_3,H; and Ph,Ph) and with nitriles ($RC\equiv N$) (such as

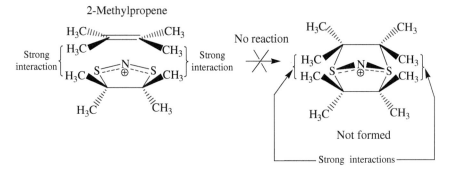

FIGURE 4.32   Cycloaddition of $[S=N=S]^+$ to 2,3-dimethyl-2-butene. Based on Brooks et al. [12].

R=CF$_3$, CH$_3$, Bu^t, Ph, and p-O$_2$NC$_6$H$_4$) are shown in Figure 4.33. Each reaction is effected by a symmetry-allowed, synchronous or nonsynchronous interaction of the HOMO of the alkyne or nitrile (1,3-dipolarophile) with the LUMO of NS$_2^+$ (reverse electron demand). The direct formation of 6 π aromatic rings in the cycloadditions is

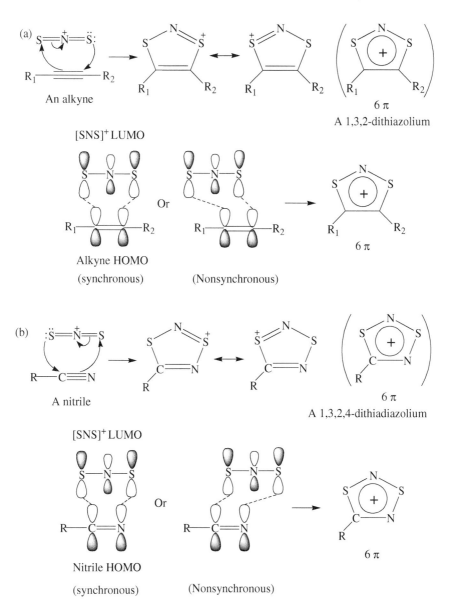

FIGURE 4.33 Thermally symmetry-allowed cycloadditions of [S=N=S]$^+$ to alkynes (a) and nitriles (b) leading to the formations of aromatic CNS-containing hereocycles.

unusual. It serves as an additional thermodynamic driving force for the reactions. Since each of the heterocyclic products is aromatic, it does not undergo further cycloaddition reactions.

Kinetics Kinetic studies of cycloadditions of NS_2^+ with various alkynes and nitriles [13] show that an electron-withdrawing group on C≡C or C≡N slows down the reaction. For example, the rate constant for the cycloaddition of $CF_3C≡CH$ (CF_3 is electron-withdrawing) is 20 times smaller than that for the cycloaddition of HC≡CH. The rate constant for the cycloaddition of $CF_3C≡N$ is 100 times smaller than that for the cycloaddition of $CH_3C≡N$. The observed effect of an electron-withdrawing group on rate constant of the cycloaddition of NS_2^+ with alkynes and nitriles is just opposite to the effect on the Diels–Alder reactions discussed in Section 4.3. It can be understood in terms of the electron-withdrawing group that attracts π electrons in C≡C or C≡N, lowering the energy level of its HOMO. As a result, the $HOMO_\pi$–LUMO (NS_2^+) energy gap has increased and the cycloaddition reaction is slowed down. Conversely, an electron-donating group on C≡C or C≡N accelerates the cycloaddition by decreasing the $HOMO_\pi$–LUMO(NS_2^+) energy gap. For example, the rate constant for the cycloaddition of PhC≡CPh (Ph is electron-donating) is about 20 times greater than that for the cycloaddition of HC≡CH. The rate constant for the cycloaddition of $(CH_3)_2NC≡N$ [$(CH_3)_2N$ is strongly electron-donating] is about 100 times greater than that for the cycloaddition of $CH_3C≡N$.

Quantitatively, the energy level of $HOMO_\pi$ is equal to $-IP$ (the negative of ionization potential) of the alkyne or nitrile. It has been found [13] that for over 15 different alkynes and nitriles, the common logarithm of relative rate constant ($\lg k_r$) decreases almost linearly as a function of the ionization potential (IP, eV) of the alkyne or nitrile, formulated as follows:

$$\lg k_r = -0.86\,IP + 10.5 \quad \left(R^2 = 0.86\right) \tag{4.1}$$

k_r is the relative rate constant of a cycloaddition to that for the CH_3CN reaction. Since $E(HOMO_\pi) = -IP$, the relationship between $\lg k_r$ and $E(HOMO_\pi)$ can be established as follows (Eq. 4.2) on the basis of Equation 4.1.

$$\lg k_r = 0.86\,E(HOMO_\pi) + 10.5 \tag{4.2}$$

Equation 4.2 shows that $\lg k_r$ increases linearly as a function of $E(HOMO_\pi)$. Increase in $E(HOMO_\pi)$ makes the $HOMO_\pi$–LUMO(NS_2^+) energy gap smaller accelerating the cycloaddition. Therefore, the relationship described by Equation 4.2 has confirmed the proposed reverse-electron-demand FMO interactions (Figs. 4.28, 4.29, 4.30, and 4.33) for the cycloadditions of NS_2^+ to alkenes (C=C), alkynes (C≡C), and nitriles (C≡N).

4.5   OTHER PERICYCLIC REACTIONS

A **pericyclic (electrocyclic) reaction** can also occur to an acyclic conjugated diene (4 π) or triene (6 π) in which both ends of the π bond interact intramolecularly, resulting in a ring-closure (cyclization). The reaction is concerted and takes place via certain molecular rotations. It can be reversed. The backward process involves a ring-opening mechanism following the opposite direction of the same pathway.

4.5.1   Conjugated Trienes

The intramolecular cyclization of an acyclic conjugated triene (such as 1,3,5-hexatriene) proceeds via a 6 π [(4n + 2) with n = 1] transition state (Fig. 4.34). According to the Huckel's rule, the reaction is thermally allowed. In fact, it takes place readily upon heating and is reversible [2]. The ring-closure occurs via the HOMO of 1,3,5-hexatriene (Fig. 4.34). The two p orbitals in the terminal carbon atoms have the same phase in the HOMO. They undergo rotations in different directions (defined as **disrotatory**). As a result, as the old π system is being transformed into a new π system (conjugated diene), a σ bond is being formed between the two terminal carbons simultaneously by overlap of two p orbital lobes in the same phase. Because the transition state is aromatic possessing a relatively low energy, rotations of the two

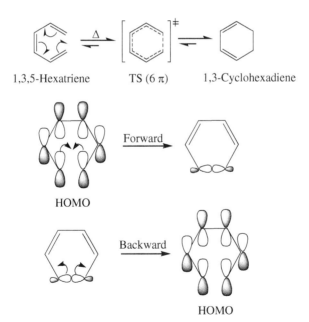

1,3,5-Hexatriene   TS (6 π)   1,3-Cyclohexadiene

FIGURE 4.34   Mechanism for intramolecular ring-closure (cyclization) of 1,3,5-hexatriene to 1,3-cyclohexadiene and the reversal ring-opening process.

terminal carbon atoms in the conjugated triene is thermally accessible. Correspondingly, the backward process occurs via disrotatory of the two sp^3 carbons (σ bonded) in 1,3-cyclohexadiene. It leads to breaking of the C—C σ bond in the ring. In addition, the disrotatory generates two p orbitals in the same phase, consistent with the HOMO of 1,3,5-hexatriene.

The disrotatory involved in the pericyclic reaction can be confirmed by examining the thermal ring-closure of 2,4,6-octatriene [4].

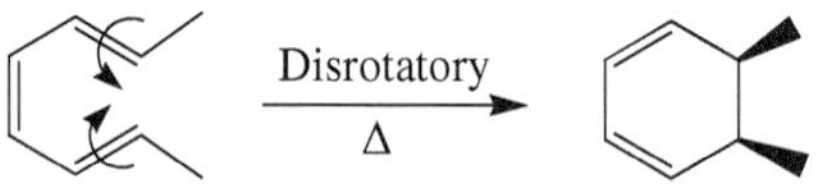

| 2,4,6-Octatriene | *cis*-Configuration |

The reaction gives *cis*-5,6-dimethyl-1,3-cyclohexadiene due to disrotatory of C2 and C7 in 2,4,6-octadiene. The *trans*-molecules are not formed.

Mechanism of the backward ring-opening reactions can be further confirmed by examining the product of the reaction of the following *trans*-5,6-dimethyl-1,3-cyclohexadiene [4].

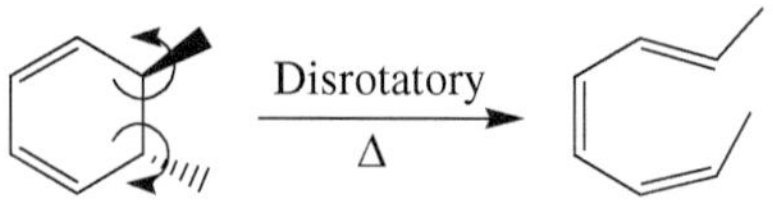

Only one stereoisomer of 2,4,6-octatriene is formed showing that the ring-opening reaction takes place via disrotatory of the two sp^3 carbons in the reactant molecule.

4.5.2 The Cope Rearrangement

Similar to 1,3,5-hexatriene which undergoes a thermal ring-closure on the two terminal carbons, 1,5-hexadiene can also have a σ bond connection between the terminal carbons. Because the molecule contains a saturated C_3—C_4 σ bond, as the C1 and C6 are being connected by a σ bond and a new π system is being formed, the C_3—C_4 σ bond is breaking simultaneously. The overall reaction is shown below.

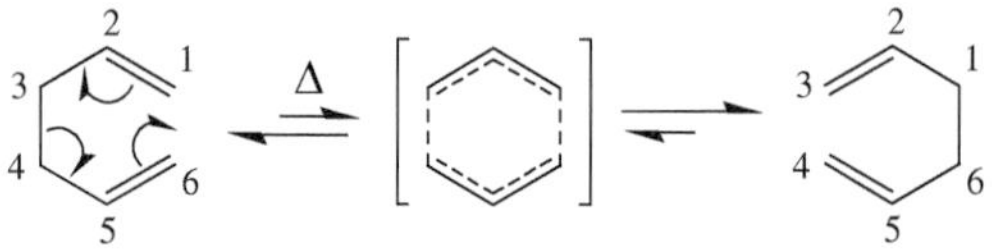

| 1,5-Hexadiene | TS ($6\,\pi$) |

It proceeds via a $6\,\pi$ cyclic transition state and is thermally allowed. This type of reactions is called the Cope rearrangement.

For the Cope rearrangement of 1,5-hexadiene, the product and reactant have the same identity. In order to demonstrate the bond-making and bond-breaking process involved in the rearrangement, an experiment is performed on a deuterium isotope-labeled 1,5-hexadiene as shown in Figure 4.35. In the reactant, the deuterium atoms are bonded to vinyl carbons ($=CD_2$). In the product, the deuterium atoms are located at allylic carbons ($-CD_2$), while the vinyl carbons are occupied by hydrogen atoms ($=CH_2$). The change is now observable [2].

Figure 4.35 shows that the Cope rearrangement of the deuterium isotope-labeled 1,5-hexadiene takes place via its HOMO. Since the diene is not conjugated, the two π bonds in the molecule are separated. Therefore, the HOMO can have two different phases. As the terminal carbon atoms (the $=CD_2$ carbons) undergo conrotatory (rotations in the same direction) or disrotatory (rotations in different directions), a σ bond between the two terminal $=CD_2$ carbons and a new π system are formed. This disrupts the H_2C-CH_2 σ bond in the reactant molecule. Simultaneously, the two σ bonded, saturated CH_2 carbons start conrotatory or disrotatory leading to breaking of the H_2C-CH_2 bond. The overall rearrangement is completed. In Chapter 10, we will give more accounts for this type of rearrangement.

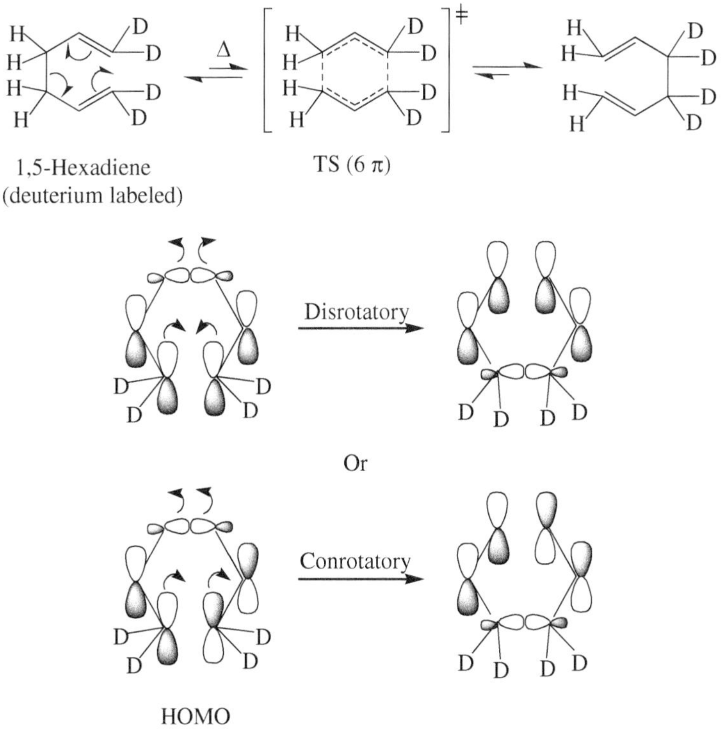

FIGURE 4.35   Mechanism for the Cope rearrangement of deuterium isotope-labeled 1,5-hexadiene.

4.5.3   Conjugate Dienes

The intramolecular cyclization of a conjugate diene (such as 1,3-butadiene) proceeds via a 4 π [(4n) with n = 1] transition state (Fig. 4.36). According to the Huckel's rule, the reaction is photochemically allowed, but thermally forbidden. In fact, it can only take place in UV light. One electron is excited from HOMO to LUMO by the UV light. Then the ring-closure occurs via the LUMO of 1,3-butadiene. The two p orbitals in the terminal carbon atoms have the same phase in the LUMO. They undergo rotations in different directions (disrotatory). As a result, as the old π system is being transformed into a new π system (cyclobutene), a σ bond is being formed simultaneously between the two terminal carbons in 1,3-butadiene. Because the transition state is antiaromatic possessing a relatively high energy, the molecular rotation is not thermally accessible. However, the photons of the UV light are energetic sufficiently to bring the molecule to the transition state [15], resulting in the photochemical ring-closure reaction. The disrotatory mechanism can be confirmed by examining the photochemical ring-closure of (E,E)-2,4-hexadiene.

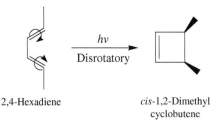

2,4-Hexadiene   cis-1,2-Dimethyl
                                 cyclobutene

Only cis-1,2-dimethylcyclobutene is formed via disrotatory. The alternative trans-configuration is not observed.

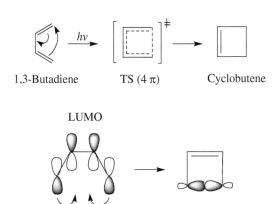

1,3-Butadiene   TS (4 π)   Cyclobutene

LUMO

FIGURE 4.36   Photochemical ring-closure of 1,3-butadiene to cyclobutene.

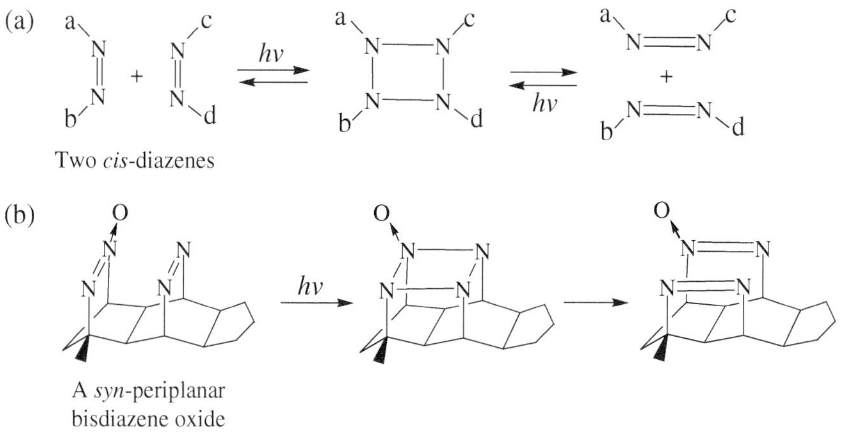

FIGURE 4.37   The 4π-cycloaddition between the N=N bonds. (a) Pathway for a reaction between two cis-diazenes and (b) Reaction pathway for a syn-periplanar bisdiazene oxide. (b) Based on Exner and Prinzbach [16].

4.5.4   The 4π-Cycloaddition Between the N=N Bonds

A photochemical cycloaddition can take place between two N=N π bonds (such as those in cis-diazenes) to lead to the formation of a metastable saturated N_4 ring via a 4π transition state (Fig. 4.37a). Similar to the photochemically symmetry-allowed cycloaddition of two ethylene molecules, the photochemical cycloaddition between two N=N bonds occurs via the HOMO–SOMO interaction. The N_4 ring intermediate undergoes a facile dissociation to make two different N=N molecules. The reaction is reversible, and the backward reaction follows the same 4π-cycloaddition mechanism. The reversibility for the overall reaction should be determined by the relative stability between reactants and products. Such a photochemical 4π-cycloaddition between the N=N bonds have been found to take place in a syn-periplanar bisdiazene oxide (Fig. 4.37b) [16]. On irradiation with UV light, the bisdiazene oxide undergoes metathesis giving another bisdiazene oxide. The reaction takes place via a highly unstable N_4 ring intermediate which has been observed by a low temperature matrix characterization [16].

4.5.5   The 4π-Cycloaddition of Buckminsterfullerene C_{60}

C_{60} is highly unsaturated with many C=C double bonds and is electrophilic. Research has shown that a C=C π bond in C_{60} can undergo a photochemical cycloaddition with a C=C π bond in 2,5-dimethyl-2,4-hexadiene (a 4π-cycloaddition) to form a saturated C_4 ring (Fig. 4.38) [17]. The reaction does not take place in the absence of light. However, upon irradiation with UV light, a rapid reaction occurs giving the cycloaddition product in 60% yield.

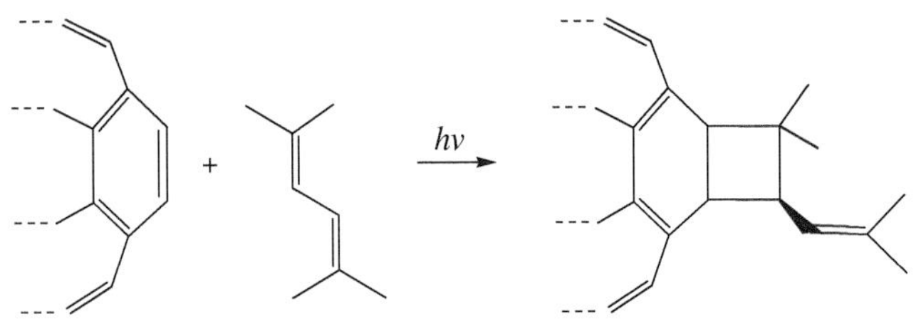

C_{60}

FIGURE 4.38   The 4π-cycloaddition of C_{60} with 2,5-dimethyl-2,4-hexadiene. Based on Vassilikogiannakis et al. [17].

4.6   DIELS–ALDER CYCLOADDITIONS IN WATER: THE GREEN CHEMISTRY METHODS

As described in Section 1.12 (the Green Chemistry Methodology), many organic chemical reactions can be conducted in (or on) water more efficiently than in organic media (or with neat organic reactants) in terms of faster conversions and higher product yields. As a result, the green chemistry methodology has been developed to minimize the utilization of hazardous and toxic organic solvents and to reduce the burden of organic synthesis to the environment. **The principal molecular basis for the green methodology is to form a strong hydrogen bond with the transition state of the reaction occurring in the organic–water interface to stabilize the transition state and lower the activation energy.** Such green methodology has found effective and efficient applications in Diels–Alder reactions.

Research has shown that the Diels–Alder cycloaddition of 1,3-cyclopentadiene with an ethylene derivative below is many times faster performed in water than in organic media [18].

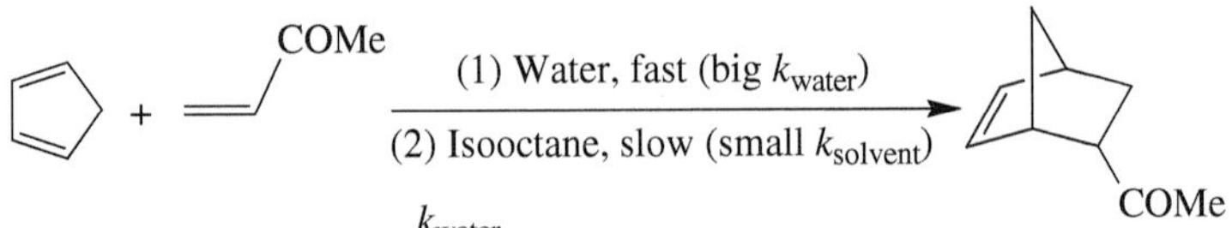

For the above reaction, the $k_{water}/k_{solvent}$ ratios at room temperature for isooctane and acetonitrile are 740 and 290, respectively (k_{water} and $k_{solvent}$ are rate constants for the reactions performed in water and in an organic solvent, respectively). Figure 4.39 shows the FMO interactions for the cycloaddition. It is effected by the interaction of HOMO of the diene (4π) and the LUMO of the dienophile (the ethylene derivative, 2π, the π-electrons in carbonyl disregarded). When the reaction is performed in water, the carbonyl of the ethylene derivative (both the unreacted molecule and the molecule in the transition state) can form a hydrogen bond with water in the

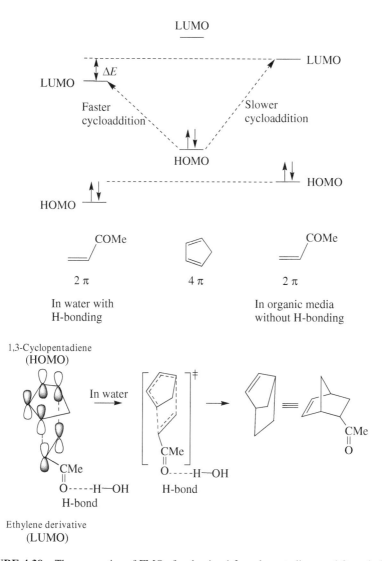

FIGURE 4.39   The energetics of FMOs for the 4 π 1,3-cyclopentadiene and 2 π ethylene derivative in water and in organic media, and their HOMO–LUMO interaction and H-bonding for the Diels–Alder cycloaddition in water.

organic–water interface where the reaction takes place. The H-bonding is equivalent to an electron-withdrawing entity connecting to the carbonyl, lowering the energy levels for both HOMO and LUMO of the ethylene derivative. As a result, the HOMO–LUMO gap between the reactants has decreased (ΔE) relative to that for the reactions conducted in the organic media (without H-bonding). This lowers the activation energy (stabilization of the transition state) for the reaction, making the reaction in water much faster than those performed in organic media.

Figure 4.40 shows the Diels–Alder cycloaddition between a nitrogen-containing 1,3-dipole (C=N$^+$–C:$^-$) (4 π) with styrene (PhCH=CH$_2$, 2 π, the π-electrons in aromatic ring disregarded) [18]. The C=N$^+$–C:$^-$ moiety of the 1,3-dipolar molecule contains 4 π electrons (2 electrons formally contributed from the C=N bond and the other 2 electrons from the carbon lone pair) which are delocalized to form a three-center, four-electron π bond. Its reaction with the C=C π bond in styrene is a thermally allowed 6π-cycloaddition. Research has shown that this reaction, when performed in water, is much faster than in organic solvents (such as acetonitrile). It has been found that the ratio of the rate constant in water (k_{water}) to the rate constant in acetonitrile ($k_{solvent}$), $k_{water}/k_{solvent}$, at room temperature is 15 (Fig. 4.40) [18].

Figure 4.41 shows the FMOs of the 1,3-dipole (4 π molecule) and styrene (2 π molecule) and their interactions giving a Diels–Alder cycloaddition. Since the 4 π 1,3-dipolar molecule contains electronegative nitrogen atoms and cyano groups, its HOMO and LUMO energies are greatly lowered, while the electron-donating Ph in the 2 π styrene enhances the energies for its FMOs. As a result, the 4 π HOMO is lower than the 2 π HOMO, while the 4 π LUMO is located between HOMO and LUMO of the 2 π molecule. This makes the cycloaddition *reverse electron demand*, namely, that the reaction is initiated by the interaction of the 2 π HOMO and 4 π LUMO (Fig. 4.41).

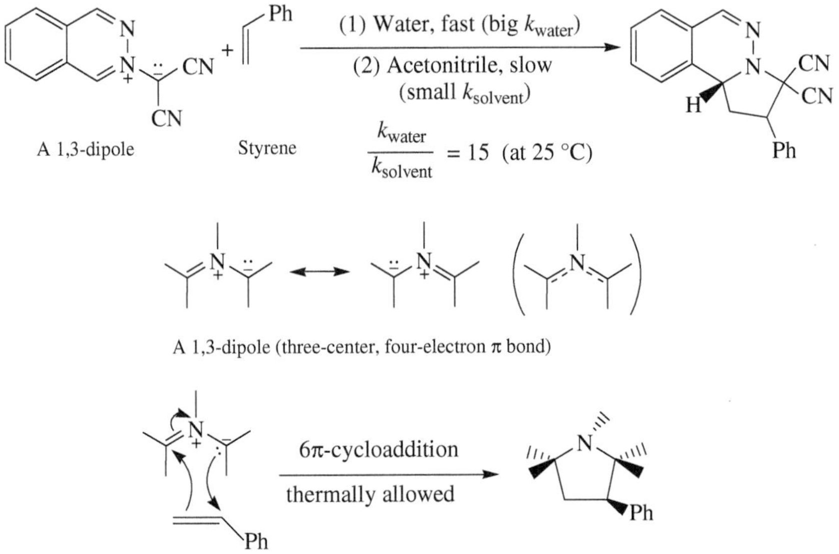

FIGURE 4.40   Comparison of the 6 π Diels–Alder cycloadditions of a 1,3-dipole with styrene in water and in acetonitrile (an organic solvent). Based on Butler et al. [18].

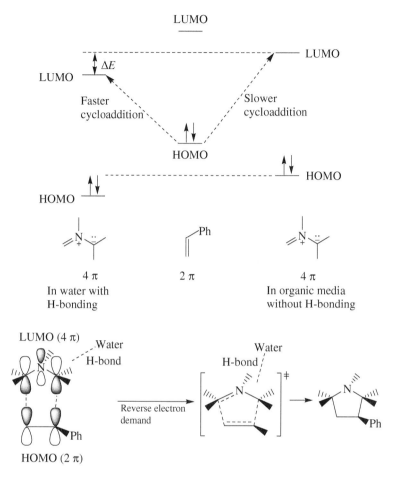

FIGURE 4.41   The energetics of FMOs for a 4 π 1,3-dipole and the 2 π styrene in water and in organic media, and their HOMO–LUMO interaction and H-bonding for the Diels–Alder cycloaddition in water.

When the reaction in Figure 4.41 is performed in water, all the nitrogen atoms in the 4 π 1,3-dipolar molecule (both the unreacted molecule and the molecule in the transition state) can form hydrogen bonds with water in the organic–water interface where the reaction takes place. The H-bonding, equivalent to an electron-withdrawing entity, lowers the energy levels for both HOMO and LUMO of the 4 π molecule, while there is no H-bonding on the 2 π molecule. As a result, the interacting HOMO–LUMO gap between the 2 π and 4 π reactant molecules has decreased (ΔE) relative to that for the reactions conducted in the organic media (without H-bonding). This lowers the activation energy (stabilization of the transition state) and makes the reaction in water much faster than those performed in organic media.

These examples demonstrate that for regular Diels–Alder reactions that take place via the $HOMO_{4\pi}$–$LUMO_{2\pi}$ interactions (normal electron demand), the H-bonding on the 2 π molecule (lowering $LUMO_{2\pi}$) will increase the reaction rate [on the other hand, the H-bonding on the 4 π molecule (lowering $HOMO_{4\pi}$) will decrease the reaction rate]; for the reverse electron demand, Diels–Alder reactions which take place via the $HOMO_{2\pi}$–$LUMO_{4\pi}$ interactions, the H-bonding on the 4 π molecule (lowering $LUMO_{4\pi}$) will increase the reaction rate [the H-bonding on the 2 π molecule (lowering $HOMO_{2\pi}$) will decrease the reaction rate].

Research has also shown that quadricyclane and dimethylazodicarboxylate undergo a cycloaddition resembling the Diels–Alder reaction [19].

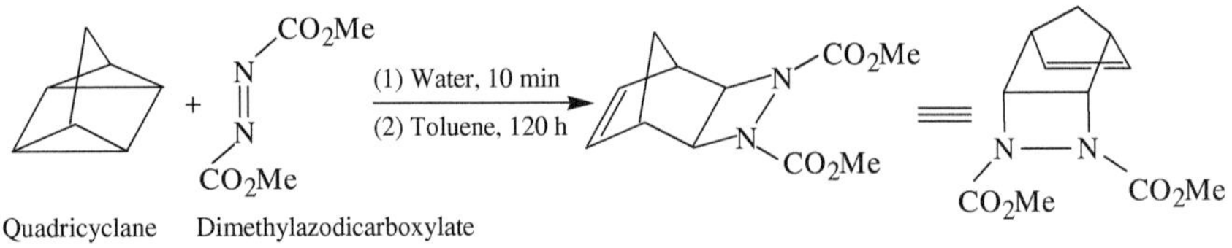

Quadricyclane   Dimethylazodicarboxylate

It only takes 10 min for the above reaction to complete in water as the reaction mixture is vigorously agitated at 23 °C. However, it takes 48 h for the reaction to complete when it is conducted by mixing the neat reactants in the absence of solvent. When the reaction is performed in toluene as a solvent, the completion time is 120 h. When the two reactants are mixed in water with vigorous agitation, the cycloaddition reaction takes place in the organic–water interface. Clearly, water acts as a catalyst to greatly speed up the reaction.

Figure 4.42 illustrates the FMO interactions between quadricyclane (C—C bonds) and dimethylazodicarboxylate (N=N) to lead to the Diels–Alder type of cycloaddition. Different from the common Diels–Alder reactions that usually occur between a 4π-molecule and a 2π-molecule, the reaction of quadricyclane and dimethylazodicarboxylate occurs on two C—C σ bonds in the three-membered carbon rings of quadricyclane. By the end of the reaction, both of the C—C bonds break and a new C=C π bond is formed. The molecular strain in the three-membered carbon rings weakens the C—C bonds and makes the cycloaddition occur. Each of the reacting C—C σ bonds in quadricyclane is formed by the overlap of two sp^3 orbitals. Therefore, a total of four sp^3 orbitals participate in the reaction. The linear combinations of the four sp^3 orbitals give two degenerate HOMOs and two degenerate LUMOs for the reacting C—C bonds. The N=N π HOMO is lower than the C–C HOMOs, while the N=N π LUMO is located between HOMOs and LUMOs of the C—C bonds. Thus, the cycloaddition reaction results from the interaction of the C–C HOMO(1) with the N=N LUMO. The two orbitals are of symmetry match.

As the HOMO(1)–LUMO interaction (overlap) starts happening (Fig. 4.42), all the four sp^3 orbitals in the reacting C—C σ bonds are being transformed into four near-p_π orbitals in the transition state, and the two C—C σ bonds are being partially broken, coincident with the partial formation of a C=C π bond in the transition state. Simultaneously, two near-p_π orbitals in quadricyclane overlap with the N=N π

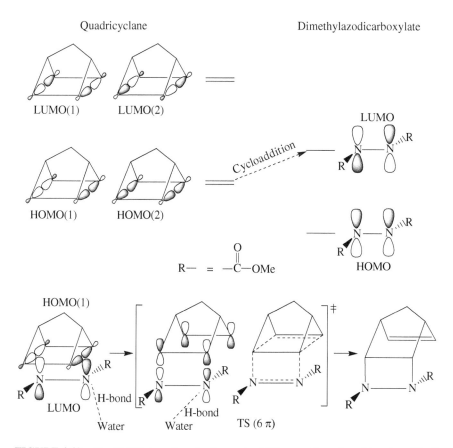

FIGURE 4.42 The FMO interactions for the cycloaddition reaction of quadricyclane (C—C bonds) with dimethylazodicarboxylate (N=N).

LUMO to lead to the partial formations of two N—C σ bonds in the transition state, and the N=N π bond is partially broken. In the transition state, each of the partially developed near-p_π orbitals in quadricyclane contributes one electron, and there are 4 π electrons in quadricyclane. On the other hand, the N=N bond in the transition state contributes 2 π electrons. Therefore, the cycloaddition possesses a 6π-transition state, analogous to the Diels–Alder reaction of a conjugated diene (C=C—C=C, 4π) with an ethylene derivative (C=C, 2π). Overall, the cycloaddition reaction of quadricyclane and dimethylazodicarboxylate (Fig. 4.42) is thermally symmetry-allowed, consistent with the experimental observation.

In dimethylazodicarboxylate both the N=N nitrogens and carbonyl groups (in both the unreacted molecule and in the transition state) can form strong hydrogen bonds with water molecules in the organic–water interface to greatly lower the energies of both HOMO and LUMO of the N=N bond (Fig. 4.42). Therefore, when the cycloaddition is performed in water, the HOMO–LUMO gap between the two reactants becomes very small to greatly reduce the activation energy and stabilize the transition state. As a result, the reaction proceeds very fast in water; while in organic solvents (such as toluene), no H-bond is formed on the N=N molecule and the reaction is very slow.

With employing the green methodology by conducting the organic reactions in water, not only utilization of organic solvents can be avoided but also the reactions can be catalyzed by water and proceed much faster than using the traditional methods.

4.7   BIOLOGICAL APPLICATIONS

4.7.1   Photochemical Synthesis of Vitamin D_2 via a Cyclic Transition State

Pericyclic reactions have practical applications in biochemically related areas. For example, vitamin D_2 (ergocalciferol) is nonenzymatically formed in the skin of animals through photochemical conversion (by UV light) of the steroid ergosterol into the hexatriene precalciferol [20]. It occurs by electrocyclic ring-opening reaction of a conjugated 1,3-cyclohexadiene ring in ergosterol via disrotatory of σ-bonded carbon atoms in the ring (Fig. 4.43). This ring-opening is analogous to the backward ring-opening process of 1,3-cyclohexadiene as shown in Figure 4.34. Precalciferol undergoes a 1,7-hydrogen shift via a cyclic 8 π transition state (photochemically symmetry allowed) to form a different hexatriene, vitamin D_2. This vitamin is also known as ergocalciferol, in recognition of its key role in calcium uptake. Vitamin D_2 can also be synthesized in industry [2] by the photochemical conversion as illustrated in Figure 4.43.

4.7.2   Ribosome-Catalyzed Peptidyl Transfer via a Cyclic Transition State: Biosynthesis of Proteins

In the living systems, DNA and RNA work together to direct biosynthesis of proteins (polypeptides). The overall process includes gene transcription (synthesis of an mRNA chain directed by the starting DNA sequence) and translation (making polypeptide bonds through tRNAs) [20]. The mRNA chain, after being synthesized, connects to multiple tRNAs via the complementary nitrogen base pairing. Each tRNA molecule links to a specific α-amino acid residue on its 3′-end ribose sugar (Fig. 4.44). Two tRNAs, along with the amino acid residues connected to them, go into the structure of a ribosome where a peptide bond is made between the two amino acid residues (Fig. 4.44).

The formation of a peptide bond between two α-amino acid residues is catalyzed by ribosome, and its detailed mechanism is illustrated in Figure 4.44 [20]. The beginning $tRNA_1$ and incoming $tRNA_2$, along with their amino acid residues, are bound to P site and A site in ribosome, respectively. With the help of the ribosome structure,

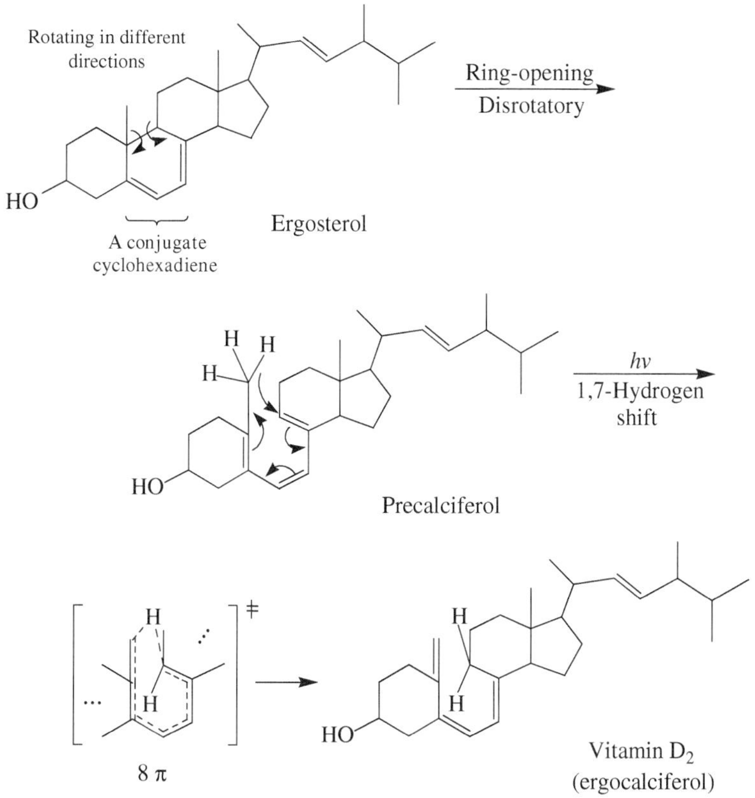

FIGURE 4.43   Chemical synthesis of Vitamin D_2 via the ring-opening process of a 1,3-cylohexadiene derivative.

the N—H bond in A site is oriented to and approaches the OH oxygen of $tRNA_1$ in P site, and an O...H...N hydrogen bond is formed. This H-bond weakens the N—H bond and enhances the nucleophilicity of the NH_2 nitrogen in A site. It attacks the electrophilic carbonyl of the amino acid residue in P site to make a peptide (C–N) bond and to cleave the C—O bond simultaneously. As the C—O bond is being cleaved, the oxygen is protonated by the adjacent OH group from the sugar. The overall conversion is concerted, involving transfer of **10 electrons** as indicated by the five arrows in the figure and leading to the formation of **a cyclic transition state that contains six partially bonded atoms in a ring**. The collapse of the transition state leads to the formation of a peptide bond and to making a dipeptidyl on $tRNA_2$ in A site. Then the free $tRNA_1$ moves out of the ribosome and $tRNA_2$ moves into the P site with the newly formed dipeptidyl from A site. The third $tRNA_3$ then moves into the vacant A site with an amino acid residue attached to it. The dipeptidyl of $tRNA_2$ in P site now is transferred to the $tRNA_3$ in A site by forming a peptide

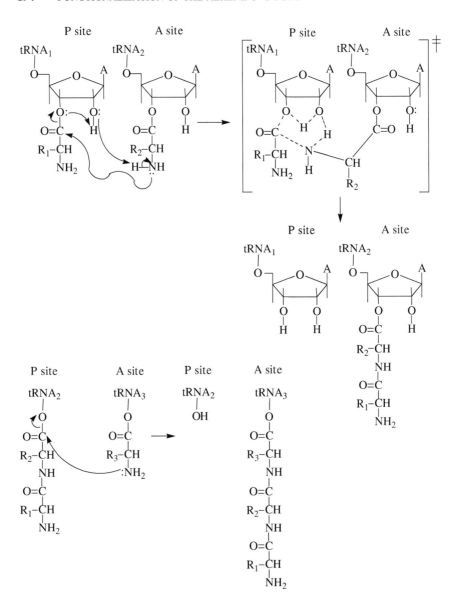

FIGURE 4.44 The ribosome-catalyzed formation of the peptide bond (C–N) in biosynthesis of proteins that takes place via a 10-electron, 6-membered cyclic transition state. The transition state contains six partially bonded atoms linked by dotted lines.

bond via a cyclic transition state in the same manner. This way a polypeptide is made eventually from the N-terminus to the C-terminus via the repeating concerted peptidyl transfers catalyzed by ribosome.

The formation of each peptide bond (the peptidyl transfer) takes place via a **stabilized ten-electron [(4n + 2), n = 2], six-membered cyclic transition state**. The reaction possesses a relatively low activation energy.

PROBLEMS

4.1 Write out the mechanism for reaction of α-pinene with CH_2I_2 in the presence of zinc. Account for the stereochemistry for the reaction.

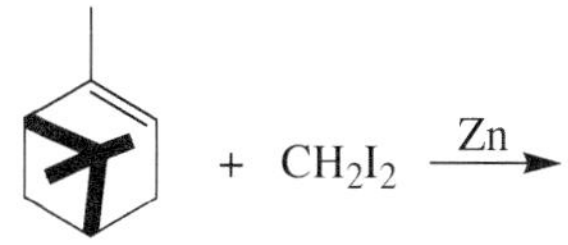

α-Pinene

4.2 Using curved arrows, show electron transfers in the Diels–Alder reaction of 1,3-butadiene ($CH_2=CHCH=CH_2$) with 3,3,3-trifluoropropyne ($HC\equiv CCF_3$). Is this reaction slower or faster than the reaction of 1,3-butadiene with propyne? Give the reason.

4.3 In the following reaction, the N=N π bond (equivalent to a C=C π bond) undergoes a cycloaddition to the conjugate diene which is analogous to the Diels–Alder reaction. Using curved arrows, show electron transfers in the reaction. Also show the frontier molecular orbital interactions that lead to the cycloaddition.

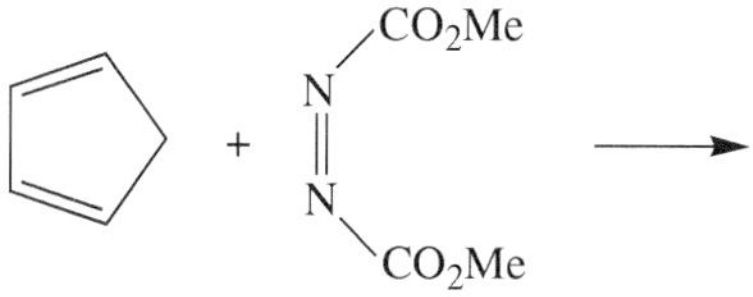

4.4 Using curved arrows, show electron transfers in the following cycloaddition reaction. How many electrons are involved in the reaction? Is the cycloaddition thermally allowed or photochemically allowed. Give the reason.

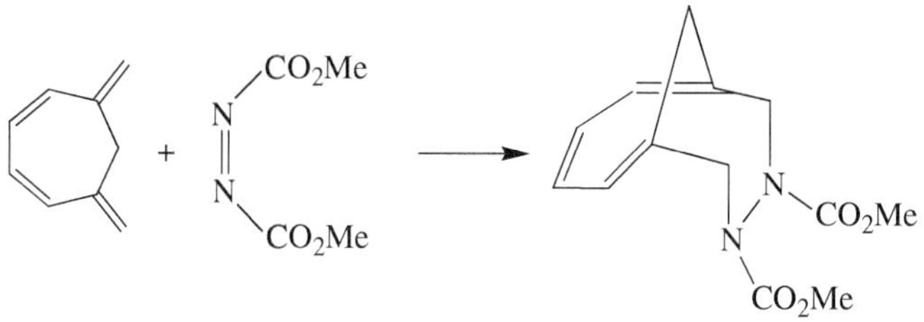

4.5 As shown below, 7-methyl-1,3,5-cycloheptatriene can undergo both 1,2-shift of hydrogen and 1,4-shift of hydrogen. One is thermally allowed. The other is photochemically allowed, but thermally forbidden.

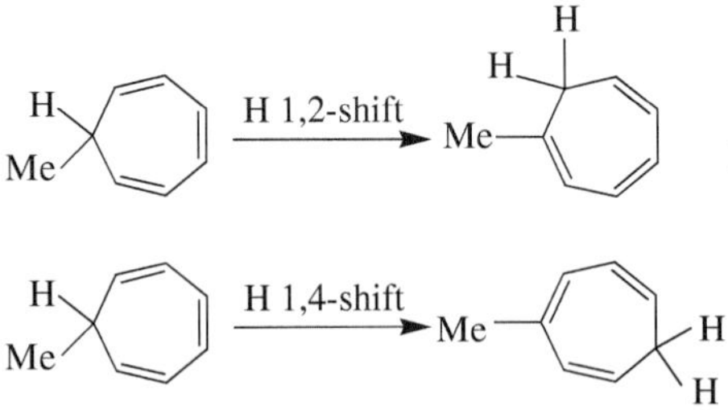

7-Methyl-1,3,5-cycloheptatriene

Using curved arrows, show electron transfers in both 1,2- and 1,4-shift of hydrogen. How many electrons are involved in each of the reactions? Which reaction is thermally allowed, and which is photochemically allowed? Give the reasons.

4.6 As shown below, 2,4,6,8-decatetraene can undergo electrocyclic reaction giving either a *cis*- or *trans*-product. One is thermally allowed. The other is photochemically allowed, but thermally forbidden.

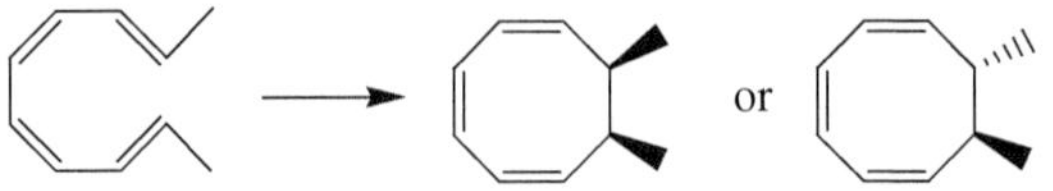

2,4,6,8-Decatetraene   *cis*   *trans*

By the aid of molecular orbital analysis, show mechanisms for formation of both *cis*- and *trans*-products. Which reaction is thermally allowed, and which is photochemically allowed? Give the reasons.

4.7 In the presence of SbF$_5$, the following chlorocyclopropane derivative undergoes a ring-opening to give an allylic cation.

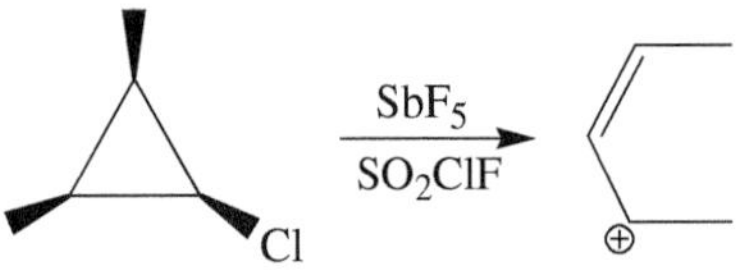

Provide a mechanism for the change and explain why the ring-opening process is thermally allowed.

4.8 Consider the following reaction.

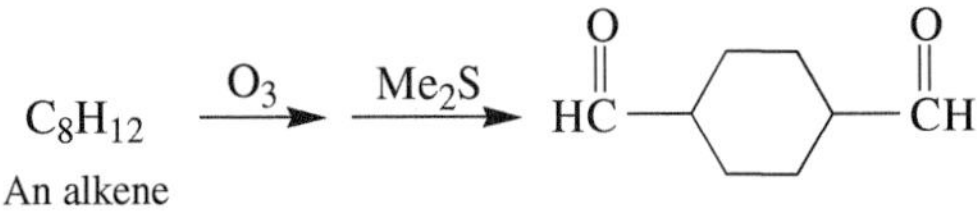

C$_8$H$_{12}$
An alkene

What is the structure of the starting alkene C_8H_{12}? Provide the mechanism for cycloaddition of O_3 to the alkene molecule by showing the electron transfers and frontier molecular orbital interactions that lead to the cycloaddition. Is the product in *cis*- or *trans*-configuration? Explain and draw the most stable chair conformation of the product.

4.9 As demonstrated in Section 4.4.3, the linear dithionitronium NS_2^+ ion undergoes general thermally symmetry allowed cycloaddition reactions with nitriles (RCN). Cyanogen $(CN)_2$ has been shown to undergo cycloaddition reactions with NS_2^+ on both of the –C≡N groups consecutively. Write out both reactions and indicate the electron transfers using curved arrows. Which cycloaddition reaction goes faster, the first or the second? Explain. NS_2^+ can also react with *p*-Ph$(CN)_2$. Is this reaction faster or slower than the $(CN)_2$ reaction? Explain.

REFERENCES

1. Bruckner, R. *Advanced Organic Chemistry: Reaction Mechanisms*, Chapter 12, pp. 477–518, Harcourt/Academic Press, Orlando, FL, USA (2002).

2. Fox, M. A.; Whitesell, J. K. *Core Organic Chemistry*, 2nd ed., Jones and Bartlett, Sudbury, MA, USA (1997).

3. Huheey, J. E.; Keiter, E. A.; Keiter, R. L. *Inorganic Chemistry: Principles of Structure and Reactivity*, 4th ed., Harper Collins, New York, NY, USA (1993).

4. Jackson, R. A. *Mechanisms in Organic Reactions*, The Royal Society of Chemistry, Cambridge, UK (2004).

5. Hoffman, R. V. *Organic Chemistry: An Intermediate Text*, 2nd ed., Wiley, Hoboken, NJ, USA (2004).

6. Alves, M. J.; Gilchrist, T. L. Generation and Diels–Alder Reactions of *t*-Butyl 2*H*-Azirine-3-carboxylate. *Tetrahedron Lett.* 1998, *39*, 7579–7582.

7. Marchand, A. P.; Ganguly, B.; Watson, W. H.; Bodige, S. G. Thermodynamic *vs.* Kinetic Control in the Diels–Alder Cycloaddition of Cyclopentadiene to 2,3-Dicyano-*p*-benzoquinone: Kinetic Control Revisited. *Tetrahedron*, 1998, *54*, 10967–10972.

8. Murakata, M.; Kawase, A.; Kimura, N.; Ikeda, T.; Nagase, M.; Koizumi, M.; Kuwata, K.; Maeda, K.; Shimizu, H. Synthesis of Tofogliflozin as an SGLT2 Inhibitor via Construction of Dihydroisobenzofuran by Intramolecular [4 + 2] Cycloaddition. *Org. Process Res. Dev.* 2019, *23*, 548–557.

9. Brown, W. H.; Foote, C. S.; Iverson, B. L.; Anslyn, E. V. *Organic Chemistry*, 6th ed., Brooks/Cole, Belmont, CA, USA (2012).

10. Martin-Vila, M.; Hanafi, N.; Jimenez, J. M.; Alvarez-Larena, A.; Piniella, J. F.; Branchadell, V.; Oliva, A.; Ortuno, R. M. Controlling π-Facial Diastereoselectivity in the 1,3-Dipolar Cycloadditions of Diazomethane to Chiral Pentenoates and Furanones: Enantioselective Stereodivergent Syntheses of Cyclopropane Hydroxy Acids and Didehydro Amino Acids. *J. Org. Chem.* 1998, *63*, 3581–3589.

11. Parsons, S.; Passmore, J. Rings, Radicals, and Synthetic Metals: The Chemistry of SNS$^+$. *Acc. Chem. Res.* 1994, *27*, 101–108.

12. Brooks, W. V. F.; Brownridge, S.; Passmore, J.; Schriver, M. J.; Sun, X. Preparation and Characterization of 1,3,2-Dithiazolidine and 1,4-Dithia-7-Azabicyclo[2.2.1]heptanes Cations, and a Mechanistic Study of the Cycloaddition Reactions of Alkenes with SNS⁺. *J. Chem. Soc., Dalton Trans.* 1996, *9*, 1997–2009.

13. Parsons, S.; Passmore, J.; Schriver, M. J.; Sun, X. Quantitative Preparations of 1,3,2-Dithiazolium, 1,3,2,4-Dithiadiazolium, and 1-Halo-1,2,4,3,5-Trithiadiazolium Cations by the General, Symmetry-Allowed Cycloadditions of SNSAsF₆ with Alkynes, Nitriles, and Thiazyl Halides. Kinetics of Cycloadditions of the 1,3-Dipolar-like SNS⁺ with Aliphatic and Aromatic Nitriles and Alkynes. *Inorg. Chem.* 1991, 30, 3342–3348.

14. Huisgen, R. *1,3-Dipolar Cycloaddition Chemistry*, Padwa, A. Ed., Vol. *1*, Chapter 1, pp. 1–176, Wiley, New York, NY, USA (1984).

15. Silbey, R. J.; Alterty, R. A.; Bawendi, M. G. *Physical Chemistry*, 4ᵗʰ ed., Wiley, Danvers, MA, USA (2005).

16. Exner, K.; Prinzbach, H. Highly Efficient Photometathesis in a Proximate, Synperiplanar Diazene–Diazene Oxide Substrate: Retention of Optical Purity, Mechanistic Implications. *Chem. Commun.* 1998, *7*, 749–750.

17. Vassilikogiannakis, G.; Chronakis, N.; Orfanopoulos, M. A New [2 + 2] Functionalization of C₆₀ with Alkyl-Substituted 1,3-Butadienes: A mechanistic Approach. Stereochemistry and Isotope Effects. *J. Am. Chem. Soc.* 1998, *120*, 9911–9920.

18. Butler, R. N.; Coyne, A. G. Water: Nature's Reaction Enforcer—Comparative Effects for Organic Synthesis "In-Water" and "On-Water". *Chem. Rev.* 2010, *110*, 6302–6337.

19. Klijn, J. E.; Engberts, J. B. F. N. Fast reactions 'on water'. *Nature*, 2005, *435*, 746–747.

20. Voet, D.; Voet, J. G.; Pratt, C. W. *Fundamentals of Biochemistry*, 5ᵗʰ ed., Wiley, Hoboken, NJ, USA (2016).

5

THE AROMATIC C—H BOND FUNCTIONALIZATION AND RELATED REACTIONS

Aromatic hydrocarbons (arenes) and their derivatives are characterized by electrophilic substitution reactions of the C—H bond, including (but not limited to) nitration effected by the nitronium NO_2^+ ion (or a nitrating agent), halogenations effected by Cl^+ or Br^+, Friedel–Crafts alkylation and acylation effected by a carbocation and the acylium ($RC^+=O \leftrightarrow RC=O^+$) ion, respectively, and sulfonation effected by SO_3 [1–3]. In each type of the reactions, the aromatic ring (electron-rich due to the conjugated π electron cloud) is attacked by a reactive electrophile (E^+, electron-deficient) giving an intermediate arenium ion (a σ-complex) [4]. It is commonly the relatively **slow**, reversible, rate-determining step. The arenium intermediate is highly energetic and unstable because of disruption of the original aromatic ($4n + 2$) π (typically, $n = 1$) system. It undergoes subsequent **fast** elimination of proton (H^+) (deprotonation) spontaneously, resulting in the formation of a functionalized aromatic compound with a new ($4n + 2$) π system established. The overall process is illustrated in Figure 5.1.

Most recent research has shown that for some electrophiles, in particular the nitronium NO_2^+ ion, a charge-transfer complex (π-complex) between the aromatic substrate (ArH) and NO_2^+, and an ion-radical pair [$ArH^{+\cdot}$, $NO_2^{\cdot}$] are formed prior to the formation of the intermediate arenium [$Ar(H)NO_2$]$^+$ ion (σ-complex) in the course of aromatic nitration [5]. The findings have furnished the chemistry of conventional electrophilic aromatic substitution (EAS) with new insights. The full mechanism will be presented. The EAS reactions with some structurally related electrophiles

Organic Mechanisms: Reactions, Methodology, and Biological Applications,
Second Edition. Xiaoping Sun.
© 2021 John Wiley & Sons, Inc. Published 2021 by John Wiley & Sons, Inc.
Companion website: www.wiley.com/go/Sun/OrgMech_2e

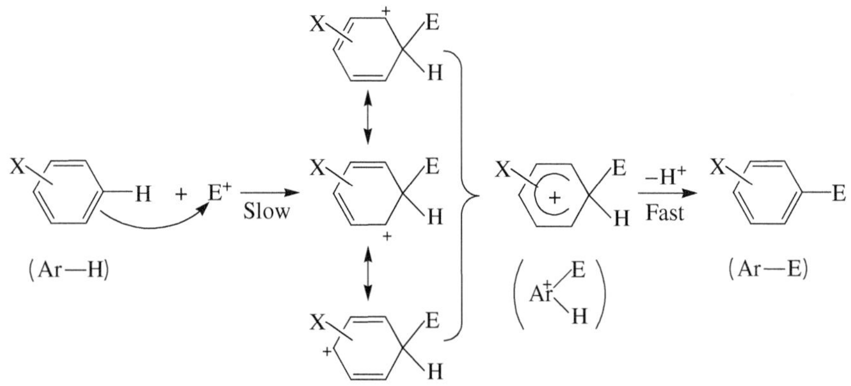

X: A substituent

E^+ = An electrophile such as NO_2^+, Cl^+, Br^+, R^+, RC^+=O, SO_3, etc.

FIGURE 5.1 General mechanism for the electrophilic aromatic substitution (EAS) reactions.

will be discussed. Various EAS reactions which possess special and sophisticated mechanistic aspects and major synthetic applications will receive much attention in this chapter. Biological applications of functionalized aromatic compounds will be described.

Nucleophilic aromatic substitution (NAS) reactions represent major significance in functional group transformations in aromatic compounds. Mechanisms and synthetic applications of this type of reactions will also be addressed in this chapter.

5.1 AROMATIC NITRATION: ALL REACTION INTERMEDIATES AND FULL MECHANISM FOR THE AROMATIC C–H BOND SUBSTITUTION BY NITRONIUM (NO₂⁺) AND RELATED ELECTROPHILES

Among various EAS processes, aromatic nitration has been of particular wide interest owing to the early unambiguous identification of the nitronium NO_2^+ ion as the reactive electrophile from nitric acid. Traditionally, it was thought that the first step of aromatic nitration involved direct addition of electrophilic NO_2^+ ion to an aromatic ring giving an arenium intermediate (Eq. 5.1) [5].

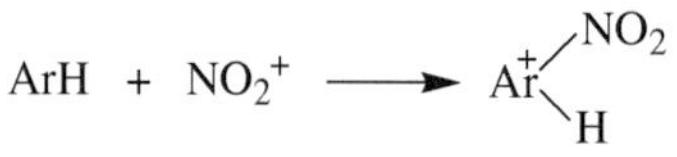

$$\text{ArH} + NO_2^+ \longrightarrow Ar\overset{+}{\underset{H}{\diagup}}\overset{NO_2}{}$$

(5.1)

However, most recent experimental and theoretical studies have revealed an alternative, **stepwise mechanism** which occurs via consecutive **single-electron transfers** from the arene substrate to NO_2^+ (Eq. 5.2) [5].

$$ArH + \underbrace{NO_2^+}_{e} \longrightarrow [ArH^{+\cdot}, NO_2^{\cdot}] \xrightarrow{\text{Fast}} Ar\overset{+}{\underset{H}{\overset{NO_2}{<}}} \tag{5.2}$$

An ion-radical pair

The formation of the intermediate ion-radical pair has been confirmed by ultrafast time-resolved absorption spectroscopy [5–7]. In this section, we give a brief account for all the observed reaction intermediates and full mechanism for the **charge-transfer** aromatic nitration.

5.1.1   Charge-Transfer Complex [ArH, NO_2^+] Between Arene and Nitronium

The electrophilic NO_2^+ and nitronium carriers (nitrating agents) NO_2–Y, such as nitric acid (–Y = –OH), acetyl nitrate (–Y = –OAc), dinitrogen pentoxide (–Y = –ONO_2), nitryl chloride (–Y = –Cl), N-nitropyridinium (–Y = –Py), and tetranitromethane [–Y = –$C(NO_3)_3$], are all electron-deficient and thus capable of serving as effective electron acceptors. The initial mixture of each nitrating agent NO_2–Y with an electron-donating arene (such as benzene, toluene, pentamethylbenzene, naphthalene, or hexamethylbenzene) exhibits characteristic UV–Vis absorption bands attributable to a single-electron transfer (charge-transfer) from the highest occupied molecular orbital (HOMO) of the arene to the lowest occupied molecular orbital (LUMO) of NO_2–Y (or NO_2^+). For a given NO_2–Y, the charge-transfer energy ($h\nu_{CT}$) is shown to be directly proportional to the ionization potential (IP) of the arene (E_{HOMO} = –IP). The observation is diagnostic of the very rapid formation of a π-complex (also called charge-transfer complex) in the first step of aromatic nitration (Eq. 5.3) [5].

$$ArH + NO_2Y \text{ (or } NO_2^+) \rightleftharpoons [ArH, NO_2Y] \text{ or } [ArH, NO_2^+]$$

A π-complex
(charge-transfer complex)

$$\tag{5.3}$$

The π-complex is weak, and formed via an **intermolecular interaction** between the arene conjugated π bond and the electrophilic NO_2Y or NO_2^+ (π–$\pi*$ interaction). In general, the intermediate π-complex formed in the course of an aromatic nitration is fleeting (reactive) and quickly transformed into an ion-radical pair [$ArH^{+\cdot}$, $NO_2^{\cdot}$], the formation of which will be accounted for below.

5.1.2   Ion-Radical Pair [ArH$^{+\cdot}$, NO$_2$$^{\cdot}$]

Picosecond time-resolved spectroscopy has enabled characterization of electron-transfer process in the intermediate π-complex [ArH, NO$_2$Y] and identification (capture) of the short-lived ion-radical pair [ArH$^{+\cdot}$, NO$_2$$^{\cdot}$]. The π-complex [ArH, NO$_2$Y], once formed, is excited (activated) subsequently by the specific irradiation ($h\nu_{CT}$) of the corresponding charge-transfer band. As a result, a single-electron transfer from ArH (electron-donor) to NO$_2$Y (electron-acceptor) takes place within the π-complex giving an intermediate ion-radical pair [ArH$^{+\cdot}$, NO$_2$$^{\cdot}$] (Eq. 5.4) [5–7].

$$
[ArH, NO_2Y] \xrightarrow{h\nu_{CT}} [ArH^{+\cdot}, NO_2^{\cdot}]Y^-
$$

A π-complex   An ion-radical pair   (5.4)

The ion-radical pair consists of a cationic arene ArH$^{+\cdot}$ radical and a nitrogen dioxide NO$_2$$^{\cdot}$ radical, which are in close contact once formed, and extremely reactive toward each other due to the facile combination of two single electrons forming a σ-bond. The short-lived ion-radical pair is observed by picosecond absorption spectroscopy.

Activation of the various π-complexes [ArH, NO$_2$Y] can also be thermally induced to effect aromatic nitration. For example, mixing naphthalene (ArH) with PyNO$_2$$^+$ in acetonitrile immediately exhibits a bright lemon color attributable to the formation of a charge-transfer complex [ArH, PyNO$_2$$^+$], which has been confirmed by UV–Vis spectroscopic characterization [5]. The color slowly fades spontaneously within an hour to afford a mixture of 1- and 2-nitronaphthalene (ArNO$_2$).

5.1.3   Arenium [Ar(H)NO$_2$]$^+$ Ion

As aforementioned, the ion-radical pair [ArH$^{+\cdot}$, NO$_2$$^{\cdot}$] is short-lived due to ultrafast combination between the two radicals via the single electrons. Picosecond time-resolved spectroscopic studies have shown that the ion-radical pair collapses rapidly with a first-order rate-law to the arenium [Ar(H)NO$_2$]$^+$ ion, a σ-complex (Eq. 5.5).

[ArH$^{+\cdot}$, NO$_2$$^{\cdot}$]   [Ar(H)NO$_2$]$^+$ (5.5)

An ion-radical pair   A σ-complex (arenium)

This first-order collapse is shown to be much faster than a diffusive separation of the ion-radical pair [5]. Therefore, the ion-radical pair [ArH$^{+\cdot}$, NO$_2$$^{\cdot}$] is transformed into a σ-complex $[Ar(H)NO_2]^+$ exclusively prior to possible separation of the two radicals, giving rise to an essentially quantitative nitration with no side products.

The σ-complex is transient and undergoes rapid elimination of a proton (H^+) (deprotonation) giving the nitroarene product (Eq. 5.6).

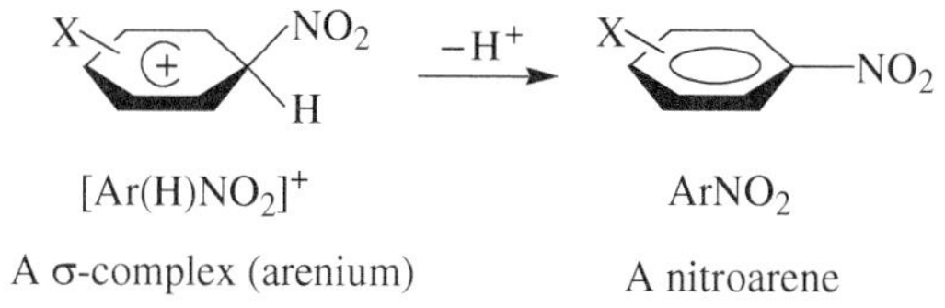

$$[Ar(H)NO_2]^+ \qquad\qquad\qquad ArNO_2$$

A σ-complex (arenium)   A nitroarene

$$(5.6)$$

The formation of a σ-complex $[Ar(H)NO_2]^+$ in the course of nitration is further confirmed by direct observation of the arenium ion formed between hexamethylbenzene (C_6Me_6) and NO_2^+ (Fig. 5.2) using ^1H and ^{13}C NMR spectroscopy [4, 8]. Since the aromatic ring in C_6Me_6 does not contain any hydrogen, the arenium ion (σ-complex) is relatively stable. The formation of a readily observable arenium ion between C_6Me_6 and NO_2^+ implies that in general, aromatic nitration and other types of EAS reactions occur through a σ-complex.

5.1.4   Full Mechanism for Aromatic Nitration

In the course of aromatic nitration effected by either the nitronium NO_2^+ ion or a nitrating agent $NO_2–Y$, three metastable reaction intermediates, a π-complex (charge-transfer complex) {$[ArH, NO_2Y]$ or $[ArH, NO_2^+]$}, an ion-radical pair $[ArH^{+\cdot}, NO_2{^\cdot}]$, and a σ-complex (arenium ion) $[Ar(H)NO_2]^+$, have been found. The full mechanism for the overall reaction is summarized in Figure 5.3.

The first step in the mechanism is the rapid formation (reversible) of a weak π-complex $[ArH, NO_2Y]$ or $[ArH, NO_2^+]$ through the intermolecular interaction of an arene (electron-donor) and a nitrating agent NO_2Y or NO_2^+ (electron-acceptor). Since the intermolecular π-complex exhibits a charge-transfer absorption which originates from a single-electron transfer from the arene to NO_2Y or NO_2^+, it is also called charge-transfer complex. The π-complex is unstable and can be readily

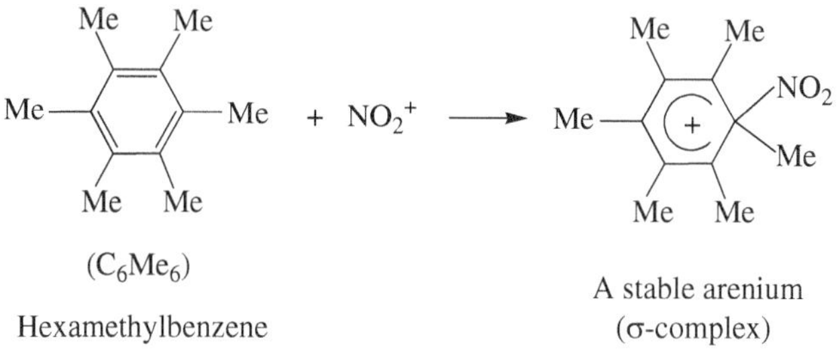

(C_6Me_6)   A stable arenium

Hexamethylbenzene   (σ-complex)

FIGURE 5.2   Formation of a stable arenium (σ-complex) between hexamethylbenzene (C_6Me_6) and nitronium (NO_2^+). Olah et al. [4, 8].

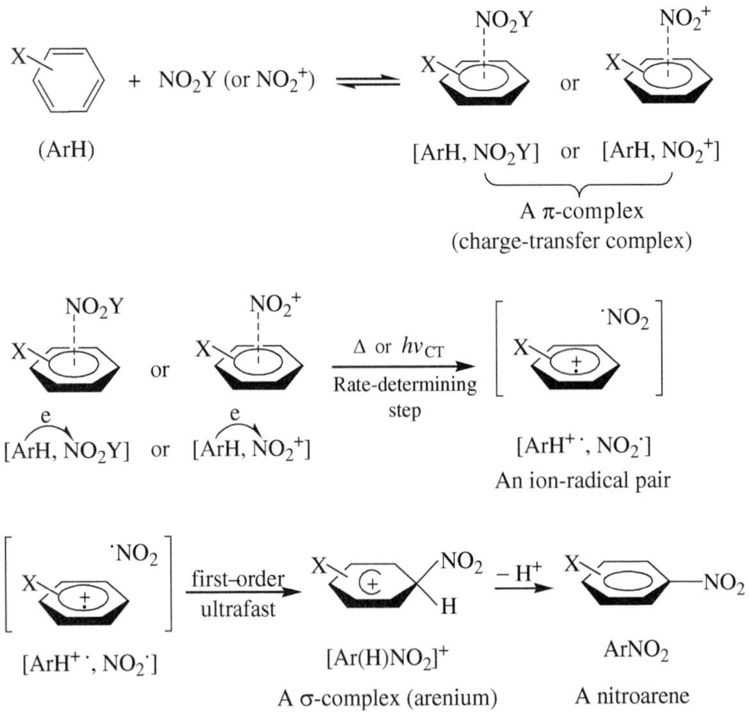

FIGURE 5.3 The full mechanism for the charge-transfer aromatic nitration.

excited (activated) thermally or photochemically. Once excited, an inner-sphere electron transfer from the arene to NO_2Y or NO_2^+ takes place within the π-complex giving an ion-radical pair $[ArH^{+\cdot}, NO_2^{\cdot}]$. This is an **irreversible, rate-determining step** which involves a single-electron transfer [5]. The ion-radical pair $[ArH^{+\cdot}, NO_2^{\cdot}]$ is transient and undergoes ultrafast (on the picosecond-timescale) first-order decay to a σ-complex (arenium ion) $[Ar(H)NO_2]^+$ due to the facile combination of the single electrons. The σ-complex $[Ar(H)NO_2]^+$ in turn transforms rapidly into the nitroarene product $ArNO_2$. Since the overall reaction involves a single-electron transfer (charge-transfer) from the arene (HOMO) to the nitrating agent (LUMO), it is referred to as charge-transfer nitration.

5.2 MECHANISMS AND SYNTHETIC UTILITY FOR AROMATIC C–H BOND SUBSTITUTIONS BY OTHER RELATED ELECTROPHILES

The dithionitronium NS_2^+ ion NS_2^+ (or $[S=N=S]^+$) is structurally analogous to NO_2^+ [9]. Similar to NO_2^+, NS_2^+ is electron-deficient and can serve as an electron-acceptor. However, different from NO_2^+, the terminal sulfur atoms in NS_2^+ are less electronegative than the central nitrogen atom. The difference in electronegativities makes the LUMO of NS_2^+ (reacting MO) sulfur based (refer to Figure 4.28 for FMOs of NS_2^+). As a result, reactions of NS_2^+ with a π bond (electron-rich) occur on a sulfur atom (Section 4.4.3). This is also true for the electrophilic substitution reaction of NS_2^+ with benzene [10].

Similar to aromatic nitration, the initial mixture of NS_2^+ with an arene (such as benzene, toluene, *t*-butylbenzene or pentamethylbenzene) exhibits characteristic UV–Vis absorption bands attributable to the rapid formation of a π-complex (charge-transfer complex) [ArH, SNS]+. Then [ArH, SNS]+ is spontaneously trans-formed into [ArSNSH]+, a substitution product formed by a net insertion of SNS+ into the aromatic C—H bond [10]. The reaction takes place via a σ-complex (Fig. 5.4).

Figure 5.4 shows the mechanism for electrophilic substitution of the dithionitro-nium NS_2^+ ion with benzene [10]. The first step is the rapid formation of an inter-mediate π-complex $[C_6H_6, SNS]^+$ through the weak interaction of benzene HOMO and SNS+ LUMO. Then the π-complex $[C_6H_6, SNS]^+$ transforms spontaneously into a transient σ-complex $[C_6H_5(H)SNS]^+$ with a C—S bond formed. Different from the σ-complex $[Ar(H)NO_2]^+$ formed in an aromatic nitration, the σ-complex $[C_6H_5(H)SNS]^+$ undergoes an intramolecular hydrogen transfer (a concerted transfer of the *ipso*-H in the benzenium ring to the terminal sulfur atom in the –SNS moiety) giving the final substitution product $[C_6H_5SNSH]^+$. The formation of an intermediate σ-complex $[C_6H_5(H)SNS]^+$ in the course of the EAS reaction is confirmed by direct observation of the arenium ion formed between hexamethylbenzene (C_6Me_6) and SNS+ (Fig. 5.5) using ^1H and ^{13}C NMR spectroscopy [10]. Overall, the $[S=N=S]^+$ ion has undergone an insertion reaction into the C—H bond of benzene following a general EAS mechanism. The reaction gives a product that is different from that of $[O=N=O]^+$ (aromatic nitration).

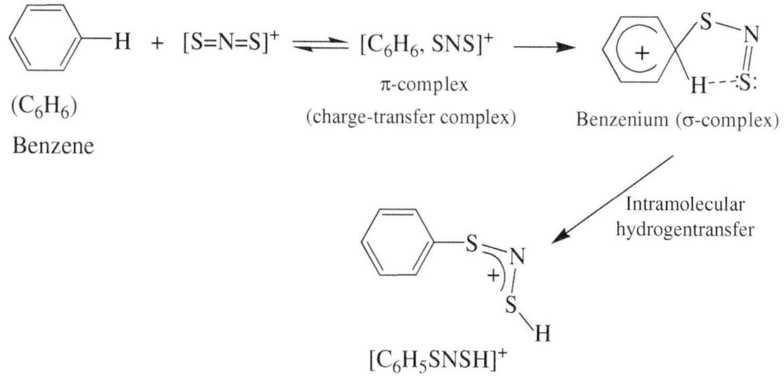

FIGURE 5.4   Mechanism for the electrophilic substitution reaction of benzene with the dithionitronium (NS_2^+) ion.

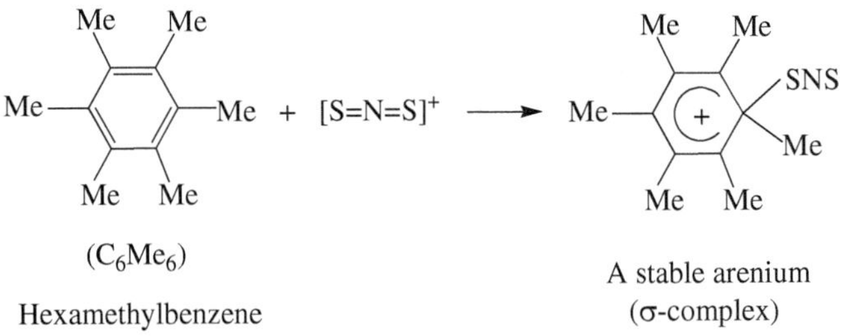

(C$_6$Me$_6$)

Hexamethylbenzene

A stable arenium
(σ-complex)

FIGURE 5.5 Formation of a stable arenium (σ-complex) between hexamethylbenzene (C$_6$Me$_6$) and dithionitronium (NS$_2^+$).

The Cl$_2$SO$^+$–$^-$AlCl$_3$ (thionyl chloride—aluminum chloride) zwitterionic adduct
Thionyl chloride has been shown to combine with aluminum chloride readily via an O→Al coordination bond forming a 1:1 zwitterionic adduct Cl$_2$SO$^+$–$^-$AlCl$_3$ (Eq. 5.7) [11].

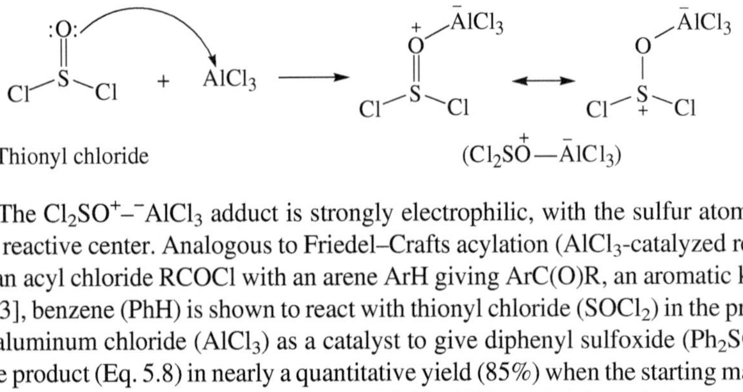

Thionyl chloride

(Cl$_2$S$\overset{+}{O}$—$\overset{-}{A}$lCl$_3$)

(5.7)

The Cl$_2$SO$^+$–$^-$AlCl$_3$ adduct is strongly electrophilic, with the sulfur atom being the reactive center. Analogous to Friedel–Crafts acylation (AlCl$_3$-catalyzed reaction of an acyl chloride RCOCl with an arene ArH giving ArC(O)R, an aromatic ketone) [2, 3], benzene (PhH) is shown to react with thionyl chloride (SOCl$_2$) in the presence of aluminum chloride (AlCl$_3$) as a catalyst to give diphenyl sulfoxide (Ph$_2$SO) as a sole product (Eq. 5.8) in nearly a quantitative yield (85%) when the starting materials are mixed in the molar ratio of PhH:SOCl$_2$:AlCl$_3$ = 2:1:1, and the reaction is performed at 70 °C with granular AlCl$_3$ added piecewise to a mixture of PhH and SOCl$_2$ [11].

$$2\ \text{PhH} + \text{SOCl}_2 \xrightarrow[70\ °C]{\text{AlCl}_3} \text{Ph}_2\text{SO} + \text{HCl} \qquad (5.8)$$

The reaction is believed to take place via the electrophilic Cl$_2$SO$^+$–$^-$AlCl$_3$ adduct (Fig. 5.6) [11].

Ph$_2$SO is an important fundamental synthetic reagent. For example, it has been used effectively in catalytic oxidation of various alkyl sulfides (RSR$'$) to sulfoxides (RSOR$'$) [12]. The above reaction represents a simple and very efficient method to synthesize this important organic reagent.

At lower temperatures (0–25 °C), the reaction of PhH with SOCl$_2$ in the presence of AlCl$_3$ is found to give a mixture of Ph$_2$SO and reduced Ph$_2$S (diphenyl sulfide) in

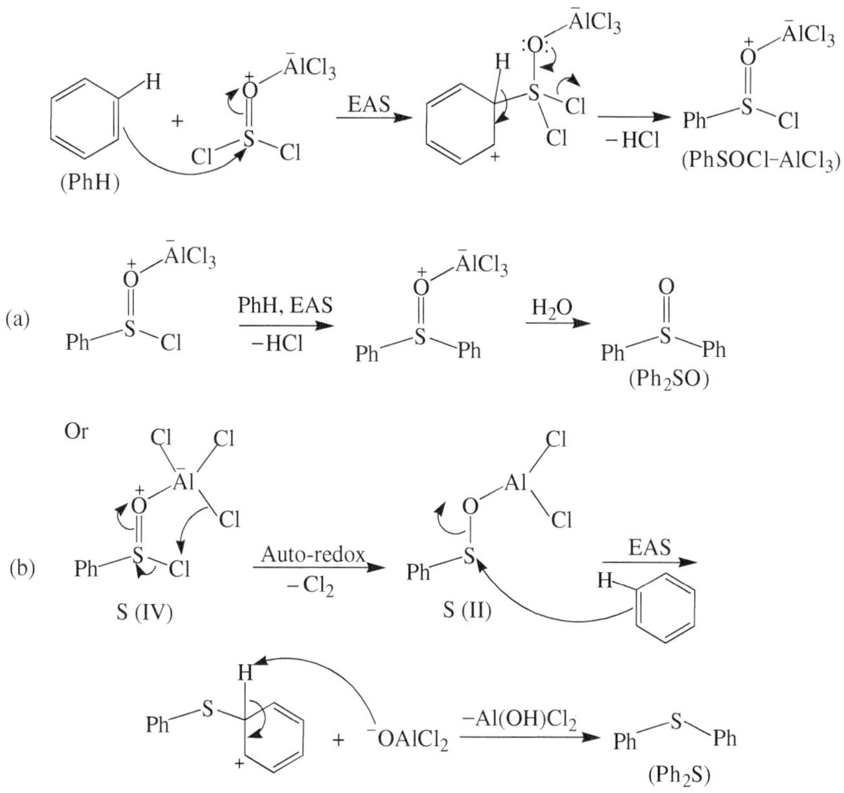

FIGURE 5.6 Possible mechanism for the AlCl$_3$-catalyzed electrophilic substitution reaction of benzene with thionyl chloride (SOCl$_2$). (a) EAS reaction and (b) Auto-redox reaction. Reprinted from Ref. [11] by permission of the publisher (Taylor & Francis, http://www.tandf.co.uk/journals).

different molar ratios, depending on the temperature and how the starting materials are mixed (Eq. 5.9) [11].

$$2\,\text{PhH} \;+\; \text{SOCl}_2 \;\xrightarrow[\text{0–25 °C}]{\text{AlCl}_3}\; \text{Ph}_2\text{SO} \;+\; \text{Ph}_2\text{S} \qquad (5.9)$$

Particularly interesting is that when the reaction is conducted at 0 °C with benzene (PhH) quickly poured at once into a mixture of SOCl$_2$ and AlCl$_3$, Ph$_2$S is formed as a major product along with minor Ph$_2$SO, with the molar ratio of Ph$_2$S/Ph$_2$SO = 1.4. In addition, chlorine (Cl$_2$) is identified as a by-product of the reaction. Clearly, the formation of Ph$_2$S containing S(II) from SOCl$_2$ containing S(IV) involves a reduction process in the sulfur.

Possible mechanism of the overall reaction is shown in Figure 5.6 [11]. First, a benzene molecule is attacked by the electrophilic zwitterionic Cl$_2$SO$^+$–$^-$AlCl$_3$

adduct resulting in an intermediate EAS product PhSOCl–AlCl$_3$. At 70 °C PhSOCl–AlCl$_3$ is attacked by a second benzene molecule (Approach A) exclusively effecting another EAS reaction in a similar manner to give Ph$_2$SO. The postulated Approach B does not occur.

At lower temperatures, especially 0 °C, the rate of Approach A decreases substantially. However, the effect of temperature on the rate of Approach B is small because its rate-determining step, the first intramolecular auto-redox process, has a relatively small entropy effect ($\Delta S^{\ddagger}$~0). Therefore, Approach B is more competitive than Approach A at lower temperatures, and becomes the major reaction at 0 °C. An Al–Cl bond in PhSOCl–AlCl$_3$ (the first EAS product) can reasonably be a chloride (Cl$^-$) donor (nucleophilic). The S–Cl chlorine atom is activated by the electronegative sulfoxide (S=O) group, possibly being electrophilic and readily attacked by the Al–Cl chloride to give Cl$_2$. S(IV) is in turn reduced to S(II). The conversion of S(IV) to S(II) in Approach B is accomplished via an intramolecular auto-redox process. High affinity of aluminum to oxygen is believed to aid cleavage of the S–O bond (removal of oxygen from sulfur) in the S(II) intermediate, making the sulfur in the S(II) intermediate electrophilic and facilitating a second EAS reaction to give Ph$_2$S. Concurrently, Approach A also takes place to a lesser extent giving Ph$_2$SO as a minor product.

When an aromatic ring bears a strong electron donating group (EDG), the reaction of the substituted benzene with SOCl$_2$ in the presence of AlCl$_3$ results in exclusive reduction of S(IV) to S(II) and only the corresponding diaryl sulfide (Ar$_2$S) is formed. For example, the AlCl$_3$-catalyzed reaction of phenol (C$_6$H$_5$OH) with SOCl$_2$ by quick addition of C$_6$H$_5$OH to a SOCl$_2$–AlCl$_3$ mixture at 25 °C gives the reduced (p-HOC$_6$H$_4$)$_2$S in 91% yield, along with the formations of a very low yield (9%) of the side product p-HOC$_6$H$_4$Cl and a by-product Cl$_2$. (**p-HOC$_6$H$_4$)$_2$SO is not found** (Eq. 5.10) [13].

$$\tag{5.10}$$

Figure 5.7 illustrates the mechanism for the AlCl$_3$-catalyzed reaction of C$_6$H$_5$OH with SOCl$_2$ (Reaction 5.10) and explains the possible reason why an EDG facilitates the auto-redox in SOCl$_2$ and promotes the formation of a diaryl sulfide [13]. Similar to the benzene reaction, the first step in the reaction mechanism for C$_6$H$_5$OH is that the aromatic ring undergoes an EAS reaction on the *para*-carbon with the electrophilic zwitterionic Cl$_2$SO$^+$–$^-$AlCl$_3$ adduct to give the intermediate p-HOC$_6$H$_4$–SOCl–AlCl$_3$ (I). The *para*-OH group in the intermediate (I) facilitates delocalization of the positive charge to the entire molecule giving rise to a resonance stabilization of the reaction intermediate, with the structure (III) being the major contributor. The positive charge delocalization decreases the electrophilicity of

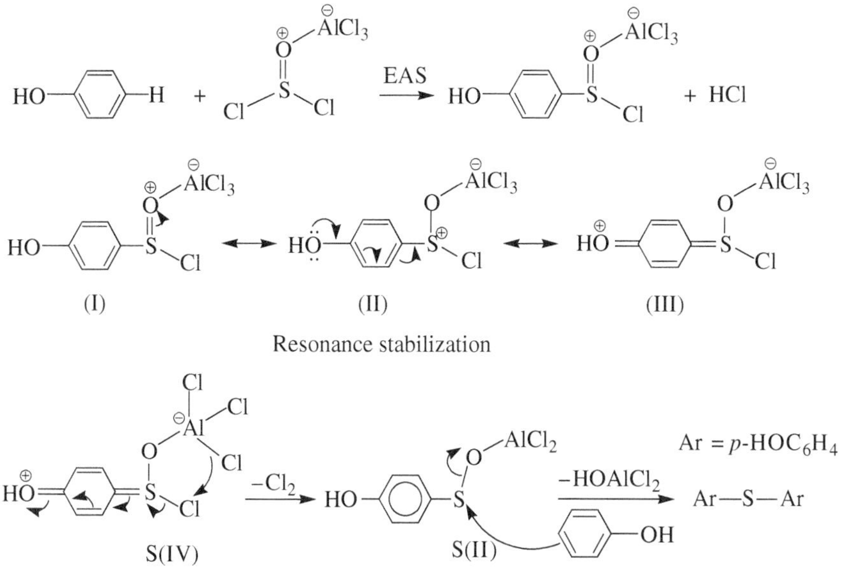

FIGURE 5.7 Possible mechanism for the $AlCl_3$-catalyzed electrophilic aromatic substitution and concurrent auto-redox reaction of phenol and thionyl chloride ($SOCl_2$). Based on Sun et al. [13].

the sulfur center so that it cannot directly react with the aromatic ring of the second phenol molecule to form the diaryl sulfoxide (p-HOC_6H_4)$_2$SO. Instead, the more favorable intramolecular auto-redox of S(IV) occurs exclusively to make S(II) and give off Cl_2. In the following step, a second phenol molecule attacks the S(II) center giving the final diaryl sulfide (p-HOC_6H_4)$_2$S product. In the presence of $AlCl_3$, Cl_2 (generated from the auto-redox) and phenol undergo very minor EAS reaction (side reaction) giving p-HOC_6H_4Cl in a very low yield.

The **p-CH₃C₆H₄SO₂Cl–AlCl₃** *(p-toluenesulfonyl chloride–aluminum chloride) zwitterionic adduct*   Benzene or toluene reacts with p-toluenesulfonyl chloride in the presence of equimolar aluminum chloride to give sulfone products (Fig. 5.8) [14]. Aluminum chloride functions as a catalyst. By nature, the reaction is EAS effected by a p-$CH_3C_6H_4SO_2Cl$–$AlCl_3$ (p-toluenesulfonyl chloride–aluminum chloride) zwitterionic adduct with the electrophilic center being on the positively charged sulfur. The zwitterionic p-$CH_3C_6H_4SO_2Cl$–$AlCl_3$ adduct is similar to Cl_2SO^{+}–$^{-}AlCl_3$ as shown in Equation 5.7. The electrophilic p-$CH_3C_6H_4SO_2Cl$–$AlCl_3$ adduct attacks an arene (such as benzene) with its sulfur atom to lead to the formation of an arenium intermediate (Fig. 5.8), which upon elimination of a proton gives a sulfone product. For the toluene reaction, the yield of *para*-sulfone is much greater than that of *ortho*-sulfone presumably due to the steric effect.

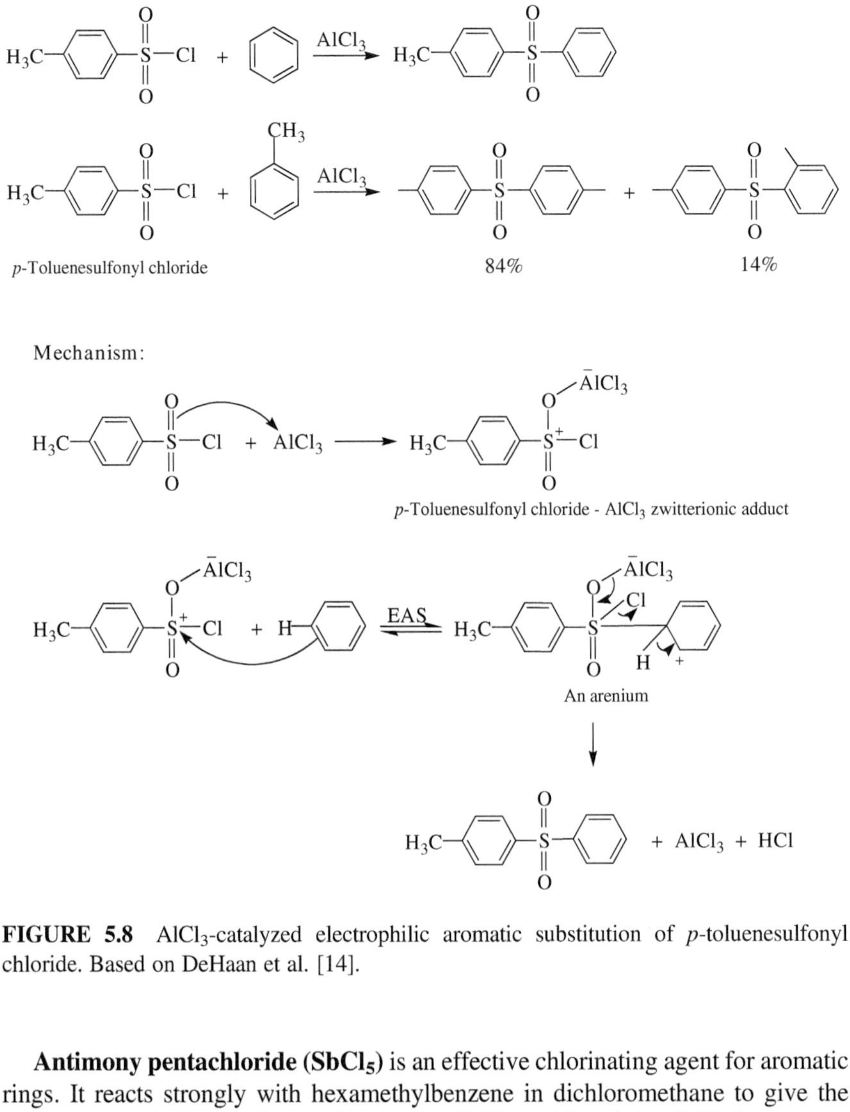

Mechanism:

FIGURE 5.8   AlCl$_3$-catalyzed electrophilic aromatic substitution of p-toluenesulfonyl chloride. Based on DeHaan et al. [14].

Antimony pentachloride (SbCl$_5$) is an effective chlorinating agent for aromatic rings. It reacts strongly with hexamethylbenzene in dichloromethane to give the chlorohexamethylbenzenium cation (as the SbCl$_6^-$ salt) and the SbCl$_3$ by-product (Fig. 5.9a). **The net result is that a positive chlorinium (Cl$^+$) ion has been added to the aromatic ring of hexamethylbenzene** [15]. Conceivably, an auto-redox occurs to SbCl$_5$, so that the Sb(V) is reduced to Sb(III) by a chloride (Cl$^-$) and the Cl$^-$ is in turn oxidized to chlorinium (Cl$^+$). The postulated auto-redox involves two SbCl$_5$ molecules (2 SbCl$_5$ ➔ SbCl$_3$ + SbCl$_6^-$ + Cl$^+$). The chloroarenium cation has been structurally characterized by X-ray crystallography. The structural data show that the carbon atom which is attached to chlorine becomes tetrahedral (sp^3 hybridized) with all the bonds on it being characteristic of typical single bonds.

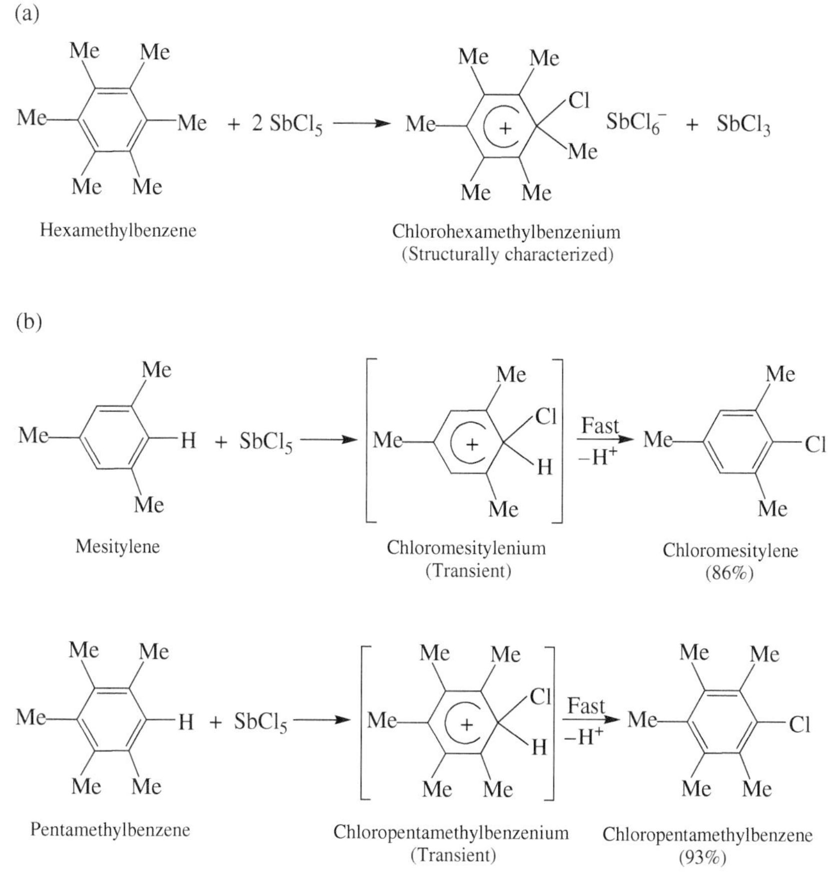

(a)

Hexamethylbenzene

Chlorohexamethylbenzenium
(Structurally characterized)

(b)

Mesitylene

Chloromesitylenium
(Transient)

Chloromesitylene
(86%)

Pentamethylbenzene

Chloropentamethylbenzenium
(Transient)

Chloropentamethylbenzene
(93%)

FIGURE 5.9   (a) Reaction of antimony pentachloride (SbCl₅) with hexamethylbenzene giving the stable chlorohexamethylbenzenium cation. Based on Rathore et al. [15]; and (b) Reaction of SbCl₅ with mesitylene and pentamethylbenzene giving chloromesitylene and chloropentamethylbenzene, respectively.

The ring is shown to be a planar cyclohexadienyl system as illustrated in Figure 5.9a [15]. Its identity is consistent with that of the arenium ions formed between hexamethylbenzene and NO_2^+ and between hexamethylbenzene and NS_2^+, respectively (Figs 5.2 and 5.5) [8, 10].

Research has also shown [15] that SbCl₅ reacts with mesitylene and pentamethylbenzene nearly quantitatively to give chloromesitylene and chloropentamethylbenzene, respectively (Fig. 5.9b). The reactions are believed to be effected by a positive chlorinium (Cl^+) ion similarly to the hexamethylbenzene reaction (Fig. 5.9a) and take place via transient intermediate chloroarenium cations, the structures of which are analogous to that of the stable chlorohexamethylbenzenium. However, these chloroarenium intermediates are not detectable due to very rapid deprotonation to the final stable EAS products.

Unexpected ring-opening in the acylium ion (–C≡O⁺)   Friedel–Crafts reaction of 2,2-difluorocyclopropanecarbonyl chloride with an arene (such as toluene) gives a major rearranged ketone product (70%), and the reaction takes place via an unexpected ring-opening process in an intermediate acylium ion (Fig. 5.10) [16]. In the first step of the reaction mechanism, a ring-intact acylium intermediate is formed. It subsequently attacks the toluene substrate to effect a minor EAS reaction (Step 2). After the minor unrearranged EAS process of the acylium takes place to give an aromatic ketone and hydrogen chloride, the nucleophilic chloride ion from the HCl by-product attacks the electrophilic CF_2 carbon in the cyclopropane ring of the acylium intermediate to effect a fast major ring-opening process, affording a ketene intermediate (Step 3). The ketene is then protonated by $H^+AlCl_4^-$ to give a second linear-chain acylium intermediate (Step 4). The linear-chain acylium then attacks toluene (Step 5) to bring about a major EAS reaction leading to the formation of the observed rearranged ketone product, and the HCl consumed in the ring-opening process (Step 3) is regenerated. The regenerated HCl is ready to induce ring-opening of another acylium ion (ring-intact) to continue the major rearranged EAS reaction. Overall, the major Friedel–Crafts ring-opening reaction is a chain-like reaction. It is initiated by a small amount of hydrogen chloride produced from the minor unrearranged EAS process of the acylium ion (the first two steps in the mechanism). The last three steps (Steps 3–5) in the mechanism keep repeating to form chain-growth steps.

5.3   THE IRON (III) CATALYZED ELECTROPHILIC AROMATIC C–H BOND SUBSTITUTION

Among the most common salts of iron (III) is iron (III) chloride ($FeCl_3$). It is a very good Lewis acid and has been widely used in organic catalysis, especially used as an effective, inexpensive catalyst for EAS reactions [17]. Although both $AlCl_3$ and $FeCl_3$ are interchangeably employed to catalyze simple EAS reactions (such as halogenation and Friedel–Crafts acylation) of small arenes (such as benzene and toluene), by comparison with $AlCl_3$, the $FeCl_3$-catalyzed EAS reactions of complicated aromatic substrates with some special electrophiles have higher yields and better selectivity [17, 18]. Recent research has suggested that one element responsible for such differences is attributable to the participation of the iron (III) d orbital in the catalysis [19]. In this section, we will study some recently developed EAS reactions catalyzed by $FeCl_3$.

EAS reactions with aldehydes   Research has shown that in the presence of $FeCl_3$ in catalytic amounts (typically 5–10 mol%), benzene and substituted benzenes (ArH) react with aldehydes (RCHO) (2:1 reactions) to give diaryl methanes (Ar_2CHR)

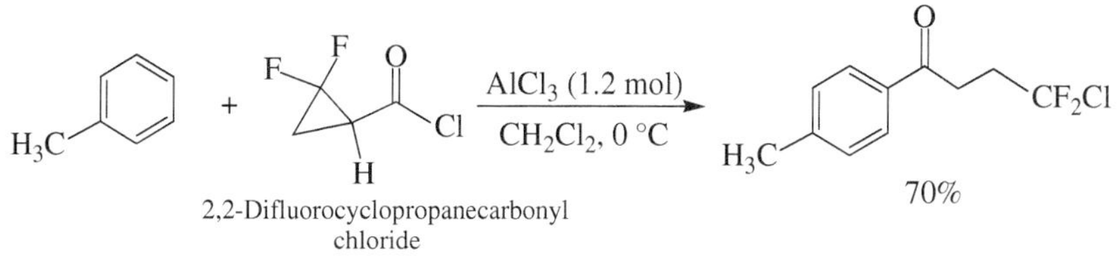

Mechanism:

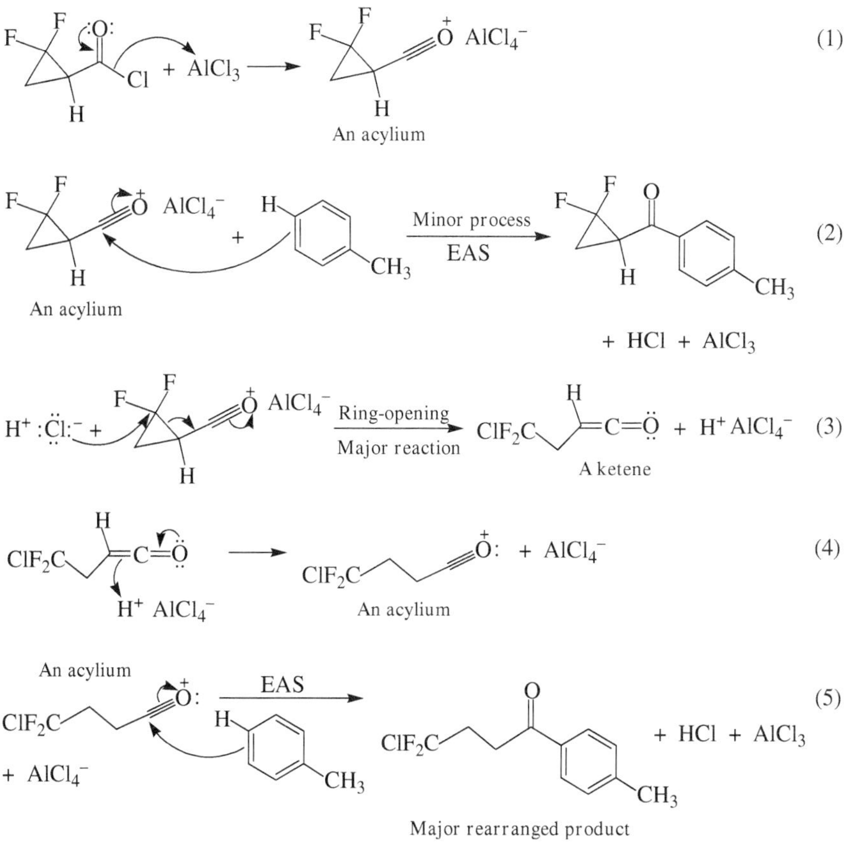

FIGURE 5.10 Unexpected ring-opening process in the AlCl$_3$-catalyzed Friedel–Crafts reaction of toluene with 2,2-difluorocyclopropanecarbonyl chloride. Based on Dolbier et al. [16].

(Fig. 5.11) [17]. Most likely, the reactions take place via an RCHO–FeCl$_3$ adduct analogous to the electrophilic zwitterionic Cl$_2$SO$^+$–$^-$AlCl$_3$ adduct formed between SOCl$_2$ and AlCl$_3$ (Eq. 5.7). Conceivably, the aromatic ring is attacked by the electrophilic carbonyl in RCHO–FeCl$_3$ to effect a favorable EAS reaction. The

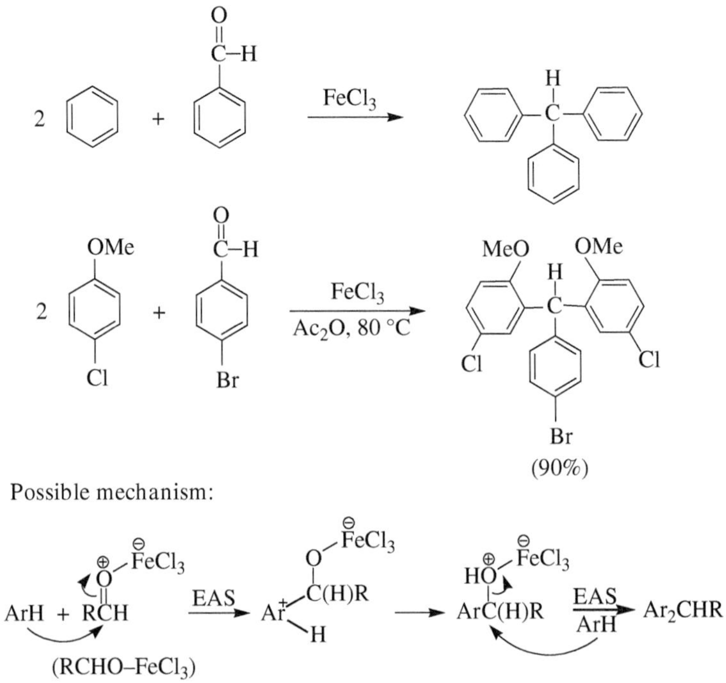

Possible mechanism:

FIGURE 5.11 FeCl$_3$-catalyzed EAS reactions of arenes with aldehydes. Based on Bauer and Knolker [17].

intermediate can further react with a second aromatic substrate molecule to afford the final diaryl methane (Ar$_2$CHR) product.

EAS reactions with alcohols Figure 5.12 shows the intramolecular EAS reactions of the benzyl alcohol derivatives catalyzed by FeCl$_3$ (5–10 mol%) [17]. Clearly, the reactions take place via carbocation intermediates generated by abstraction of hydroxide (OH$^-$) from the alcohols by FeCl$_3$. For both reactions, the carbocation intermediates are stabilized by the aromatic rings attached to the positive carbons, and the intramolecular EAS reactions occur between aromatic rings and electrophilic carbocationic centers to lead to the formations of stable five-membered rings. For the reaction in Figure 5.12b, the carbocation intermediate also experiences a resonance stabilization. An intramolecular EAS reaction occurs via the second resonance contributor to give the final bicyclic product.

A regular secondary alcohol (such as cycloheptanol) can also react with an arene in the presence of both FeCl$_3$ and AgSbF$_6$ to lead to an EAS product in a high yield (Eq. 5.11) [17].

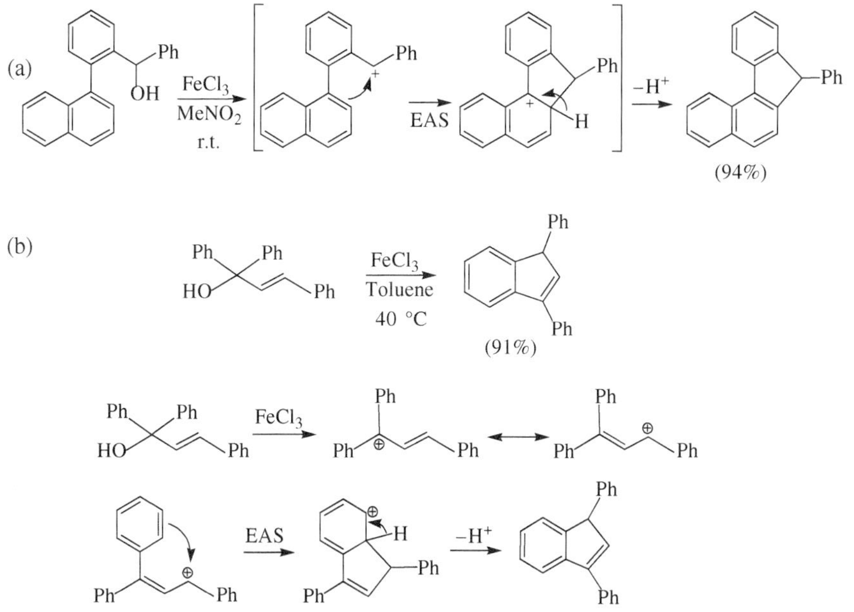

FIGURE 5.12 FeCl$_3$-catalyzed intramolecular EAS reactions of benzyl alcohol derivatives. (a) and (b) show reaction mechanisms for different benzyl alcohols.

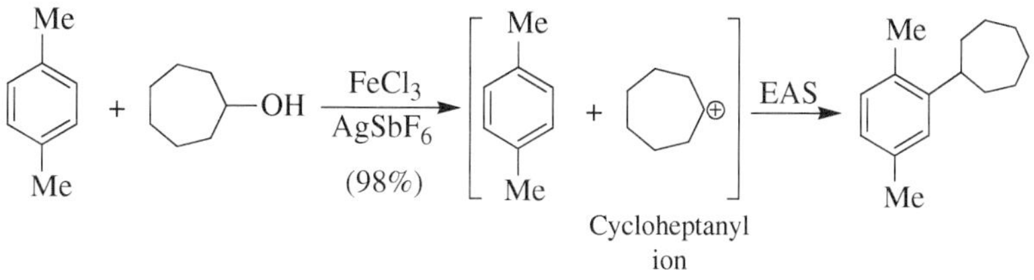

$$(5.11)$$

Reaction 5.11 takes place via the intermediate cycloheptanyl ion (a secondary carbocation) which is generated by abstraction of a hydroxide from the alcohol and the carbocation is substantially stabilized by AgSbF$_6$ (SbF$_6^-$ is an extremely weak base).

EAS reactions with alkenes   Research has shown that FeCl$_3$ catalyzes efficient reactions of alkenes with aromatic compounds to lead to alkylation of the aromatic rings (Eq. 5.12).

$$\text{ArH} + \text{>C=C<} \xrightarrow{\text{FeCl}_3} \quad -\underset{\text{Ar}}{\overset{|}{C}}-\underset{\text{H}}{\overset{|}{C}}- \qquad (5.12)$$

By the nature, Reaction 5.12 belongs to EAS with respect to the aromatic substrate. The reaction can also be regarded as an electrophilic addition of the aromatic

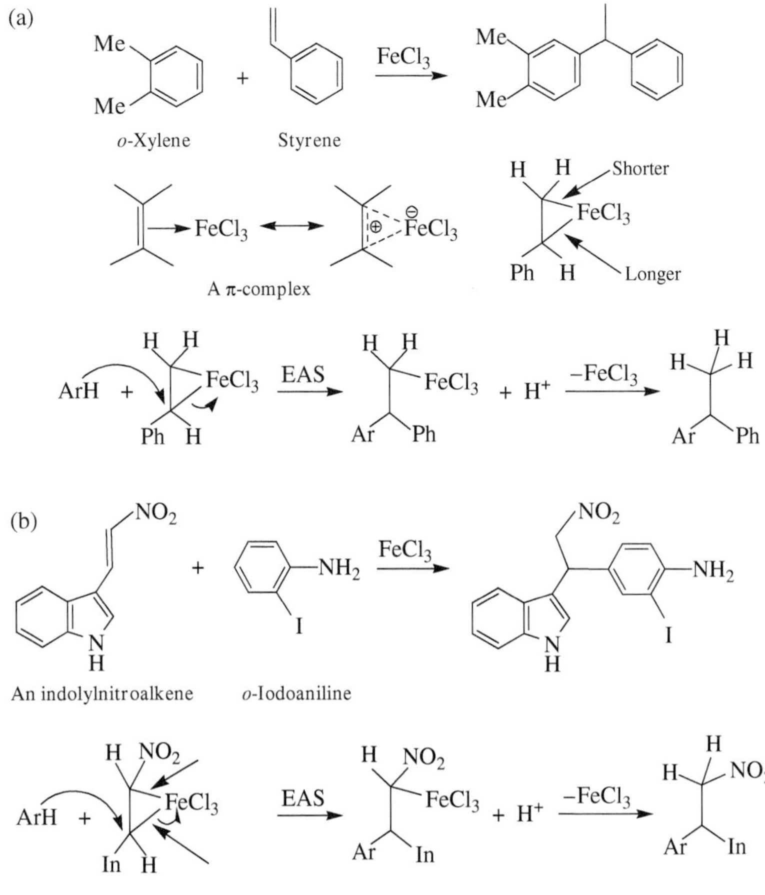

(a)

o-Xylene   Styrene

A π-complex

(b)

An indolylnitroalkene   *o*-Iodoaniline

FIGURE 5.13 FeCl₃-catalyzed regioselective EAS reactions of alkenes and possible mechanism. (a) Reaction of styrene with o-xylene and (b) Reaction of an indolylnitroalkene with o-iodoaniline. (a,b) Based on Bauer and Knolker [17].

C–H bond to an alkene following Markovnikov's rule with the aromatic ring connected to the C=C carbon containing less hydrogens. Figure 5.13 shows examples of the FeCl₃-catalyzed reactions of alkenes with aromatic compounds.

Figure 5.13a shows reaction of 1,2-dimethylbenzene (*o*-xylene) with phenylethene (styrene) in the presence of FeCl₃ in a catalytic amount. The aromatic ring of *o*-xylene is selectively added to the C_1-carbon (the doubly bonded carbon with phenyl) of styrene [17]. In general, iron (III) can readily form a π-complex with an alkene C=C bond [20]. While the π electrons are donated into an iron (III) d orbital to make a coordination bond (indicated by an arrow between C=C and FeCl₃), the d_π–$p_\pi*$ back bonding between iron (III) and the C=C bond is believed to further stabilize the π-complex. The stabilization of the alkene–iron (III) π-complex assures its EAS reaction with *o*-xylene. If the alkene is unsymmetrical, the C–Fe bond on the carbon bearing the electron donating Ph group is longer than the C–Fe bond on the carbon with two hydrogens (without EDG). This facilitates the delocalization of the positive charge making the intermediate π-complex more stable (c.f. the C–Hg bonds in the alkene–Hg(II) complex in Section 3.1.3, Figure 3.10). The π-complex carries partial carbocationic feature. Thus, the EAS reaction preferably occurs on the C_1-carbon of styrene with a longer C–Fe bond, explaining the observed regioselectivity for the reaction.

Figure 5.13b shows the $FeCl_3$-catalyzed reaction of an indolylnitroalkene with *o*-iodoaniline (with an amino $–NH_2$ group attached to an aromatic ring). The reaction gives a regiospecific product with the alkylation of the aromatic ring of iodoaniline happening on the indolylnitroalkene C=C carbon without a nitro ($–NO_2$) group. In addition, the functional groups are intact in the reaction [18]. Research has also shown that when the reaction is catalyzed by $AlCl_3$, a complex mixture is produced [18]. Similar to the above *o*-xylene–styrene reaction (Fig. 5.13a), the $FeCl_3$-catalyzed reaction of the indolylnitroalkene with *o*-iodoaniline (Fig. 5.13b) should also occur via a nitroalkene–iron (III) π-complex in which the C–Fe bond on the electron withdrawing nitro carbon is shorter than the C–Fe bond on the other carbon. As a result, the positive charge can be better delocalized to stabilize the π-complex. Thus, the EAS reaction preferably occurs on the C=C carbon without the nitro group. The indolylnitroalkane product made in this reaction is an important bioactive compound and possesses biological and pharmacological activity [18].

Fecl₃-catalyzed EAS reaction of chlorobenzene with SOCl₂   Recent research has shown that the $FeCl_3$-catalyzed EAS reaction of chlorobenzene (C_6H_5Cl) with $SOCl_2$ at 0 °C gives a mixture of $(4\text{-}ClC_6H_4)_2SO$ (28%, minor) and reduced $(4\text{-}ClC_6H_4)_2S$ (57%, major); while when the reaction is catalyzed by $AlCl_3$ at 0 °C, the overwhelming major product is $(4\text{-}ClC_6H_4)_2SO$ (98%) along with the formation of tiny $(4\text{-}ClC_6H_4)_2S$ (1%) [19]. Clearly, $FeCl_3$ has facilitated dramatically the auto-redox of sulfur in $SOCl_2$ in the course of its EAS reaction. This effect is thought owing to participation of the iron (III) d orbital in the reaction.

Similar to the $AlCl_3$-catalyzed EAS reaction with $SOCl_2$ which takes place via the zwitterionic $Cl_2SO^+–^-AlCl_3$ adduct (Eq. 5.7), the $FeCl_3$-catalyzed EAS reaction with $SOCl_2$ is believed to take place via an analogous zwitterionic $Cl_2SO^+–^-FeCl_3$ adduct (Fig. 5.14) [19]. The EAS reaction gives an intermediate $ArSOCl–FeCl_3$ (Ar = $4\text{-}ClC_6H_4$) first. Then the intermediate undergoes EAS with a second chlorobenzene molecule (a very minor process) to give $(4\text{-}ClC_6H_4)_2SO$ (only 28%). Alternatively, the major reaction for the $ArSOCl–FeCl_3$ intermediate is an auto-redox to convert S(IV) to S(II), followed by an EAS with a second chlorobenzene molecule to give the major reduced $(4\text{-}ClC_6H_4)_2S$ (57%) product.

Figure 5.15 shows that the symmetry of the empty $p_\pi*$ orbital (the antibonding π∗ orbital) in the S=O bond domain of the $ArSOCl–FeCl_3$ intermediate matches symmetry of the filled d_π orbital in the iron (III) atom. Therefore, the d_π and $p_\pi*$ orbitals can overlap effectively in sideways to form a $d_\pi \to p_\pi*$ back bond between the iron and oxygen atoms in $ArSOCl–FeCl_3$ [19]. As a result, the d_π electrons in iron

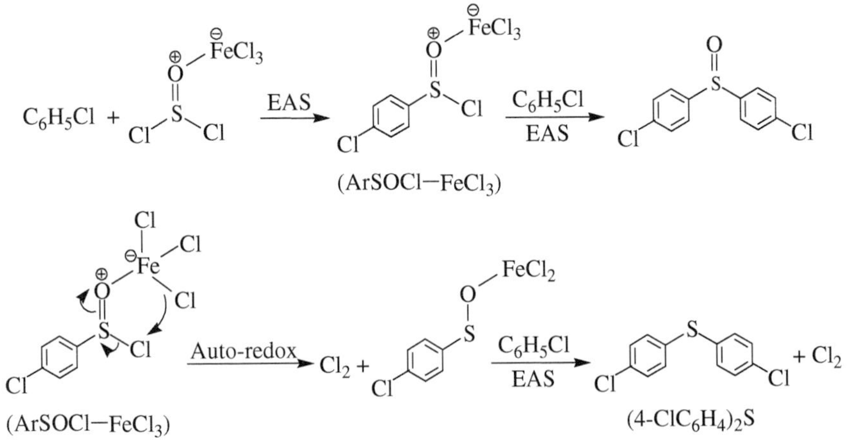

FIGURE 5.14 Mechanism for the $FeCl_3$-catalyzed reaction of chlorobenzene with $SOCl_2$. Based on Sun et al. [19].

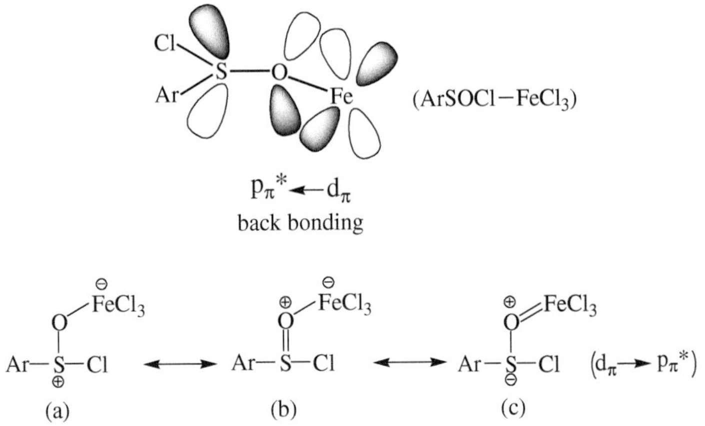

FIGURE 5.15 The d_π–$p\pi*$ back bonding between Fe(III) and the S=O bond in the $ArSOCl$–$FeCl_3$ adduct and possible resonance structures of $ArSOCl$–$FeCl_3$. The d_π (d_{yz}) orbital is perpendicular to the $FeCl_3$ molecule plane (the *xy*-plane). Reprinted from Ref. [19] by permission of the publisher (Taylor & Francis, http://www.tandf.co.uk/journals).

flow into the S=O $p_\pi*$ orbital to partially neutralize the positive charge in sulfur decreasing its electrophilicity. The resonance structure (C) in Figure 5.15 results from the $d_\pi \rightarrow p_\pi*$ back bonding and further demonstrates the electron transfer from iron to sulfur. Consequently, the sulfur center in $ArSOCl$–$FeCl_3$ cannot be attacked efficiently by the nucleophilic aromatic ring in chlorobenzene to make Ar_2SO (Ar = p-ClC_6H_4) as a major product. Instead, the intramolecular auto-redox process in $ArSOCl$–$FeCl_3$, as shown in Figure 5.14, becomes more prominent (much faster than the EAS of ArH with $ArSOCl$–$FeCl_3$) to give Ar_2S as the major product. This is believed to account for the role of $FeCl_3$ in facilitating the auto-redox processes in the $FeCl_3$-catalyzed EAS reactions of $SOCl_2$ [19].

5.4   THE ELECTROPHILIC AROMATIC C—H BOND SUBSTITUTION REACTIONS VIA S_N1 AND S_N2 MECHANISMS

The aromatic C—H bond alkylation possesses important synthetic utility. It follows the EAS mechanism on the arene substrate via an arenium intermediate (σ-complex). The reaction involves an S_N1 or S_N2 step on the alkylating agent. In this section, we will use some key examples to demonstrate the effective combinations of EAS and S_N steps involved in the arene alkylation processes.

5.4.1   Reactions Involving S_N1 Steps

Friedel–Crafts alkylation reactions   Figure 5.16 shows the Friedel–Crafts reaction of benzene with (R)-2-chlorobutane [1]. The reaction gives a racemic mixture of (R)- and (S)-2-phenylbutane, indicating that it takes place via an S_N1 step on (R)-2-chlorobutane, a chiral alkylating agent. First, the chloroalkane and aluminum chloride combine to form an adduct. As a result, the C—Cl bond is activated. The subsequent cleavage of the C—Cl bond results in an intermediate secondary carbocation, which is attacked by benzene from top and bottom equally giving a racemic mixture. The overall reaction follows an EAS mechanism on benzene. An S_N1 step is involved in the chloroalkane.

Figure 5.17 shows Friedel–Crafts reaction of benzene with 1-chlorobutane. In the course of the reaction, a hydrogen rearrangement takes place resulting in an intermediate secondary carbocation which has the same identity to the carbocation in the reaction of (R)-2-chlorobutane in Figure 5.16. Then the attack of benzene on the carbocation occurs in both sides equally giving a racemic mixture. An S_N1 step is involved in the Friedel–Crafts alkylation.

In the presence of a catalytic amount of $AlCl_3$, benzene undergoes Friedel–Crafts alkylation reaction with a stereospecific cyclohexyl chloride giving only one stereoisomer (Eq. 5.13).

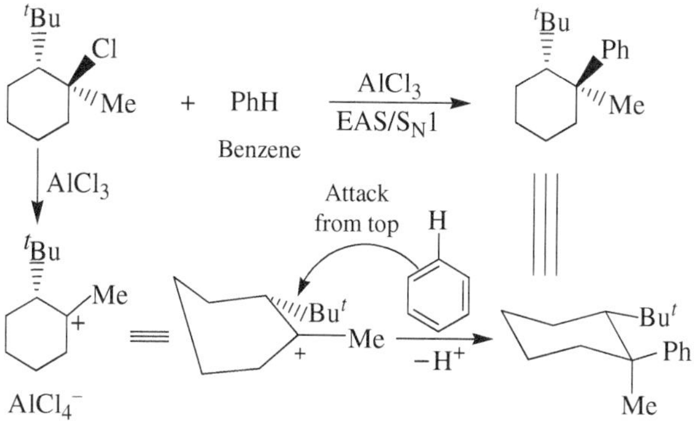

$$(5.13)$$

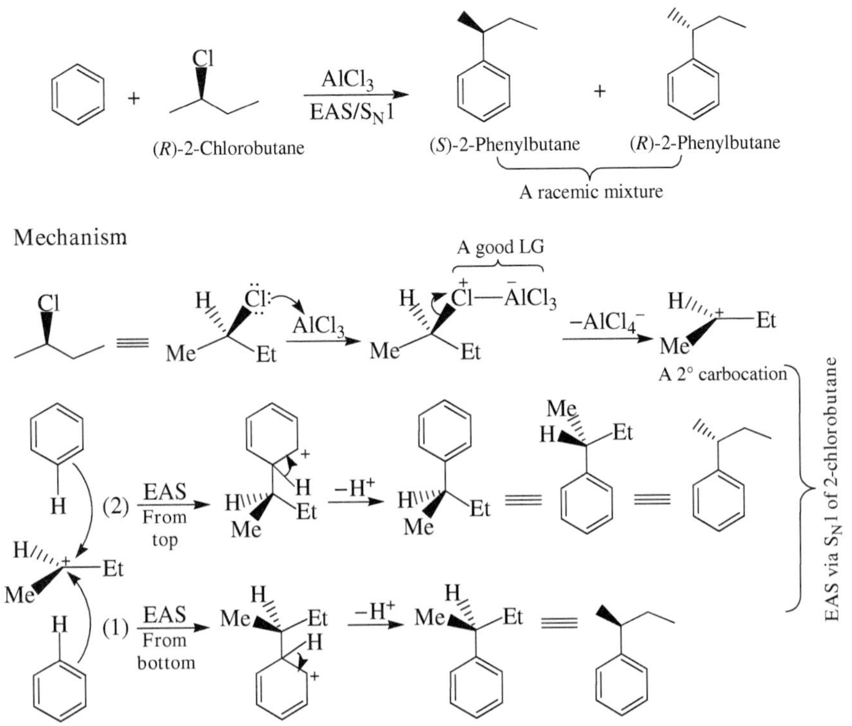

FIGURE 5.16 Mechanism for $AlCl_3$-catalyzed Friedel–Crafts reaction of benzene with (R)-2-chlorobutane.

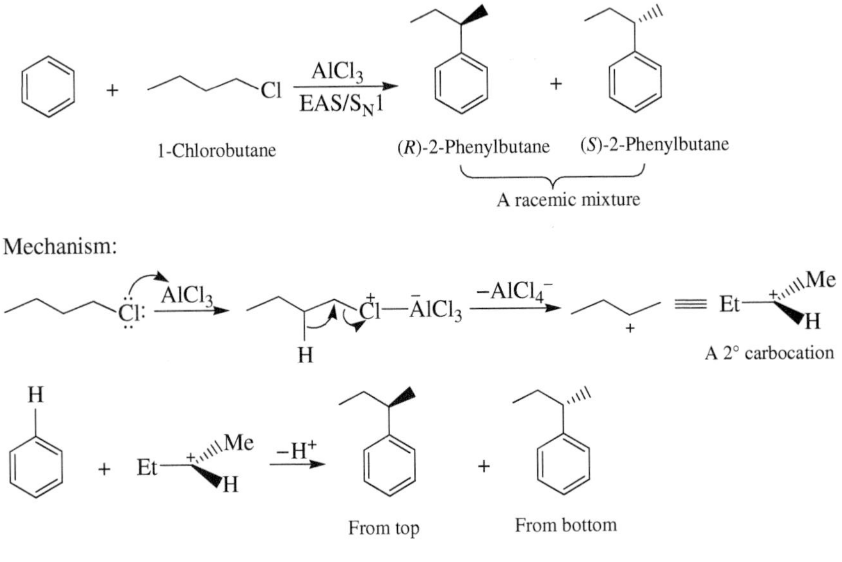

FIGURE 5.17 Mechanism for $AlCl_3$-catalyzed Friedel–Crafts reaction of benzene with 1-chlorobutane.

The reaction takes place via a cyclic tertiary carbocation, of which the two faces have different steric environments. The bulky *tert*-butyl group preferably stays in an equatorial position and generates strong steric hindrance in the bottom face of the molecule plane. Thus, the nucleophilic benzene molecule attacks the carbocation, preferably, from the top face giving a single stereoisomer in which the *tert*-butyl and phenyl groups stay in different sides of the cyclohexane ring (*trans* to each other) and both stay in the preferred equatorial positions of the most stable chair conformation.

Dichloromethane (CH_2Cl_2) contains two chloro (–Cl) groups on the carbon atom. The molecule can undergo two consecutive Friedel–Crafts alkylation reactions via S_N1 steps on the sp^3 carbon giving the diphenylmethane product (Fig. 5.18). Combination of the first chloro group in CH_2Cl_2 with aluminum chloride affords the intermediate chloromethyl $^+CH_2Cl$ cation. The carbocation is stabilized by a lone pair of electrons in chlorine (resonance stabilization) making its formation possible. Then the electrophilic $^+CH_2Cl$ attacks the first benzene molecule effecting an EAS reaction to give $PhCH_2Cl$, and the reaction has involved an S_N1 step on CH_2Cl_2. Afterwards, $PhCH_2Cl$ undergoes another Friedel–Crafts alkylation reaction with a

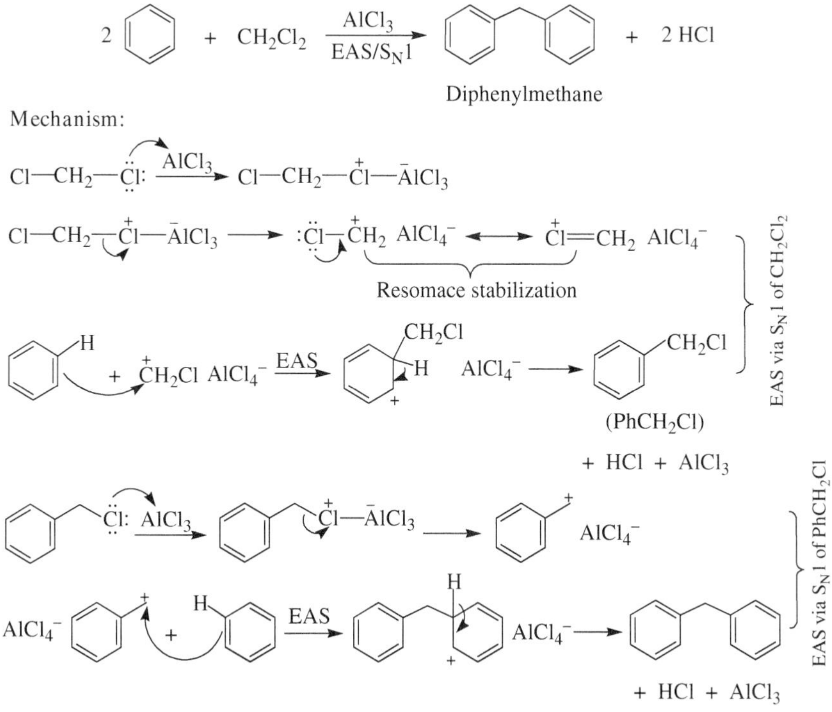

FIGURE 5.18   Mechanism for $AlCl_3$-catalyzed Friedel–Crafts reaction of benzene with dichloromethane.

second benzene molecule via the benzylic cation ($PhCH_2^+$). The reaction involves an S_N1 step on $PhCH_2Cl$ giving the final diphenylmethane product.

Other reactions   Bisphenol A is used in the production of epoxy and polycarbonate resins (plastics). In industry, it is synthesized by reaction of phenol with acetone in the presence of an acid catalyst such as sulfuric acid (Fig. 5.19) [1–3]. The first step in the reaction mechanism is the protonation of acetone. Then the protonated carbonyl group, a strong carbon electrophile, attacks a phenol molecule on its *para*-carbon effecting an EAS reaction. The intermediate alcohol product is subsequently protonated to lead to the formation of an electrophilic intermediate tertiary carbocation, which attacks a second phenol molecule on its *para*-carbon to effect another

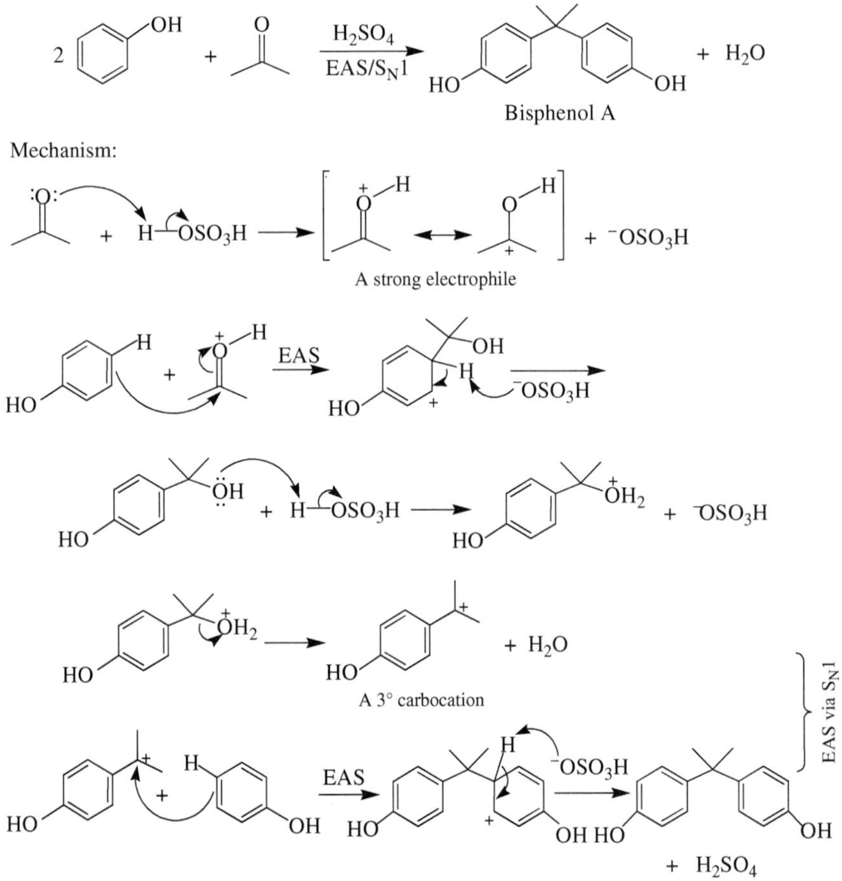

FIGURE 5.19   Mechanism for acid-catalyzed electrophilic substitution reaction of phenol with acetone. Bruckner [1]; Fox and Whitesell [2].

EAS reaction and to finally give the bisphenol A product. The second step of EAS reaction involves an S_N1 step on the intermediate alcohol product.

In the presence of an acid catalyst, benzene can react with an alkene resulting in aromatic C–H alkylation (Fig. 5.20). The reaction follows an EAS mechanism through an intermediate tertiary carbocation which is produced from protonation of a C=C bond. The reaction of an arene (such as benzene) with an alkene giving an alkylarene can also be catalyzed by a Lewis acid (such as $AlCl_3$ and $FeCl_3$) as described in Figure 3.13 of Section 3.1.5 and in Figure 5.13.

Figure 5.21 shows a ring-closure reaction taking place via a combination of an EAS step on the aromatic ring and an S_N1 step on the alcohol C–OH carbon [1]. After a tertiary carbocation is formed by protonation of the alcohol, it preferably attacks the aromatic ring intramolecularly giving a stable five-membered ring. The intramolecular ring-closure is facilitated by a favorable entropy effect, while attack of the intermediate carbocation on the aromatic ring of a different molecule (intermolecular reaction) would be thermodynamically and kinetically less favorable.

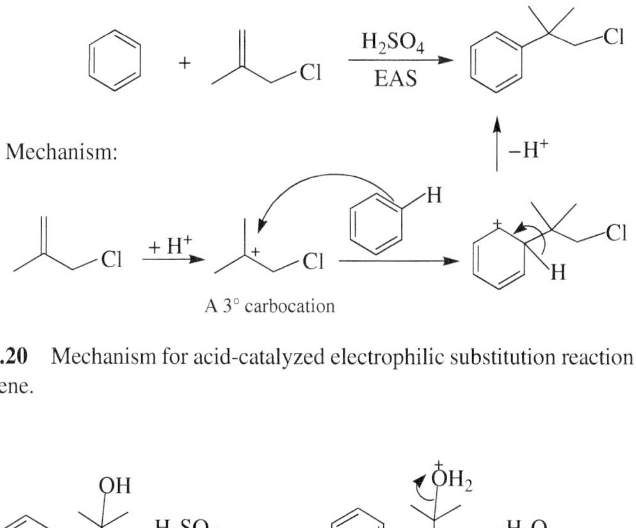

FIGURE 5.20   Mechanism for acid-catalyzed electrophilic substitution reaction of benzene with an alkene.

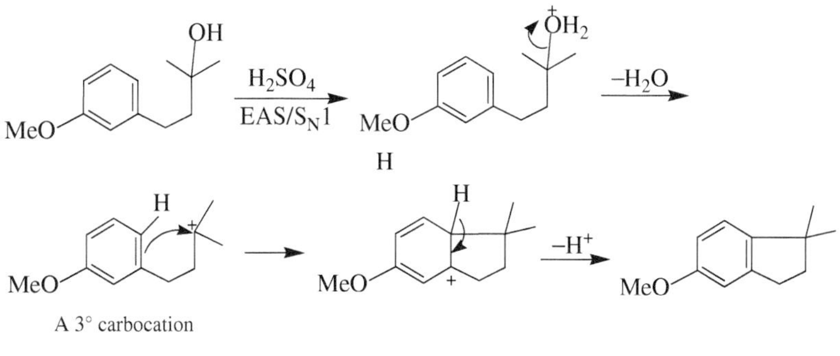

FIGURE 5.21   Mechanism for an intramolecular electrophilic aromatic substitution reaction via an S_N1 step. Based on Bruckner [1].

5.4.2 Reactions Involving S$_N$2 Steps

The reaction shown in Figure 5.22 is a nucleophilic substitution that takes place on a sp^3-hybridized carbon of an ester molecule [1]. In the presence of a catalytic amount of aluminum chloride, a methoxy (–OMe) group is substituted by a phenyl ring and benzene is alkylated. Different from the S$_N$1 reaction described in Figure 5.16 which gives a racemic mixture of two products, this reaction gives only one product that has the opposite configuration to that of the reactant ester molecule. The result shows that the reaction follows an S$_N$2 mechanism on the ester molecule.

The –OMe group in the ester molecule is a poor leaving group. In the absence of a catalyst, the reaction will not take place. In the presence of a catalytic amount of aluminum chloride, the methoxy oxygen atom can readily combine with AlCl$_3$ via a strong O→Al coordination bond to form an intermediate 1:1 adduct. As a result, the O—C bond is activated by a positive charge and a good leaving group (LG) is created. As the O—C bond is being cleaved and the central carbon atom becomes electrophilic, the benzene molecule (nucleophilic) will attack the carbon atom simultaneously from the opposite side of the methoxy group. As a result, an S$_N$2 reaction happens to the ester molecule, while the benzene molecule undergoes an EAS reaction producing an intermediate benzenium ion. Then the *ipso*-H in the benzenium

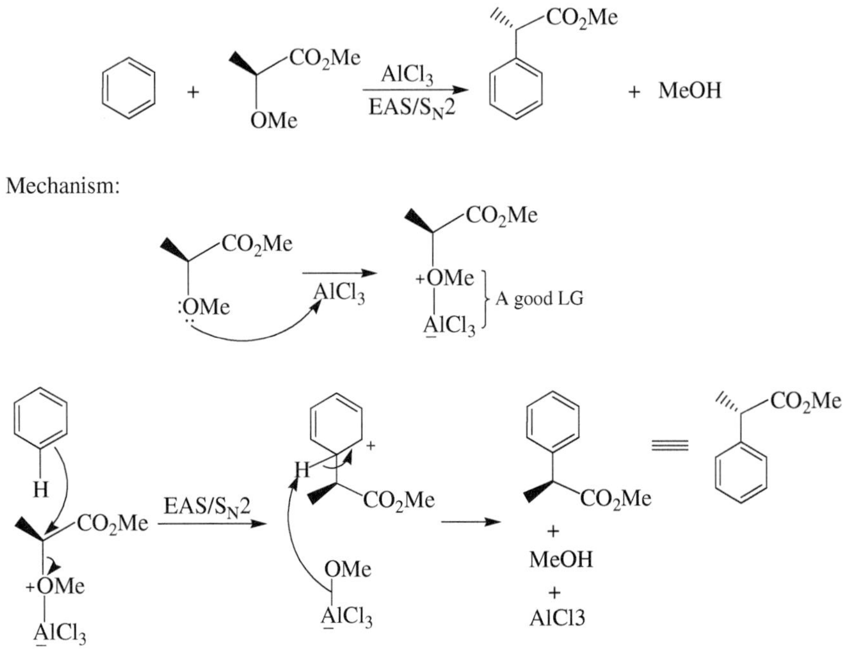

FIGURE 5.22 Mechanism for an electrophilic aromatic substitution reaction of benzene via an S$_N$2 step. Based on Bruckner [1].

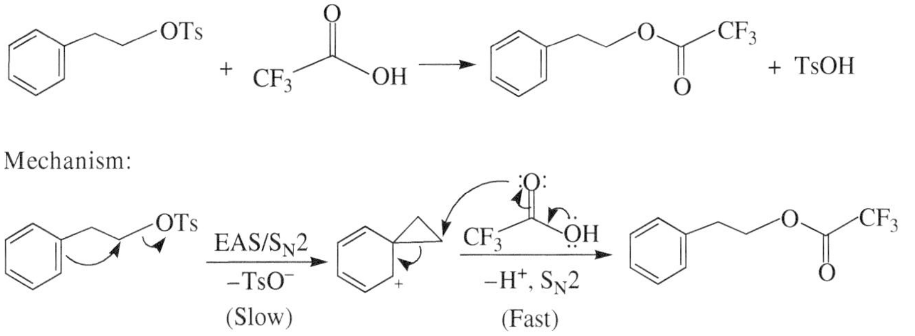

Mechanism:

(Slow)   (Fast)

FIGURE 5.23   Mechanism for an intramolecular electrophilic aromatic substitution reaction via an S_N2–like reaction. Based on Bruckner [1].

ion is eliminated as a proton (deprotonation) giving the final substitution product. The proton (H^+), once eliminated from the benzenium ring, combines concurrently with the methoxy group in MeO–$^-$AlCl$_3$ (generated from the S_N2 process) to form methanol. The AlCl$_3$ catalyst is regenerated.

Figure 5.23 shows a functional group transformation facilitated by a phenyl ring. The reaction is found to be first-order in PhCH$_2$CH$_2$OTf, and the rate is independent of concentration of CF$_3$CO$_2$H (zero-order in the acid) [1]. The triflate (–OTf) group is commonly considered a super good leaving group. Therefore, the terminal carbon bearing the –OTf group can be effectively attacked intramolecularly by the benzene ring, even though benzene is only very weakly nucleophilic. This gives rise to an S_N2-like mechanism on the terminal sp^3 carbon whose rate is directly proportional to the molar concentration of PhCH$_2$CH$_2$OTf because the reaction takes place intramolecularly, while an EAS reaction happens to the phenyl ring. The first step is the rate-determining step due to the formation of a metastable benzenium (σ-complex). This accounts for the observed overall first-order rate law for the reaction. In the intermediate benzenium cation, the *ipso* C$_{benzenium}$—C$_{sp3}$ bond is highly activated because its bonding electron pair has strong tendency to move back to the ring, restoring the 6 π aromatic system. As the bonding electron pair is transferred to the benzenium ring and the *ipso* C$_{benzenium}$—C$_{sp3}$ bond is being broken, the sp^3 carbon will be attacked simultaneously by a nucleophilic trifluoroacetic acid molecule (an S_N2 reaction) to give the final ester product.

5.5   SUBSTITUENT EFFECTS ON THE ELECTROPHILIC AROMATIC SUBSTITUTION REACTIONS

Figure 5.1 indicates that the formation of an intermediate arenium ion (σ-complex) is the rate-determining step for an EAS reaction. In general, if the aromatic ring bears an EDG, part of the positive charge in the arenium ring can delocalize to the EDG. As a result, the arenium intermediate will be stabilized. This lowers the activation energy and accelerates the EAS reaction, namely, that the EAS reaction of an arene bearing an EDG has a greater rate constant than that for the reaction of benzene with the same electrophile. Conversely, if the aromatic ring bears an electron withdrawing group (EWG) which is usually electronegative, electrostatic repulsion will be encountered between the positively charged arenium ring and the electronegative EWG. The atom in an EWG that is bonded to an arenium ring is positively charged, such as the nitro group $-NO_2$, with the nitrogen atom being positively charged. As a result, the arenium intermediate will be destabilized. This enhances the activation energy and slows down the EAS reaction, namely, that the EAS reaction of an arene bearing an EWG has a smaller rate constant than that for the reaction of benzene with the same electrophile. In addition, the EDG and EWG also dictate the regiochemistry of EAS reactions electronically. The effects of a substituent group (EDG or EWG) on the regiochemistry is called **directing effects**. In this section, we give an account for directing effects of EDG and EWG on the basis of analysis of charge distributions in the intermediate arenium ions formed in EAS reactions.

Figure 5.24 shows on the left side a simplified model of the charge distribution in the benzenium ion (σ-complex). This charge distribution is obtained by superimposing the three resonance structures of Figure 5.1, where X = H. According to this model, the positive charges only appear on the *ortho-* and *para*-carbons to the carbon that bears the electrophile (E), and the *meta*-carbons are electrically neutral. However, refined theoretical studies have shown that each of the *meta*-carbons also bears a small fraction (+0.10) of the positive charge [1]. A more subtle model of the charge distribution in the benzenium ion is shown on the right side of Figure 5.24. The following discussion on the directing effects of EDG and EWG will be based on the more subtle (refined) model for the charge distribution. It is assumed that introducing a substituent group to the ring will not alter the charge distribution.

Common electron-donating and electron-withdrawing groups (EDG's and EWG's) and their relative abilities to donate or withdraw electrons are presented in Figure 5.25.

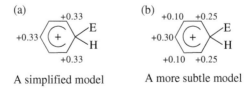

A simplified model　　　A more subtle model

FIGURE 5.24 Simplified and more subtle models for the charge distribution in a benzenium ion (σ-complex) (a) A simplified model and (b) a more subtle model. Reprinted from Ref. [1], Page 177, with permission from Elsevier.

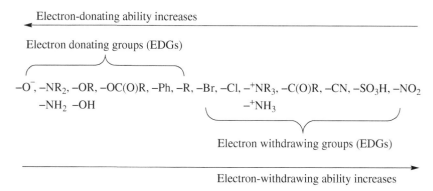

FIGURE 5.25 Common electron-donating and electron-withdrawing groups (EDG's and EWG's) and their relative electron-donating or electron-withdrawing abilities.

5.5.1 *Ortho-* and *para-*Directors

Figure 5.26 shows the intermediate *ortho-*, *para-*, and *meta-*arenium ions (σ-complexes) which bear an EDG in the *ortho-*, *para-*, and *meta-*carbon, respectively, in the arenium. In the *ortho-* and *para-*arenium ions, an EDG is attached to a carbon that bears 25% (+0.25) and 30% (+0.30) of a full positive charge (+1), respectively. The arenium intermediates are strongly stabilized by the EDG due to delocalization of the corresponding portion (25% or 30%) of the positive charge to the EDG. This gives rise to low activation energies for the formations of the *ortho-* and *para-*arenium ions (the rate-determining steps for the formations of the *ortho-* and *para-*products, respectively). As a result, the formations of the *ortho-* and *para-*products are fast. In the *meta-*arenium, an EDG is attached to a carbon that bears only 10% (+0.10) of a full positive charge (+1), and thus the arenium intermediate is only slightly stabilized by the EDG. This gives rise to a relatively high activation energy for the formation of the *meta-*arenium ion (the rate-determining step for the formation of the *meta-*product). As a result, the formation of the *meta-*product is slow. Figure 5.26 also shows the energy profiles for the formations of *ortho-*, *para-*, and *meta-*arenium ions, the rate-determining steps for the formations of *ortho-*, *para-*, and *meta-*products, respectively, which bear an EDG. The product distribution is kinetically controlled. Therefore, an EDG is in general an *ortho-* and *para-*director, making the *ortho-* and *para-*products the major products.

For alkyl groups (EDGs), steric hindrance on the *ortho-*carbon becomes more appreciable as the size of an alkyl is getting larger. It increases the percentage of the *para-*product and decreases the percentage of the *ortho-*product (Fig. 5.27) [2]. The bulky tertiary butyl group [$-C(CH_3)_3$] has strong steric hindrance on the *ortho-*carbon. In general, it is only a *para-*director for all types of EAS reactions (if the *para-*carbon is unsubstituted).

A strong EDG (such as $-OH/-OR$) that bears a lone pair of electrons in the atom that is attached to an aromatic ring is usually only a *para-*director for substitution of

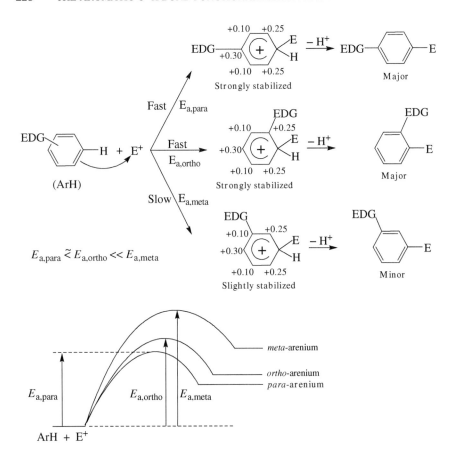

FIGURE 5.26   The intermediate *ortho*-, *meta*-, and *para*-arenium ions which bear an electron donating group (EDG), their relative stabilities, and energy profiles for their formations.

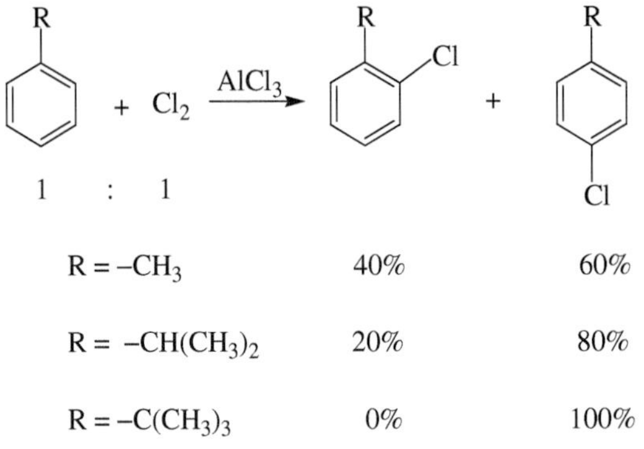

R	ortho	para
R = −CH$_3$	40%	60%
R = −CH(CH$_3$)$_2$	20%	80%
R = −C(CH$_3$)$_3$	0%	100%

FIGURE 5.27   Chlorination of alkylbenzenes: Directing effects of different alkyl groups.

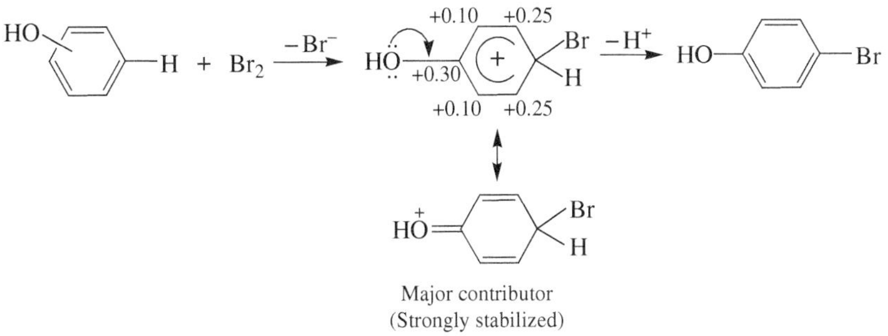

FIGURE 5.28   Reaction of phenol with bromine: *Para*-directing effect of the hydroxyl (–OH) group.

the first aromatic C—H bond by an electrophile (if the *para*-carbon is unsubstituted). After the C_{para}—H bond is substituted, the following EAS will take place on an *ortho*-carbon. For example, reaction of phenol with Br_2 gives exclusively *p*-bromophenol (Fig. 5.28) [2]. In the intermediate *para*-arenium, the active lone pair of electrons in oxygen can reach all the p_π carbons (positively charged) via hyperconjugation effect to delocalize the full positive charge from the arenium ring to the oxygen atom. As a result, a "quinone-like" resonance structure that contains a positively charged oxygen in the *para*-carbon would be a major contributor relative to other structures as shown in Figure 5.1, and this "quinone-like" structure is strongly stabilized. This makes the reaction on the *para*-carbon particularly kinetically favorable, giving the *p*-bromophenol as a sole product. Introduction of a bromo group to phenol giving *p*-bromophenol deactivates the aromatic ring toward further EAS reactions. This makes the bromination on each of the *ortho*-carbons in *p*-bromophenol much slower than the first bromination on the *para*-carbon. Thus, bromination of phenol can be controlled stepwise to obtain 4-bromophenol (*p*-bromophenol), 2,4-dibromophenol, and 2,4,6-tribromophenol, respectively by using different amounts of Br_2, because each successive reaction is slower.

5.5.2   *Meta*-Directors

Figure 5.29 shows the intermediate *meta*-, *ortho*-, and *para*-arenium ions each of which bears an EWG. In the *meta*-arenium, an EWG is attached to a carbon that bears only 10% (+0.10) of a full positive charge (+1), and the arenium intermediate is slightly destabilized by the EWG primarily due to a small electrostatic repulsion between the slightly positively charged carbon (+0.10) and the electronegative EWG. This gives rise to a high activation energy for the formation of the *meta*-arenium ion (the rate-determining step). As a result, the formation of the *meta*-product is slow. In the *ortho*- and *para*-arenium ions, an EWG is attached to a carbon that bears 25% (+0.25) and 30% (+0.30) of a full positive charge (+1), respectively. The arenium intermediates are strongly destabilized by the EWG primarily due to a

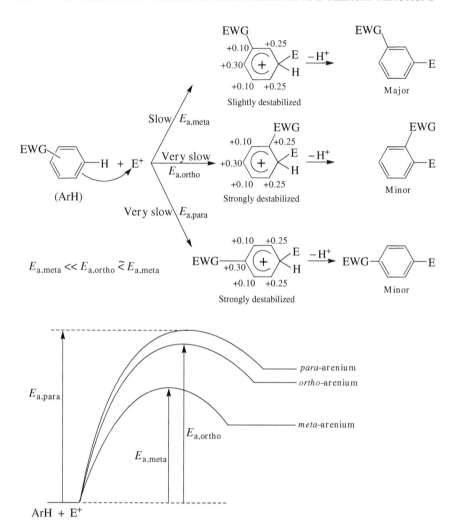

FIGURE 5.29 The intermediate *ortho*-, *meta*-, and *para*-arenium ions which bear an electron withdrawing group (EWG), their relative stabilities, and energy profiles for their formations.

great electrostatic repulsion between the substantially positively charged carbon (+0.25 or +0.30) and the electronegative EWG. This gives rise to **very** high activation energies for the formations of the *ortho*- and *para*-arenium ions (the rate-determining steps). As a result, the formations of the *ortho*- and *para*-products are **very** slow. Figure 5.29 also shows the energy profiles for the formations of *meta*-, *ortho*-, and *para*-arenium ions, the rate-determining steps for the formations of *meta*-, *ortho*-, and *para*-products, respectively, which bear an EWG. Although reactions on all the carbon atoms are slow due to the presence of a deactivating EWG in the aromatic ring to destabilize the intermediate arenium, the formation of the meta-product is relatively much faster than the formations of *ortho-* and *para*-products. This is because the $E_{a,meta}$ (activation energy for the formation of the *meta*-arenium) is relatively much lower than $E_{a,ortho}$ and $E_{a,para}$ (activation energies for the formations of the *ortho*-arenium and *para*-arenium, respectively). Once again, the product distribution is kinetically controlled. Therefore, an EWG is in general a *meta*-director, making the *meta*-product the major product.

The nitro group ($-NO_2$) is a strong EWG (Fig. 5.25). Nitration reaction of nitrobenzene gives a mixture of *meta-*, *ortho-*, and *para*-dinitrobenzene in the molar ratio of 93:6:1 (Eq. 5.14) [2].

$$\text{(5.14)}$$

$$93\% \qquad 6\% \qquad 1\%$$

Chloro ($-Cl$) and bromo ($-Br$) groups are weak EWGs. However, they both are *para-* and *ortho*-directors. This exception to the above general model of Figure 5.29 is owing to the activity of the lone-pair of electrons in each of the halogen atoms. The hyperconjugation effect of the electron-pair on the arenium ion can neutralize the positive charge distributed to the carbon which is connected to the halo group. As a result, the positive charge can be delocalized to the halo group stabilizing the arenium. Since the *para-* and *ortho*-carbons in the arenium bear greater positive charges (+0.30 and +0.25, respectively) than does the *meta*-carbon (+0.10), the *para-* and *ortho*-arenium ions which bear a halo group are more strongly stabilized than is the *meta*-arenium. This makes a halo group a *para-* and *ortho*-director. *In general, a substituent group that contains a lone pair of electrons in the atom that is attached to an aromatic ring is a para- and ortho-director primarily owing to the hyperconjugation effect of the electron-pair.*

5.6   ISOMERIZATIONS EFFECTED BY THE ELECTROPHILIC AROMATIC SUBSTITUTION REACTIONS

Sometimes, a defunctionalization (removal of a functional group) from an aromatic ring can be effected by proton. It is an *ipso*-substitution (with the electrophile being a proton H^+) via an intermediate arenium ion. This may be followed by a re-functionalization. The latter is an EAS reaction with Y^+ (generated from the first *ipso*-substitution by proton) on a different carbon via another arenium ion, resulting in a net isomerization reaction (Eq. 5.15).

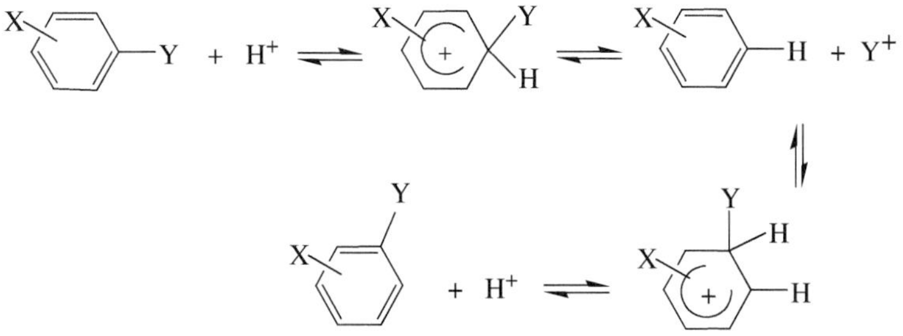

$$(5.15)$$

If the product is thermodynamically more stable than the reactant, the above proton-catalyzed isomerization proceeds irreversibly.

Figure 5.30 shows the facile isomerization of 1,2,4-tri(t-butyl)benzene to 1,3,5-tri (t-butyl)benzene which is catalyzed by $H^+AlCl_4^-$ (a proton source, produced by interaction of a catalytic amount of $AlCl_3$ with trace water—wet $AlCl_3$. See Eq. 5.16) [1].

$$AlCl_3 + H_2O \text{ (trace)} \xrightarrow[-Al(OH)Cl_2]{} HCl \xrightarrow{AlCl_3} H^+ AlCl_4^- \qquad (5.16)$$

The reactant 1,2,4-tri(t-butyl)benzene contains bulky t-butyl groups on two adjacent carbons in the aromatic ring. They interact strongly (electrostatic repulsion) giving rise to molecular constraint and destabilization of the molecule. Defunctionalization (removal of a t-butyl group) from the C_1-carbon will relieve the molecular

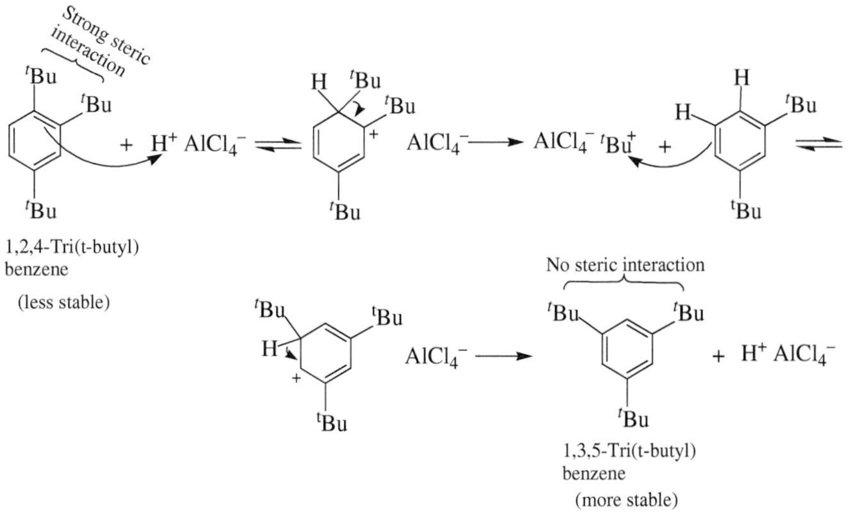

FIGURE 5.30   Mechanism for proton-catalyzed isomerization of 1,2,4-tri(t-butyl)benzene to 1,3,5-tri(t-butyl)benzene. Based on Bruckner [1].

constraint, which serves as the thermodynamic driving force for the *ipso*-substitution of *t*-butyl by H^+. The *ipso*-substitution is followed by a favorable EAS on a *meta*-carbon by a *t*-butyl cation (generated from the first step of the *ipso*-substitution) giving 1,3,5-tri(*t*-butyl)benzene, which is thermodynamically more stable than the reactant because of the relief of the steric interactions of *t*-butyl groups in the product.

Reaction of CH_3Cl with *p*-xylene in the presence of $AlCl_3$ at 25 °C gives 1,2,4-trimethylbenzene. When further heated to 100 °C *in situ*, 1,2,4-trimethylbenzene (a kinetic isomer) is transformed into 1,3,5-trimethylbenzene (a thermodynamic isomer) in the yield of ~60% (Eq. 5.17) [2].

$$(5.17)$$

The by-product HCl generated in the first reaction at 25 °C combines with $AlCl_3$ to give a small amount of $H^+AlCl_4^-$ (Eq. 5.16) which catalyzes the isomerization of 1,2,4-trimethylbenzene to 1,3,5-trimethylbenzene at 100 °C (Reaction 5.17). Similar to the reaction in Figure 5.30, the isomerization in Reaction 5.17 takes place via an *ipso*-substitution. When the second step of de-methylation (loss of methyl) occurs, the methyl lost from the arenium most likely combines with a chloride in $AlCl_4^-$ simultaneously giving a CH_3Cl–$AlCl_3$ adduct. This way, the formation of a highly unstable methyl ion is avoided. Finally, the intermediate 1,3-dimethylbenzene undergoes an EAS reaction with the CH_3Cl–$AlCl_3$ adduct (in which methyl is electrophilic) to give the final 1,3-trimethylbenzene product.

Analogous to the above isomerizations, heating a mixture of 2,4-dichlorocumene [made by chlorination of cumene (isopropylbenzene)] and benzene in the presence of a catalytic amount of $H^+AlCl_4^-$ (generated from $AlCl_3$ and trace water as shown in Equation 5.16) results in removal of the isopropyl group from 2,4-dichlorocumene giving *m*-dichlorobenzene, and the isopropyl group is subsequently transferred to a benzene molecule forming cumene (isopropylbenzene) (Fig. 5.31) [2]. The dealkylation reaction of 2,4-dichlorocumene is initiated by *ipso*-substitution of the isopropyl group by H^+, relieving the steric interaction between the chloro and isopropyl

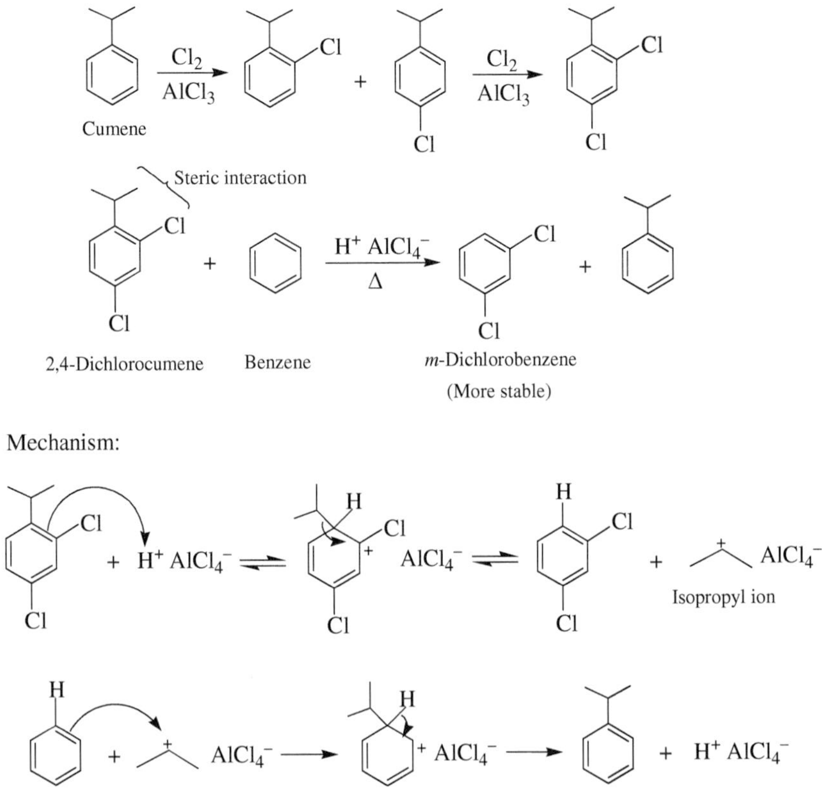

Mechanism:

FIGURE 5.31   Mechanism for proton-catalyzed dealkylation of 2,4-dichlorocumene giving *m*-dichlorobenzene.

groups in the reactant molecule. Since the isopropyl cation generated from the initial *ipso*-substitution is highly reactive toward an aromatic C–H bond, in the presence of benzene, the isopropyl cation will attack the benzene molecule favorably (both kinetically and thermodynamically) giving cumene. The EAS reaction of benzene with the isopropyl cation is free from steric hindrance and almost irreversible, driving the overall reaction to completion.

While *ortho*- and *para*-dichlorobenzene are readily made by chlorination of benzene, it is not easy to make *meta*-dichlorobenzene because the chloro group is the *ortho*- or *para*-director, but not a *meta*-director. The reaction presented in Figure 5.31 provides a general method to synthesize *meta*-dichlorobenzene, with an alkyl benzene, such as cumene, serving as a mediator which is unconsumed. The overall net reaction is a double chlorination of benzene giving *m*-dichlorobenzene.

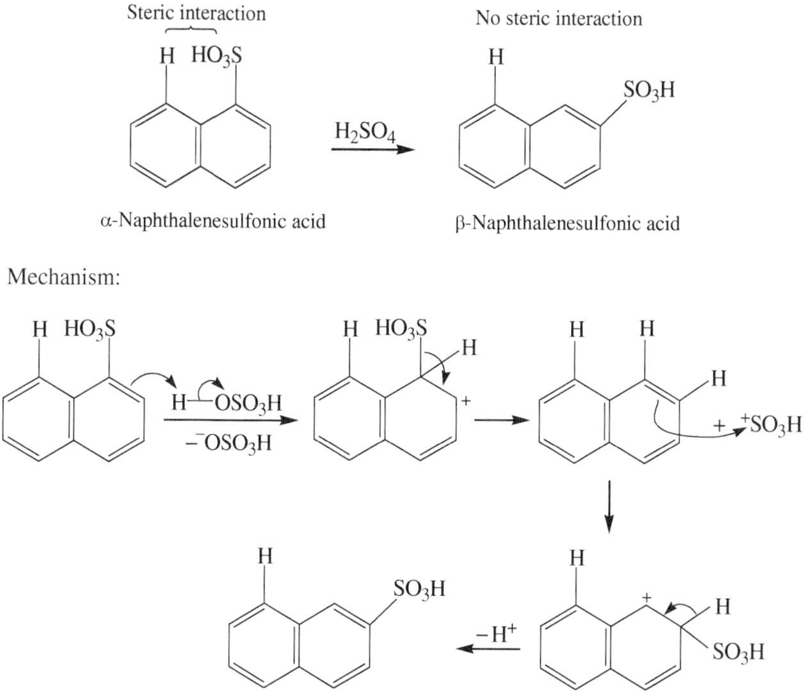

FIGURE 5.32   Mechanism for acid-catalyzed isomerization of α-naphthalenesulfonic acid to β-naphthalenesulfonic acid. Based on Bruckner [1].

In the presence of sulfuric acid, α-naphthalenesulfonic acid isomerizes to the thermodynamically more stable β-naphthalenesulfonic acid (Fig. 5.32) [1]. The reaction is also initiated by an *ipso*-substitution with H^+ (defunctionalization) releasing a reactive $^+SO_3H$ cation. Then $^+SO_3H$ preferably attacks the β-carbon (less sterically hindered than the α-carbon) of naphthalene produced from the *ipso*-substitution to give the more stable β-naphthalenesulfonic acid (re-functionalization). The overall isomerization is both kinetically and thermodynamically favorable.

5.7   ELECTROPHILIC SUBSTITUTION REACTIONS ON THE AROMATIC CARBON–METAL BONDS: MECHANISMS AND SYNTHETIC APPLICATIONS

The aryl Grignard (Ar–Mg–X, X = Cl, Br, or I) and aryllithium (Ar–Li) compounds are readily made from aryl halides. They are strong nucleophiles, while the corresponding Ar–H is only weakly nucleophilic. Therefore, aryl Grignard and aryllithium reagents can effectively react with many weak electrophiles to bring about electrophilic substitutions on the aromatic ring, which cannot be achieved by weakly nucleophilic aromatic C–H bond. In addition, some functional groups that are attached to an aromatic ring and contain a carbonyl (C=O) unit can direct metallation (a process of making a carbon–metal bond) selectively on the *ortho*-carbon, affording a synthetically significant methodology.

5.7.1 Aryl Grignard and Aryllithium Compounds

A remarkable example is the functionalization of mesitylene (1,3,5-trimethylbenzene), which can be very effectively achieved using the aryl-metal compounds (Fig. 5.33) [1]. First, mesitylene is brominated, followed by reaction of the resulting aryl bromide (ArBr) with magnesium or lithium in diethyl ether to make aryl Grignard reagent (ArMgBr) or aryllithium (ArLi). Then each of the aryl-metal compounds will be treated with different electrophiles, such as carbon dioxide, acetone, methyl iodide, and D$_2$O, to give a carboxylic acid, a tertiary alcohol, a

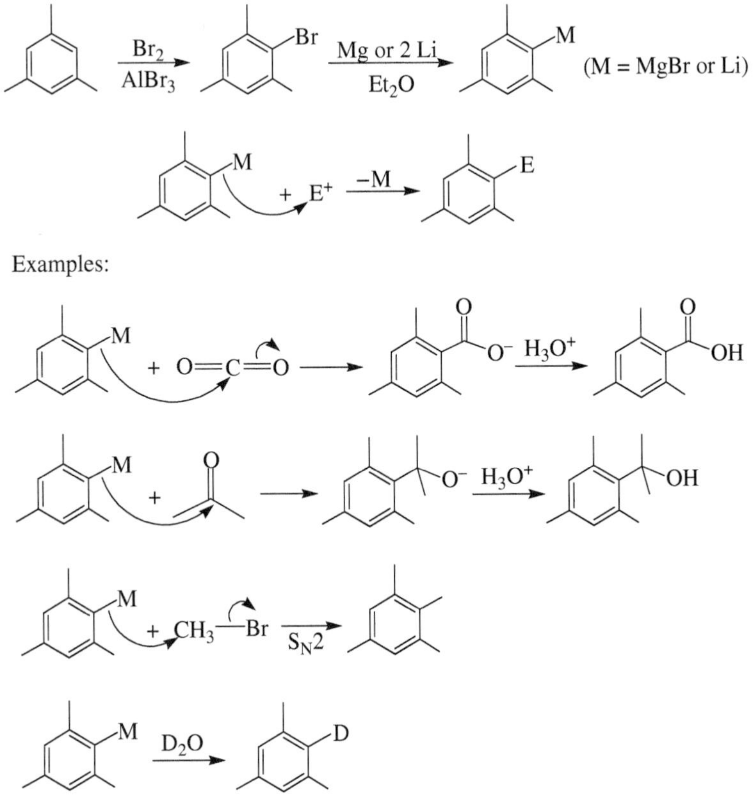

Examples:

FIGURE 5.33 Metals (magnesium and lithium) facilitated electrophilic aromatic C–H bond substitutions. Based on Bruckner [1].

tetramethylbenzene, and a deuterated mesitylene, respectively. Introducing each of these groups (carboxyl, hydroxyl, methyl, and deuterium) to mesitylene is not easy to fulfill. With the aid of the aryl Grignard and aryllithium compounds, the above synthetic goals can be readily achieved.

5.7.2 *Ortho*-Metallation-Directing Groups (*o*-MDGs): Mechanism and Synthetic Applications

Figure 5.34 shows that some functional groups containing a carbonyl (C=O) unit, such as the –C(O)NR$_2$ and –OC(O)NR$_2$ groups, can direct a lithiumation (making a C—Li bond) using *sec*-butyllithium (^{s}BuLi) on the *ortho*-carbon of the aromatic ring, such that the C$_{ortho}$–H hydrogen is substituted by a lithium atom forming an aryl-lithium (ArLi) compound [1]. Such a substitution is unusual. It cannot happen to the C—H bond in a regular aryl ring. With an oxygen lone-pair of electrons, the carbonyl group in –C(O)NR$_2$ or –OC(O)NR$_2$, an *ortho*-metallation-directing group (*o*-MDG), can strongly interact with the positively charged lithium (–C=O: →Li–^{s}Bu). This interaction weakens the C$_{sec}$—Li bond in ^{s}BuLi making the ^{s}Bu–group highly basic and nucleophilic. The C$_{ortho}$—H bond in the aromatic ring is activated by the adjacent carbonyl group, becoming slightly acidic. As the active C$_{sec}$—Li bond in ^{s}BuLi is being held in place by an *o*-MDG toward the C$_{ortho}$—H bond in the aromatic ring, the highly basic ^{s}Bu– will attack the C$_{ortho}$—H hydrogen effectively to weaken the C$_{ortho}$—H bond and to make the *ortho*-carbon nucleophilic.

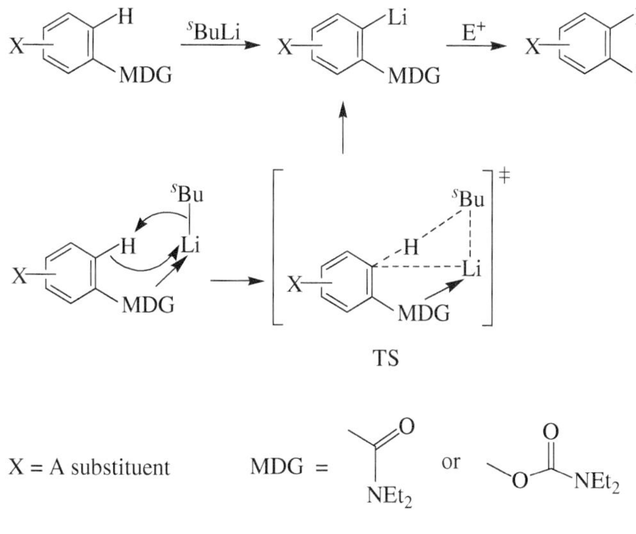

FIGURE 5.34   Metal directing groups (MDG's) facilitated lithiumation of arenes. Reprinted from Ref. [1], Page 201, with permission from Elsevier.

Simultaneously, the nucleophilic *ortho*-carbon in the aromatic ring attacks the lithium in ^{S}BuLi, which has been held in place, to make a C_{ortho}—Li bond. After the aryllithium compound (Ar–Li) is formed, it can react with an electrophile in the same way as that shown in Figure 5.33, to give Ar–E. As a result, the *ortho*-carbon has been selectively functionalized. Examples of selective *ortho*-carbon functionalization facilitated by the aryl C—Li bond in aryllithium compounds are shown in Figure 5.35.

The EAS reactions facilitated by *ortho*–MDGs have recently found interesting applications in synthesis of useful polycyclic aromatic hydrocarbon derivatives (Fig. 5.36) [21]. The $Et_2NCOO–$ group in the polycyclic arene (Fig. 5.36) directs lithiumation (the formation of a C—Li bond) by ^{S}BuLi on the *ortho*-carbon as depicted in Figure 5.35. Then the intermediate aryllithium compounds react with CH_3I (source of the electrophilic methyl) to give the targeted products in almost quantitative yields. In the second reaction in Figure 5.36, the lithiumation as well as the following methylation are regioselective on one specific *ortho*-carbon as reaction occurring on the other *ortho*-carbon would be strongly sterically hindered by the adjacent ring structure. The polycyclic aromatic products have found utility as fluorescence materials as well as in certain medicinal areas such as some related compounds possessing anticancer activity [21].

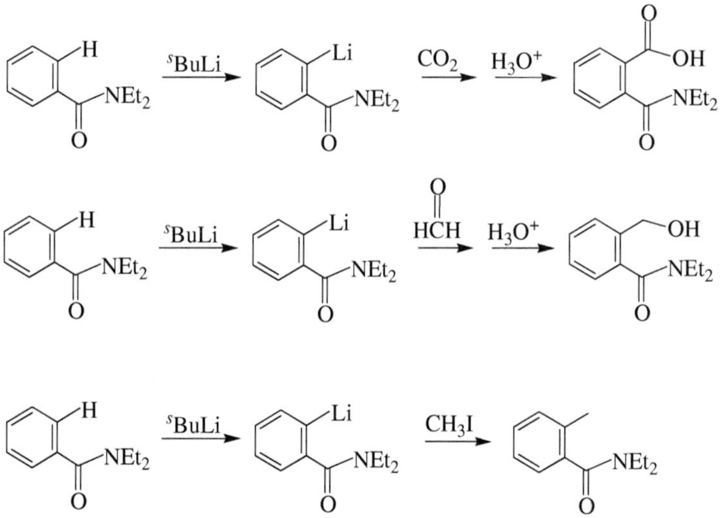

FIGURE 5.35 Reactions of substituted arenes with ^{S}BuLi followed by substitutions with various electrophiles.

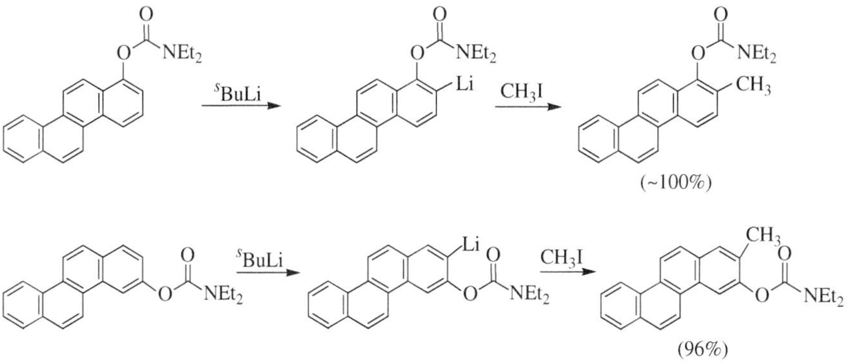

(~100%)

(96%)

FIGURE 5.36 *ortho*-Metal directing group facilitated lithiumation and subsequent methylation of polycyclic arenes. Based on Kancherla et al. [21].

5.8 NUCLEOPHILIC AROMATIC SUBSTITUTION VIA A BENZYNE (ARYNE) INTERMEDIATE: FUNCTIONAL GROUP TRANSFORMATIONS ON AROMATIC RINGS

In an electrophilic substitution reaction, an aromatic C—H bond is substituted by an electrophile giving a functionalized aromatic compound. When a halo (–X) group [X = Cl (chloro), Br (bromo), or I (iodo)] is attached to a benzene (arene) ring, the C—X bond can be substituted by a strong nucleophile (also a strong base) such as hydroxide (OH^-), cyanide (CN^-), and amide (NH_2^-). As a result, one type of functionalized aromatic compound (halobenzene) is transformed into another (phenol, benzonitrile, or aniline). This type of reactions is called **nucleophilic aromatic substitution (NAS)**.

The nucleophilic substitution reaction of a halobenzene (haloarene) by a strong nucleophile (also a base) $:B^-$ (such as OH^-, CN^-, or NH_2^-) takes place via an intermediate **benzyne (aryne)** which is developed from the benzene (arene) ring in the course of the reaction, with a C≡C triple bond formed in the ring (Fig. 5.37). Different from the C≡C bond in an alkyne, a linear molecule with the triply bonded carbons sp-hybridized and each of the two π bonds formed by overlap of p orbitals in sideways, the triply bonded carbons in benzyne are sp^2-hybridized [1]. The two sp^2 orbitals are pointing away from the ring. They have distorted sideway overlap outside of the ring, forming a very weak π bond. Therefore, benzyne (aryne) is highly reactive and vulnerable of a nucleophilic attack by a second $:B^-$ molecule. The overall reaction is stepwise, initiated by a strong-base ($:B^-$) induced β-elimination of hydrogen halide (HX) (E2-like reaction), followed by a nucleophilic addition of $:B^-$ to the intermediate benzyne (aryne).

The stepwise NAS mechanism via a benzyne intermediate is confirmed by reaction of hydroxide (OH^-) with the C-14 isotope-labeled chlorobenzene (Fig. 5.38) [1]. In the reactant, the chloro group is attached to a C-14 (a carbon enriched by

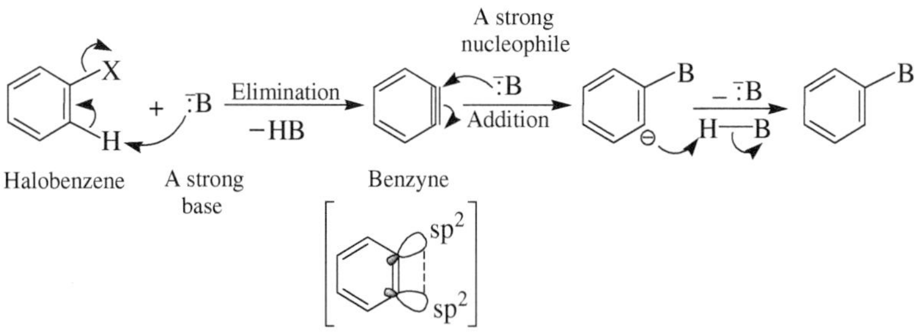

FIGURE 5.37 Nucleophilic aromatic substitution (NAS) of a halobenzene via the benzyne intermediate. Based on Bruckner [1].

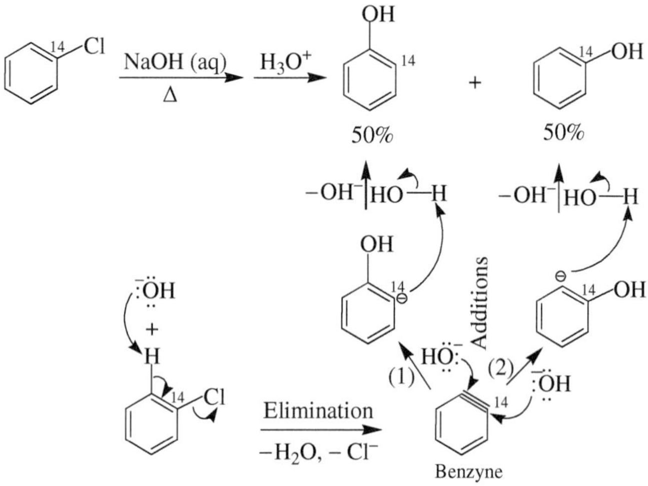

FIGURE 5.38 Mechanism for nucleophilic substitution of the carbon-14 isotope-labeled chlorobenzene with hydroxide: Proof for NAS via benzyne. Based on Bruckner [1].

the C-14 isotope). The reaction gives two types of phenol products in an equal amount. In one type of the phenol product, the hydroxyl group is attached to a regular carbon (with a natural abundance for each of the isotopes), and in the other type of the phenol product, the hydroxyl group is attached to a C-14. The results demonstrate that the nucleophilic substitution of chlorobenzene by hydroxide takes place via an intermediate benzyne in which the triple bond is formed on the C-14 ($^{14}C\equiv C$), the same as expected from a β-elimination because the chloro group in the original chlorobenzene is attached to C-14. In the second step, hydroxide (nucleophile) attacks equally the two triply bonded $^{14}C\equiv C$ carbons, the C-14 and regular carbon, in the benzyne molecule, giving an equal amount of two products as observed.

In industry, phenol is made (one method) by NAS reaction of chlorobenzene with sodium (or potassium) hydroxide at high temperature as shown in Equation 5.18 [1].

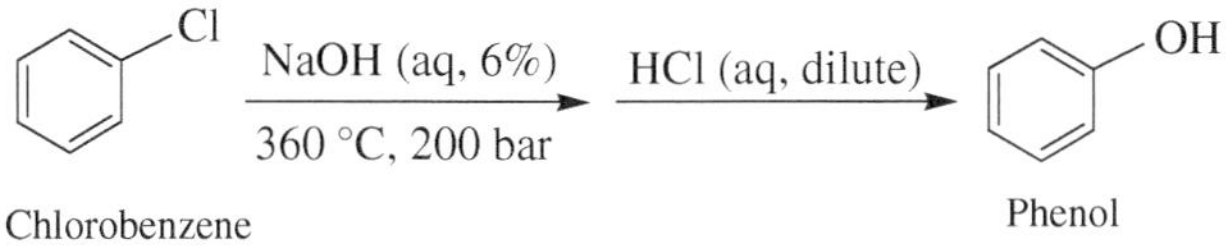

$$\text{Chlorobenzene} \xrightarrow[\text{360 °C, 200 bar}]{\text{NaOH (aq, 6\%)}} \xrightarrow{\text{HCl (aq, dilute)}} \text{Phenol} \tag{5.18}$$

Reaction of *o*-chlorotoluene with sodium hydroxide gives roughly an equal mixture of 2-methylphenol (*o*-cresol) and 3-methylphenol (*m*-cresol) (Fig. 5.39) [3]. It further demonstrates that the NAS reaction proceeds via an intermediate aryne formed by a strong-base (OH$^-$) initiated β-elimination of an HCl molecule that takes place between the *ortho*- and *meta*-carbons. Then either the triply bonded *o*-carbon or *m*-carbon will be attacked by the nucleophilic hydroxide (OH$^-$) giving two structural isomers of methylphenol.

Benzonitrile and aniline can also be made by similar NAS reactions of halobenzene with nucleophilic (also basic) cyanide (CN$^-$) and amide (NH$_2^-$), respectively (Fig. 5.40). Research has shown that reaction of NH$_2^-$ with C14-labelled iodobenzene on the C–I carbon gives two aniline products, one bearing ^{14}C—NH$_2$ bond (47%) and the other bearing regular C—NH$_2$ bond (53%) in roughly equal amounts. The results support the formation of a benzyne intermediate during the reaction (Fig. 5.40). The C-14 and regular carbon in the ^{14}C≡C bond of the benzyne can be attacked by NH$_2^-$ roughly equally giving the two aniline products in about equal amounts. With each above nucleophile (base), reaction of a halotoluene will give a mixture of two structural isomers comparable to the reaction in Figure 5.39.

Aryldiazonium (Ar–$^+$N=N) salts undergo NAS reactions following an S$_N$1-like mechanism (Fig. 5.41). For the first reaction in Figure 5.41, an aryldiazonium is heated in water (a nucleophile) and as a result, the –$^+$N=N group is substituted by the hydroxyl –OH group giving a phenol. For the second reaction, an

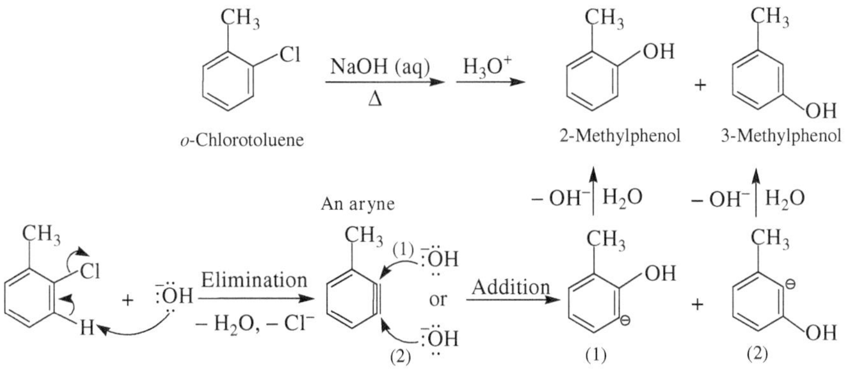

FIGURE 5.39   Nucleophilic substitution reaction of *o*-chlorotoluene with hydroxide.

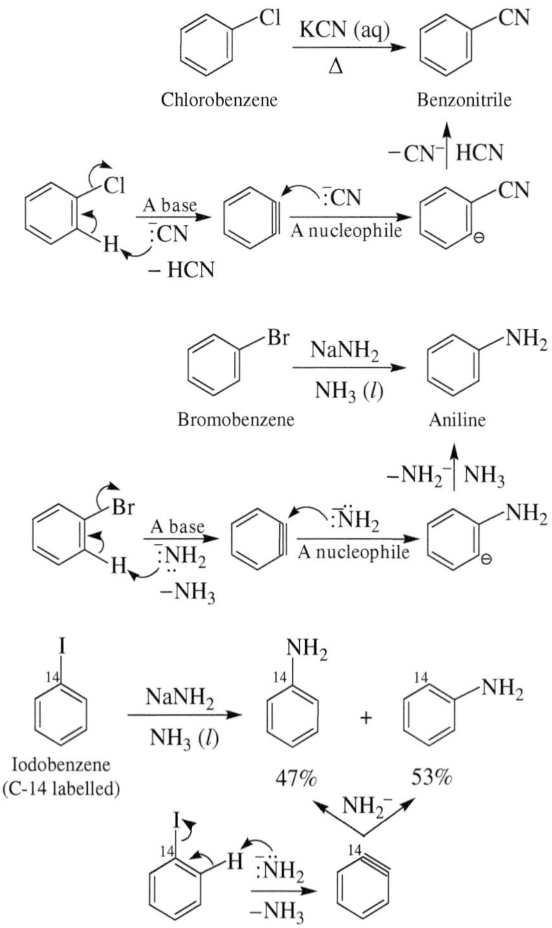

FIGURE 5.40 Nucleophilic substitution reactions of halobenzenes with cyanide and amide.

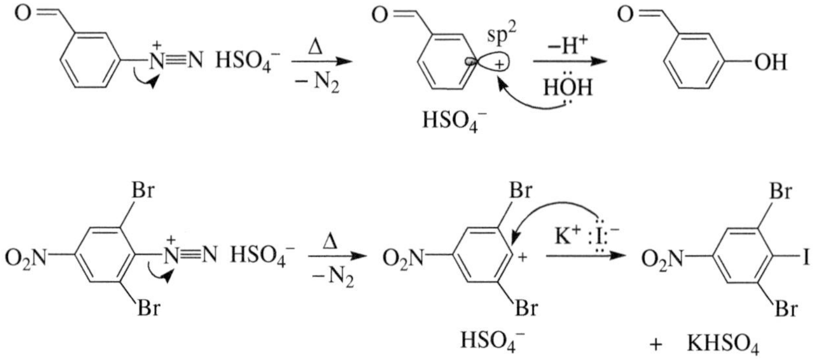

FIGURE 5.41 Nucleophilic substitution reactions of aryldiazonium (Ar–$^+$N=N) following an S_N1-like mechanism.

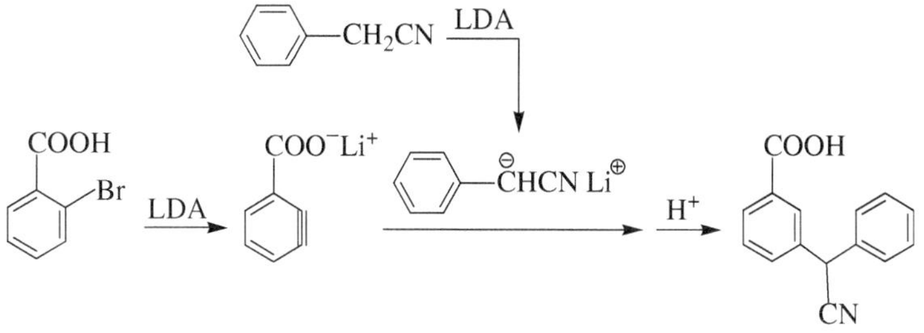

FIGURE 5.42   Reaction of 2-bromobenzoic acid with benzyl nitrile ($C_6H_5CH_2CN$) in the presence of excess LDA. Based on Wang et al. [22].

aryldiazonium is treated with potassium iodide (a nucleophile) by heating, and the $-^+N{=}N$ group is substituted by the iodo $-I$ group. For both reactions, the first step is a thermal elimination of the $-^+N{=}N$ group from an aromatic ring resulting in an intermediate cation in which a positive charge is held in an sp^2 orbital of a carbon in the aromatic ring. The cationic intermediate is highly reactive and is readily attacked by a nucleophile completing the nucleophilic substitution reaction.

Research has shown that in the presence of excess lithium diisopropyl amide (LDA, an extremely strong base), 2-bromobenzoic acid undergoes the NAS reaction with $ArCH_2CN$ via an aryne intermediate and the nucleophilic $ArCH^-CN$ carbanion (Fig. 5.42) [22]. With its negatively charged carbon, $ArCH^-CN$ attacks the *meta* $C{\equiv}C$ carbon in the aryne to give the final product after aqueous work-up. Only a *meta*-product is formed most likely owing to the steric hindrance of the $-COO^-Li^+$ group on the *ortho*-carbon of the aryne preventing the nucleophilic attack on it. In addition, a rearranged 2-cyanobenzoic acid product is also formed.

5.9   NUCLEOPHILIC AROMATIC SUBSTITUTION VIA AN ANIONIC MEISENHEIMER COMPLEX

A halobenzene containing an EWG undergoes nucleophilic substitution which pro-
ceeds via an anionic Meisenheimer complex [1]. The general mechanism is shown in Figure 5.43. The intermediate Meisenheimer anion is stabilized by the EWG via delocalization of the negative charge from the aromatic ring to the EWG. As a result, the formation of the anionic intermediate effected by an initial nucleophilic attack is kinetically favorable with a relatively low activation energy. The charge distributions to the *ortho-*, *meta-*, and *para*-carbons in the Meisenheimer complex are com-
parable to those for an arenium ion (Section 5.5, Figure 5.24). The *ortho-* and *para*-
carbons receive the major portions of the negative charge. Therefore, if these carbons are occupied by EWGs, the NAS via a Meisenheimer intermediate will be particularly facilitated. *If the aromatic ring of a haloarene does not contain any EWGs or contain one or more EDGs, the NAS reaction will occur via an aryne intermediate.*

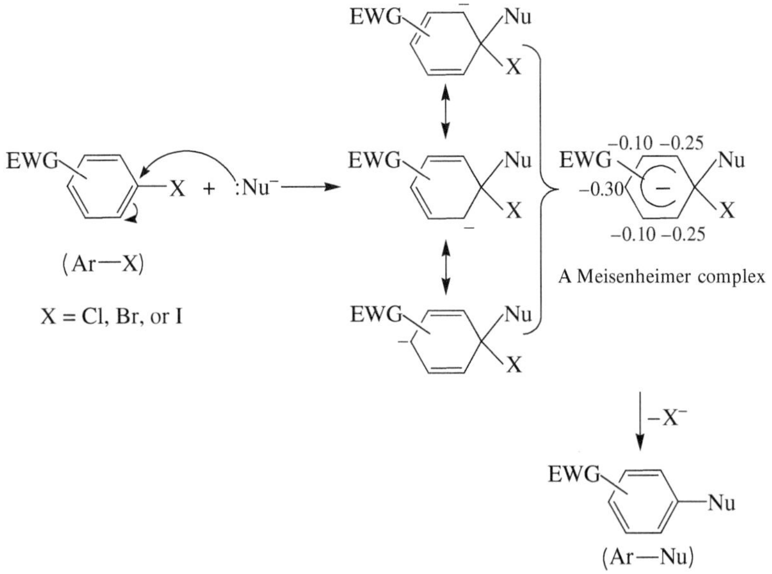

FIGURE 5.43   Nucleophilic aromatic substitution via the Meisenheimer complex.

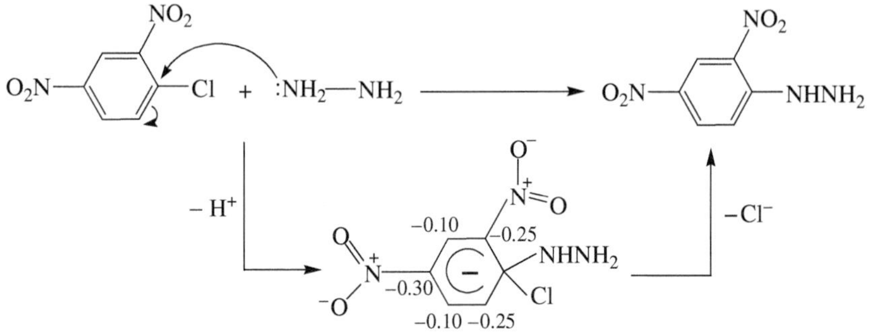

FIGURE 5.44   Reaction of 2,4-dinitrochlorobenzene with hydrazine giving 2,4-dinitrophenylhydrazine.

Figure 5.44 shows reaction of 2,4-dinitrochlorobenzene with nucleophilic hydrazine (NH_2NH_2). In the reactant, the *ortho-* and *para-*carbons to the chloro group are occupied by strongly electron-withdrawing nitro ($-NO_2$) groups. In the Meisenheimer intermediate, the negative charge is largely (55%) distributed to the *ortho-* and *para-*carbons and is very well stabilized by the nitro groups. As a result, the reaction possesses a relatively low activation energy. The chloro group can be substituted readily by the nucleophilic NH_2NH_2 giving 2,4-dinitrophenylhydrazine. Only one isomer of dinitrophenylhydrazine is formed from the reaction, which excludes the possibility for the reaction to proceed via an aryne intermediate. Instead, due to the effective stabilization of the negative charge by the electron-withdrawing nitro groups, the reaction proceeds via a more favorable, relatively lower-energy Meisenheimer complex.

Similarly, 2,4-dinitrochlorobenzene reacts with water (a weak nucleophile) upon relatively mild heating (100 °C) to give 2,4-dinitrophenol (Eq. 5.19) [1].

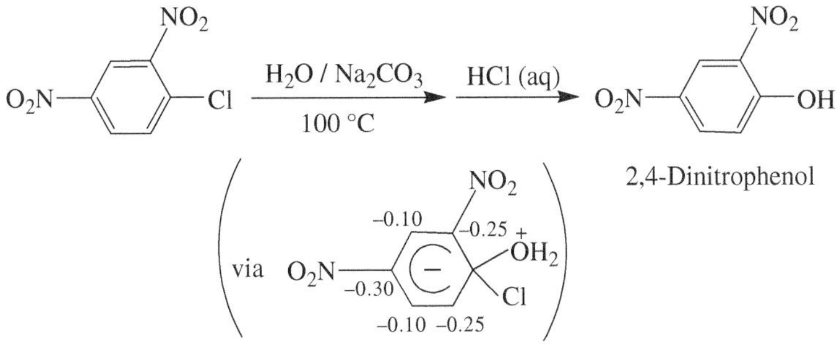

$$
\begin{array}{c}
\text{(Eq. 5.19)}
\end{array}
$$

2,4-Dinitrophenol   (5.19)

The intermediate Meisenheimer complex is greatly stabilized by the two nitro groups in the *ortho*- and *para*-carbons. Thus, even weakly nucleophilic water can effect the NAS reaction. Only one isomer of dinitrophenol is formed from the reaction, which excludes the possibility for the reaction to proceed via an aryne intermediate.

The highly polar C—F bonds in 2,4-difluoronitrobenzene make the fluorine-bonded carbon atoms electrophilic. In liquid ammonia, the compound undergoes a facile NAS reaction, which almost exclusively occurs on the C—F bond of the *ortho*-carbon (Fig. 5.45) [23]. The reaction proceeds via a Meisenheimer complex which is stabilized by an intramolecular hydrogen bond formed between the $-NH_3^+$ and $-NO_2$ groups. Such stabilization by an intramolecular hydrogen bond would be impossible for the *para*-substitution. Therefore, the substitution of the C—F bond of the *ortho*-carbon by ammonia is much faster than that of the *para* C—F bond and the corresponding *ortho*-substitution product is almost the sole product.

Oxygen nucleophiles, such as alkoxide and phenoxide ions, react readily with 4-nitrofluorobenzene in liquid ammonia to give the corresponding substitution product (Eq. 5.20) [23].

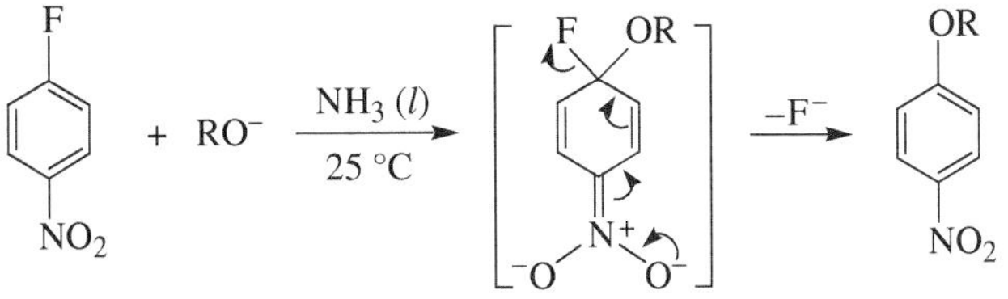

$$
\begin{array}{c}
\text{(Eq. 5.20)}
\end{array}
$$
(5.20)

There is little solvolysis product formed as the reaction of 4-nitrofluorobenzene with ammonia (the C—F bond substitution) is too slow to compete with the nucleophilic C—F bond substitution by anionic O-nucleophiles.

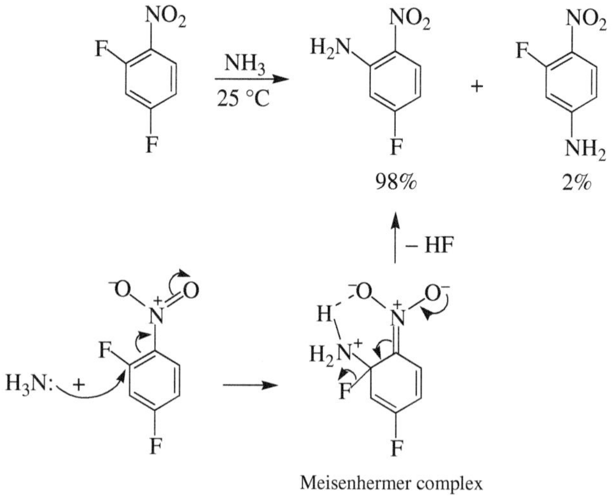

FIGURE 5.45 Nucleophilic substitution of 2,4-difluoronitrobenzene with liquid ammonia. Based on Ji et al. [23].

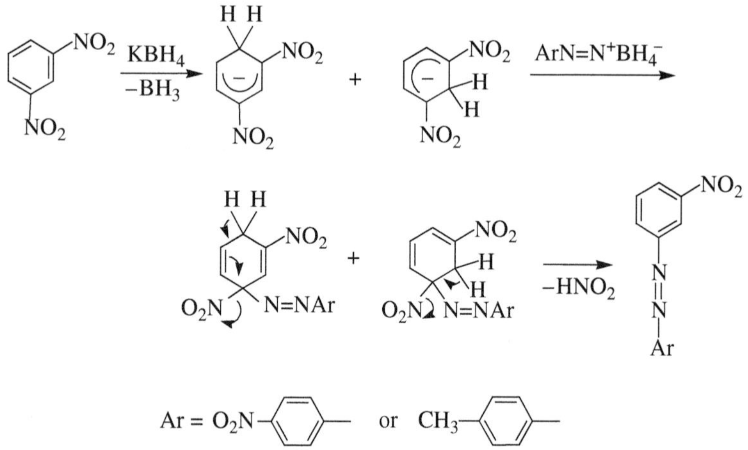

FIGURE 5.46 The azocoupling reaction between 1,3-dinitrobenzene and aryldiazonium via isomeric Meisenheimer anions.

Treatment of 1,3-dinitrobenzene with reductive and nucleophilic potassium borohydride (KBH_4), followed by reacting with substituted phenyldiazonium tetrafluoroborates ($ArN_2^+BF_4^-$) leads to an azocoupling reaction (Fig. 5.46) [24]. BH_4^- attacks an electrophilic carbon which is *ortho* to a nitro group in 1,3-dinitrobenzene to give two isomeric Meisenheimer anions (also called σ-complexes) identified by NMR spectroscopy. Then the ArN_2^+ ion react with each of the

Meisenheimer anions on a C–NO_2 carbon. This will be followed by a rapid elimination of HNO_2 to give the azocoupling product.

5.10   BIOLOGICAL APPLICATIONS OF FUNCTIONALIZED AROMATIC COMPOUNDS

Combretastatin A-4 (CA4), a derivative of *cis*-stilbene (Fig. 5.47), has been found significantly cytotoxic owing to its inhibitory effects on tubulin polymerization and cell growth. It has been used as a chemotherapeutic drug. Modifications of CA4 have been performed by substituting the C=C moiety by a variety of spacers of the two aromatic rings, such as a carbonyl group (C=O), a methylene group (CH_2), a reduced CH_2CH_2 group, and a divalent sulfur (S) atom (Fig. 5.47). These structurally similar compounds have shown to have analogous biological activities to those possessed by CA4 [25]. Studies have shown that double bond reduction of CA4 yields compounds with only moderately reduced biological activity, as does a methylene spacer group. Moreover, if the hydroxyl group in CA4 is missing or replaced by another group (such as methoxy), the effect on the biological activity is only minimal [25, 26].

The **diaryl sulfide** analogue (Fig. 5.48) of CA4 has particular high activity in inhibitory effects on the cancer cell growth. Its synthesis involves a series of nucleophilic substitution reactions as shown in Figure 5.48 [26].

The first NAS reaction is the substitution of the $-^+N\equiv N$ group from an $Ar-^+N\equiv N$ (aryldiazonium) by a strongly nucleophilic iodide anion, which has been discussed in Section 5.8 (Fig. 5.41). Then the resulting aryl iodide reacts with an ArSH in neocuproine (a nitrogen base) in the presence of a catalytic amount of copper (I) iodide (CuI). The aryl iodide forms an adduct with CuI readily via a lone pair of electrons in the iodine (Fig. 5.48). Thus, the aromatic C—I bond is activated by the positive charge on iodine. Analogous to the $Ar-^+N\equiv N$ reaction, the cleavage of the activated C—I bond results in the formation of a highly reactive Ar^+ intermediate, which is subsequently attacked by the strongly nucleophilic ArS^-. This gives rise to an overall S_N1-like substitution to produce the biologically very active diaryl sulfide.

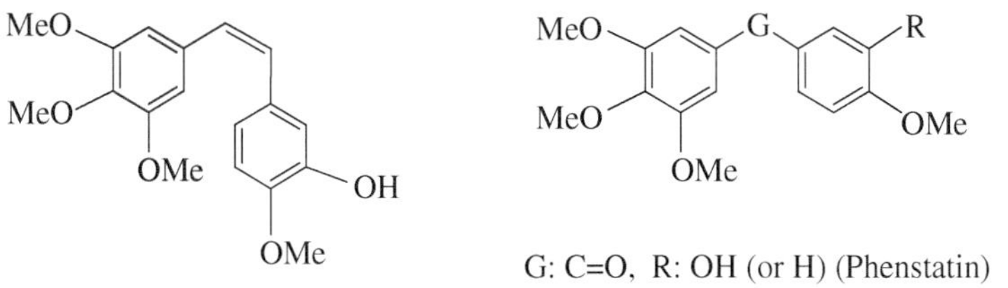

Combretastatin A-4 (CA4)
(A derivative of *cis*-stilbene)

G: C=O, R: OH (or H) (Phenstatin)
G: CH_2, R: OH (or H) (Phenstatin)
G: CH_2CH_2, R: OH (or H)
G: S, R: H (Diaryl sulfide)

FIGURE 5.47   Structures of Combretastatin A-4 (CA4) and modified analogues.

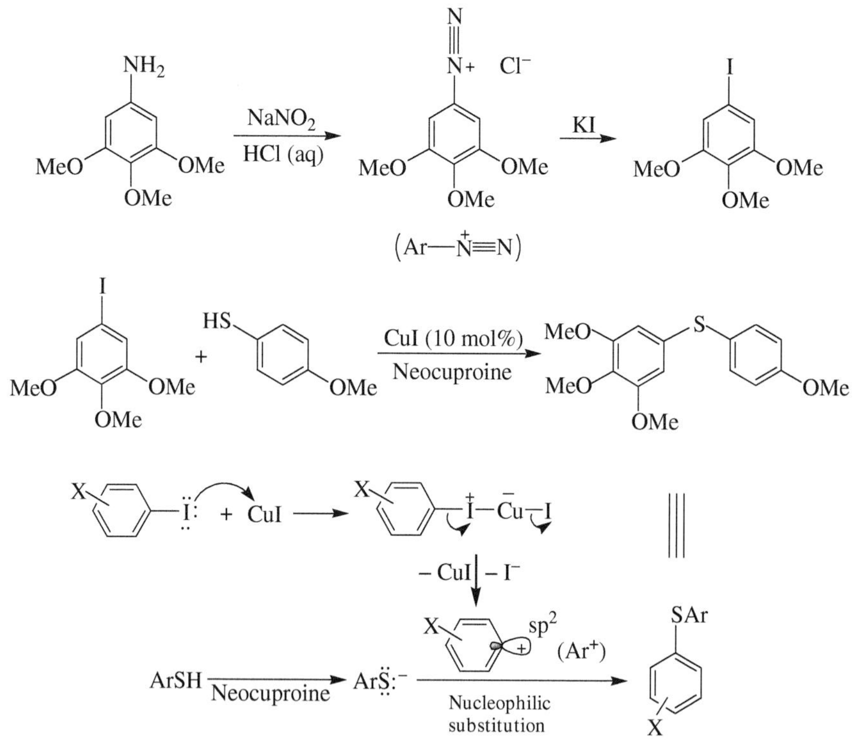

FIGURE 5.48 Synthesis of the diaryl sulfide analogue of CA4 by nucleophilic aromatic substitution reactions and the reaction mechanism. Based on Barbosa et al. [26].

Aryl-sulfoxide-containing aromatic nitrogen mustards (Fig. 5.49) have been used as hypoxia-directed bioreductive cytotoxins (anti-cancer drugs) to treat solid tumors. The sulfoxide is first reduced by NADPH *in vivo* to sulfide. It is believed that the lone pair of electrons in sulfur of the aryl sulfide activates the nitrogen atom in a side chain by enhancing its nucleophilicity via a conjugation effect through the aromatic ring. As a result, an intramolecular nucleophilic substitution takes place in a side chain to give an aziridinium with a positive charge formed on the tetravalent nitrogen. The highly reactive (electrophilic) aziridinium is responsible for the cytotoxicity.

A particularly effective cytotoxic nitrogen mustard of this class is an aryl-sulfoxide-containing polycyclic system which is synthesized by NAS of 9-chloroacridine (Fig. 5.50) [25]. The C–Cl carbon in the neutral aromatic 9-chloroacridine molecule is essentially unreactive toward any nucleophiles. However, after the nitrogen atom in the ring is protonated by HCl, delocalization of the positive charge resulted from protonation makes the C–Cl carbon strongly electrophilic so that it can be attacked by a nucleophilic $ArNH_2$ to form an Meisenheimer-complex-like intermediate

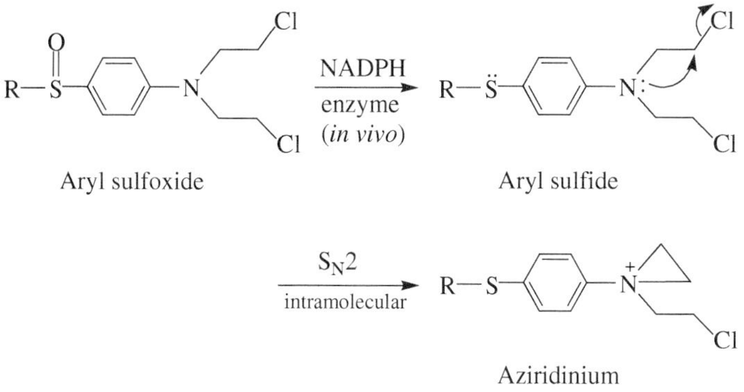

$R = n\text{-}C_3H_7,\ C_6H_5,\ p\text{-}O_2NC_6H_4,\ p\text{-}H_2NC_6H_4,$ and $p\text{-}(C_2H_4Cl)_2NC_6H_4.$

FIGURE 5.49   The aryl-sulfoxide-containing nitrogen mustard and its *in vivo* reactions.

(Fig. 5.50). Upon elimination of the *ipso*-chloride, the reactive intermediate is transformed into the biomedically active aryl-sulfoxide-containing nitrogen mustard.

Other biomedically active aromatic compounds include **ibuprofen**, an anti-inflammatory agent that functions as a pain-relieving and fever-reducing drug. Its precursor 4-isobutylacetophenone can be made from benzene by consecutive Friedel–Crafts acylation reactions (Fig. 5.51). Then a series of reactions on the carbonyl group of 4-isobutylacetophenone are performed giving racemic ibuprofen as shown in Figure 5.51. Only the (*S*)-enantiomer is biologically active and possesses medical effect. The opposite (*R*)-enantiomer is biologically inactive, but it is not harmful. Therefore, in practice, the drug is synthesized as a racemic mixture with both the biomedically active (*S*)-enantiomer and inactive (*R*)-enantiomer staying together without further separation.

2,6-Di(*t*-butyl)-4-methylphenol, alternatively known as butylated hydroxytoluene (BHT), is used as an antioxidant in foods to retard spoilage. BHT is synthesized in industry from 4-methylphenol by reaction with 2-methylpropene in the presence of phosphoric acid as a catalyst (Eq. 5.21) [3].

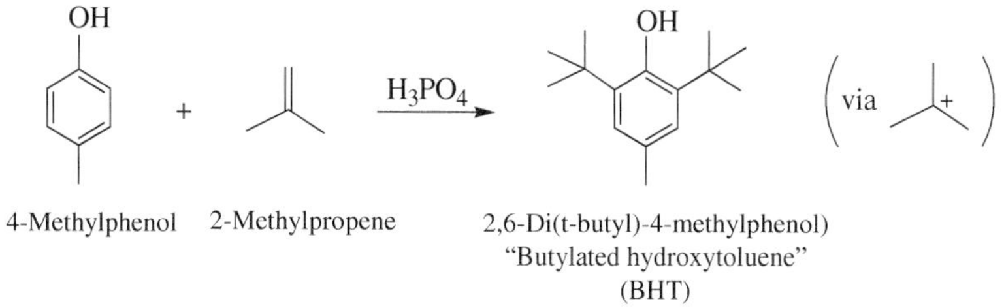

4-Methylphenol   2-Methylpropene   2,6-Di(t-butyl)-4-methylphenol)
                                          "Butylated hydroxytoluene"
                                                    (BHT)

(5.21)

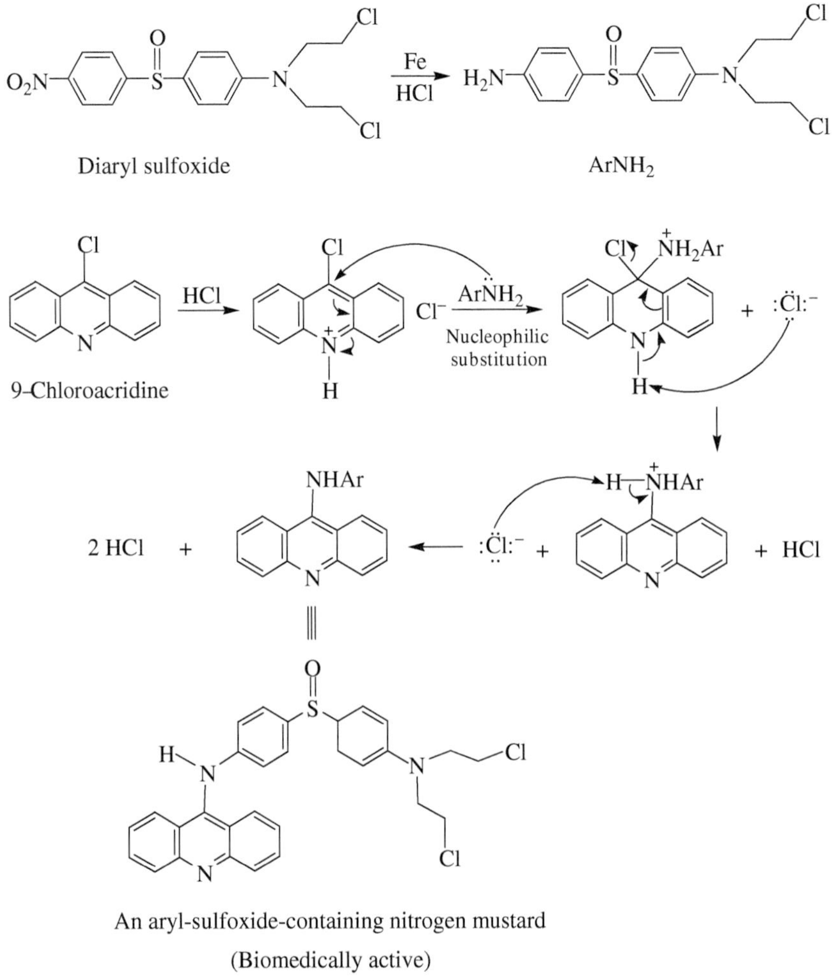

FIGURE 5.50   Synthesis of a biomedically active aryl-sulfoxide-containing polycyclic system and the reaction mechanism. Based on Sun et al. [25].

The EAS reaction takes place via a *t*-butyl cation formed by the initial protonation of 2-methylpropene.

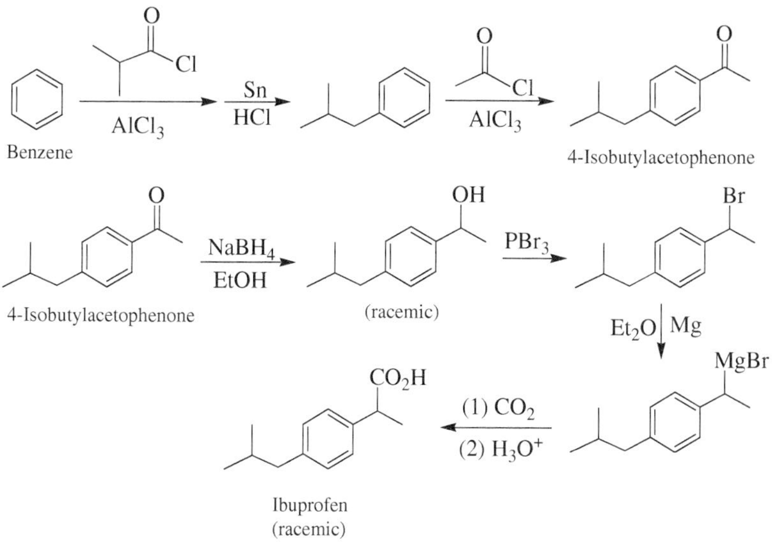

FIGURE 5.51 Chemical synthesis of the inflammatory ibuprofen and the reaction mechanism.

PROBLEMS

5.1 Using nitration of benzene by NO_2^+ as an example, draw reaction profile for the aromatic charge transfer nitration. Include all the intermediates in the profile. What is the rate-determining step?

5.2 In the presence of $AlBr_3$, reaction of benzene with elemental bromine giving bromobenzene is shown to take place through the intermediate brominium Br^+ ion which is formed by interaction of Br_2 with $AlBr_3$. The following schemes represent two alternative mechanisms for the reaction.

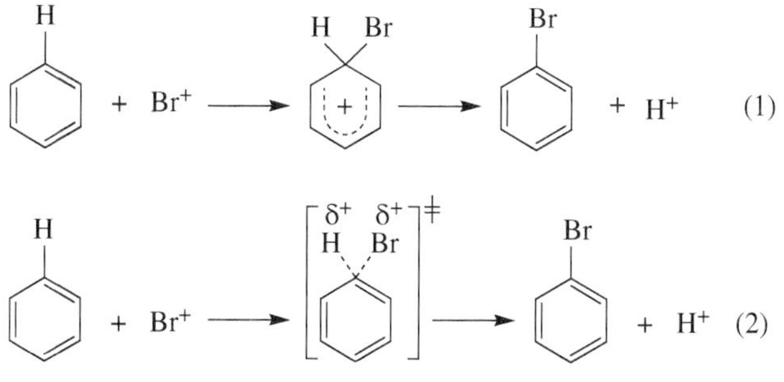

Which mechanism is true for the reaction? Explain your answer by comparison of energetics for both mechanisms.

5.3 Provide a mechanism to account for each of the following reactions.

(1)

(2)

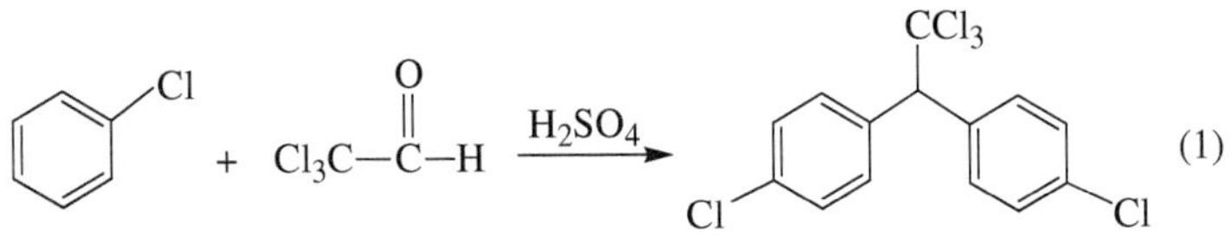

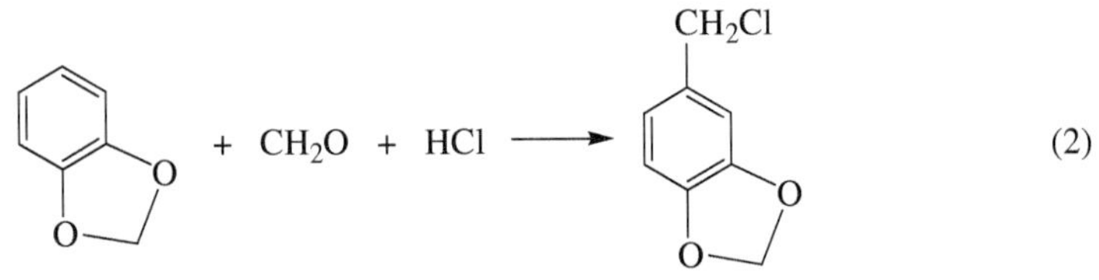

5.4 Propose a mechanism for the following reaction.

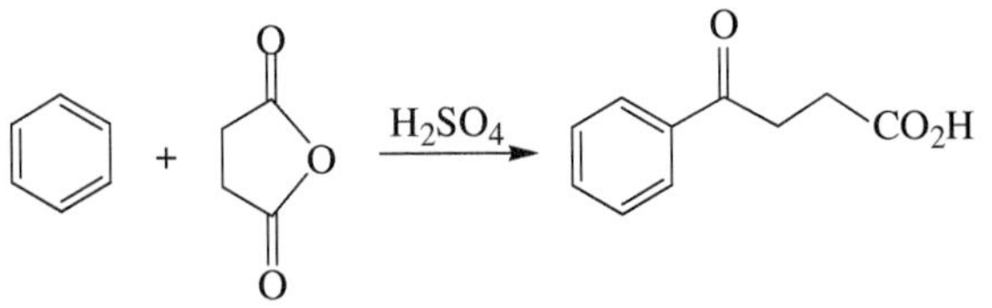

5.5 Give the major product and show a mechanism for each of the following reactions.

(1)

(2)

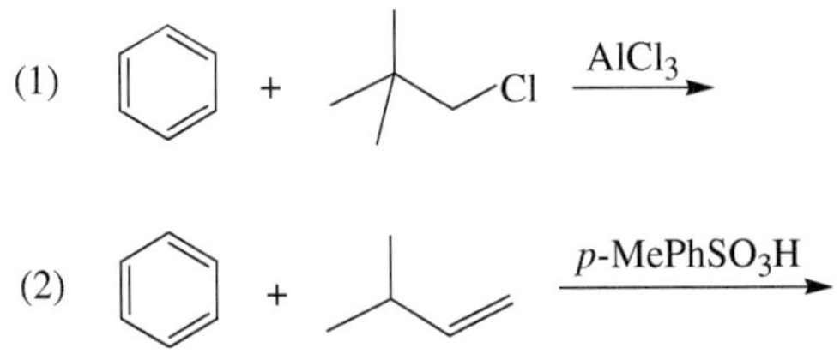

5.6 As shown below, the AlCl₃ catalyzed chlorination of cumene gives a dichlorination product, which upon heating undergoes isomerization. Propose a detailed mechanism to account for the isomerization process.

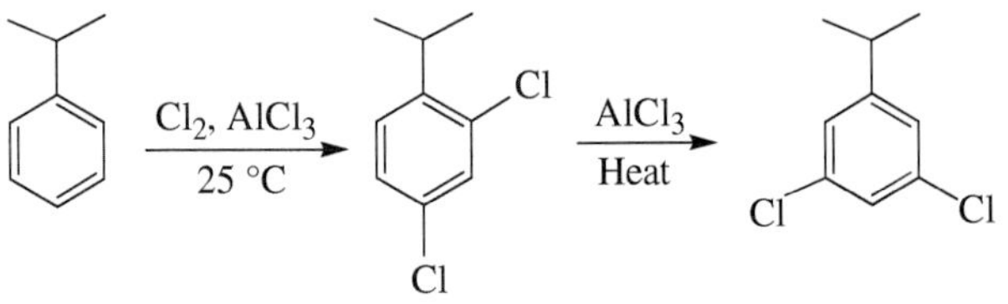

5.7 Give product for the following reaction and suggest a mechanism to account for the reaction.

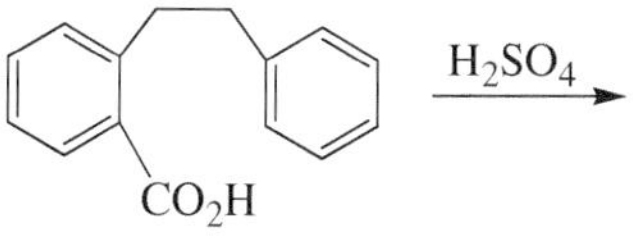

5.8 Provide a mechanism to account for the following reaction. What are the functions of the substituent groups? If all the substituents were changed to methyl groups, would the reaction still be able to take place? Explain your answer.

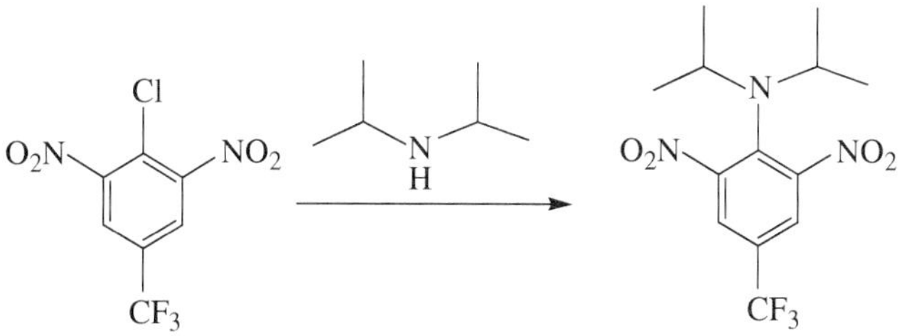

5.9 In the presence of a base, one of the stereoisomers of the following oxime undergoes a ring closure. Suggest a mechanism to account for the reaction and identify the configuration of the oxime.

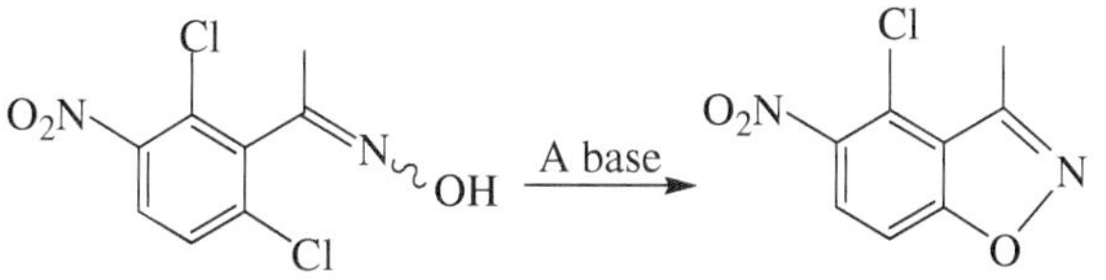

An oxime

Note: ∿ is a bond whose stereochemistry is uncertain

5.10 Describe the mechanism for each of the following conversions and give any necessary reagents.

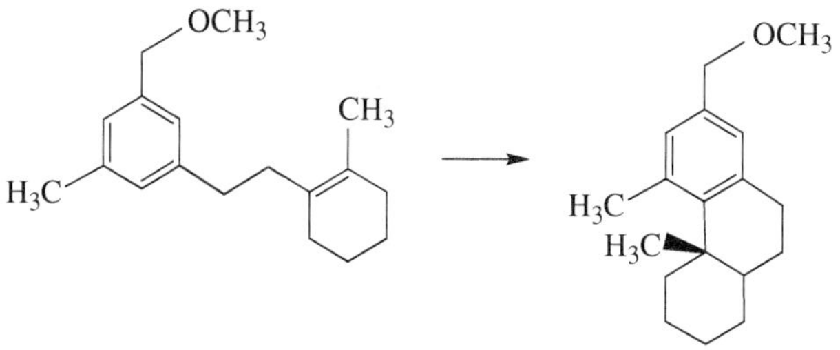

(1)

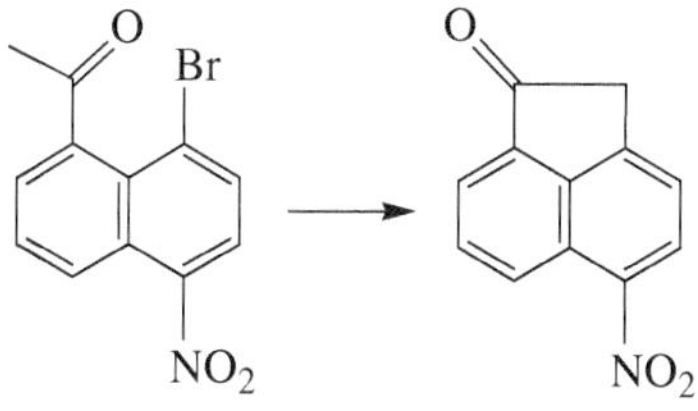

(2)

REFERENCES

1. Bruckner, R. *Advanced Organic Chemistry: Reaction Mechanisms*, Chapter 5, pp. 169–219, Harcourt/Academic Press, Orlando, FL, USA (2002).

2. Fox, M. A.; Whitesell, J. K. *Core Organic Chemistry*, 2nd ed., Jones and Bartlett, Sudbury, MA, USA (1997).

3. Brown, W. H.; Foote, C. S.; Iverson, B. L.; Anslyn, E. V. *Organic Chemistry*, 6th ed., Brooks/Cole, Belmont, CA, USA (2012).

4. Olah, G. A.; Malhotra, R.; Narang, S. C. *Nitration: Methods and Mechanisms*, VCH, New York, NY, USA (1989).

5. Kochi, J. K. Inner-Sphere Electron Transfer in Organic Chemistry. Relevance to Electrophilic Aromatic Nitration. *Acc. Chem. Res.* 1992, *25*, 39–47.

6. Kim, E. K.; Bockman, T. M.; Kochi, J. K. Electron-Transfer Mechanism for Aromatic Nitration via the Photoactivation of EDA Complexes. Direct Relationship to electrophilic Aromatic Substitution. *J. Am. Chem. Soc.* 1993, *115*, 3091–3104.

7. Gwaltney, S. R.; Rosokha, S. V.; Head-Gordon, M.; Kochi, J. K. Charge-Transfer Mechanism for Electrophilic Aromatic Nitration and Nitrosation via the Convergence of (ab Initio) Molecular-Orbital and Marcus–Hush Theories with Experiments. *J. Am. Chem. Soc.* 2003, *125*, 3273–3283.

8. Olah, G. A.; Lin, H. C.; Forsyth, D. A. Charge Distribution in Benzenium and Nitrobenzenium Ions Based on 13C Nuclear Magnetic Resonance Studies and Their Relevance to the Isomer Distribution in Electrophilic Aromatic Substitutions. *J. Am. Chem. Soc.* 1974, *96*, 6908–6911.

9. Parsons, S.; Passmore, J. Rings, Radicals, and Synthetic Metals: The Chemistry of SNS. *Acc. Chem. Res.* 1994, *27*, 101–108.

10. Brownridge, S.; Passmore, J.; Sun, X. The Electrophilic Substitution Reaction of the Dithionitronium Cation [SNS]+ with Benzene. *Can. J. Chem.* 1998, *76*, 1220–1231.

11. Sun, X.; Haas, D.; Sayre, K.; Weller, D. Formation of Diphenyl Sulfoxide and Diphenyl Sulfide via the Aluminum Chloride Facilitated Electrophilic Aromatic Substitution of Benzene with Thionyl Chloride, and Novel Reduction of Sulfur (IV) to Sulfur (II). *Phosphorus Sulfur Silicon Relat. Elem.*, 2010, *185*, 2535–2542.

12. Arterburn, J. B.; Nelson, S. L. Rhenium-Catalyzed Oxidation of Sulfides with Phenyl Sulfoxide. *J. Org. Chem.* 1996, *61*, 2260–2261.

13. Sun, X.; Haas, D.; McWilliams, S.; Smith, B.; Leaptrot, K. Investigations on the Lewis-Acid-Catalyzed Electrophilic Aromatic Substitution Reactions of Thionyl Chloride and Selenyl Chloride, the Substituent Effects, and the Reaction Mechanisms. *J. Chem. Res.* 2013, *37*, 736–744.

14. DeHaan, F. P.; Ahn, P. Y.; Kemnitz, C. R.; Ma, S. K.; Na, J.; Patel, B. R.; Ruiz, R. M.; Villahermosa, R. M. Electrophilic Aromatic Substitution. 14. [1] A Kinetic, NMR, and Raman Study of the Aluminum Chloride-Catalyzed Reaction of *p*-Toluenesulfonyl Chloride with Benzene and Toluene in Dichloromethane. *J. Chem. Kinet.* 1998, *30*, 367–372.

15. Rathore, R.; Hecht, J.; Kochi, J. K. Isolation and X-ray Structure of Chloroarenium Cations as Wheland Intermediates in Electrophilic Aromatic Chlorination. *J. Am. Chem. Soc.* 1998, *120*, 13278–13279.

16. Dolbier, Jr., W. R.; Cornett, E.; Martinez, H.; Xu, W. Friedel–Crafts Reactions of 2,2-Difluorocyclopropanecarbonyl Chloride: Unexpected Ring-Opening Chemistry. *J. Org. Chem.* 2011, *76*, 3450–3456.

17. Bauer, I.; Knolker, H.-J. Iron Catalysis in Organic Synthesis. *Chem. Rev.* 2015, *115*, 3170–3387.

18. Zanwar, M. R.; Kavala, V.; Gawande, S. D.; Kuo, C.-W.; Huang, W.-C.; Kuo, T.-S.; Huang, H.-N.; He, C.-H.; Yao, C.-F. FeCl$_3$ Catalyzed Regioselective C-Alkylation of Indolylnitroalkenes with Amino Group Substituted Arenes. *J. Org. Chem.* 2014, *79*, 1842–1849.

19. Sun, X.; Haas, D.; Lockhart, C. Iron(III) Chloride (FeCl$_3$)–Catalyzed Electrophilic Aromatic Substitution of Chlorobenzene with Thionyl Chloride (SOCl$_2$) and the Accompanying Auto-Redox in Sulfur to Give Diaryl Sulfides (Ar$_2$S): Comparison to Catalysis by Aluminum Chloride (AlCl$_3$). *Phosphorus Sulfur Silicon Relat. Elem.*, 2017, *192*, 376–380.

20. Huheey, J. E.; Keiter, E. A.; Keiter, R. L. *Inorganic Chemistry: Principles of Structure and Reactivity*, 4th ed., Harper Collins, New York, NY, USA (1993).

21. Kancherla, S.; Lotentzen, M.; Snieckus, V.; Jorgensen, K. B. Directed *ortho*-Metalation and Anionic *ortho*-Fries Rearrangement of Polycyclic Aromatic *O*-Carbamates: Regioselective Synthesis of Substituted Chrysenes. *J. Org. Chem.* 2018, *83*, 3590–3598.

22. Wang, A.; Maguire, J. A.; Biehl, E. Preparation of 2-Cyanobenzoic Acids from the Reaction of Bromobenzoic Acids with Arylacetonitriles and LDA. *J. Org. Chem.* 1998, *63*, 2451–2455.

23. Ji, P.; Atherton, J. H.; Page, M. I. The Kinetics and Mechanisms of Aromatic Nucleophilic Substitution Reactions in Liquid Ammonia. *J. Org. Chem.* 2011, *76*, 3286–3295.

24. Blockhina, N. I.; Atroshchenko, Y. M.; Gitis, S. S.; Blokhin, I. V.; Grudtsyn, Y. D.; Andrianov, V. F.; Kaminskii, A. Y. Reactions of Aromatic Nitrocompounds. LXXII. Anionic σ-Complexes of Nitroarenes in the Azocoupling Reaction. *Russ. J. Org. Chem.* 1998, *34*, 533–535.

25. Sun, Z.-Y.; Botros, E.; Su, A.-D.; Kim, Y.; Wang, E.; Baturay, N. Z.; Kwon, C.-H. Sulfoxide-Containing Aromatic Nitrogen Mustards as Hypoxia-Directed Bioreductive Cytotoxins. *J. Med. Chem.* 2000, *43*, 4160–4168.

26. Barbosa, E. G.; Bega, L. A. S.; Beatriz, A.; Sarkar, T.; Hamel, E.; Amaral, M. S. d.; Lima, D. P. d. A Diaryl Sulfide, Sulfoxide, and Sulfone Bearing Structural Similarities to Combretastatin A-4. *Eur. J. Med. Chem.* 2009, *44*, 2685–2688.

6

NUCLEOPHILIC SUBSTITUTIONS ON SP³-HYBRIDIZED CARBONS: FUNCTIONAL GROUP TRANSFORMATIONS

6.1 NUCLEOPHILIC SUBSTITUTION ON MONO-FUNCTIONALIZED SP³-HYBRIDIZED CARBON

By the nature, many organic reactions are those that take place between an electron-deficient reagent (electrophile—an electron-pair acceptor) and an electron-rich reagent (nucleophile—an electron-pair donor). They interact to form a covalent bond as a result of transfer of one pair of electrons from the nucleophile to the electrophile. A nucleophilic substitution reaction is one of these types of reactions. In Section 5.7, nucleophilic aromatic substitution (NAS) which takes place on an aromatic ring (sp^2-carbon) bearing a good leaving group is discussed. The substitution by a nucleophile can also take place on a sp^3-hybridized (tetrahedral) carbon that bears one (*usually only one*) functional group as a good leaving group. In this type of reactions, a nucleophile displaces the leaving group (–LG) on the electrophilic sp^3 carbon. As a result, one functionalized organic compound is transformed into another (Eq. 6.1):

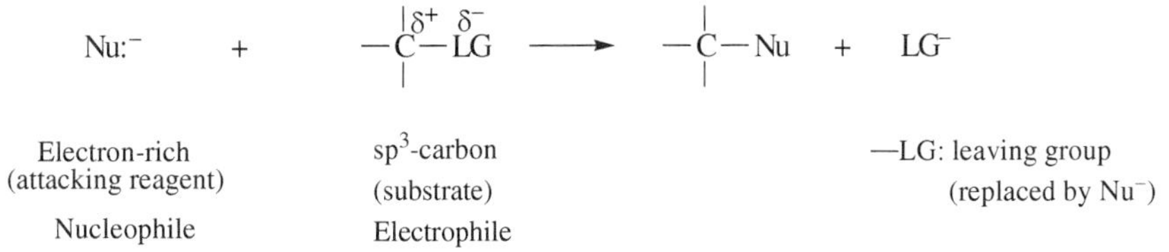

$$\text{Nu:}^- \quad + \quad -\overset{|}{\underset{|}{C}} \overset{\delta^+}{-} \overset{\delta^-}{LG} \quad \longrightarrow \quad -\overset{|}{\underset{|}{C}} - \text{Nu} \quad + \quad LG^-$$

Electron-rich (attacking reagent)	sp^3-carbon (substrate)	—LG: leaving group (replaced by Nu⁻)
Nucleophile	Electrophile	

$$(6.1)$$

Organic Mechanisms: Reactions, Methodology, and Biological Applications,
Second Edition. Xiaoping Sun.
© 2021 John Wiley & Sons, Inc. Published 2021 by John Wiley & Sons, Inc.
Companion website: www.wiley.com/go/Sun/OrgMech_2e

Very often, the ease of departure for a leaving group (–LG) is related to the bond dissociation energy (BDE) of the C—LG bond. Approximately, for structurally similar –LG groups, the smaller the BDE, the easier (qualitatively, less energy required) to cleave the C—LG bond, therefore, the better is the leaving group –LG. For example, halo groups (chloro –Cl, bromo –Br, and iodo –I) are among the most common good leaving groups. The order of standard BDEs is BDE (C–Cl, 327 kJ/mol) > BDE (C–Br, 285 kJ/mol) > BDE (C–I, 213 kJ/mol). Therefore, the increase of relative ease of leave for halo groups follows the order of –Cl < –Br < –I. On the other hand, the hydroxyl group –OH is a poor leaving group. However, the protonated hydroxyl –OH₂⁺ ("water") is a good leaving group. Further discussion on various good and poor leaving groups will be presented in Section 6.2.

There are two mechanistic extremes for nucleophilic substitution reactions which take place on sp³-hybridized carbons. In the first extreme, the nucleophile attacks the electrophilic carbon in the substrate, and the formation of the new Nu—C bond and the breaking of the old C—LG bond occur simultaneously. The reaction proceeds via a single transition state. This mechanism is called the S_N2 **mechanism**. The following examples (Reactions 6.2 and 6.3) are S_N2 reactions.

$$CH_3CH_2\overset{..}{\underset{..}{O}}{:}^- \quad + \quad H_3C\!-\!Br \quad \longrightarrow \quad CH_3CH_2\!-\!O\!-\!CH_3 \quad + \quad :\overset{..}{\underset{..}{Br}}{:}^-$$

Nucleophile   Electrophile   –Br: leaving group

              (sp³-carbon)   (replaced by $CH_3CH_2O^-$)

$$(6.2)$$

$$N\!\!\equiv\!\!C{:}^- \quad + \quad H_3C\!-\!Br \quad \longrightarrow \quad H_3C\!-\!C\!\!\equiv\!\!N \quad + \quad :\overset{..}{\underset{..}{Br}}{:}^-$$

Nucleophile   Electrophile   –Br: leaving group

              (sp³-carbon)   (replaced by CN^-)

$$(6.3)$$

In the second extreme, the C—LG bond in the substrate dissociates first giving a carbocation intermediate. Then the strongly electrophilic carbocation is attacked by the nucleophile to give a substitution product. This stepwise mechanism is called the S_N1 **mechanism**. The following example is an S_N1 reaction (Reaction 6.4):

$$(CH_3)_3C\!-\!Cl \quad \rightleftharpoons \quad (CH_3)_3C^+ \quad \xrightarrow[-H^+]{H_2O} \quad (CH_3)_3C\!-\!OH \qquad (6.4)$$
$$+ \; Cl^-$$

The nucleophilic substitution reactions have the following two general results:

(1) A new carbon–heteroatom bond is formed (e.g., Reactions 6.2 and 6.4), or

(2) A new carbon–carbon bond is formed (effected by a carbon nucleophile, a species often containing a negatively charged carbon atom, e.g., Reaction 6.3)

6.2   FUNCTIONAL GROUPS WHICH ARE GOOD AND POOR LEAVING GROUPS

The strength of the C—LG bond (represented approximately by BDE) on a sp^3 carbon can be directly and conveniently correlated to the basicity of the corresponding LG^- anion. Thus, we may estimate the relative strength of the C—LG bond (or relative ease of leave for the –LG group) on the basis of our general knowledge about the acidity–basicity for the conjugated acid–base ($HLG–LG^-$) pair. *In general, the weaker the basicity of LG^-, the weaker the C—LG bond, and therefore the better is the leaving group –LG.* In other words, weak bases are usually good leaving groups, and strong bases are poor leaving groups [1].

Triflic acid (HOTf) is a well-known superacid (acidity much stronger than sulfuric acid) with $pK_a = -15$. Therefore, its conjugate base triflate (TfO^-) is an extremely weak base (Fig. 6.1a). It is a *super good leaving group* and can even be displaced by a very weak nucleophile via S_N2 mechanism (referred to Fig. 5.23 in Section 5.4.2).

Tosylate (TsO^-) and mesylate (MsO^-) groups (Fig. 6.1b) are analogous to triflate. The acidity of their conjugate acids [$pK_a = -6.5$ (TsOH) and $pK_a = -2$ (MsOH)] is relatively comparable to that of sulfuric acid ($pK_a = -3$), but much weaker than triflic acid. Therefore, the two groups are very weak bases (but much stronger than triflate). *They are very good leaving groups (but relatively poorer than triflate).*

Hydrogen halides HX (X = Cl, Br, and I) are strong acids with the pK_a values being –7 (HCl), –9 (HBr), and –10 (HI). Correspondingly, their conjugate bases chloride (Cl^-), bromide (Br^-), and iodide (I^-) are weak bases. *They are all good leaving groups (but poorer than tosylate and mesylate).* On the basis of their relative basicity ($Cl^- > Br^- > I^-$, determined by the pK_a's of HX), the increase in ease of leaving is $–Cl < –Br < –I$, consistent with the order of the corresponding BDEs for the C—X bonds (see Section 6.1).

Another type of good leaving groups is the divalent oxygen –O– in the epoxide rings presumably due to the molecular strain in the cyclic structure. An epoxide can react with strong nucleophiles leading to breaking of a C—O bond in the ring (Reaction 6.5):

Carbon electrophile

Nu^- : OH^-, OR^-, R-MgX, etc. (strong nucleophiles)

$$(6.5)$$

In contrast, hydrogen fluoride (HF) is a weak acid with $pK_a = 3.2$. Correspondingly, the basicity of its conjugate base fluoride (F^-) is relatively strong (much stronger than all the other halides). Therefore, the fluoro group (–F) is much more difficult to leave than the other halo groups. In fact, *–F is a poor leaving group.*

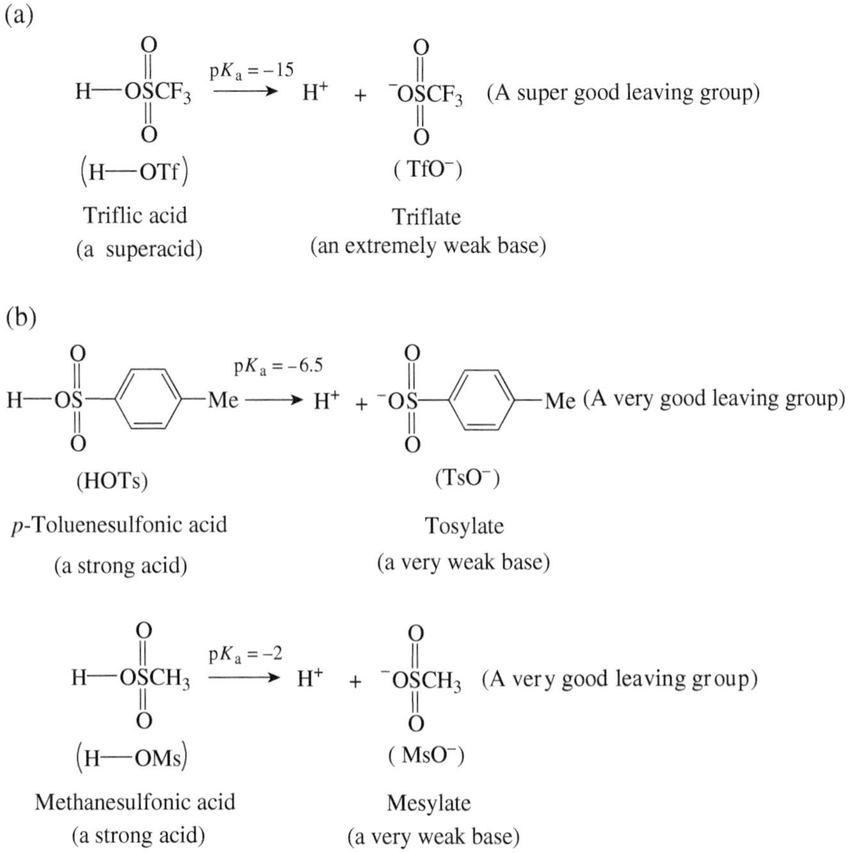

(a)

$$H—O\overset{\overset{\displaystyle O}{\|}}{\underset{\underset{\displaystyle O}{\|}}{S}}CF_3 \xrightarrow{\;pK_a = -15\;} H^+ + \;^-O\overset{\overset{\displaystyle O}{\|}}{\underset{\underset{\displaystyle O}{\|}}{S}}CF_3 \quad \text{(A super good leaving group)}$$

$\left(H—OTf\right)$ (TfO$^-$)

Triflic acid   Triflate

(a superacid)   (an extremely weak base)

(b)

$$H—O\overset{\overset{\displaystyle O}{\|}}{\underset{\underset{\displaystyle O}{\|}}{S}}\!\!-\!\!\!\left\langle\;\right\rangle\!\!-\!\!Me \xrightarrow{\;pK_a = -6.5\;} H^+ + \;^-O\overset{\overset{\displaystyle O}{\|}}{\underset{\underset{\displaystyle O}{\|}}{S}}\!\!-\!\!\!\left\langle\;\right\rangle\!\!-\!\!Me \;\text{(A very good leaving group)}$$

(HOTs)   (TsO$^-$)

p-Toluenesulfonic acid   Tosylate

(a strong acid)   (a very weak base)

$$H—O\overset{\overset{\displaystyle O}{\|}}{\underset{\underset{\displaystyle O}{\|}}{S}}CH_3 \xrightarrow{\;pK_a = -2\;} H^+ + \;^-O\overset{\overset{\displaystyle O}{\|}}{\underset{\underset{\displaystyle O}{\|}}{S}}CH_3 \quad \text{(A very good leaving group)}$$

$\left(H—OMs\right)$ (MsO$^-$)

Methanesulfonic acid   Mesylate

(a strong acid)   (a very weak base)

FIGURE 6.1   Super and very good leaving groups.

Other poor leaving groups include –SR(H), –NR$_2$(H$_2$), –NO$_2$, and –CN presumably attributable to their relatively strong basicity. *None of these poor leaving groups can be displaced by any common nucleophiles.*

For the same rationale, the hydroxyl (–OH) and alkoxide (–OR) groups are poor leaving groups because the corresponding anions OH$^-$ and OR$^-$ are strong bases due to very weak acidity of their conjugated acids [pK_a = 15 (HOH) and pK_a = 16–18 (HOR)]. This gives rise to a strong C—O bond in each case. Neither of the groups can be displaced by any common nucleophiles. Upon treated with a strong acid (such as H$_2$SO$_4$ with pK_a = –3), the protonated hydroxyl (–$^+$OH$_2$) and protonated alkoxide [–$^+$O(H)R] become good leaving groups as both water (HOH) and an alcohol (ROH) are only weakly basic. This is because of high acidity of the protonated –$^+$OH$_2$ and –$^+$O(H)R each of which has a pK_a ~ –2. Therefore, alcohols (R–OH) and ethers (R$'$– OR) can undergo nucleophilic substitution and elimination reactions in strongly acidic media, which will be discussed later in this chapter and in Chapter 7. The –OH and –OR can also be activated by strong Lewis acids such as BF_3 and $AlCl_3$ and as a result, the adducts [such as $-^+O(AlCl_3)R$] may become super good leaving groups. They can even be displaced by a very weak nucleophile via S_N2 mechanism (referred to Section 5.4.2, Fig. 5.22).

6.3   GOOD AND POOR NUCLEOPHILES

In Section 1.9.2, we introduced briefly the quality of nucleophiles. In general, we can relate nucleophilicity of a species to its basicity. Bases always contain a lone pair of electrons which can react with a proton in an acid molecule to form a covalent bond. In addition, with the lone pair of electrons many bases can react with an electrophilic sp^3-hybridized carbon to effect nucleophilic substitution reactions. When reacting with an electrophilic carbon, they are termed nucleophiles. Table 6.1 presents a quantitative scale of nucleophilicity (n) for a variety of anions on the basis of their S_N2 reactions. In general, strong bases tend to be strong nucleophiles. That is because both react with a positively charged species. This trend is followed very well when the central atom of a nucleophile (base) is the same [1]. The case can be clearly seen, for example, for the oxygen-based nucleophiles (bold entries in Table 6.1). The substantial difference in nucleophilicity of OH^- ($n = 4.2$) and that of PhO^- ($n = 3.5$) and $CH_3CO_2^-$ ($n = 2.7$) is due to resonance stabilizations of PhO^- and $CH_3CO_2^-$ because of the negative charge delocalizations (Fig. 6.2), while OH^- lacks stabilization, with the negative charge residing on oxygen. Resonance stabilizations of PhO^- and $CH_3CO_2^-$ reduce their basicity by spreading out the negative charge in each of the anions and therefore their nucleophilicities are lower than OH^-.

TABLE 6.1   Quantitative Strengths of Various Nucleophiles

Nucleophile (Nu^-)	Nucleophilicity $[n = \lg(k_N/k_O)]^a$	pK_a of conjugate acid (HNu)
HS^-	5.1	7.0 (H_2S)
CN^-	5.1	9.2 (HCN)
I^-	5	−10 (HI)
RNH_2	4.5	10–11 (RNH_3^+)
OH^-	**4.2**	**15.7 (H_2O)**
Br^-	3.5	−9 (HBr)
PhO^-	**3.5**	**9.9 (PhOH)**
Cl^-	2.7	−7 (HCl)
$CH_3CO_2^-$	**2.7**	**4.8 (CH_3CO_2H)**
F^-	2.0	3.2 (HF)
H_2O	**0**	**−1.7 (H_3O^+)**

Reproduced from Ref. [1], with permission from The Royal Society of Chemistry.
^{a} Nucleophilicity (n) is defined as $\lg(k_N/k_O)$, where k_N/k_O is the relative rate of attack by the nucleophile on a substrate compared with the attack by water on the same substrate.

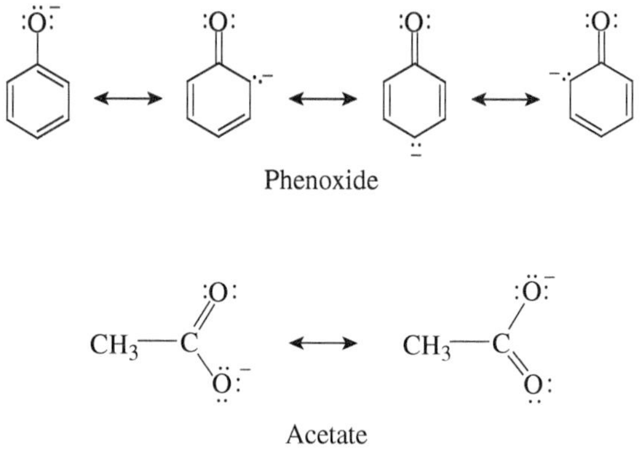

Phenoxide

Acetate

FIGURE 6.2   Possible resonance structures for phenoxide and acetate, accounting for the negative charge delocalization.

Table 6.1 also indicates that as the size of an anion, such as I^- and HS^-, becomes very large, its basicity (the reactivity toward a very small proton) decreases because overlap of the very small 1s orbital of a proton with very large valence orbitals in these anions is less effective. However, the nucleophilicity of the anion (the reactivity toward an electrophilic carbon) increases as the relatively large carbon atom would prefer to bonding to a large anion so that the size of valence orbitals would better match to produce a more effective overlap [1]. In addition, small anions such as OH^- are highly solvated, whereas larger anions such as I^- and HS^- are less solvated. This is particularly true for the anions in protic solvents. The solvation on the lone pairs of electrons in small anions reduces their reactivity toward electrophilic carbon. All this explains why the weakly basic I^- [$pK_a(HI) = -10$, $n = 5$] and HS^- [$pK_a(H_2S) = 7$, $n = 5.1$] are more nucleophilic than the strongly basic OH^- [$pK_a(H_2O) = 15.7$, $n = 4.2$]. Similarly, the nucleophilicity of RS^- is stronger than that of RO^-, while the basicity of RS^- is weaker than that of RO^-. The analogous size effect on the nucleophilicity of halides is observed such that the nucleophilicity for halides increases in the order of F^- ($n = 2.0$) < Cl^- ($n = 2.7$) < Br^- ($n = 3.5$) < I^- ($n = 5$).

For the anions with the negatively charged atoms belonging to the elements of the same period, both nucleophilicity and basicity increase as one moves from the right to the left in the period. For example, the nucleophilicity and basicity for the following anions increase in the order of $F^- < OH^- < NH_2^- < CH_3^-$. In general, carbanions (a negatively charged trivalent carbon $R_1R_2R_3C:^-$) are extremely strong nucleophiles and extremely strong bases and hence do not exist in protic solvents. If any of the R groups in the carbanion is unsaturated (e.g., a phenyl group), the nucleophilicity and basicity decrease because of the delocalization of the negative charge to the unsaturated group (resonance stabilization).

Since nucleophilicity is the reactivity of a species toward an electrophilic tetrahedral carbon which is surrounded by a few groups, it is also affected by bulkiness (steric hindrance) of the nucleophile. Basicity of a species is not influenced by steric hindrance as a proton in an acid is always protruding, vulnerable to the attack by a base. For example, the nucleophilicity of different types of alkoxides (as the sodium salts) decreases in the order of CH_3O^- (methoxide) > $CH_3CH_2O^-$ (ethoxide) > $(CH_3)_2CHO^-$ (*i*-propoxide) > $(CH_3)_3CO^-$ (*t*-butoxide) as the oxygen atom becomes more sterically hindered, while the basicity increases in the same order of CH_3O^- < $CH_3CH_2O^-$ < $(CH_3)_2CHO^-$ < $(CH_3)_3CO^-$ due to the electron donating effect of the methyl groups. Similarly, the nucleophilicity of different types of amides decreases in the order of NH_2^- (amide) > $(CH_3CH_2)_2N^-$ (diethyl amide) > $[(CH_3)_2CH]_2N^-$ (diisopropylamide) as the nitrogen atom becomes more sterically hindered, while the basicity increases in the same order of NH_2^- < $(CH_3CH_2)_2N^-$ < $[(CH_3)_2CH]_2N^-$ due to the electron donating effect of the methyl groups. NH_2^- is an extremely strong base and an extremely strong nucleophile. CH_3O^- is a very strong base and a very strong nucleophile. However, $[(CH_3)_2CH]_2N^-$ is essentially a nonnucleophilic extremely strong base, while $(CH_3)_3CO^-$ is a strong base with relatively weak nucleophilicity. These nonnucleophilic and weakly nucleophilic bases play important roles in some types of organic reactions which will be discussed in the following chapters.

6.4 S$_N$2 REACTIONS: KINETICS, MECHANISM, AND STEREOCHEMISTRY

The S$_N$2 reactions are of major mechanistic and synthetic significance in organic chemistry. This type of reactions is concerted, bimolecular, and possesses a second-order rate law (first order in the nucleophile and first order in the organic substrate containing an electrophilic sp^3 carbon) as shown in Equation 6.6:

$$Nu{:}^- + R{-}X \xrightarrow[k]{S_N2} Nu{-}R + X^-$$

$$\text{(6.6)}$$

$$\text{Rate} = d[NuR]/dt = k[Nu^-][RX] \qquad k = A\,\exp(-E_a/RT)$$

The activation energy (E_a) is determined by the energy level of the transition state. If the sp^3 carbon attached to the functional group in the substrate is asymmetric (a chiral center), the reaction is stereospecific, namely that the product will possess a different configuration than the starting substrate. First, we will demonstrate mechanism, kinetics, and stereochemistry of S$_N$2 reactions using the reaction of sodium azide (NaN$_3$) with a (2S)-2-triflyloxyester in CH$_3$CN (an aprotic solvent), which gives a (2R)-2-azidoester (Fig. 6.3).

6.4.1 Mechanism and Stereochemistry for S$_N$2 Reactions

Figure 6.3 shows that as azide approaches (attacks) the electrophilic sp^3-hybridized asymmetric carbon atom (a chiral center) in (2S)-2-triflyloxyester from the opposite side of the –OTf group (a super good leaving group), the formation of a new N—C

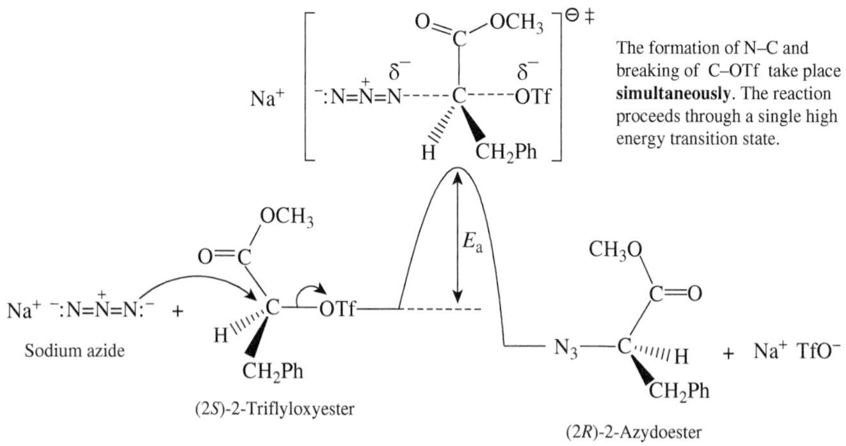

FIGURE 6.3   Reaction of sodium azide and (2S)-2-triflyloxyester in CH$_3$CN (an aprotic solvent): the S$_N$2 mechanism.

$$\text{Rate} = d[(R)\text{-ester}]/dt = k[N_3^-][(S)\text{-ester}]$$

$$k = A\text{-exp}(-E_a/RT)$$

bond and breaking of an old C—OTf bond occur **simultaneously** [2]. At the same time, the –H, –C(O)OCH$_3$, and –CH$_2$Ph groups on the functionalized sp^3 carbon flip from the left side toward the right side. The reaction proceeds via a single transition state in which the old C—OTf bond is being partially broken, coincident with the partial formation of a new N—C bond, and the initial negative charge carried over from the nucleophile is spread out. The –H, –C(O)OCH$_3$, and –CH$_2$Ph groups have moved toward the "midregion," forming roughly a trigonal-planar configuration (more rigorously speaking, an early transition state). The transition state possesses the maximum energy level in the reaction profile. It collapses (dissociates) spontaneously and rapidly to lead to full breaking of the old C—OTf bond in the reactant and the concurrent complete formation of the new N—C bond in the product. Simultaneously, the –H, –C(O)OCH$_3$, and –CH$_2$Ph groups move to the right side, giving an ester product with the opposite configuration. The reaction is overall a second-order process, first order in azide (nucleophile) and first order in the (S)-ester substrate. At a given temperature, the rate constant (k) for the reaction is determined by the stabilization (E_a) of the transition state.

The stereochemistry for S$_N$2 reactions is further demonstrated by examining the configuration change in a cyclohexane derivative during an S$_N$2 reaction (Fig. 6.4) [3]. The substrate is in *trans*-configuration, with both bromo and methyl groups staying in the favorable equatorial positions in the predominant chair-conformation. HS$^-$ attacks the C–Br carbon from the opposite side of the bromo group to effect an S$_N$2 reaction. As a result, the –SH group is introduced onto the axial position in the

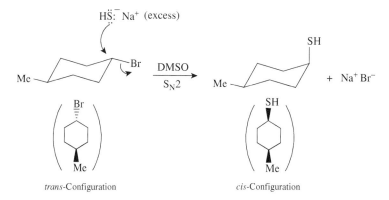

FIGURE 6.4   Reaction of *trans*-1-bromo-4-methylcyclohexane and sodium hydrogen sulfide in DMSO: stereochemistry of the S$_N$2 reaction. Modified from Bruckner [3].

product giving a *cis*-configuration. The *trans*-product is not observed. Since the thiol product (RSH) is also nucleophilic, excess HS$^-$ is used in the reaction to prevent the starting bromoalkane from reacting with RSH. Instead, all the limiting reactant is converted to the thiol RSH product.

The S$_N$2 reactions can be employed to effect interconversion of enantiomers of chiral compounds. Equation 6.7 shows the overall process of quantitative conversion of (*R*)-1-phenyl-2-propanol ($\alpha = +33.0°$) to its enantiomer (*S*)-1-phenyl-2-propanol ($\alpha = -33.2°$), in which the inversion of the absolute configuration is accomplished by an S$_N$2 reaction of an alkyl tosylate with ethanoate (acetate) (as the sodium salt) [1].

$$
\begin{array}{ccccc}
\text{Ph}\diagdown\diagup\text{OH} & \xrightarrow{\text{TsCl}} & \text{Ph}\diagdown\diagup\text{OTs} & \xrightarrow[\substack{\text{Ac}_2\text{O}\\ \text{S}_N2}]{\text{:OCCH}_3} & \text{Ph}\diagdown\diagup\text{OCCH}_3 & \xrightarrow{\text{OH}^-\text{(aq)}} & \text{Ph}\diagdown\diagup\text{OH}
\end{array}
$$

(*R*)-1-Phenyl-2-propanol	(*R*)-Configuration	(*S*)-Configuration	(*S*)-1-Phenyl-2-propanol
$\alpha = +33.0°$	$\alpha = +31.1°$	$\alpha = -7.1°$	$\alpha = -33.2°$

(6.7)

In the first step, the poor leaving group –OH in the starting alcohol is converted to tosylate (–OTs), a very good leaving group, by tosyl chloride (TsCl) with the initial configuration of the chiral center retained. As shown in Table 6.1, acetate (CH$_3$CO$_2$$^-$) is a moderately strong nucleophile with the reactive center being on an oxygen atom. It can react with a primary or a secondary alkyl tosylate (R–OTs) to form an ester. The reaction follows the S$_N$2 mechanism, with the nucleophilic acetate attacking the organic substrate from opposite side of tosylate (–OTs) effecting inversion of the original (R)-configuration in the alkyl tosylate to the opposite (S)-configuration in the ester. Saponification (hydrolysis in basic solution) of the (S)-ester affords the final (S)-alcohol.

6.4.2   Steric Hindrance on S_N2 Reactions

A remarkable feature for S_N2 reactions is that the rate constants are strongly subject to the steric hindrance in the substrate. Figure 6.5 shows that as the electrophilic α-carbon (the carbon atom that bears a functional group) is getting more crowded by attaching to methyl groups, the pathway along which the nucleophile approaches the electrophilic carbon becomes more sterically hindered by the methyl groups, and as a result, the relative rate constant (k_{rel}) for the S_N2 reaction is getting smaller [3]. From halomethane (CH_3X) to haloethane (CH_3CH_2X), increase in one methyl group on the α-carbon results in decrease in k_{rel} by ~30 folds. In 2-halo-2-methylpropane (a 3° haloalkane), the pathway for a S_N2 reaction is almost completely blocked by the three methyl groups on the α-carbon, and k_{rel} becomes zero essentially. Instead, the reaction follows the S_N1 mechanism exclusively. We will study the S_N1 mechanism in Section 6.6. The results indicate that in general, the sterically inactive (unbulky) methyl and primary substrates [RCH_2X (R = H or linear alkyl)] are the best for S_N2 reactions. In fact, they do not undergo S_N1 reactions because of the high instability of methyl and primary carbocations (CH_3^+ and a primary carbocation cannot be actually formed due to instability). On the other hand, the S_N2 reactions do not occur to tertiary substrates (R_3CX) due to a strong steric hindrance in the substrates. Instead, a tertiary substrate undergoes S_N1 reactions readily owing to stabilization of the intermediate tertiary carbocation.

Figure 6.6 shows that methyl groups present on the β-carbon of haloalkanes also have substantial steric hindrance to the S_N2 reactions. When all the R groups are methyl groups, the S_N2 reaction is almost completely hindered. The haloalkane substrate can undergo an S_N1 reaction with skeletal rearrangement observed (Reaction 6.8).

$$CH_3\!-\!\underset{\underset{CH_3}{|}}{\overset{\overset{CH_3}{|}}{C}}\!-\!CH_2\!-\!X \xrightarrow[-X^-]{S_N1} CH_3\!-\!\underset{\oplus}{\overset{\overset{CH_3}{|}}{C}}\!-\!CH_2\!-\!CH_3 \xrightarrow{Nu:^-} CH_3\!-\!\underset{\underset{Nu}{|}}{\overset{\overset{CH_3}{|}}{C}}\!-\!CH_2\!-\!CH_3$$

$$\text{A 3° carbocation}$$

$$(6.8)$$

Skeletal rearrangement of carbocations will be discussed in Chapter 10.

The effect of the steric hindrance on rate constants for S_N2 reactions can be further accounted for by examining the transition states for the reactions (Fig. 6.7). If the groups attached to the central electrophilic carbon are sterically active (bulky), they will interact with the incoming nucleophile and with the leaving group strongly in the early transition state and late transition state, respectively. The steric interactions destabilize the transition state by enhancing its energy, slowing down the reaction. As the number of methyl groups increases in the central

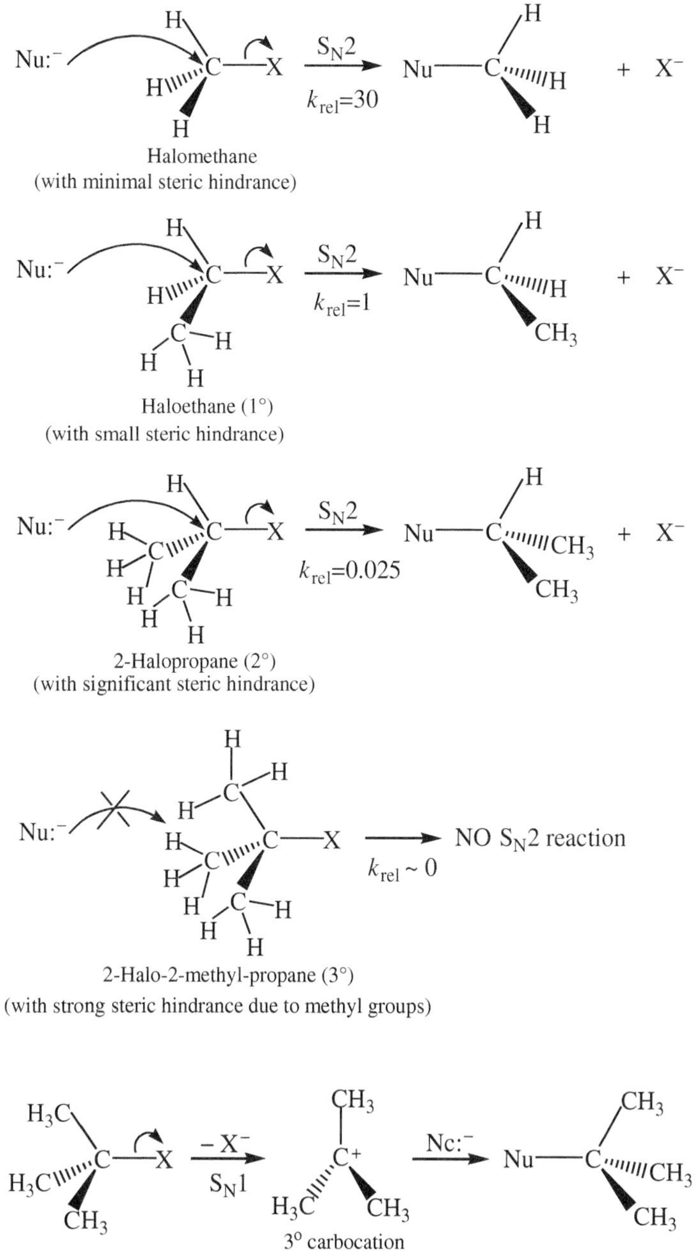

FIGURE 6.5   Effect of steric hindrance on relative rate constants for the S_N2 reactions of methyl, primary, secondary, and tertiary alkyl halides.

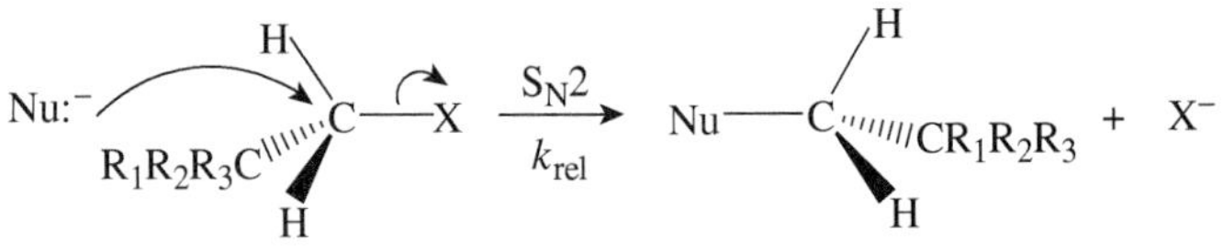

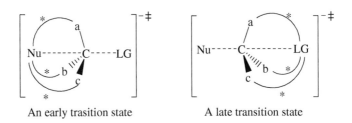

$R_1 = R_2 = R_3 = H$ $k_{rel} = 1$

$R_1 = R_2 = H, R_3 = CH_3$ $k_{rel} = 0.4$

$R_1 = H, R_2 = R_3 = CH_3$ $k_{rel} = 0.03$

$R_1 = R_2 = R_3 = CH_3$ $k_{rel} \sim 0$

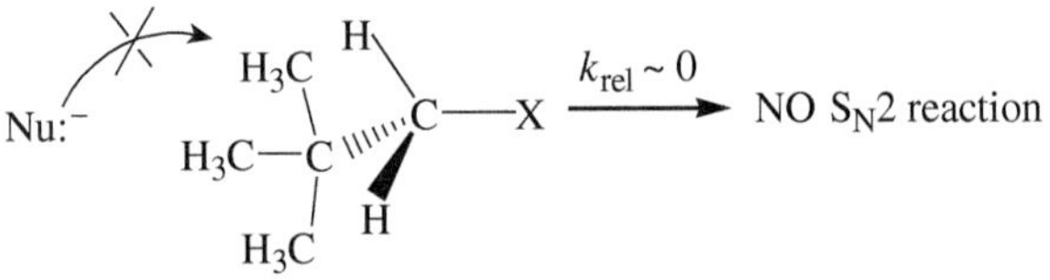

FIGURE 6.6 Effect of steric hindrance on relative rate constants for the S_N2 reactions of $R_1R_2R_3CCH_2X$.

An early trasition state   A late transition state

* Steric interaction, which destabilizes the transition state.

FIGURE 6.7 Destabilization of the S_N2 transition states by steric interactions.

electrophilic carbon of haloalkanes, the steric interactions in the transition state are getting greater, making the energy of the S_N2 transition state for a tertiary haloalkane greater than that for a secondary haloalkane, and the energy of the S_N2 transition state for a secondary haloalkane is greater than that for a primary haloalkane and for a halomethane. All this accounts for the trend in the relative rate constants (k_{rel}) for S_N2 reactions of different types of organic substrates.

6.4.3   Effect of Nucleophiles

The rate law for an S$_N$2 reaction indicates that the reaction rate is determined by reactivities of both the nucleophile and the electrophilic organic substrate. The electrophilicity of the carbon bearing a leaving group in the substrate is low as the carbon atom in the C—LG bond is only slightly positively charged. It would require a strong nucleophile to react with to have the S$_N$2 reaction proceed in an appreciable rate. *In general (exceptions exist), only a strong nucleophile (typically, an anion) can effect an S$_N$2 reaction. A weak nucleophile (typically, a neutral molecule containing a lone pair of electrons) can only effect S$_N$1 (but not S$_N$2) reactions unless there is a super good (or a very good) leaving group.* This can be further understood by examining the transition states for the related S$_N$2 reactions (Fig. 6.8).

Figure 6.8 shows that hydroxide (a strong nucleophile) undergoes a S$_N$2 reaction with bromomethane giving methanol. However, there is essentially no reaction (extremely slow reaction) between the weakly nucleophilic water and bromomethane. The rate constant for a concerted reaction is directly determined by the energy level (stabilization) of the transition state. For a strong nucleophile (an anion), the initial negative charge is dispersed in the transition state making its energy (E_a) relatively low. This gives rise to a relatively large rate constant [$k = A \exp(-E_a/RT)$]. When a weak nucleophile (a neutral molecule) approaches the electrophilic carbon, the transition state would have a charge separation making it highly unstable (energetic) with a large E_a, as creating a partial positive charge on an electronegative oxygen atom would require much energy input and greatly

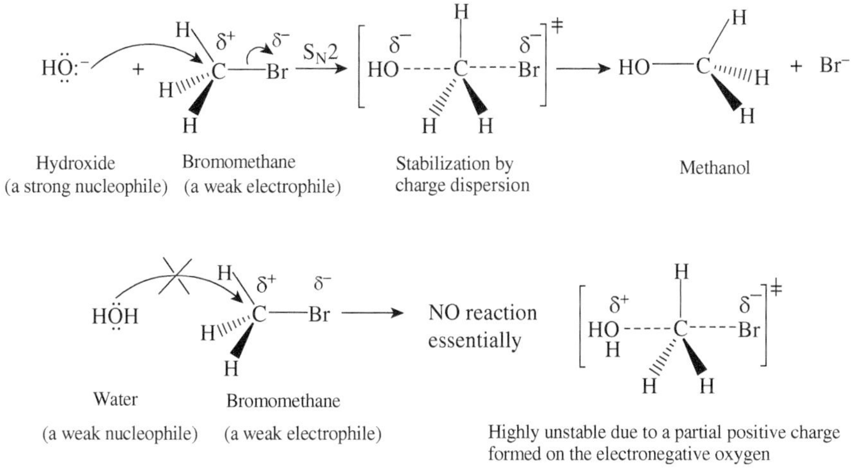

Hydroxide (a strong nucleophile)   Bromomethane (a weak electrophile)   Stabilization by charge dispersion   Methanol

Water (a weak nucleophile)   Bromomethane (a weak electrophile)   NO reaction essentially   Highly unstable due to a partial positive charge formed on the electronegative oxygen

FIGURE 6.8   The S$_N$2 reactions of hydroxide and water with bromomethane: comparison of relative stability of the transition states.

enhance the energy level of the transition state. This will result in an extremely slow reaction.

Similarly, a primary haloalkane such as bromoethane undergoes a S_N2 reaction with a strongly nucleophilic alkoxide such as isopropoxide (Williamson ether synthesis), but essentially not with a weakly nucleophilic alcohol such as isopropanol (Reaction 6.9).

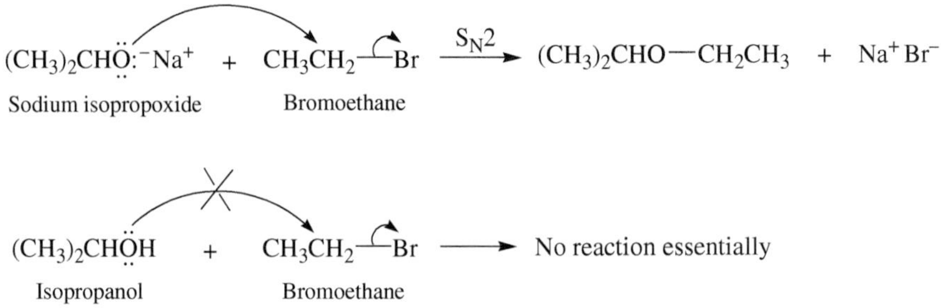

$$(CH_3)_2CH\ddot{O}:^-Na^+ \ + \ CH_3CH_2-Br \ \xrightarrow{S_N2} \ (CH_3)_2CHO-CH_2CH_3 \ + \ Na^+Br^-$$

Sodium isopropoxide   Bromoethane

$$(CH_3)_2CH\ddot{O}H \ + \ CH_3CH_2-Br \ \longrightarrow \ \text{No reaction essentially}$$

Isopropanol   Bromoethane

$$(6.9)$$

For the substrate molecules in which the electrophilic sp^3 carbon is attached to a very good or a super good leaving group, their electrophilicity is much elevated. As a result, the substrates can undergo S_N2 reactions with even very weak nucleophiles. (Reactions 6.10 and Reactions 6.11. Also see Figs. 5.22 and 5.23 in Section 5.4.2.)

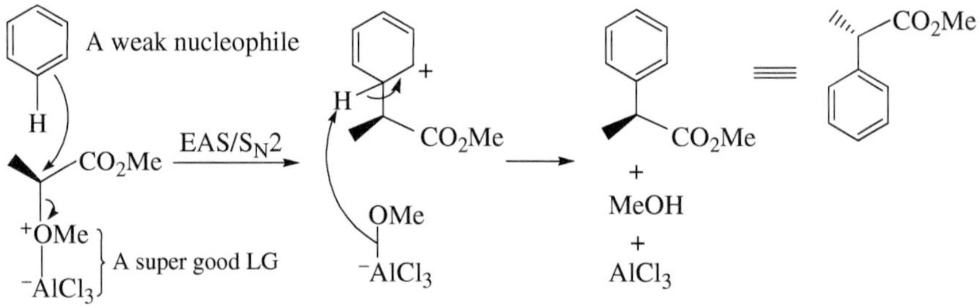

$$(6.10)$$

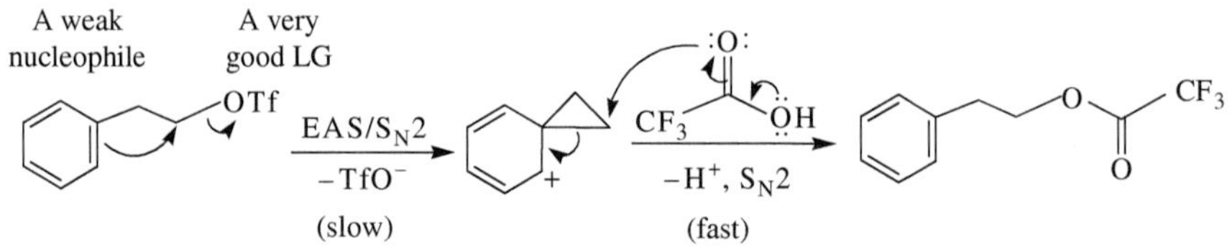

$$(6.11)$$

6.4.4   Solvent Effect

Usually, S$_N$2 reactions proceed faster in **polar** *aprotic solvents* (whose molecules cannot function as hydrogen bond donors) than in *protic solvents* (whose molecules function as hydrogen bond donors). Nonpolar aprotic solvents (e.g., *n*-hexane) are poor S$_N$2 solvents. The situation is true if the nucleophile is an anion, which is generally required for S$_N$2 reactions because most of the common strong nucleophiles are anionic. In general, a protic solvent is more polar than an aprotic solvent. Therefore, an anionic nucleophile is better stabilized by a more polar protic solvent than by a less polar aprotic solvent (Fig. 6.9) [4]. Hydrogen bonding between the nucleophile and a protic solvent is particularly important in stabilization of the nucleophile, while there is no hydrogen bond with an aprotic solvent. Since the negative charge in the transition state for the reaction is dispersed (Fig. 6.3), the interactions of the transition state with either type of solvents are relatively small so that its energy level is only very slightly affected by the solvent. As a result, the activation energy for the reaction performed in a polar aprotic solvent (E_{a2}) is smaller than that for the same reaction performed in a protic solvent (E_{a1}) (Fig. 6.9). This accounts for the difference in S$_N$2 reaction rates in protic and polar aprotic solvents.

The following reactions of 1-bromohexane CH$_3$(CH$_2$)$_4$CH$_2$Br, a primary haloalkane, with sodium cyanide (NaCN), a strong nucleophile, in different media have well demonstrated how the nature of the solvents affects the S$_N$2 reactions (Reaction 6.12) [4].

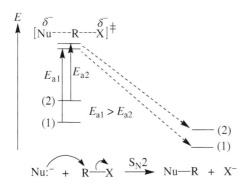

(1) S$_N$2 reaction in protic solvent (more polar): slower

(2) S$_N$2 reaction in aprotic solvent (less polar): faster

FIGURE 6.9   Energy profiles for the S$_N$2 reactions in (1) a more polar protic solvent and (2) a less polar aprotic solvent.

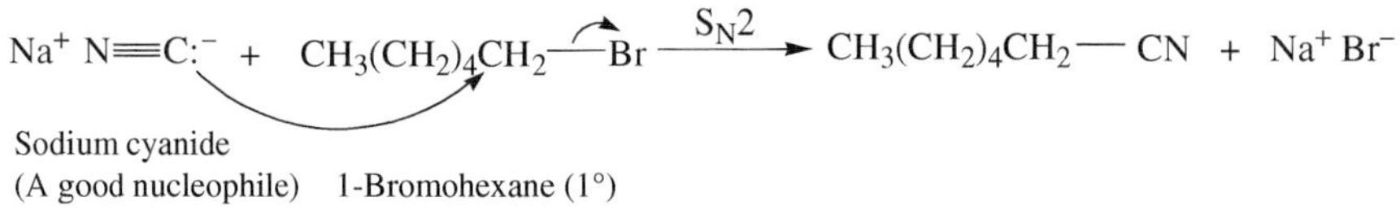

$$\text{Na}^+ \text{ N}{\equiv}\text{C:}^- \;+\; \text{CH}_3(\text{CH}_2)_4\text{CH}_2{-}\text{Br} \xrightarrow{\text{S}_N2} \text{CH}_3(\text{CH}_2)_4\text{CH}_2{-}\text{CN} \;+\; \text{Na}^+ \text{Br}^-$$

Sodium cyanide
(A good nucleophile)   1-Bromohexane (1°)

$$(6.12)$$

In a H_2O–CH_3OH mixture, a protic medium, it takes 20 h for the reaction to complete and product yield is 71%. In dimethyl sulfoxide (DMSO), an aprotic solvent, it only takes 20 min for the reaction to complete and product yield is 91%. In both media, the reactions follow the S_N2 mechanism because the substrate is a 1° haloalkane and the nucleophile (CN^-) is strong. The nature of the solvents does not affect the reaction mechanism but affects the relative reaction rates. An aprotic solvent (such as DMSO) facilitates an S_N2 reaction and makes it very fast (20 min) giving high yield (91%) of product, while a protic medium (H_2O–CH_3OH mixture) slows down an S_N2 reaction so that the reaction in H_2O–CH_3OH takes a long time (20 h) to complete giving lower yield (71%) of product.

Although amines (and ammonia) are neutral molecules, they are strong nucleophiles with the nucleophilicity comparable to that of hydroxide (Table 6.1). Therefore, they undergo S_N2 reactions with methyl and primary haloalkanes and other substrates bearing a good leaving group as shown below:

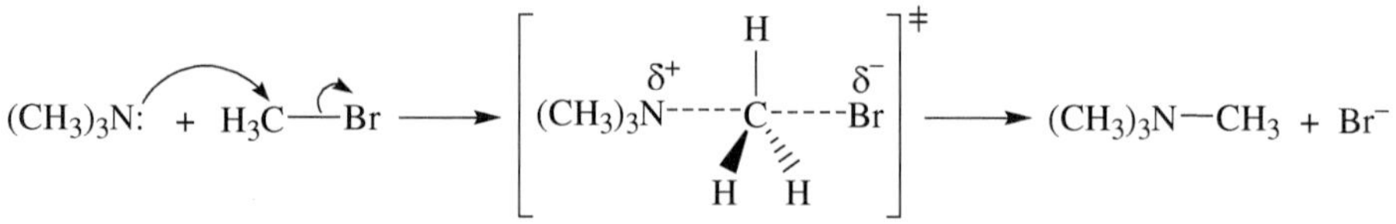

$$(\text{CH}_3)_3\text{N:} \;+\; \text{H}_3\text{C}{-}\text{Br} \longrightarrow \left[(\text{CH}_3)_3\overset{\delta^+}{\text{N}}{-}{-}{-}{-}\overset{\overset{\text{H}}{\mid}}{\text{C}}{-}{-}{-}{-}{-}\overset{\delta^-}{\text{Br}} \right]^{\ddagger} \longrightarrow (\text{CH}_3)_3\text{N}{-}\text{CH}_3 \;+\; \text{Br}^-$$

When an S_N2 reaction is effected by a neutral nucleophile such as an amine, the reaction proceeds faster in a protic solvent than in an aprotic solvent. This is because the transition state for the reaction has a charge separation. It is better stabilized by a more polar protic solvent than by a less polar aprotic solvent. In addition, a protic solvent molecule can form a hydrogen bond on the transition state, further stabilizing it and accelerating the reaction.

6.4.5   Effect of Unsaturated Groups Attached to the Functionalized Electrophilic Carbon

The S_N2 reactions are substantially accelerated by an unsaturated group, such as phenyl (C_6H_5-) and vinyl ($CH_2{=}CH-$), which is attached to the functionalized electrophilic carbon in the organic substrate (Fig. 6.10). This effect is due to stabilization of the transition state by an unsaturated group via conjugation effect. The negative charge can be further spread out into the π system, lowering the energy of the transition state.

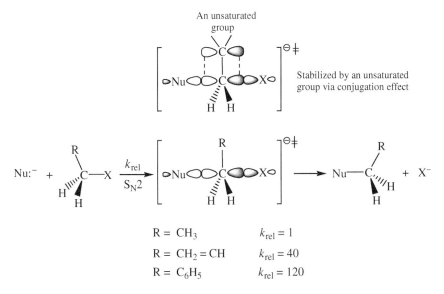

$$R = CH_3 \qquad k_{rel} = 1$$
$$R = CH_2 = CH \qquad k_{rel} = 40$$
$$R = C_6H_5 \qquad k_{rel} = 120$$

FIGURE 6.10 Stabilization of the S$_N$2 transition state by an unsaturated group in the central electrophilic carbon.

6.5 ANALYSIS OF THE S$_N$2 MECHANISM USING SYMMETRY RULES AND MOLECULAR ORBITAL THEORY

6.5.1 The S$_N$2 Reactions of Methyl and Primary Haloalkanes RCH$_2$X (X = Cl, Br, or I; R = H or an Alkyl Group)

Halo groups (–Cl, –Br, and –I) are among the most common good leaving groups. In this section, the S$_N$2 reactions of methyl and primary haloalkanes are studied using symmetry rules and molecular orbital (MO) theory.

A halomethane molecule CH$_3$X (X = Cl, Br, or I) possesses the C$_{3v}$ symmetry. Its principal threefold axis is along the C—X bond. This direction is defined as z-axis (Fig. 6.11a) [5]. The hydrogen and halogen atoms in CH$_3$X are considered four ligands coordinating to the central carbon atom. According to MO theory, the linear combination of the four ligand orbitals forms four MOs called *ligand group orbitals* (LGOs). Then bonding in CH$_3$X can be treated by further linear combination of LGOs with valence orbitals in the central carbon atom (2s, 2p$_x$, 2p$_y$, and 2p$_z$) based on symmetry match, forming a set of MOs.

The LGOs in CH$_3$X can be established by using the C$_{3v}$ character table and symmetry rules, and they are formulated as follows [5, 6]:

$$LGOs = 2a_1 + e$$

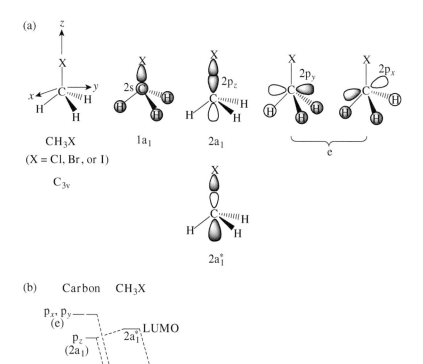

FIGURE 6.11 (a) Molecular orbitals (MOs) in CH_3X (X = Cl, Br, or I) which are formed by linear combination of ligand group orbitals (LGOs = $2a_1$ + e) and 2s, $2p_x$, $2p_y$, and $2p_z$ in the central carbon based on symmetry-match. The four MOs on top are occupied bonding orbitals. Each of them has a counterpart of an unoccupied antibonding MO. Only the LUMO ($2p_z$-based LUMO $2a_1^*$ along the C—X bond) is displayed. From Sun [5]. © 2003 MDPI. (b) Molecular orbital diagram of CH_3X, showing correlations of LGOs to the carbon atomic orbitals and qualitative energy levels of related orbitals. From Sun [6]. © 2003 MDPI. CC BY- 4.0. Public Domain.

where a_1 (singlet) correlates to 2s or $2p_z$ orbital and e (doublet) correlates to $2p_x$ and $2p_y$ orbitals in the central carbon. As a result, four bonding MOs are formed in the entire molecule (Fig. 6.11a). The predominant MO responsible for the formation of the C—X bond is $2a_1$ [highest occupied molecular orbital (HOMO)]. It is a carbon $2p_z$-based bonding orbital along the C—X bond. Its counterpart antibonding orbital $2a_1*$ is LUMO (lowest unoccupied molecular orbital). A qualitative MO diagram for CH_3X is shown in Figure 6.11b [6].

In an S_N2 reaction of CH_3X, the carbon atom functions as the electrophilic center (electron-pair acceptor). Therefore, the $2p_z$-based LUMO ($2a_1*$) should be the reacting orbital. It is an antibonding orbital along the C—X bond, responsible for cleavage of the C—X bond upon being filled with a pair of electrons. A vulnerable lobe of $2a_1*$ is identified along z-axis, just being positioned in the opposite side of the halogen atom. When a nucleophile (Nu^-) is approaching the central carbon from the opposite side of the C—X bond, the filled Nu^- orbital has the maximum overlap with $2a_1*$ (Fig. 6.12a) [5]. This effects electron transfer into $2a_1*$, which leads to breaking of the C—X bond and the formation of a new Nuc—C bond. The nucleophilic attack from any other orientations would generate less effective orbital overlap and thus, it would be kinetically unfavorable. The correlations of frontier molecular orbitals (FMOs) between reactants and products for S_N2 reaction of CH_3X are shown in Figure 6.12b [6].

The S_N2 reactions of primary haloalkanes RCH_2X (X = Cl, Br, or I; R = alkyl group) can be studied in an analogous manner to that described above for halomethanes. RCH_2X possesses the C_s symmetry with the X–C–R moiety contained in the $_h$ plane. The z-axis is perpendicular to the X–C–R defined $_h$ plane, while the x-axis (or y-axis) could be defined along the C—X bond (Fig. 6.13a). The LGOs in RCH_2X are formulated as follows based on the C_s character table [5, 6]:

$$\text{LGO's} = 3a' + a''$$

where a' (singlet) correlates to 2s, $2p_x$, or $2p_y$ orbital and a'' (singlet) correlates to 2 p_z orbital in the valence shell of the central carbon atom according to the C_s character table. Analogous to CH_3X, the predominant MO responsible for the formation of the C—X bond in RCH_2X is $2a'$, a carbon $2p_x$-based bonding MO (HOMO) along the C—X bond. Its counterpart LUMO $2a'*$ is formed in RCH_2X along its C—X bond (x-axis) as well (Fig. 6.13a). Conceivably, this $2p_x$-based $2a'*$ orbital (LUMO) should be the reacting orbital in an S_N2 reaction of RCH_2X, responsible for the cleavage of the C—X bond. The nucleophilic attack from the opposite side of the C—X bond (along x-axis) generates the maximum overlap between the Nu^- orbital and $2a'*$ (Fig. 6.13b) that effects electron transfer into $2a'*$ leading to breaking of the C—X bond and the formation of a new Nu—C bond [5, 6]. The nucleophilic attack from any other orientations would have less effective orbital overlap and thus, it would be kinetically unfavorable.

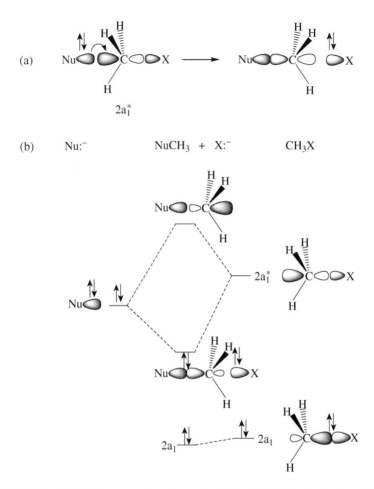

FIGURE 6.12 (a) Maximum overlap of a nucleophile orbital with the $2p_z$-based LUMO $2a_1^*$ in the opposite side of the leaving group –X in CH_3X (X = Cl, Br, or I) results in electron transfer to $2a_1^*$ breaking the C—X bond and effecting an S_N2 reaction. From Sun [5]. © 2003 MDPI. (b) Correlations of frontier molecular orbitals for the S_N2 reaction of CH_3X. From Sun [6]. © 2003 MDPI. CC BY- 4.0. Public Domain.

6.5.2   Reactivity of Dichloromethane CH_2Cl_2

Dichloromethane (CH_2Cl_2) is chosen to be the prototype for disubstituted functionalized molecules. The molecule possesses the C_{2v} symmetry. Its principal twofold axis is located between the two C—Cl bonds and also between the two C—H bonds. This direction is defined as z-axis (Fig. 6.14). The x and y orientations are defined accordingly. As a result, there is not a C—Cl (or C—H) bond along any of the x, y, or z axis. The MO treatment based on LGOs and x, y, and z basis functions in the central atom (LGOs = $2a_1 + b_1 + b_2$) does not generate any p-orbital based bonding or

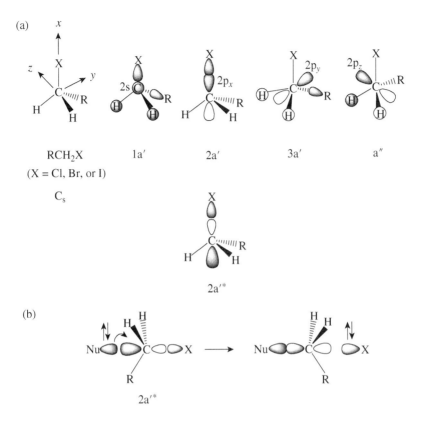

FIGURE 6.13   (a) Molecular orbitals (MOs) in RCH$_2$X (X = Cl, Br, or I; R = alkyl group) which are formed by linear combination of ligand group orbitals (LGO's = 3a' + a'') and 2s, 2p$_x$, 2p$_y$, and 2p$_z$ in the central carbon based on symmetry match. The four MOs on top are occupied bonding orbitals. Each of them has a counterpart of an unoccupied antibonding MO. Only the LUMO (2p$_x$-based LUMO 2a'* along the C—X bond) is displayed. (b) Maximum overlap of a nucleophile orbital with the 2p$_x$-based LUMO 2a'* in the opposite side of the leaving group –X results in electron transfer to 2a$_1$* breaking the C—X bond and effecting an S$_N$2 reaction.

antibonding MO along a C—Cl bond (Fig. 6.14). The lack of p-orbital based antibonding MO along a C—Cl bond implies there is not a single antibonding MO mainly responsible for breaking of the C—Cl bond. Consequently, the nucleophilic attack on any of the antibonding orbitals cannot effect a C—Cl bond breaking [5]. This explains why a disubstituted carbon like CH$_2$Cl$_2$ does not undergo an S$_N$2 reaction readily.

According to the MO model, the key to electrophilicity of a functionalized sp^3 hybridized carbon lies in presence of a carbon 2p-orbital based antibonding MO along the C—X bond (X = Cl, Br, or I—a good leaving group). This antibonding MO, upon accepting a pair of electrons from a nucleophile, gives rise to dissociation

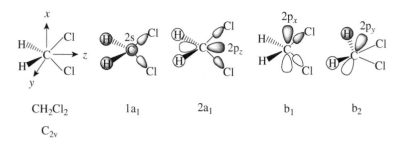

$$CH_2Cl_2 \qquad\qquad 1a_1 \qquad\qquad 2a_1 \qquad\qquad b_1 \qquad\qquad b_2$$

$$C_{2v}$$

FIGURE 6.14   Occupied bonding molecular orbitals (MOs) in CH_2Cl_2 which are formed by linear combination of ligand group orbitals (LGOs = $2a_1 + b_1 + b_2$) and 2s, $2p_x$, $2p_y$, and $2p_z$ in the central carbon based on symmetry match. Each of the bonding MOs has a counterpart of an unoccupied antibonding MO which is omitted. There is no 2p-based MO along any of the C—H or C—Cl bonds.

of the C—X bond and the formation of a new Nu—C bond [5]. The presence or absence of such an antibonding MO in various functionalized organic molecules can be identified readily by a symmetry analysis. The subsequent overall orbital interactions in an S_N2 reaction can be established by MO analysis.

6.6 S_N1 REACTIONS: KINETICS, MECHANISM, AND PRODUCT DEVELOPMENT

6.6.1   The S_N1 Mechanism and Rate Law

An S_N1 reaction is a stepwise process which takes place via an intermediate carbocation (Fig. 6.15). The reaction profile shows that the carbocation is highly energetic, resulting in a high activation energy (E_a) for the first step (the formation of a carbocation from reversible dissociation of RX). On the other hand, the highly energetic, unstable carbocation is extremely reactive, resulting in a very low activation-energy for the second step (nucleophilic attack on the carbocation to form the RNu product). Therefore, the first step (forward) is slow (*rate-determining step*), and the second step is fast ($k_1 \ll k_2$). The overall reaction rate is only determined by the molar concentration of the substrate, but is independent of the concentration of the nucleophile. This case can be confirmed by examining the rate law for the S_N1 reaction.

From the S_N1 reaction in Figure 6.15, the rate equations for RNu and R^+ can be written as

$$\text{Rate} = d[\text{RNu}]/dt = k_2[R^+][Nu^-] \qquad (6.13)$$

$$d[R^+]/dt = 0 = k_1[\text{RX}] - k_{-1}[R^+][X^-] - k_2[R^+][Nu^-] \qquad (6.14)$$

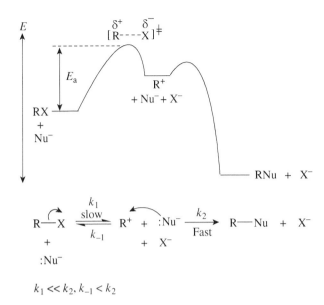

FIGURE 6.15   Reaction profile for the S$_N$1 mechanism.

Since RNu is only formed in the bimolecular k_2 step, its rate equation (Eq. 6.13) is second order. Equation 6.14 shows the rate of net increase in concentration of the carbocation intermediate, which is equal to rate of its formation $k_1[RX]$ (unimolecular dissociation) minus rate of its consumption ($k_{-1}[R^+][X^-] + k_2[R^+][Nu^-]$) (bimolecular attacks of both X^- and Nu^- on the carbocation). The steady-state assumption is applicable to the reactive carbocation intermediate ($d[R^+]/dt = 0$).

From Equation 6.14, concentration of the carbocation intermediate can be written as

$$[R^+] = \{k_1/(k_{-1}[X^-] + k_2[Nu^-])\}[RX] \approx (k_1/k_2[Nu^-])[RX] \qquad (6.15)$$

Since the k_2 step is fast and $[X^-] \ll [Nu^-]$, we have $k_{-1}[X^-] \ll k_2[Nu^-]$ and

$$k_{-1}[X^-] + k_2[Nu^-] \approx k_2[Nu^-]$$

Combination of Equations 6.13 and 6.15 leads to

$$\text{Rate} = d[RNu]/dt = k_2(k_1/k_2)([RX]/[Nu^-])[Nu^-]$$

Therefore,

$$\text{Rate} = d[RNu]/dt = k_1[RX] \qquad (6.16)$$

Equation 6.16 is the rate law for the S_N1 reaction in Figure 6.15. It clearly shows that the reaction is only first order in the substrate, and the rate of the overall reaction is determined by the first step (k_1). Since the nucleophile is only involved in the second step (k_2) but not in the first one, the rate of the reaction is independent of the concentration and nature of the nucleophile.

6.6.2 Solvent Effect

While polar aprotic solvents accelerate rates of S_N2 reactions (Section 6.4.4 and Fig. 6.9), S_N1 reactions are better facilitated by protic solvents. The transition state for rate-determining step of an S_N1 reaction (dissociation of RX to a carbocation, Fig. 6.15) has a charge separation. Therefore, it is better stabilized by a more polar protic solvent than by a less polar aprotic solvent. In addition, a protic solvent molecule can be hydrogen bonded to the X^- anion, further stabilizing the anion. This lowers the energy of the transition state. As a result, the formation of a carbocation in a protic solvent is faster (with a smaller activation energy) than in an aprotic solvent. This accounts for why protic solvents are usually employed for S_N1 reactions. In fact, many S_N1 reactions of haloalkanes are readily effected by solvation of protic solvents such as water, alcohols, and carboxylic acids (Fig. 6.16). In the reactions of Figure 6.16, 2-chloro-2-methylpropane (a tertiary haloalkane) is mixed with excess water, ethanol, and acetic acid, respectively. Each of the solvents also functions as a nucleophile, attacking the carbocation intermediate to effect the S_N1 reaction.

The reaction presented in Figure 6.16b represents a useful method for esterification to make a tertiary alkyl ester by the S_N1 reaction of a carboxylic acid with a tertiary alkyl halide. Very often, an ester is made by acid-catalyzed reaction of a carboxylic acid with an alcohol. The type of reactions does not work well for tertiary alcohols presumably due to steric hindrance of the tertiary alkyl group in the alcohol. The above S_N1 reaction provides an alternative way to make the ester product.

6.6.3 Effects of Carbocation Stability and Quality of Leaving Group on the S_N1 Rates

Different from the S_N2 reactions which are strongly subject to steric hindrance (Section 6.4.2 and Fig. 6.5), the rates for S_N1 reactions are determined by relative stability of the carbocation intermediates. In general, the rate constants for S_N1 reactions of different types of substrates to the same nucleophile increases in the order of $CH_3X < CH_3CH_2X$ (1°) $\ll (CH_3)_2CHX$ (2°) $< (CH_3)_3CX$ (3°) $<$ $PhCH_2X$. In fact, methyl and primary alkyl halides essentially do not undergo S_N1 reactions because the corresponding methyl and primary carbocations do not exist due to high instability. Such a trend is illustrated by the data for S_N1 reactions of water (solvolysis) with different types of bromoalkanes (RBr) at 25 °C (Fig. 6.17) [3].

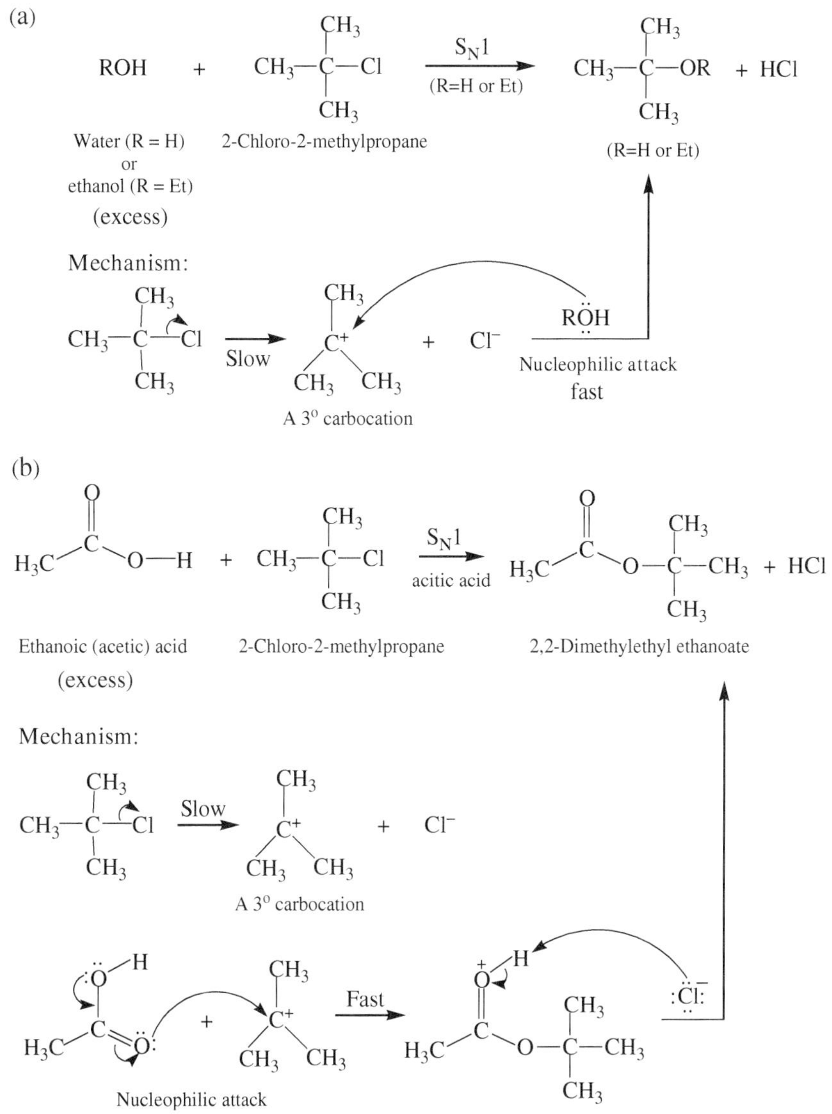

(a)

(b)

FIGURE 6.16   The $S_N 1$ reactions of 2-chloro-2-methylpropane in protic solvents.

The energy profiles in Figure 6.18 show dependence of the activation energy for the formation of a carbocation (rate-determining step for an $S_N 1$ reaction) from RX on the relative energy level (stability) of the carbocation. According to Equation 1.55 (the Bell–Evans–Polanyi principle), the activation energy (E_a) for the formation of a

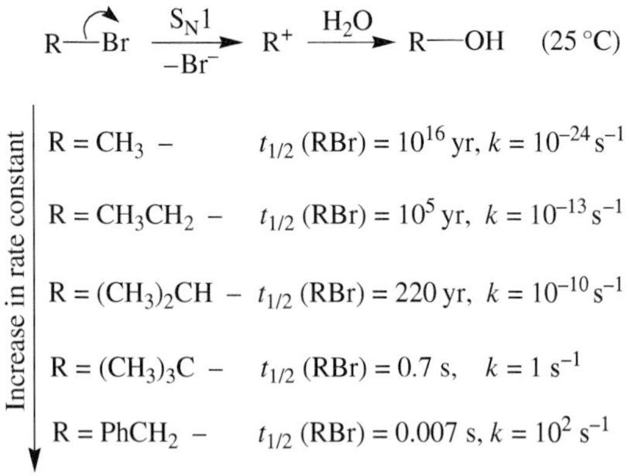

$$R—Br \xrightarrow[-Br^-]{S_N1} R^+ \xrightarrow{H_2O} R—OH \quad (25\,°C)$$

Increase in rate constant

$R = CH_3 –$ $t_{1/2}\,(RBr) = 10^{16}\,yr, k = 10^{-24}\,s^{-1}$

$R = CH_3CH_2 –$ $t_{1/2}\,(RBr) = 10^5\,yr,\ k = 10^{-13}\,s^{-1}$

$R = (CH_3)_2CH –$ $t_{1/2}\,(RBr) = 220\,yr,\ k = 10^{-10}\,s^{-1}$

$R = (CH_3)_3C –$ $t_{1/2}\,(RBr) = 0.7\,s,\quad k = 1\,s^{-1}$

$R = PhCH_2 –$ $t_{1/2}\,(RBr) = 0.007\,s, k = 10^2\,s^{-1}$

FIGURE 6.17   The S_N1 rate constants for hydrolysis of various bromoalkanes. Modified from Bruckner [3].

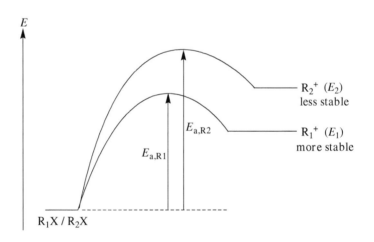

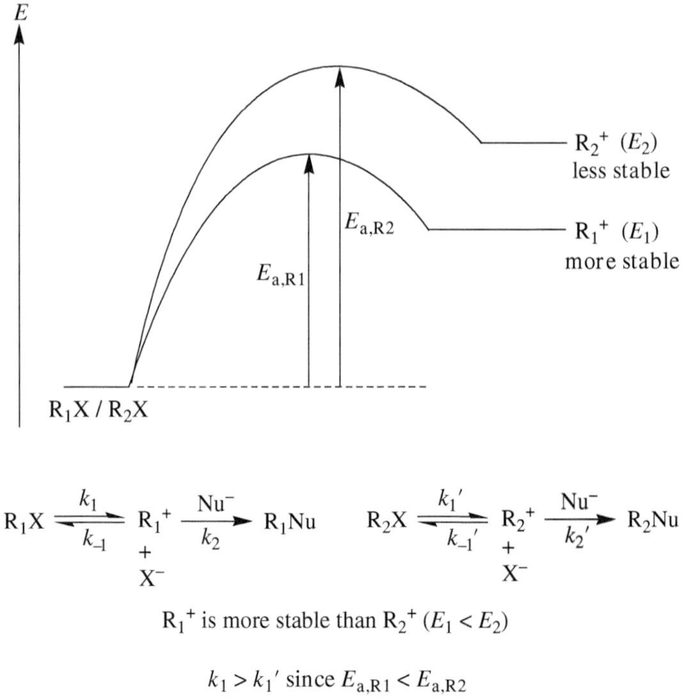

$$R_1X \underset{k_{-1}}{\overset{k_1}{\rightleftharpoons}} \underset{+}{R_1^+} \xrightarrow{\underset{k_2}{Nu^-}} R_1Nu \qquad R_2X \underset{k_{-1}'}{\overset{k_1'}{\rightleftharpoons}} \underset{+}{R_2^+} \xrightarrow{\underset{k_2'}{Nu^-}} R_2Nu$$

$$X^- \qquad\qquad\qquad\qquad\qquad X^-$$

R_1^+ is more stable than R_2^+ $(E_1 < E_2)$

$k_1 > k_1'$ since $E_{a,R1} < E_{a,R2}$

FIGURE 6.18   Energy profiles for rate-determining steps of the S_N1 reactions of different haloalkanes.

carbocation from RX increases linearly as a function of the energy level of the carbocation ($E_{carbocation}$) formulated as

$$E_a = aE_{carbocation} + b \quad (\text{a and b are constants})$$

Therefore, as a carbocation intermediate becomes more stabilized, the activation energy for the formation of the carbocation from RX is getting smaller, giving rise to a greater rate constant (Fig. 6.18). Since relative stability of the different types of carbocations increases (correspondingly, the energy decreases) in the order of CH_3^+, $CH_3CH_2^+$ (1°), $(CH_3)_2CH^+$ (2°), $(CH_3)_3C^+$ (3°), and $PhCH_2^+$ (Section 1.9.1), the rate constants for S$_N$1 reactions of the corresponding RX increase in the same order. As hydrogen atoms in $PhCH_2^+$ are replaced by phenyl groups successively, the carbocations will be more stabilized. As a result, the rate constants for S$_N$1 reactions for Ph_2CHX and Ph_3CX further increase.

For functionalized alkanes with different leaving groups, such as RCl, RBr, and RI, the substrate that contains a better leaving group undergoes faster nucleophilic substitution reaction with a given nucleophile. This is true for both S$_N$1 and S$_N$2 reactions. Figure 6.19 shows reaction profiles for both S$_N$1 and S$_N$2 reactions of RCl, RBr, and RI with a certain nucleophile. As described earlier in Section 6.1, the order of standard BDEs for the C—X bonds is BDE (C–Cl, 327 kJ/mol) > BDE (C–Br, 285 kJ/mol) > BDE (C–I, 213 kJ/mol). The energy levels of RCl, RBr, and RI are placed accordingly in the reaction profiles. For the S$_N$1 reactions, the formation of the carbocation R$^+$ intermediate from RX (the rate-determining step) has late transition states for all the haloalkanes. Therefore, all the transition

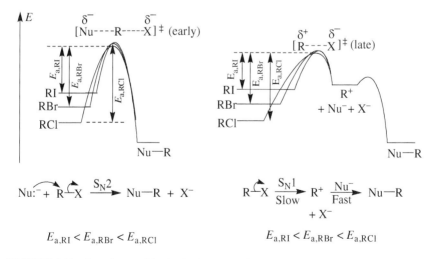

FIGURE 6.19 Reaction profiles for both S$_N$1 and S$_N$2 reactions of RCl, RBr, and RI with a certain nucleophile.

states resemble the carbocation and their energy levels are very close. Hence, the difference in activation energies for the reactions is determined almost exclusively by the relative energy levels of the starting haloalkanes ($E_{a,RI} < E_{a,RBr} < E_{a,RCl}$). For the S_N2 reactions, they all have early transition states, which resemble the starting haloalkanes. Therefore, the activation energies for reactions of different RX are also mainly determined by the energy levels of RX, and the order is the same as that for the S_N1 reactions. Therefore, for both types of reactions with a certain nucleophile, the rate constants increase in the order of RCl < RBr < RI.

6.6.4 Product Development for S_N1 Reactions

While stability of the carbocation intermediate determines the rate for an S_N1 reaction, the product development is determined primarily by the structure of the carbocation. Figure 6.20 shows general situations of product development for S_N1 reactions. As the C—X bond in the haloalkane $R_1R_2R_3C$–X substrate dissociates, first, a contact $R_1R_2R_3C^+/X^-$ ion-pair is formed in the solvent cage. If all the R groups are alkyls (or two alkyls and one hydrogen), the corresponding tertiary or secondary carbocation $R_1R_2R_3C^+$ intermediate is stable enough to be formed. However, its stability does not allow sufficient time for the X^- anion to diffuse out of the solvent cage before the carbocation is attacked by the nucleophile to complete the overall reaction. Instead, once the contact $R_1R_2R_3C^+/X^-$ ion-pair is formed, the carbocation will immediately react with the nucleophile. Clearly, the nucleophile approaches the carbocation more readily from the opposite side of the halide to give the configuration-inversed enantiomer as the major product. This is because the halide side is substantially hindered by the X^- anion. The attack of nucleophile on the carbocation from the same side of X^- is slower, giving the corresponding configuration-retained enantiomer as the minor product. Usually, when a carbocation is stabilized by an unsaturated group such as phenyl/aryl or allyl, its reaction with the nucleophile is much slower. The X^- anion will have sufficient time to diffuse out of the solvent cage before the carbocation is attacked by the nucleophile. This results in the formation of a separated $R_1R_2R_3C^+...X^-$ ion-pair when one of the R groups is an unsaturated group, with the environment of both faces of the carbocation essentially the same. Thus, the subsequent nucleophilic attacks on the carbocation from both directions possess essentially equal probability giving rise to the formation of a racemic mixture.

The situations presented in Figure 6.20 are further demonstrated by the following examples. Reaction of (S)-$R_1R_2R_3CBr$ (R_1 = i-propyl, R_2 = methyl, R_3 = H) with excess ethanol (EtOH) gives a mixture of two ethers, (R)-$EtOCR_1R_2R_3$ (major) and (S)-$R_1R_2R_3COEt$ (minor). It proceeds via a contact $R_1R_2R_3C^+/Br^-$ ion-pair. The attack of EtOH on $R_1R_2R_3C^+$ from the opposite site of Br^- giving (R)-$EtOCR_1R_2R_3$ occurs faster than the attack from the same side of Br^- giving (S)-$R_1R_2R_3COEt$. As a result, the configuration inversion of the product is greater than the configuration retention.

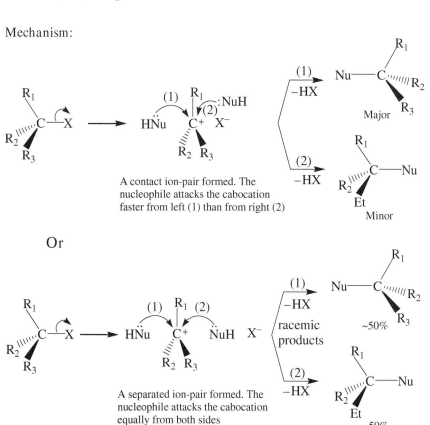

FIGURE 6.20   Mechanism and product development of the S$_N$1 reaction.

Similarly, reaction of ethanol with enantiomerically pure (R)- or (S)-R$_1$R$_2$R$_3$CBr (R$_1$ = H, R$_2$ = methyl, R$_3$ = n-hexyl) has been reported to lead to the enantiomeric R$_1$R$_2$R$_3$COEt ethers with 66% inversion and 34% retention of the original config-uration (Fig. 6.21). On the other hand, reaction of EtOH with enantiomerically pure (R)- or (S)-PhCH(Br)CH$_3$ [(1-bromoethyl)benzene] gives racemic PhCH(OEt)CH$_3$

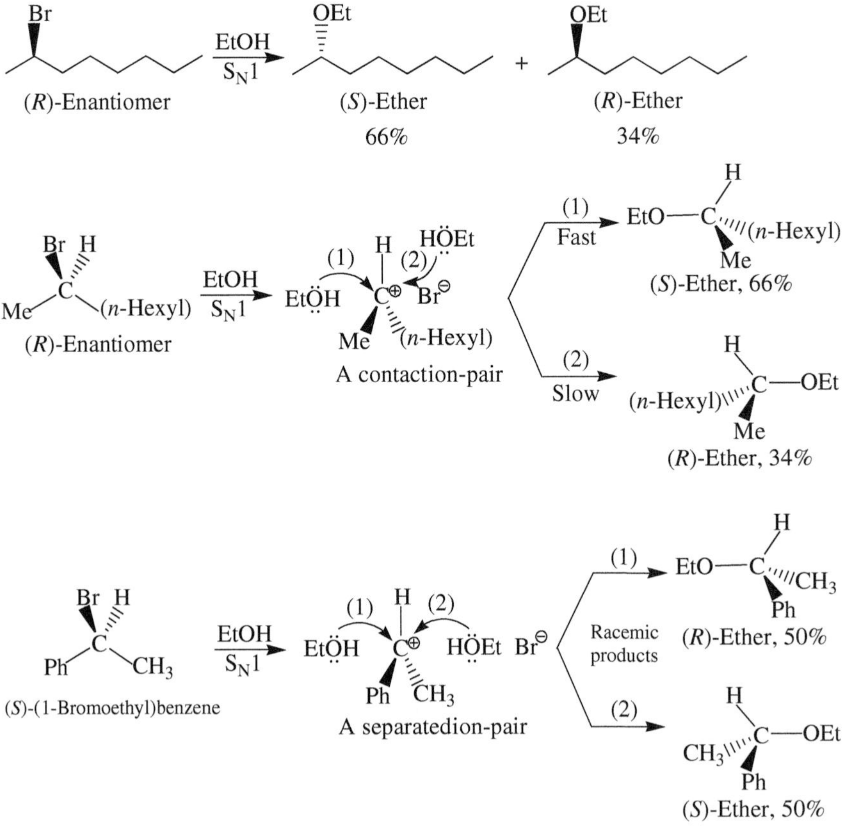

FIGURE 6.21   Stereoselectivity of the S_N1 reactions. Modified from Bruckner [3].

(1-ethoxybenzene) products (Fig. 6.21) [3]. The first reaction in Figure 6.21 takes place via a contact $(n\text{-hexyl})\text{MeHC}^+/\text{Br}^-$ ion-pair, with the formation of the configuration-inversed ether faster. The second reaction in Figure 6.21 takes place via a separated $\text{PhHCH}_3\text{C}^+\ldots\text{Br}^-$ ion-pair due to stabilization of the carbocation by a phenyl group. Then the carbocation is attacked by EtOH essentially equally from both sides to give (R)- and (S)-PhCH(OEt)CH_3 in equal amounts. The stereochemistry feature for the S_N1 reaction of PhCH(Br)CH_3 is comparable to that for the electrophilic additions of hydrogen halides (HX) to PhCH=CH_2 (Fig. 3.5) which takes place via the same prochiral PhHCH_3C^+ carbocation (with three different groups on the central trivalent carbon) and gives racemic products.

6.7   COMPETITIONS BETWEEN S_N1 AND S_N2 REACTIONS

A nucleophilic substitution reaction that takes place on an sp³-hybridized carbon can follow either the S_N1 or S_N2 mechanism, which is mainly determined by the nature of the organic **substrate** (R-LG), the **nucleophile** (strong or weak nucleophiles), and

TABLE 6.2   Factors that Determine the Reaction Mechanisms: Competition Between S_N2 and S_N1

	S_N2 mechanism	S_N1 mechanism
Substrate	Methyl, 1°, 2°	2°, 3°, benzylic, allylic
Nucleophile	Strong nucleophiles (mostly anions)	Weak (poor) nucleophiles (generally neutral molecules)
Solvent	Polar aprotic solvents (e.g., DMSO, THF, & acetone)	Protic solvents (e.g., H_2O & alcohols)

the **solvent** (protic or polar aprotic). The factors that determine the reaction mechanisms are summarized in Table 6.2.

For haloethane CH_3CH_2X (a 1° substrate) and halomethane CH_3X, because the central electrophilic carbon is attached to two or three small hydrogen atoms, steric hindrance to the nucleophilic attack from the opposite side of the leaving group is small (or minimal for the methyl substrate) (Fig. 6.5). As a result, the electrophilic carbon bearing a leaving group can be readily attacked by a strong nucleophile effecting an S_N2 reaction. On the other hand, the formation of a primary carbocation or methyl cation would require a very high activation energy due to its instability. In fact, these carbocations have not been observed experimentally. It makes an S_N1 reaction for a 1° or methyl substrate extremely slow (Fig. 6.17). *In general, the nucleophilic substitution reaction of a primary (1°) or a methyl substrate always follows the S_N2 mechanism.* Mostly, the reaction can only be effected by a strong nucleophile (mostly anions). The type of solvents can change the reaction rate (Section 6.4.4), but does not change the mechanism.

In 2-halo-2-methylpropane (*t*-butyl halide) $(CH_3)_3CX$ (a 3° substrate), the central electrophilic carbon is protected by three methyl groups, producing strong steric hindrance to the nucleophilic attack from the opposite side of the leaving group. This steric hindrance (characteristic of a tertiary-butyl group or any tertiary substrate) makes an S_N2 reaction extremely slow (Fig. 6.5). On the other hand, because a tertiary carbocation, such as $(CH_3)_3C^+$, is relatively stabilized, an S_N1 reaction for a tertiary substrate can take place in a relatively fast rate (Fig. 6.17). *In general, the nucleophilic substitution reaction of a tertiary (3°) substrate always follows the S_N1 mechanism.* The reaction can be effected by a weak nucleophile (mostly neutral molecules). The type of solvents can change the reaction rate, but does not change the mechanism.

For a secondary (2°) substrate such as 2-halopropane $(CH_3)_2CHX$, as the number of methyl groups on the central electrophilic carbon increases, the steric hindrance to the nucleophilic attack from the opposite side of the leaving group is getting more severe. This disfavors an S_N2 reaction, but does not completely prevent it from happening (Fig. 6.5). On the other hand, the stability of a secondary (2°) carbocation allows the S_N1 reaction of a 2° substrate to take place in an appreciable rate (see Fig. 6.17). In addition, a primary substrate which contains an unsaturated group (such as phenyl and vinyl) on the electrophilic carbon can also have both S_N1 and S_N2 reactions. *In general, the nucleophilic substitution reactions of a secondary*

(2°) substrate and a primary (1°) benzylic or allylic substrate can follow either the S_N2 or S_N1 mechanism, depending on the nature of the nucleophile and solvent. Usually, when a reaction is performed with a strong nucleophile (mostly, an anion) in an aprotic solvent, it follows the S_N2 mechanism. If a reaction is performed with a weak nucleophile (mostly, a neutral molecule) in a protic solvent, it follows the S_N1 mechanism.

Now, let us look at nucleophilic substitution reactions of (R)-2-bromobutane, a secondary haloalkane, performed in different conditions (Fig. 6.22).

When (R)-2-bromobutane is treated with excess water (H$_2$O), a weak nucleophile, and a protic solvent, because a weak nucleophile usually effects an S_N1 (but S_N2) reaction and a protic solvent also facilitates an S_N1 reaction, the overall reaction follows the S_N1 mechanism via a contact $R_1R_2R_3C^+/Br^-$ (R_1 = Me, R_2 = Et, R_3 = H) ion-pair with Br$^-$ staying at **front** side of the carbocation in solvent cage. Then the secondary carbocation intermediate is attacked by H$_2$O from both faces, with the attack from the **back** face faster giving the configuration-inversed (S)-2-butanol as the major enantiomer. The front face of the carbocation is hindered by bromide. Thus, the attack of water on the carbocation from the front goes more slowly giving the configuration-retained (R)-2-butanol as the minor enantiomer.

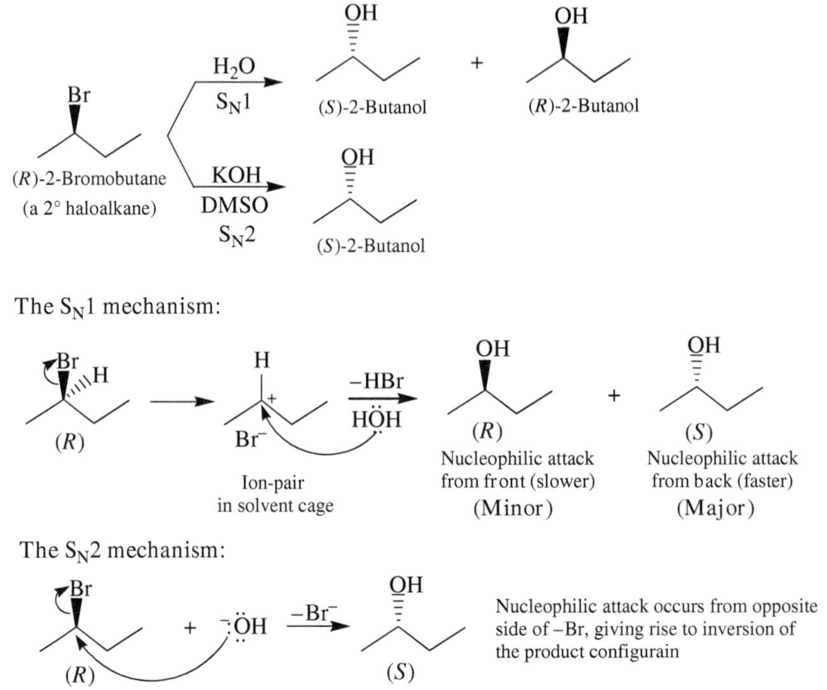

FIGURE 6.22 The S_N1 and S_N2 reactions of (R)-2-bromobutane performed in different conditions.

When (*R*)-2-bromobutane is treated with KOH (a strong nucleophile) in DMSO (a polar aprotic solvent), because a strong nucleophile can effect an S_N2 reaction and a polar aprotic solvent facilitates an S_N2 reaction but disfavors an S_N1 reaction, the overall reaction follows the S_N2 mechanism giving a single product which possesses the opposite configuration of the substrate.

Although reaction of hydroxide with a secondary haloalkane in a polar *aprotic solvent* follows the S_N2 mechanism, reactions of a secondary haloalkane (or other secondary substrates) with other strong bases (e.g., alkoxides) in general result in the formation of major elimination products. Usually, only *weakly basic*, strong nucleophiles (e.g., I^-, HS^-, RS^-, N_3^-, RCO_2^-, and CN^-) effect S_N2 reactions for secondary substrates. Hydroxide, a strongly basic nucleophile, is a remarkable exception.

Equation 6.7 (Section 6.4.1) shows that reaction of (*R*)-2-tosylpropylbenzene (a secondary alkyl tosylate) with acetate (a moderately strong nucleophile) in acetic anhydride (a polar aprotic solvent) follows the S_N2 mechanism. It leads to a complete inversion of the original configuration in the product.

(*R*)-2-Tosylpropyl-   (*S*)-1-Phenyl-2-propyl
benzene   acetate

Upon treated by excess acetic acid (a weak nucleophile and a protic solvent), (*R*)-2-tosylpropylbenzene undergoes an S_N1 reaction. It gives a mixture of two enantiomers.

(*R*)-2-Tosylpropyl-   (*S*)-1-Phenyl-2-propyl   (*R*)-1-Phenyl-2-propyl
benzene   (Excess)   acetate   acetate

When a secondary haloalkane is treated by a strong nucleophile in a protic solvent, both S_N1 and S_N2 reactions can take place concurrently, leading to a mixed rate law for the overall reaction. This is because a strong nucleophile facilitates an S_N2 reaction, and a protic solvent favors an S_N1 reaction. For example, hydrolysis of 2-bromopropane (Me_2CHBr) to 2-propanol (Me_2CHOH) in the presence of sodium hydroxide in a mixed water–ethanol medium has been found to follow the following pattern of a rate law (Eq. 6.17) [1]:

$$Rate = -d[Me_2CHBr]/dt = k_1[Me_2CHBr] + k_2[Me_2CHBr][OH^-] \qquad (6.17)$$

The rate law is established by plotting the ratio of $\{Rate/[Me_2CHBr]\}$ (initial rate) versus $[OH^-]$ which are obtained from several experiments. A linear relationship has been found as follows:

$$Rate/[Me_2CHBr] = k_1 + k_2[OH^-]$$

The k_1 and k_2 values are obtained from the intercept $(=k_1)$ and slope $(=k_2)$ of the straight line, respectively. Multiplying $[Me_2CHBr]$ on both sides leads to Equation 6.17.

The kinetic study shows that the overall reaction of Me_2CHBr with OH^- in a protic medium follows both S_N1 and S_N2 mechanisms at the same time (Reactions 6.18 and 6.19).

$$S_N1: \quad Me_2CH\!-\!Br \xrightarrow[slow]{-Br^-} Me_2CH^+ \xrightarrow[fast]{OH^-} Me_2CH\!-\!OH \qquad (6.18)$$

$$S_N2: \quad H\ddot{O}\!:\!^- + Me_2CH\!-\!Br \longrightarrow Me_2CH\!-\!OH + Br^- \qquad (6.19)$$

The first term $k_1[Me_2CHBr]$ in the rate law (Eq. 6.17) is resulted from the S_N1 component (Reaction 6.18), which is first order in Me_2CHBr and zero order in OH^-. The second term $k_2[Me_2CHBr][OH^-]$ in the rate law (Eq. 6.17) originates from the S_N2 component (Reaction 6.19), which is first order in Me_2CHBr and first order in OH^-. k_1 and k_2 can be regarded as the experimental rate constants for the pure S_N1 and S_N2 reactions, respectively. As shown above, they are determined from the $\{Rate/[Me_2CHBr]\}$ versus $[OH^-]$ relationship.

A primary benzylic substrate can undergo ether an S_N1 or S_N2 reaction. When the attacking nucleophile is weak, an S_N1 reaction takes place via $PhCH_2^+$. For example, reaction of benzyl chloride with water giving benzyl alcohol follows the S_N1 mechanism (Fig. 6.23a).

On the other hand, the sp^3 carbon in a benzyl substrate $PhCH_2$–X (X is a good leaving group, typically a halogen atom) is sterically unhindered. When the attacking nucleophile is strong, an S_N2 reaction takes place. For example, reaction of benzyl chloride with hydroxide giving benzyl alcohol follows the S_N2 mechanism (Fig. 6.23b).

6.8   SOME USEFUL S_N1 AND S_N2 REACTIONS: MECHANISMS AND SYNTHETIC PERSPECTIVES

Many nucleophilic substitution reactions that take place on sp^3 carbons have important synthetic applications. They follow the S_N1 or S_N2 mechanism depending on various factors as described above. In this section, some useful examples are presented. Their mechanistic and synthetic perspectives are discussed.

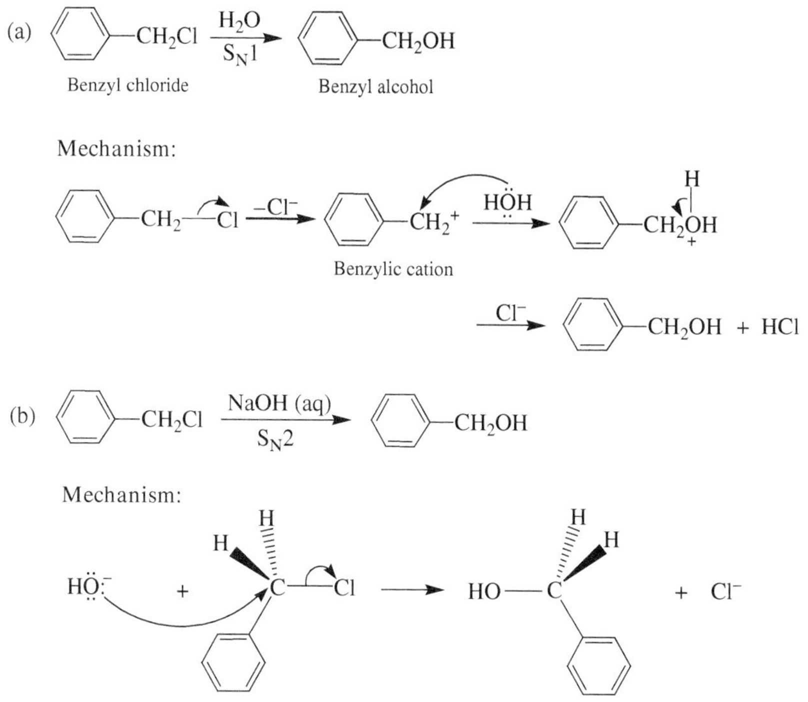

(a)
Benzyl chloride   Benzyl alcohol

Mechanism:

Benzylic cation

(b)

Mechanism:

FIGURE 6.23   The S_N1 (a) and S_N2 (b) hydrolysis of benzyl chloride.

6.8.1   Nucleophilic Substitution Reactions Effected by Carbon Nucleophiles

The first type of synthetically important carbon nucleophiles we will go over in this section is Grignard reagents (RMgX) and related alkyllithium (RLi). They are made by reactions of haloalkanes (or haloarenes) with the corresponding metals as shown in Reactions 6.20 and 6.21.

$$R\!-\!X \;+\; Mg \xrightarrow{\;Et_2O \text{ or THF}\;} R\!-\!MgX \tag{6.20}$$

(X = Cl, Br, or I; R = alkyl or aryl)

$$R\!-\!X \;+\; 2\,Li \xrightarrow{\;Et_2O \text{ or THF}\;} R\!-\!Li \;+\; LiX \tag{6.21}$$

(X = Cl, Br, or I; R = alkyl or aryl)

The formation of a Grignard reagent follows a radical mechanism as shown below:

If the C–X carbon in RX is chiral, the original configuration of the alkyl group will be retained in the RMgX product as the second step of radical coupling is extremely fast such that the carbon–magnesium bond is formed in the same side of the R—X bond. The carbon–metal bond in R–M (M = MgX or Li) is highly active and functions as a strong nucleophile although the molecule is electrically neutral. It can react with an epoxide to give an alcohol after an aqueous work-up (Reaction 6.22)

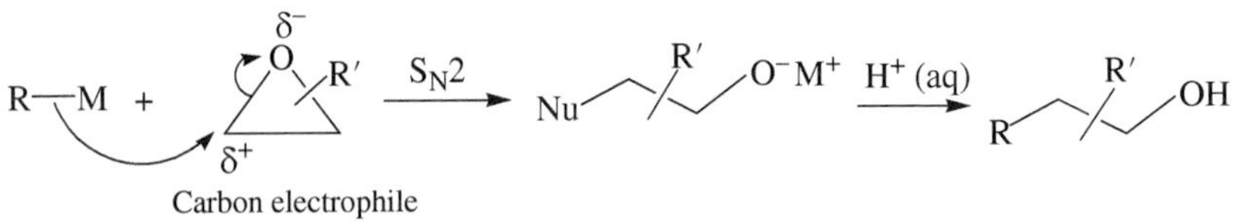

$$
\text{R} - \text{M} + \quad \underset{\text{Carbon electrophile}}{\overset{\delta^-}{\underset{\delta^+}{\bigtriangleup}}\hspace{-1em}\text{R}'} \quad \xrightarrow{\text{S}_N2} \quad \text{Nu}\diagdown\diagup\overset{\text{R}'}{\diagdown}\diagup\text{O}^-\text{M}^+ \quad \xrightarrow{\text{H}^+ \text{ (aq)}} \quad \overset{\text{R}'}{\underset{\text{R}}{\diagdown}}\diagup\text{OH}
$$

(R = alkyl or aryl, M = MgX or Li)

(6.22)

Figure 6.24 shows an example of stereospecific synthesis of an alcohol by an S$_N$2 reaction of a Grignard reagent with ethylene oxide. It is noticed that the Grignard reagent retains the initial *trans*-configuration of the iodoalkane substrate. As a result, the resulting alcohol product is in a *trans*-configuration.

If the epoxide molecule is unsymmetrical, the Grignard reagent will preferably attack the less crowded (sterically less hindered) carbon in the epoxide ring which contains more H atoms and less alkyl groups. The reaction in Figure 6.25a demonstrates the case. This is true in general for nucleophilic substitution reactions of unsymmetrical epoxides with strong nucleophiles. On the other hand, the acid-catalyzed nucleophilic substitution reactions of unsymmetrical epoxides with weak nucleophiles usually have the opposite regioselectivity, namely that the nucleophile (such as an alcohol) will preferably attack the carbon in the protonated epoxide ring which contains fewer number of H atoms (more alkyl groups) as shown in Figure 6.25b. The protonated epoxide carries partial carbocationic character due to delocalization of the positive charge. Of the three resonance contributors to the protonated epoxide in Figure 6.25b, because a 2° carbocation is more stable than a 1° carbocation, the first Lewis structure is the more important contributor. As a result, in the overall reaction, the nucleophile preferably attacks the methyl carbon in the ring.

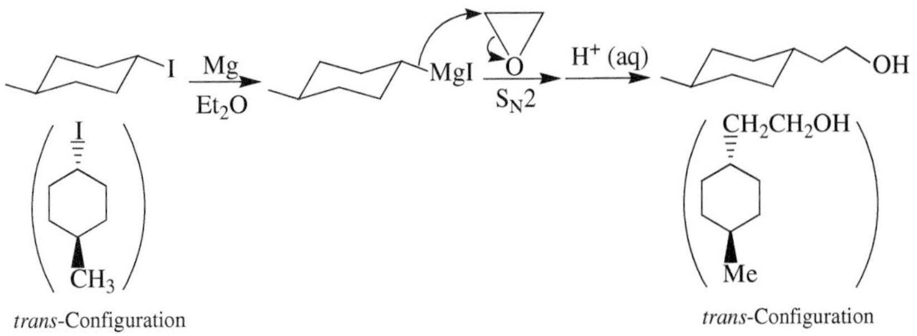

trans-Configuration   *trans*-Configuration

FIGURE 6.24   The S$_N$2 reaction of a Grignard reagent with ethylene oxide.

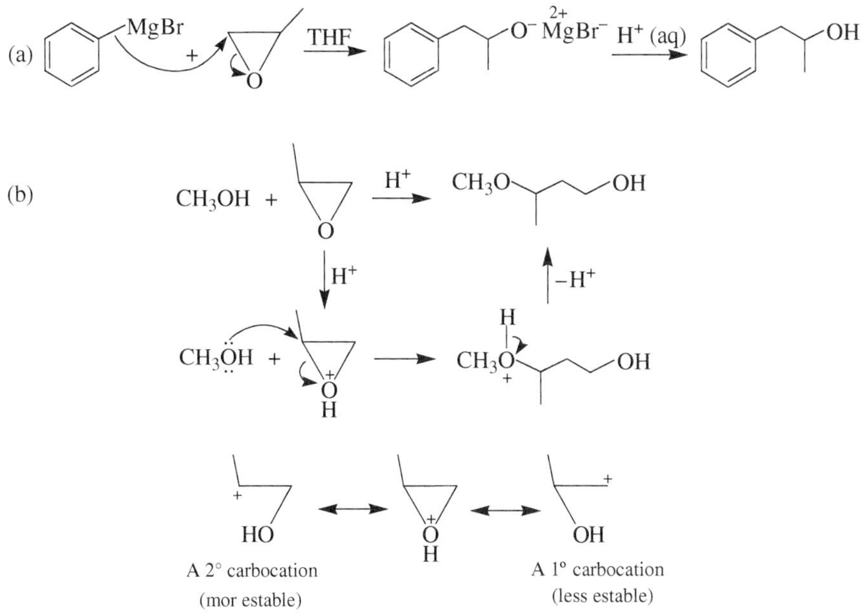

FIGURE 6.25 (a) Regiospecific S_N2 reaction of phenylmagnesium bromide with an unsymmetrical epoxide; and (b) regioselectivity of the acid catalyzed S_N2 reaction of methanol with an unsymmetrical epoxide.

The second type of synthetically useful carbon nucleophiles is the alkyne anions ($RC\equiv C{:}^-$) that are formed usually by reaction of a 1-alkyne ($RC\equiv CH$) with sodium amide ($NaNH_2$) (an extremely strong base) in liquid ammonia [7]. $RC\equiv C{:}^-$ is an extremely strong carbon nucleophile. It can react with methyl and primary haloalkanes or other analogous substrates containing a good leaving group to form larger alkynes (Fig. 6.26). Example 3 in Figure 6.26 shows the synthetic approach to a specialty ketone containing a *cis* C=C bond by consecutive S_N2 reactions of alkyne anions [2]. First, the alkyne precursor to the targeted *cis*-unsaturated ketone is made by the consecutive S_N2 reactions. Catalytic hydrogenation of the alkyne affords a *cis*-alkene (refer to Section 3.5.2). The final step is acid-catalyzed hydrolysis of a ketal to lead to a ketone. Chemistry of ketals will be discussed in Chapter 8.

The cyanide ($^-{:}C\equiv N$) ion (typically as the sodium or potassium salt) is another strong carbon nucleophile. It can react with methyl, primary, and secondary haloalkanes to form nitriles (Reaction 6.23).

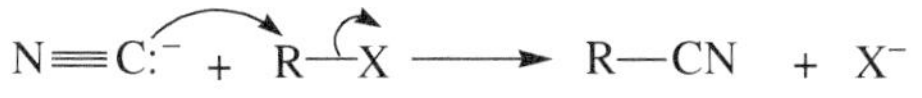

Cyanide   A nitrile   (6.23)

R = methyl, primary, or secodary alkyl;
X = Cl, Br, or I

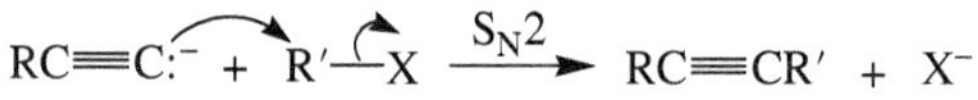

$$RC{\equiv}C{:}^- + R'{-}X \xrightarrow{S_N2} RC{\equiv}CR' + X^-$$

Examples:

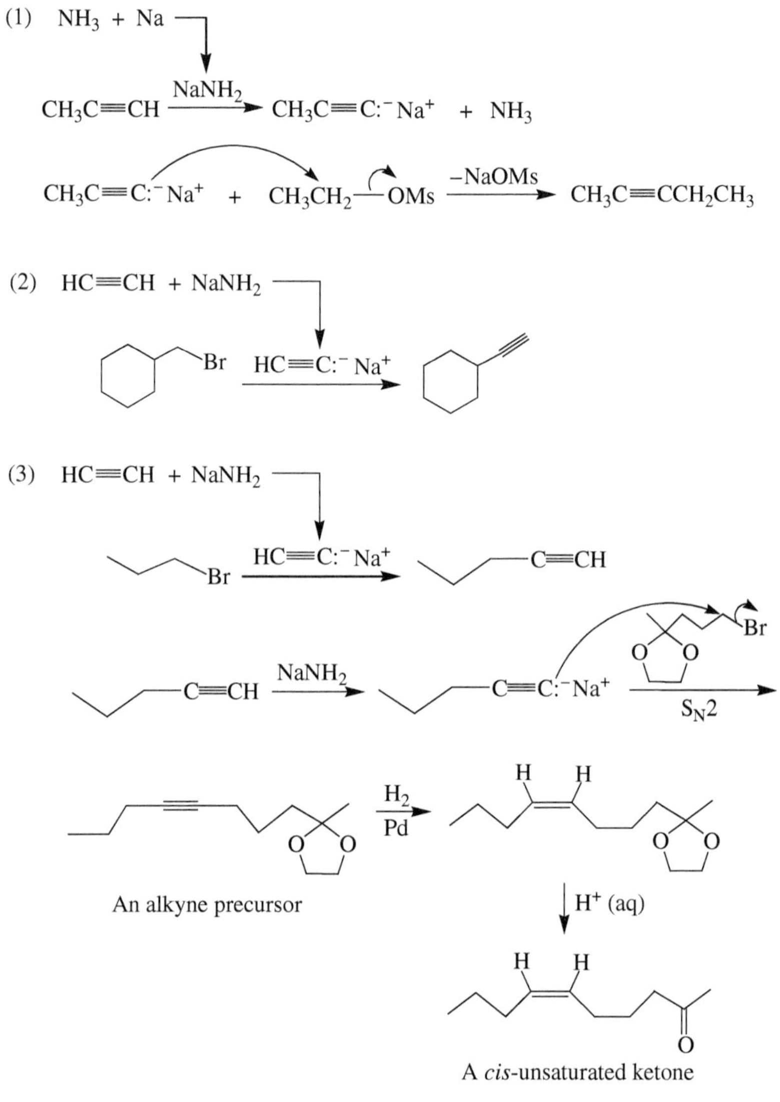

(1) NH₃ + Na

$$CH_3C{\equiv}CH \xrightarrow{NaNH_2} CH_3C{\equiv}C{:}^- Na^+ + NH_3$$

$$CH_3C{\equiv}C{:}^- Na^+ + CH_3CH_2{-}OMs \xrightarrow{-NaOMs} CH_3C{\equiv}CCH_2CH_3$$

(2) HC≡CH + NaNH₂

HC≡C:⁻ Na⁺

(3) HC≡CH + NaNH₂

HC≡C:⁻ Na⁺ ⟶ C≡CH

$$\text{—C{\equiv}CH} \xrightarrow{NaNH_2} \text{—C{\equiv}C{:}^- Na^+} \xrightarrow{S_N2}$$

An alkyne precursor

$$\xrightarrow[Pd]{H_2}$$

↓ H⁺ (aq)

A *cis*-unsaturated ketone

FIGURE 6.26   The S_N2 reactions of alkyne anions with primary haloalkanes.

Many secondary substrates give quite a lot of elimination products with the very basic nucleophile.

Enolates are strong carbon nucleophiles derived by treatment of a carbonyl compound (such as a ketone, aldehyde, or ester) with a strong base as shown below:

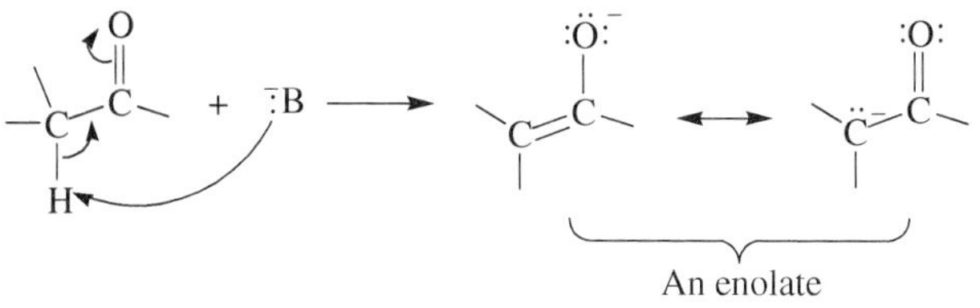

An enolate

They undergo S$_N$2 reactions with methyl and primary haloalkanes resulting in alkylation on the α-carbon of various carbonyl compounds. One example for alkylation of carbonyl compounds via an enolate is shown in Equation 6.24.

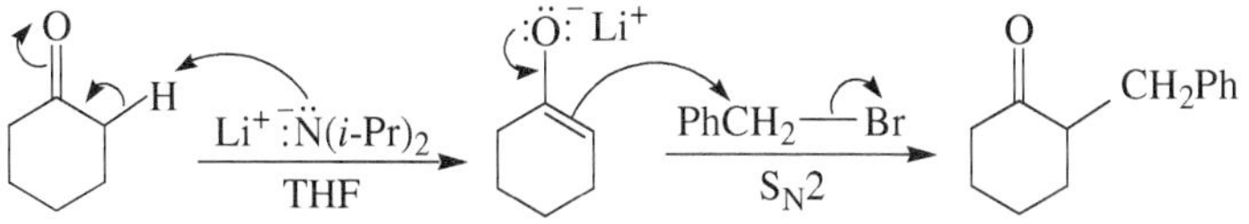

$$(6.24)$$

The chemistry of enolates will be further discussed in Chapter 9.

6.8.2   Synthesis of Primary Amines

A primary amine (RNH$_2$) can be prepared by nucleophilic substitution reaction of excess ammonia (NH$_3$) with a primary or secondary haloalkane (R–X, X = Br, or I) (Eq. 6.25).

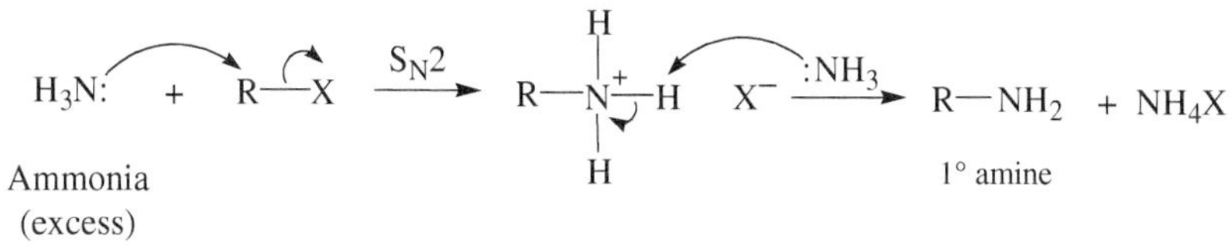

$$(6.25)$$

The RNH$_2$ product is also strongly nucleophilic and can possibly react with RX to give R$_2$NH, a secondary ammine. However, when NH$_3$ is in excess, the reaction of RX with NH$_3$ will be much faster than the reaction with RNH$_2$. As a result, all the RX can be converted to RNH$_2$ in the end of the reaction, and the formation of R$_2$NH will be minimal.

An alternative method to make a primary amine is accomplished by use of phthalimide (Gabriel synthesis), instead of using ammonia, as shown in Figure 6.27. Phthalimide is not nucleophilic because the lone pair of electrons on the nitrogen

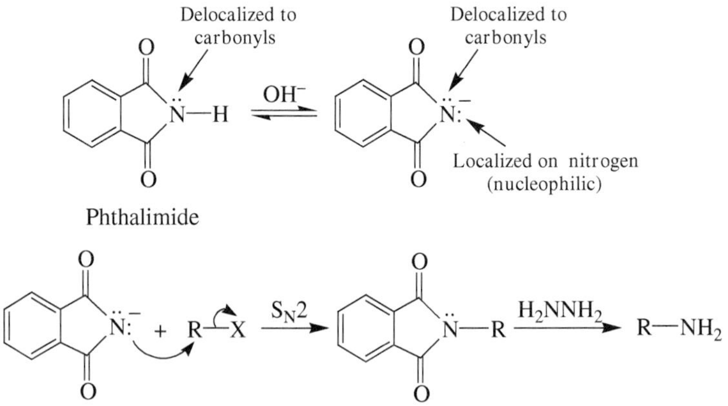

FIGURE 6.27   The Gabriel synthesis.

atom is delocalized deeply into the two carbonyl groups. The N—H bond is weakly acidic. It can be deprotonated by the hydroxide base giving a nitrogen anion. One lone pair of electrons in the nitrogen atom of the anion is delocalized to the carbonyl groups. The second lone pair of electrons is localized on the nitrogen atom, making the anion strongly nucleophilic. It undergoes an S_N2 reaction with a primary haloalkane. The intermediate product is not nucleophilic because of delocalization of the lone pair of electrons in nitrogen. The alkyl group in the intermediate is then displaced by hydrazine (H_2NNH_2), a very strong nucleophile, to give a primary amine RNH_2.

6.8.3   Synthetic Utility of Triphenylphosphine: A Strong Phosphorus Nucleophile

Triphenylphosphine (Ph_3P:) is a very useful strong phosphorus nucleophile in organic synthesis. It reacts with methyl or a primary haloalkane (RCH_2X, R = H or a linear alkyl, X = Br or I), and the reaction follows the S_N2 mechanism giving alkyltriphenylphosphonium halide ($Ph_3P^+-CH_2R\ X^-$) (Fig. 6.28a). The inductive effect of the positively charged phosphorus makes its adjacent C—H bond slightly acidic. Therefore, subsequent treatment of $Ph_3P^+-CH_2R$ with n-BuLi or $LiNR_2$ (an extremely strong base) affords a zwitterion $Ph_3P^+-^-CHR$ (phosphonium ylide or phosphorane). Because the phosphorus atom in phosphorane has empty d orbitals, it is reasonable to write a resonance structure $Ph_3P=CHR$ that contains a P=C double bond and a total of five bonds in phosphorus. The carbon atom in phosphorane is nucleophilic, while the phosphorus atom is electrophilic. The molecule undergoes a facile reaction with a ketone or aldehyde (nucleophilic addition and substitution) to give an alkene (Wittig reaction) (Fig. 6.28b).

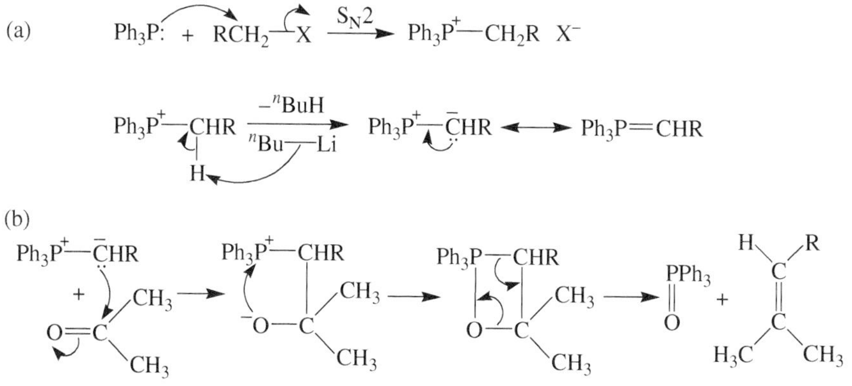

(a)

$$Ph_3P: + RCH_2-X \xrightarrow{S_N2} Ph_3\overset{+}{P}-CH_2R \quad X^-$$

$$Ph_3\overset{+}{P}-\underset{\underset{H}{|}}{C}HR \xrightarrow[{}^nBu-Li]{-{}^nBuH} Ph_3\overset{+}{P}-\overset{..}{C}HR \longleftrightarrow Ph_3P=CHR$$

(b)

FIGURE 6.28   The S_N2 reaction of triphenylphosphine (Ph_3P) with a primary haloalkane (a), followed by a Wittig reaction (b).

6.8.4   Neighboring Group-Assisted S_N1 Reactions

If the β-carbon to a functional group as a good leaving group bears a nucleophilic heteroatom, the nucleophilic heteroatom will assist an S_N1 reaction on the electrophilic carbon which is attached to the leaving group. Different from the regular S_N1 reactions which give two enantiomeric products if the substrates are chiral, such a neighboring group-assisted S_N1 reaction gives one product with its configuration the same as that for the substrate if the nucleophilic carbon in the substrate is asymmetric. The example in Figure 6.29 shows a neighboring group-assisted S_N1 reaction.

In (S)-2-bromo-n-propylmethyl ether, one β-carbon to the bromo group is attached to a nucleophilic oxygen atom. When the compound is treated with water, the bromo group is displaced by hydroxyl, but only one enantiomeric product (S)-2-hydroxy-n-propylmethyl ether is formed with the original (S)-configuration on the electrophilic carbon retained. Without the oxygen atom, the nucleophilic substitution of a chiral secondary bromoalkane [such as (S)-2-bromopentane] by water would give two enantiomeric products. The function of the neighboring nucleophilic oxygen atom is that it attacks the C–Br carbon in the first step (an S_N2-like mechanism). The attack occurs from the opposite side of the bromo group giving a highly electrophilic oxonium intermediate (a carbocation equivalent) with the oxygen atom pointing back. In the second step, a water molecule mainly attacks the asymmetric C–O carbon (a major reaction) in the oxonium intermediate (an epoxide analogue). The attack occurs from the front, restoring the original (S)-configuration in the product (another S_N2-like mechanism) (two inversions make a retention). The oxonium intermediate is unsymmetrical, with the middle C_{sec}—O bond being longer than the terminal C_{prim}—O bond. This is because that of the contributing resonance structures, the secondary carbocation is more important (more stable) than the primary carbocation. It explains why the nucleophilic attack preferably takes place on the

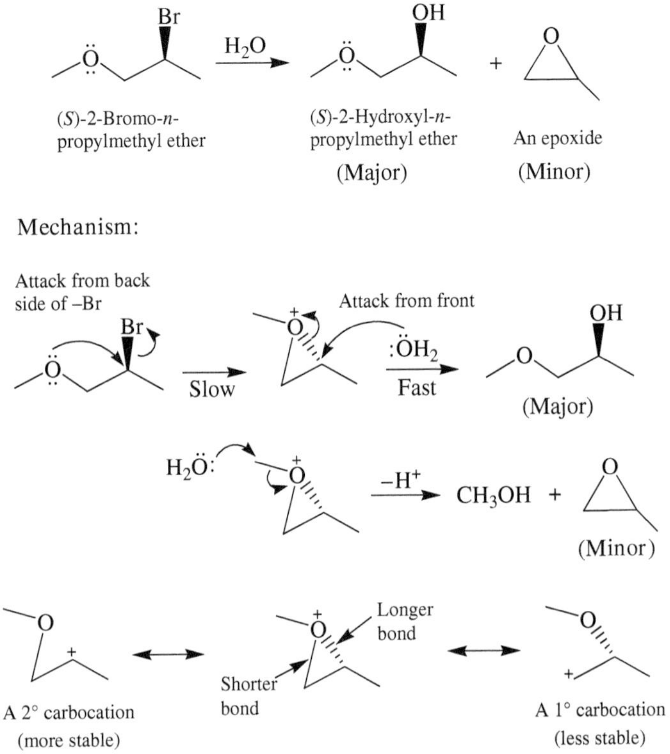

(S)-2-Bromo-n-
propylmethyl ether

(S)-2-Hydroxyl-n-
propylmethyl ether
(Major)

An epoxide
(Minor)

Mechanism:

FIGURE 6.29 Neighboring group-assisted nucleophilic substitution reaction of (S)-2-bromo-n-propylmethyl ether with water.

secondary asymmetric C–O carbon. The overall mechanistic pattern of the reaction, on the other hand, is similar to that of a regular S$_N$1 reaction. Therefore, the reaction has a first-order rate law only in the substrate. The oxonium intermediate is more stable than a regular secondary carbocation. Thus, the formation of the oxonium (the rate-determining step) is faster than the formation of a secondary carbocation from a haloalkane. This accounts for the fact that this neighboring group assisted S$_N$1 reaction is faster than the reaction of 2-bromopropane (or any other secondary bromoalkanes) with water.

In addition to the formation of the major (S)-2-hydroxy-n-propylmethyl ether product, the methyl group attached to the positive oxygen in the oxonium intermediate can also be attacked concurrently by water giving a minor competing reaction. This nucleophilic attack also follows an S$_N$2-like mechanism and leads to methanol and an epoxide in a significant amount.

Toxicity of the well-known mustard gas (ClCH$_2$CH$_2$SCH$_2$CH$_2$Cl) is due to its rapid reaction with water, producing harmful HCl to destruct human body tissues. The chemistry provides another example of neighboring group-assisted nucleophilic

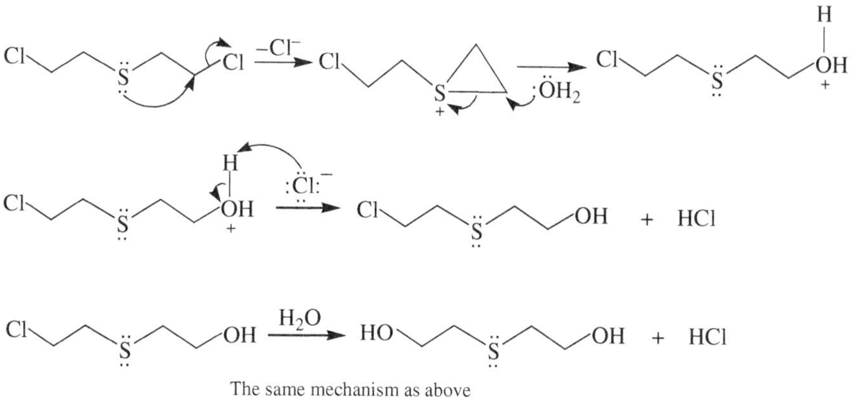

FIGURE 6.30   Neighboring group-assisted hydrolysis of mustard gas.

substitution reactions. Figure 6.30 shows mechanism of the sulfur-assisted nucleophilic substitution reaction of mustard gas with water [3, 7]. The first step (rate-determining step), nucleophilic attack of sulfur on the C–Cl carbon giving a cationic intermediate, is very fast accounting for the overall rapid reaction. Once the cationic intermediate is formed, it is attacked readily by water completing the overall substitution reaction.

Reaction of (2R,3S)-3-bromo-2-butanol with HBr leads to displacement of the –OH group (via the protonated –$^+$OH$_2$) from the alcohol by a bromide Br$^-$ ion. As a result, (meso)-2,3-dibromobutane is obtained (Fig. 6.31), and it is the only stereoisomer observed [1]. The stereochemical outcome demonstrates that the reaction is a neighboring bromo group-assisted nucleophilic substitution. Because of involvement of the neighboring –Br group in the reaction, the second bromo group which displaced the –OH group from the alcohol has maintained the orientation possessed by –OH in the original alcohol. The full mechanism is presented in Figure 6.31.

Figure 6.32 shows that reactions of both trans- and cis-2-acetyloxycyclohexyl tosylate react with acetate in acetic acid to give a trans-diester. trans-2-Acetyloxycyclohexyl tosylate reacts about 1000 times faster than the cis-isomer. In the most thermodynamically favorable conformation of trans-2-acetyloxycyclohexyl tosylate, both TsO– and the neighboring AcO– group stay in equatorial positions. The nucleophilic carbonyl oxygen in AcO– can readily attack the electrophilic carbon bearing –OTs (a very good leaving group) from the opposite side of the leaving group to displace –OTs. As a result, a reactive oxonium intermediate is formed. It is subsequently attacked very readily by AcO$^-$ from the opposite side of the positive oxygen to cleave the C—O$^+$= bond. As a result, a configuration-retained trans-diester is produced. Because AcO– is in the opposite side of TsO– in trans-2-acetyloxycyclohexyl tosylate, the acetyloxy group can effect a neighboring group-assisted S$_N$1 reaction giving a configuration-retained product. Since the oxonium intermediate experiences resonance stabilization, its

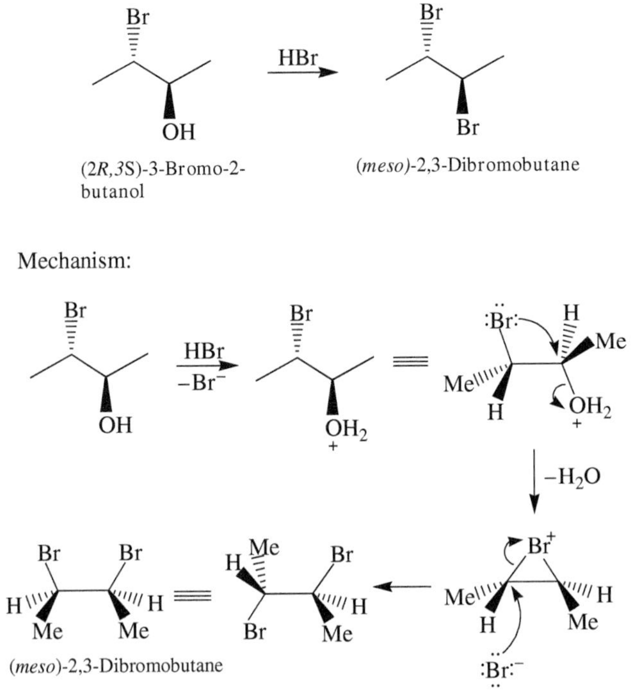

(2R,3S)-3-Bromo-2-
butanol

(meso)-2,3-Dibromobutane

Mechanism:

(meso)-2,3-Dibromobutane

FIGURE 6.31 Mechanism and stereochemistry for neighboring group-assisted nucleophilic substitution of (2R,3S)-3-bromo-2-butanol with HBr. Modified from Jackson [2].

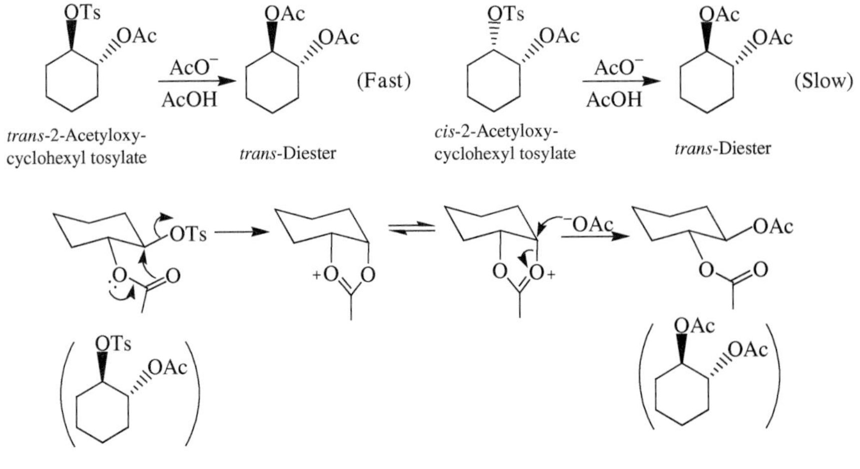

trans-2-Acetyloxy-
cyclohexyl tosylate

trans-Diester

cis-2-Acetyloxy-
cyclohexyl tosylate

trans-Diester

FIGURE 6.32 Neighboring group assisted nucleophilic substitution of trans-2-acetyloxycyclohexyl tosylate.

formation (rate-determining step) is very fast, making the overall nucleophilic substitution reaction fast. In *cis*-2-acetyloxycyclohexyl tosylate, because TsO– and the neighboring AcO– group stay in the same side of the ring, AcO– cannot displace tosylate and effects a neighboring group-assisted S_N1 reaction. Instead, *cis*-2-acetyloxycyclohexyl tosylate undergoes an S_N2 reaction with AcO⁻. Relatively, it is slow as acetic acid, a protic solvent, disfavors an S_N2 reaction.

6.8.5 Nucleophilic Substitution Reactions of Alcohols Catalyzed by Solid Bronsted Acids: A Green Chemistry Approach

Alcohols (ROH) can act as effective and efficient alkylating agents via nucleophilic substitution reactions, which play an important role in organic synthesis. Since the hydroxyl –OH group is a poor leaving group, protic (Bronsted) acids (HA) are usually employed as catalysts for the alcohol nucleophilic substitution reactions. The mineral montmorillonites (monts) contain various metal ions, such as Na, Al, Fe, and Mg. The Na^+ ion in monts can be readily replaced by H^+ upon treated with strong protic acids (e.g., hydrochloric acid) [8]. Then the proton-exchanged mont (H-mont) can act as a **solid Bronsted acid** to catalyze the alcohol nucleophilic substitution reactions that take place via carbocation intermediates as illustrated below (Eq. 6.26).

$$HNu + ROH \xrightarrow{H-mont} [HNu + R^+ + H_2O] \xrightarrow{-H^+} R-Nu + H_2O \qquad (6.26)$$

The effectiveness of the proton-exchange by HCl (aq) to make Bronsted acid sites in mont has been confirmed by infrared spectroscopy: The IR spectrum of the H-mont after being treated with pyridine exhibits a peak at $1543\ cm^{-1}$ due to the protonated pyridine, while the spectrum of the Na^+-mont after being treated with pyridine has no peak at $1543\ cm^{-1}$ [8].

The H-mont (a solid Bronsted acid) catalyzed nucleophilic substitution reactions using alcohols instead of alkyl halides as the electrophiles can minimize or eliminate the formation of waste by-products (halide salts). In addition, the solid catalyst can be readily recovered and reused. Thus, the methodology is environmentally benign. It represents a general green chemistry approach in organic synthesis.

Recent research has shown that the H-mont catalyzed reaction of diphenylmethanol (Ph_2CHOH) and allyltrimethylsilane ($CH_2=CHCH_2SiMe_3$) affords allyldiphenylmethane ($CH_2=CHCH_2CHPh_2$) in 96% of the yield (Reaction 6.27) [8].

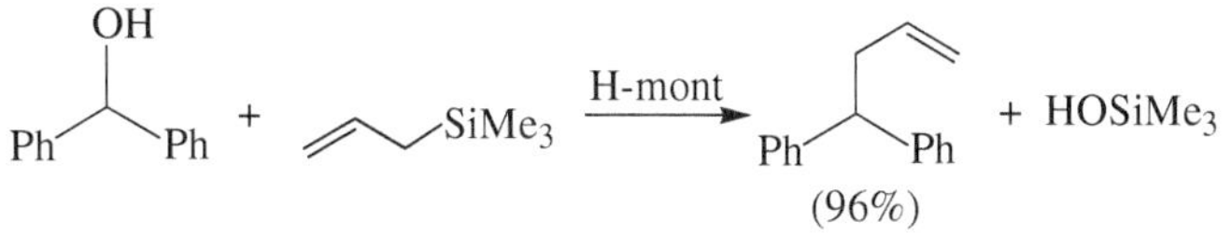

$$(6.27)$$

The reaction occurs on the surface of the solid catalyst where the alcohol is protonated (Fig. 6.33a). Then a carbocation intermediate (Ph_2CH^+) is formed as a water

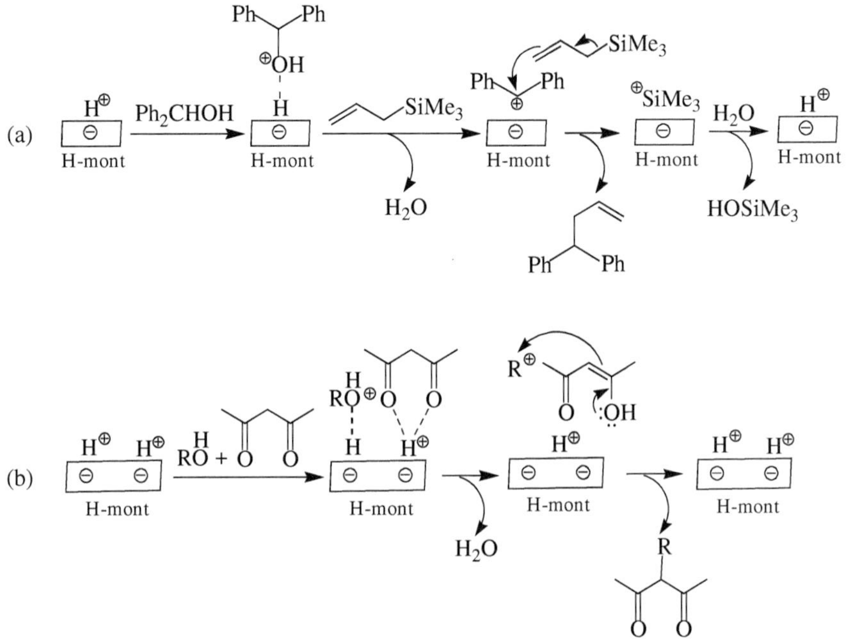

FIGURE 6.33 Mechanisms for the alcohol nucleophilic substitution reactions catalyzed by the proton-exchanged montmorillonite (H-mont, solid Bronsted acid). (a) Reaction of diphenylmethanol (Ph_2CHOH) and allyltrimethylsilane ($CH_2=CHCH_2SiMe_3$) giving allyldiphenylmethane ($CH_2=CHCH_2CHPh_2$); and (b) Reaction of an aliphatic alcohol (ROH) with a 1,3-dicarbonyl compound.

molecule leaves. The nucleophilic $CH_2=CHCH_2SiMe_3$ attacks Ph_2CH^+ subsequently to give the product $CH_2=CHCH_2CHPh_2$. Then the H_2O produced in the previous step reacts with $^+SiMe_3$ to give a by-product $HOSiMe_3$ and to regenerate H^+ on the catalyst surface.

In the presence of H-mont, a secondary aliphatic alcohol reacts with $CH_3C(O)$ $CH_2C(O)CH_3$, a 1,3-dicarbonyl compound, on the surface of the catalyst to lead to almost quantitative alkylation on the α-carbon (Reaction 6.27) [8].

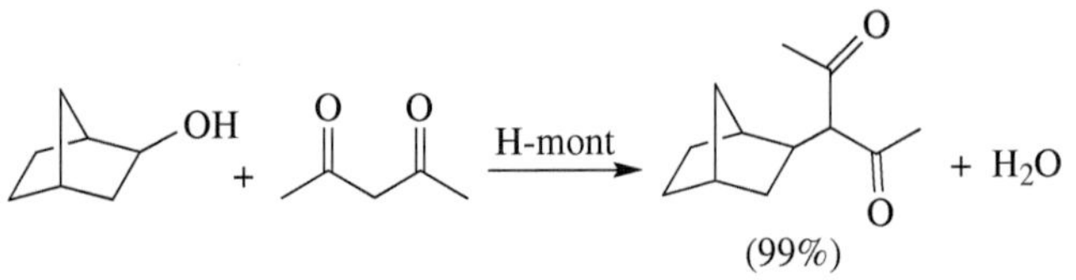

$$(6.28)$$

$$(99\%)$$

In this reaction, both the alcohol (ROH) and the 1,3-dicarbonyl compound are protonated on the surface of the catalyst (Fig. 6.33b). The protonation of the 1,3-dicarbonyl compound leads to the formation of a nucleophilic enol which attacks the carbocation intermediate produced from the alcohol resulting in the alkylation on the α-carbon.

6.9 BIOLOGICAL APPLICATIONS OF NUCLEOPHILIC SUBSTITUTION REACTIONS

6.9.1 Biomedical Applications

Busulfan (1,4-butanediol dimethanesulfonate—$MsOCH_2CH_2CH_2CH_2OMs$) is a pharmaceutical compound primarily used in the treatment of chronic myelogenous leukemia, a cancer in bone marrow [9]. Although the compound has been used in clinic for over 50 years, it was not until the last decade or so that the mechanism regarding the drug's cytotoxicity was fully understood.

Busulfan belongs to an alkylating antineoplastic agent, a drug that fights tumors by attaching an alkyl group to DNA to prevent its replication. Busulfan readily undergoes an S_N2 reaction with a nucleophilic nucleotide (Fig. 6.34) in a DNA molecule on either of its electrophilic carbons which are attached to mesylates (very good leaving groups). When reacting with a DNA nucleotide, Busulfan prefers to react with the N_7 nitrogen of Guanine or the N_3 nitrogen of Adenine [9]. The bifunctionality of Busulfan allows it to form an intrastrand crosslink (within a single DNA chain) between adjacent nucleotides (Fig. 6.35) in a DNA chain, which is accomplished by the S_N2 reactions with the nucleotides on both of the functionalized carbons in Busulfan. Experimental evidence suggests that Busulfan primarily forms intrastrand crosslinks between a guanine and adenine in the $5'$-GA-$3'$ sequence of a DNA strand (Fig. 6.35) [9]. Some guanine–guanine intrastrand crosslinks have been detected, but no adenine–adenine or adenine–guanine crosslink in the $5'$-AG-$3'$ sequence has been found. It has been demonstrated that interstrand crosslinks (between two companion DNA chains) are not formed by Busulfan as the overall length of the four-carbon chain (6 Å) of Busulfan is not large enough to span the 8 Å distance between the N_7 positions of guanines on opposite strands of a DNA molecule.

The intrastrand crosslinks formed between nucleotides of a DNA molecule via S_N2 reactions of Busulfan are believed to be primarily responsible for Busulfan's cytotoxicity. The crosslinks inhibit the activity of the cellular DNA polymerase, terminating the DNA replication and leading to eventual cell death.

If one mesylate group in Busulfan is displaced by a nucleophile such as hydroxide before the drug molecule reaches a DNA molecule, a monoadduct of Guanine–$CH_2(CH_2)_2CH_2OH$ or Adenine–$CH_2(CH_2)_2CH_2OH$ will be formed via the S_N2 reaction of the other C–OMs carbon in Busulfan after it reaches a DNA molecule. In this case, the crosslinks between two nucleotides will not be formed. Monoadducts also affect replication of DNA by creating an allosteric inhibition of DNA polymerase, but their cytotoxicity is lower than the crosslinks. $CH_3SO_2OCH_2CH_3$, a monofunctional analogue of Busulfan has been found to form the nucleotide–CH_2CH_3 monoadducts through S_N2 reactions, but the monoadducts have significantly lower cytotoxicity than the crosslinks.

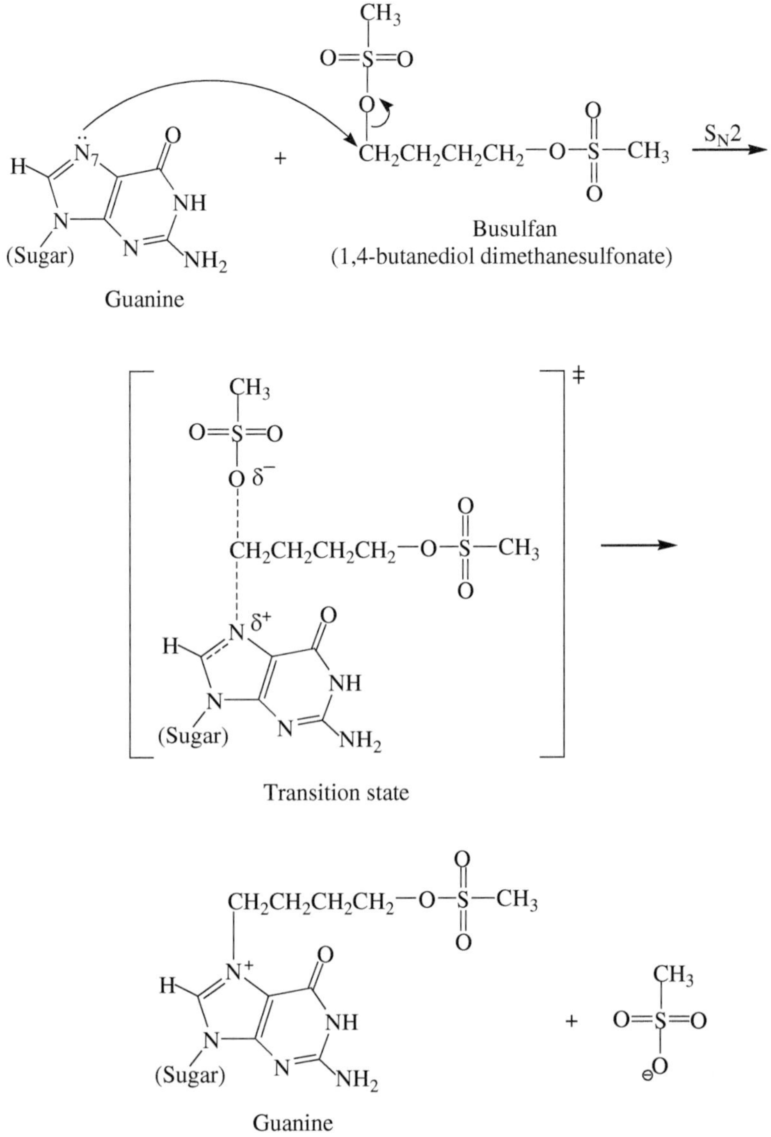

FIGURE 6.34 The S$_N$2 mechanism for alkylation of guanine in a DNA molecule by Busulfan (1,4-butanediol dimethanesulfonate).

6.9.2 Glycoside Hydrolases: Enzymes Catalyzing Hydrolytic Cleavage of the Glycosidic Bonds by the S$_N$2-Like Reactions

Monosaccharides (C$_n$H$_{2n}$O$_n$, n = 3–6) are simple sugars and mainly possess cyclic structures, typically six-membered and five-membered rings when n = 5 and 6. The six-carbon sugars (C$_6$H$_{12}$O$_6$), such as D-glucose, are among the most important

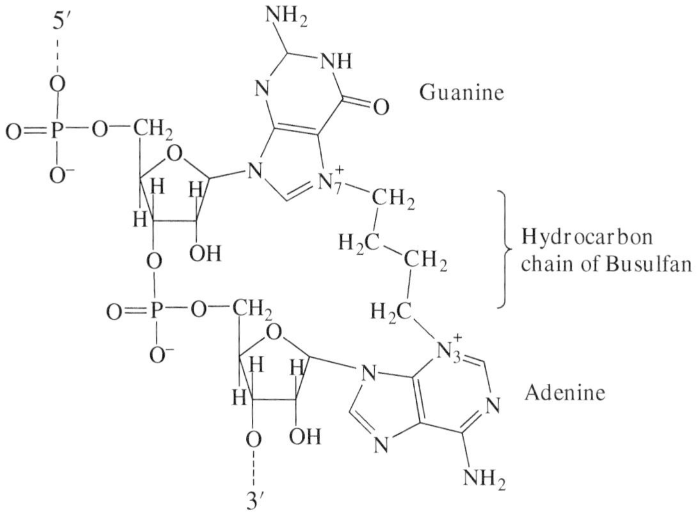

FIGURE 6.35 The intrastrand crosslink between nitrogen bases of a DNA chain by Busulfan. Modified from Iwamoto et al. [9].

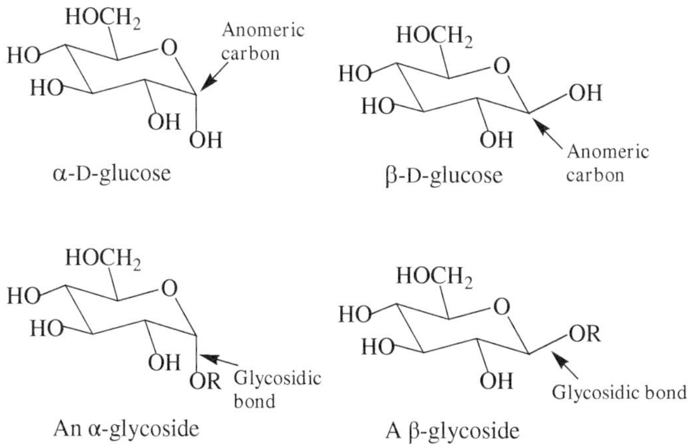

FIGURE 6.36 Cyclic structures of α- and β-D-glucose and α- and β-glycoside. The α-configuration: the anomeric OH (OR) and CH$_2$OH groups are *trans* each other. The β-configuration: the anomeric OH (OR) and CH$_2$OH groups are *cis* each other. Modified from Voet et al. [10].

monosaccharides, and they assume the chair-conformation in their naturally occurring six-membered ring structures (Fig. 6.36). When the hydroxyl –OH group in the anomeric carbon (the carbon connecting two oxygen atoms) of a cyclic sugar molecule is condensed with an alcohol (ROH) to form an –OR group on the anomeric carbon, the derivative is called **glycoside**. The bond linking the –OR oxygen to the anomeric carbon in the sugar ring is referred to as a **glycosidic bond** (Fig. 6.36) [10]. The –R group can possibly be a simple organic group (such as methyl or ethyl) as well as another monosaccharide ring. When two monosaccharide rings are linked by a glycosidic bond, the resulting molecule is referred to as a disaccharide. Common examples of disaccharides are sucrose and lactose. Linkage of many monosaccharide rings by glycosidic bonds gives polysaccharides, such as cellulose and α-amylose (starch).

Enzymes are protein molecules and catalyze biological reactions. The catalytic (active) site of an enzyme usually consists of functional groups from side chains of some α-amino acid residues in the enzyme (protein) molecule (Section 1.11). **Hydrolases** are a class of enzymes that catalyze hydrolysis of biomolecules. Glycoside (glycosyl) hydrolases are the enzymes that catalyze hydrolysis (hydrolytic cleavage) of glycosidic bonds in saccharide (sugar) molecules [11, 12]. By the nature, *the cleavage of a glycosidic bond by hydrolysis is in general nucleophilic substitution following the S_N2-like mechanism.*

Figure 6.37 shows the general mechanisms for the hydrolytic cleavage of a glycosidic bond in a β-glycoside (β-configuration: the CH_2OH and OR groups are *cis* each other) catalyzed by different glycoside hydrolases [11, 12]. In general, the active site of a glycoside hydrolase consists of a basic group (B^-) and an acidic group (HA). They are usually a carboxylate ($-COO^-$) and a carboxyl ($-COOH$), respectively, from the side chain of glutamate (Glu) or aspartic acid (Asp) [11]. The mechanism (a) leads to a **configuration-retained** product. The mechanism (b) gives a **configuration-inversed** sugar product in α-configuration (the CH_2OH and OH in the anomeric carbon are *trans* each other).

Figure 6.37a illustrates the mechanism for **retaining β-glycosidases**, a class of glycoside hydrolases leading to configuration-retained products. The basic B^- group (a carboxylate, which is also nucleophilic) in the enzyme active site attacks the anomeric carbon in the β-glycoside substrate from the opposite side of –OR, which is protonated by the acidic group HA (a carboxylic acid) in the active site to become a good leaving group. This nucleophilic attack leads to the concurrent formation of a $B-C_{anomeric}$ bond and departure of the OR group, and the resulting intermediate becomes an α-configuration. The process proceeds via a flattened transition state with the $B-C_{anomeric}$ bond partially formed, the $C_{anomeric}-OR$ bond partially broken, and an $O-C_{anomeric}$ double bond partially formed in the ring [11]. The partial formation of the double bond delocalizes the positive charge between the anomeric carbon and the ring oxygen, stabilizing the transition state. Overall, it is an S_N2-like mechanism, giving a configuration-inversed intermediate. In the next step, a water molecule enters the space between A^- (the conjugate base of HA) and the anomeric carbon of the intermediate in α-configuration. A hydrogen bond between A^- and H_2O is formed readily to weaken the $O-H$ bond in water, which activates the water molecule by enhancing its nucleophilicity. The activated water is able to attack the

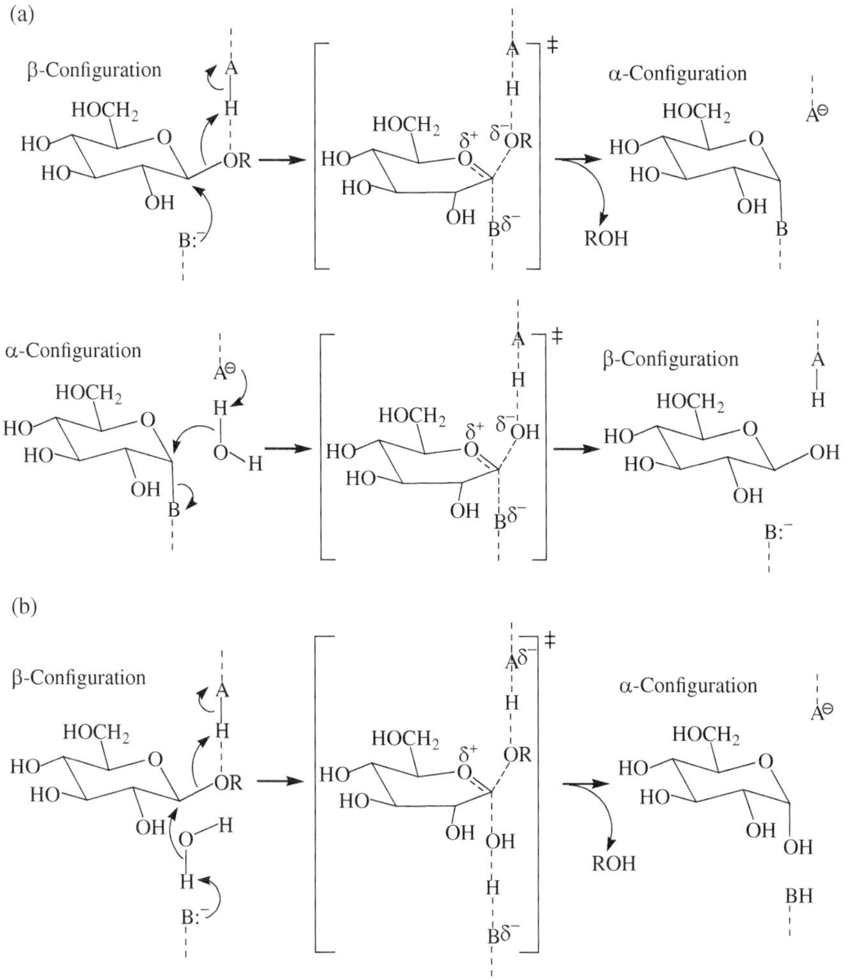

FIGURE 6.37 Mechanisms for (a) retaining β-glucosidases; and (b) inverting β-glucosidases.

anomeric carbon from the opposite side of the B—$C_{anomeric}$ bond to displace B^- and to lead to the formation of a cyclic sugar in β-configuration (configuration inversion). The process proceeds via a similar transition state and follows an S_N2-like mechanism as well. The overall nucleophilic substitution takes place via two S_N2-like steps involving double configuration inversions to give a **configuration-retained** product.

Figure 6.37b describes the mechanism for **inverting β-glycosidases**, a class of glycoside hydrolases leading to configuration-inversed products. A water molecule is inserted between the basic (nucleophilic) B$^-$ and the anomeric carbon of the β-glycoside substrate and hydrogen bonded to B$^-$. The hydrogen bonding activates the water molecule by enhancing its nucleophilicity. As a result, it can attack the anomeric carbon from the opposite side of the –OR group, which is hydrogen bonded to the acidic HA to become a good leaving group. The nucleophilic attack leads to the formation of a **configuration-inversed** product (α-configuration). The reaction proceeds via a similar transition state to that of the mechanism (a) and follows an S$_N$2-like mechanism.

In the retaining β-glycosidases (Fig. 6.37a), the distance between the basic group B$^-$ and the anomeric carbon is short. It allows B$^-$ to directly attack the anomeric carbon to bring about a nucleophilic substitution followed by a reaction of the intermediate with water to lead to configuration-retention in the product. The overall distance between B$^-$ and HA is 5.5 Å [12]. In the inverting β-glycosidases (Fig. 6.37b), the distance between the basic group B$^-$ and the anomeric carbon is large. B$^-$ cannot reach the anomeric carbon to directly initiate a nucleophilic substitution. Instead, the large distance allows a water molecule to be inserted between B$^-$ and the anomeric carbon so that H$_2$O is directly involved in the nucleophilic substitution to lead to configuration-inversion in the product. The overall distance between B$^-$ and HA for the inverting enzymes is 10 Å [12]. Examples of the retaining β-glycosidases are β-galactosidase and endoglucanase, which catalyze hydrolysis of the glycosidic bonds in lactose and cellulose, respectively. Examples of the reverting β-glycosidases are β-amylase and glucoamylase. Both catalyze hydrolysis of the glycosidic bonds in amylose [12].

The bacteria cell walls are composed of the cellulose-like polysaccharide, and its glycosidic bond is illustrated in Figure 6.38. Figure 6.38 shows the mechanism for the hydrolytic cleavage of the glycosidic bond in the bacteria cell wall polymer which is catalyzed by lysozyme, the glycoside hydrolase for this type of polysaccharides. It is a neighboring group-assisted nucleophilic substitution giving a **configuration-retained** product [11]. The first step is an intramolecular nucleophilic substitution via a nucleophilic carbonyl oxygen in the neighboring group. It is facilitated by the acidic H–A group in the active site of the enzyme to protonate the –OR oxygen in the substrate and make it a good leaving group. This is followed by an S$_N$2-like reaction with an activated water molecule situated between the substrate anomeric carbon and the basic group A$^-$. Both nucleophilic substitution steps have similar transition states to those in the glycoside hydrolases in Figure 6.37.

Glycoside hydrolases are a big class of enzymes whose mechanisms are of enduring interest owing to the ubiquity of carbohydrates in nature and their importance in human health and disease, the food, detergent, and biotechnology industries.

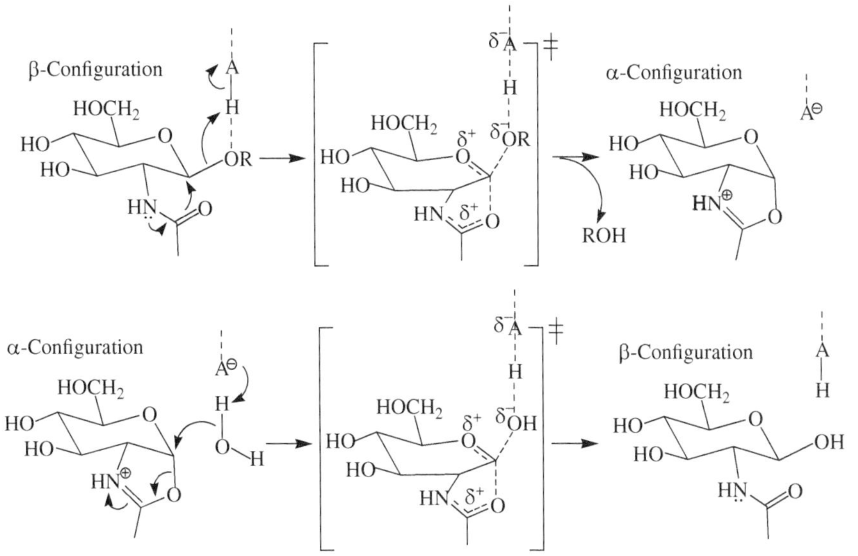

FIGURE 6.38   Mechanism for hydrolysis of bacteria walls polysaccharide by lysozyme.

6.9.3   Biosynthesis Involving Nucleophilic Substitution Reactions

Two S_N1 reactions take place during biosynthesis of the naturally occurring **geraniol**, a fragrant alcohol found in roses and used in perfumery. The process starts from the biological precursor dimethylallyl diphosphate (R–OPP) (Fig. 6.39). The diphosphate (–OPP) group is a very good leaving group and is found widely in naturally occurring substances of living organisms. Thus, it is regarded as the "biological equivalent" of a halo group (–Cl, –Br, or –I). Conceivably, R–OPP dissociates in the first step of the biosynthetic mechanism to dimethylallyl carbocation, which is subsequently attacked by a second biological precursor isopentenyl diphosphate [$CH_2=C(Me)CH_2CH_2$–OPP] to give another carbocation intermediate. In regard of R–OPP, combination of the first two steps in the overall mechanism of Figure 6.39 forms an S_N1 reaction [13]. On the other hand, the second step of the mechanism can also be considered electrophilic alkene addition for the second biological precursor $CH_2=C(Me)CH_2CH_2$–OPP. In the following step, an elimination takes place on the second carbocation through a loss of the pro-*R* hydrogen, giving the geranyl diphosphate intermediate. Then geranyl diphosphate undergoes a second S_N1 reaction via the geranyl carbocation, which, once formed, is subsequently attacked by water eventually leading to the final product Geraniol.

Figure 6.40 shows the biosynthesis of **epinephrine** from two biological precursors **norepinephrine** and **S-adenosylmethionine (SAM)** [13]. SAM contains a positively charged sulfur atom which is very easy to leave and the positive charge makes the methyl carbon electrophilic. Therefore, SAM is a very effective biological

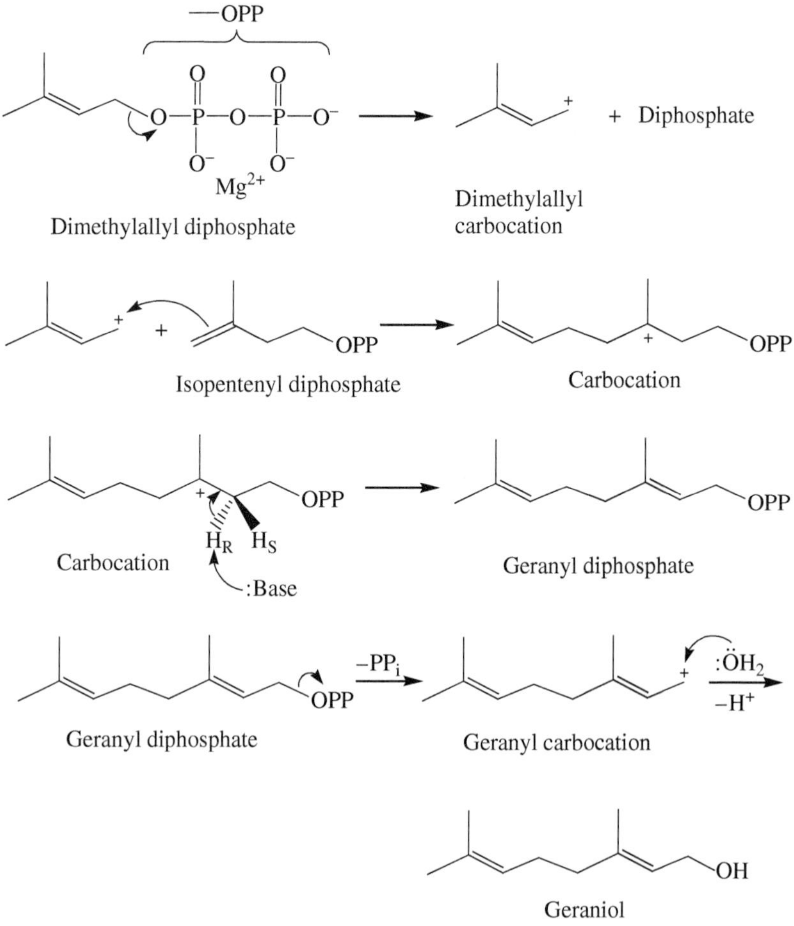

FIGURE 6.39   Biosynthesis of geraniol through S_N1 reactions.

methylating agent. In the biosynthesis, the electrophilic methyl group is attacked by the nucleophilic $-NH_2$ group in norepinephrine to lead to an S_N2 reaction. As a result, the methyl group is transferred from SAM to the nucleophile. SAM can be regarded as a biological equivalent of CH_3X (X = Cl, Br, or I).

6.9.4   An Enzyme-Catalyzed Nucleophilic Substitution of an Haloalkane

Nucleophilic substitution is one effective way by which living organisms detoxify halogenated organic compounds introduced into the environment. Enzymes that catalyze these reactions are called **haloalkane dehalogenases**. Hydrolysis of 1,2-dichloroethane to 2-chloroethanol, for example, is a biological nucleophilic substitution catalyzed by a haloalkane dehalogenase (Fig. 6.41) [14].

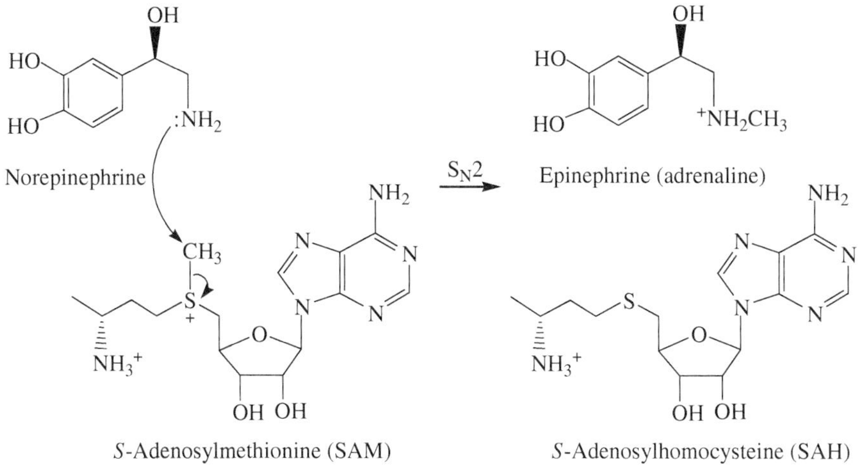

FIGURE 6.40 Biosynthesis of epinephrine from norepinephrine and S-adenosylmethionine (SAM) through an S_N2 reaction.

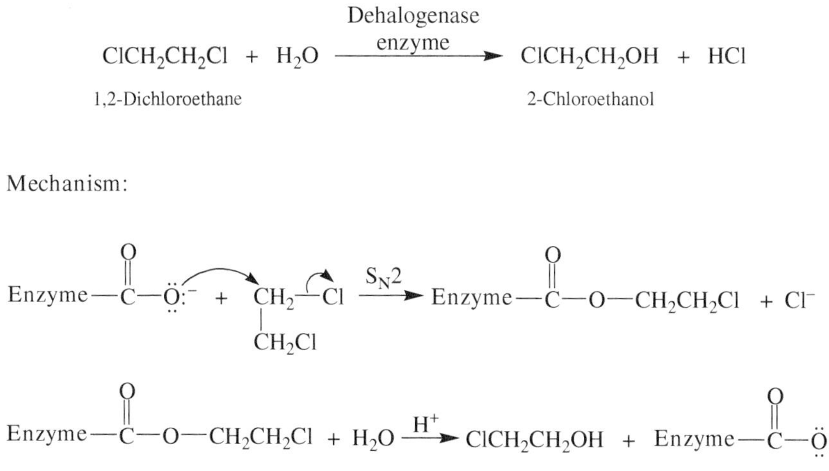

Mechanism:

FIGURE 6.41 An enzyme-catalyzed S_N2 reaction of an haloalkane. Modified from Carey [14].

The haloalkane dehalogenase enzyme is believed to act by using one of its side-chain carboxylate (moderately strongly nucleophilic) to displace a chloride from $ClCH_2CH_2Cl$ by an S_N2 reaction. The product (an ester) is then hydrolyzed to give $ClCH_2CH_2OH$ and regenerate a free enzyme which is ready to enter another reaction. The mechanism for acid-catalyzed hydrolysis of esters will receive extensive discussion in Chapter 8.

PROBLEMS

6.1 The nucleophilic substitutions including S_N1 and S_N2 reactions only take place on sp^3- hybridized carbons. If a functional group is attached to an sp^2-hybridized carbon, such as in vinyl bromide $CH_2=CHBr$, the organic compound undergoes neither S_N1 nor S_N2 reactions. Account for this lack of reactivity.

6.2 Give product(s) and show mechanism for each of the following S_N1 reactions. Work out the rate law for each reaction. Which reaction goes faster? Explain in terms of relative stability of the intermediate carbocations and activation energies of the reactions. Using molecular orbital diagram, show stabilization of an intermediate carbocation by a C—H bond.

6.3 Which compound, Me_3CBr or $Me_2(Ph)CBr$, reacts faster in S_N1 reactions? Why?

6.4 Consider the following two S_N1 reactions. Reaction (1) goes rapidly. However, the analogous Reaction (2) goes very slowly. Explain the difference.

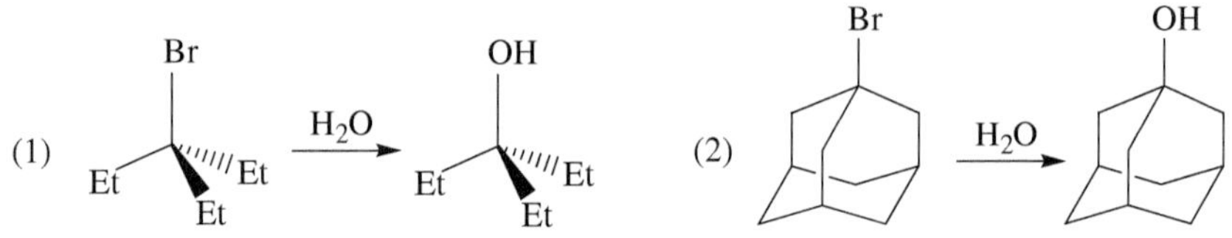

6.5 Suggest a mechanism to account for the following reaction.

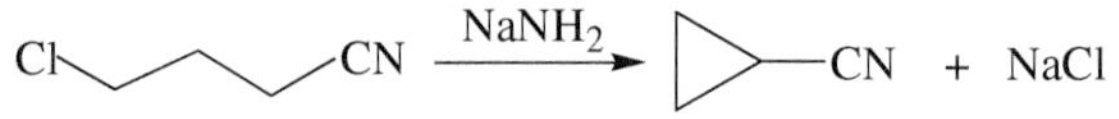

6.6 Consider the following reactions. Tertiary nitrogen atoms at the bridgehead position of bridged bicyclic amines are often more reactive as nucleophiles than are noncyclic tertiary amines. Account for this enhanced reactivity.

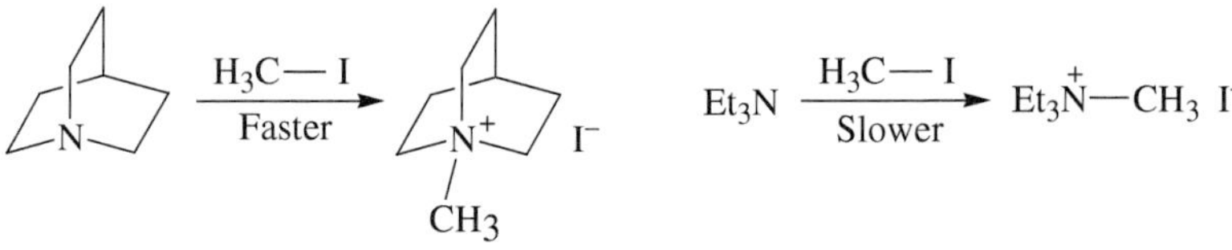

6.7 Account for the relative rates of solvolysis of the following compounds in aqueous acetic acid.

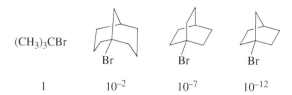

$(CH_3)_3CBr$

| 1 | 10^{-2} | 10^{-7} | 10^{-12} |

6.8 Consider solvolysis of the following bicyclic compounds (1) and (2) in acetic acid, compound (1) reacts 10^{11} times faster than does compound (2). Furthermore, solvolysis of compound (1) occurs with complete retention of configuration: The nucleophile occupies the same position as did the leaving – OSO_2Ar group.

(1)   (2)

Show the identity for the solvolysis product for each of the compounds. Account for the difference in rate of solvolysis of compounds (1) and (2). Account for the stereochemistry feature for solvolysis of compound (1).

6.9 Give product and show mechanism for the following reaction. Show absolute configuration of the product. Work out the most possible rate law for the reaction and account for the rationale.

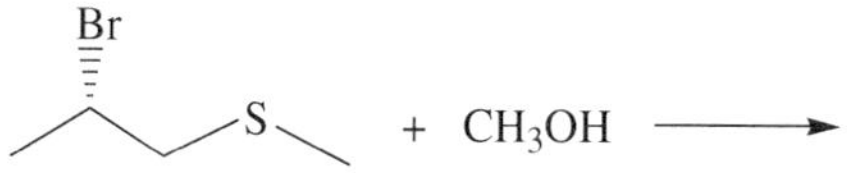

$+ CH_3OH \longrightarrow$

6.10 It is often possible to prepare symmetrical ethers by a reaction of a primary alkyl bromide with a limited amount of base in a small amount of water.

$$RCH_2-Br \xrightarrow{NaOH} RCH_2-O-CH_2R$$

What starting material would be required to prepare tetrahydrofuran (THF) by this method?
Show the detailed mechanism for conversion of this starting material to THF.

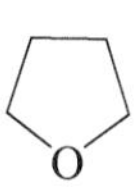

THF

6.11 Find two organic starting materials for synthesis of each of the following compounds. Show detailed mechanism for each synthesis.

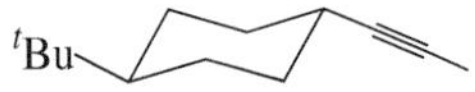

(1)

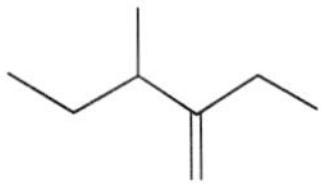

(2)

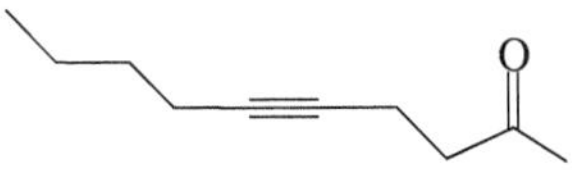

(3)

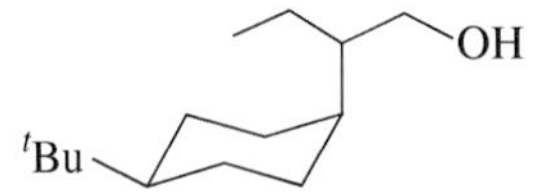

(4)

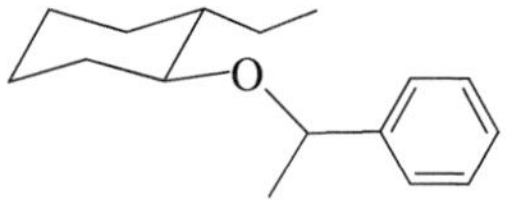

(5)

6.12 Provide a mechanism to account for the following reaction.

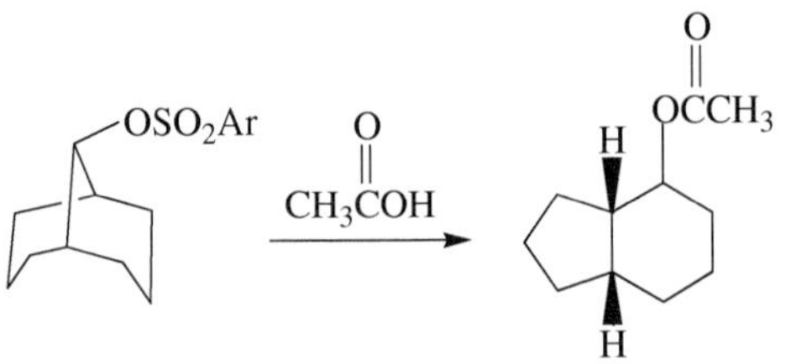

REFERENCES

1. Jackson, R. A. *Mechanisms in Organic Reactions*, The Royal Society of Chemistry, Cambridge, UK (2004).

2. Hoffman, R. V. *Organic Chemistry: An Intermediate Text*, 2nd ed., Wiley, Hoboken, NJ, USA (2004).

3. Bruckner, R. *Advanced Organic Chemistry: Reaction Mechanisms*, Chapter 2, pp. 43–83, Harcout/Academic Press, Orlando, FL, USA (2002).

4. Fox, M. A.; Whitesell, J. K. *Core Organic Chemistry*, 2nd ed., Jones and Bartlett, Sudbury, MA, USA (1997).

5. Sun, X. A Study of the S$_N$2 Mechanism by Symmetry Rules and Qualitative Molecular Orbital Theory. *Chem. Educator*, 2003, 8, 303–306.

6. Sun, X. Symmetry Analysis in Mechanistic Studies of Nucleophilic Substitution and β-Elimination Reactions. *Symmetry*, 2010, 2, 201–212.

7. Brown, W. H.; Foote, C. S.; Iverson, B. L.; Anslyn, E. V. *Organic Chemistry*, 6th ed., Brooks/Cole, Belmont, CA, USA (2012).

8. Motokura, K.; Nakagiri, N.; Mizugaki, T.; Ebitani, K.; Kaneda, K. Nucleophilic Substitution Reactions of Alcohols with Use of Montmorillonite Catalysts as Solid Bronsted Acids. *J. Org. Chem.* 2007, 72, 6006–6015.

9. Iwamoto, T.; Hiraku, Y.; Oikawa, S.; Mizutani, H.; Kojima, M.; Kawanishi, S. DNA Intrastrand Cross-Link at the 5′-GA-3′ Sequence Formed by Busulfan and its Role in the Cytotoxic Effect. *Cancer Sci.* 2004, *95*, 454–458.

10. Voet, D.; Voet, J. G.; Pratt, C. W. *Fundamentals of Biochemistry*, 5th ed., Wiley, Hoboken, NJ, USA (2016).

11. Speciale, G.; Thompson, A. J.; Davies, G. J.; Williams, S. Dissecting Conformational Contributions to Glycosidase Catalysis and Inhibition. *Curr. Opin. Struct. Biol.* 2014, *28*, 1–13.

12. Davies, G.; Henrissat, B. Structures and Mechanisms of Glycosyl Hydrolases. *Structure*, 1995, *3*, 853–859.

13. McMurry, J. *Organic Chemistry with Biological Applications*, 2nd ed., Brooks/Cole, Belmont, CA, USA (2011).

14. Carey, F. A. *Organic Chemistry*, 5th ed., McGraw Hill, Boston, New York, NY, USA (2003).

7

ELIMINATIONS

Many useful alkenes can be effectively and efficiently prepared by elimination reactions of various functionalized organic compounds, typically haloalkanes (alkyl halides), as generalized in Equation 7.1.

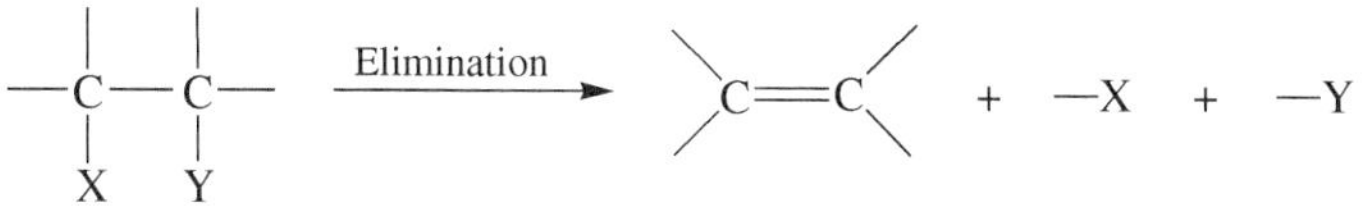

$$-\overset{\displaystyle |}{\underset{\displaystyle X}{C}}-\overset{\displaystyle |}{\underset{\displaystyle Y}{C}}- \xrightarrow{\text{Elimination}} \quad \overset{\displaystyle \diagdown}{\diagup}C = C\overset{\displaystyle \diagup}{\diagdown} \quad + \quad -X \quad + \quad -Y \tag{7.1}$$

$$(X, Y = H, LG \ \text{or} \ LG^1, LG^2)$$

This general elimination (Eq. 7.1) occurs on two adjacent carbons. When an H–LG unit (X, Y = H, LG) is eliminated (β-elimination), the reaction can follow an anti-elimination mechanism (with a staggered conformation of the substrate and –H and –LG being anti-coplanar) or a *syn*-elimination mechanism (with an eclipsed conformation of the substrate and –H and –LG being *syn*-coplanar). The anti-elimination of H–LG is bimolecular (E2 reaction) and usually requires a strong base to initiate the reaction. Very often, the E2 reaction is best represented by the bimolecular dehydrohalogenation (elimination of hydrogen halide from two adjacent carbons) of haloalkanes RX (X = Cl, Br, or I). The elimination of H–LG can also follow a stepwise mechanism via a carbocation intermediate (E1 reaction). The E1 and E2 reactions are the most important types of elimination reactions. Their mechanisms, regiochemistry, and stereochemistry are discussed in this chapter. The analysis of E2 pathway by

Organic Mechanisms: Reactions, Methodology, and Biological Applications,
Second Edition. Xiaoping Sun.
© 2021 John Wiley & Sons, Inc. Published 2021 by John Wiley & Sons, Inc.
Companion website: www.wiley.com/go/Sun/OrgMech_2e

symmetry rules and molecular orbital theory are presented on the basis of recent research. Energetics of some E1 reactions, such as acid-catalyzed dehydration of secondary and tertiary alcohols, is discussed by the aid of Bell–Evans–Polanyi principle.

Some functionalized compounds (e.g., esters and organic selenides) undergo *syn*-elimination of H–LG from two adjacent carbons. The reaction follows a unimolecular mechanism via a cyclic transition state. The study of the mechanism is aided by Huckel's $(4n + 2)$ rule and will be discussed in the chapter.

Some special bases and their effectiveness in bringing about the elimination reactions are presented. Since many bases are nucleophilic, the E1 and particularly E2 reactions are often accompanied by nucleophilic substitutions. The competitions between elimination and nucleophilic substitution reactions which occur to the same functionalized compounds will be addressed.

Elimination of LG^1/LG^2 (two functional groups) on adjacent carbons $(X, Y = LG^1, LG^2)$ is of special interest in organic chemistry. The reaction can follow an anti-elimination mechanism (with a staggered conformation of the substrate and $-LG^1$ and $-LG^2$ being anti-coplanar) or a *syn*-elimination mechanism (with an eclipsed conformation of the substrate and $-LG^1$ and $-LG^2$ being *syn*-coplanar). Both the mechanistic pathways will receive detailed discussion in the chapter.

The α-elimination (Eq. 7.2), in which an H–LG unit (X, Y = H, LG) or two functional groups $(X, Y = LG^1, LG^2)$ is eliminated from a single carbon giving a carbene, is another important type of elimination reactions.

$$
\begin{array}{c}
\overset{\displaystyle Y}{\underset{\displaystyle X}{\overset{|}{\underset{|}{—C—}}}} \xrightarrow{\text{Elimination}} \quad —\ddot{C}— \;+\; —X \;+\; —Y \\[2mm]
\hphantom{xxxxxxxxxxxxxxxxxxxxxxx}\text{A carbene}
\end{array}
\tag{7.2}
$$

$$(X, Y = H, LG \;\; \text{or} \;\; LG^1, LG^2)$$

Symmetry analysis and theoretical studies using the MO model for the α-elimination are presented in the chapter. The mechanism for the related rearrangement of carbenes to alkenes via a C—H bond elimination will be discussed.

Electronegative group facilitated E1cb reactions also possess important synthetic utility and mechanistic significance. They will be presented as well.

Elimination reactions have found many applications in biological systems. In this chapter, mechanisms for a few selected enzyme-catalyzed biological eliminations are reviewed.

7.1   E2 ELIMINATION: BIMOLECULAR β-ELIMINATION OF H/LG AND ITS REGIOCHEMISTRY AND STEREOCHEMISTRY

7.1.1   Mechanism and Regiochemistry

The E2 reaction is a concerted, bimolecular β-elimination of a functionalized alkane (R–LG, commonly, LG = Cl, Br, I, MsO, TsO, and TfO) induced by a strong base (B:⁻). Usually, the reaction requires a staggered conformation in the functionalized

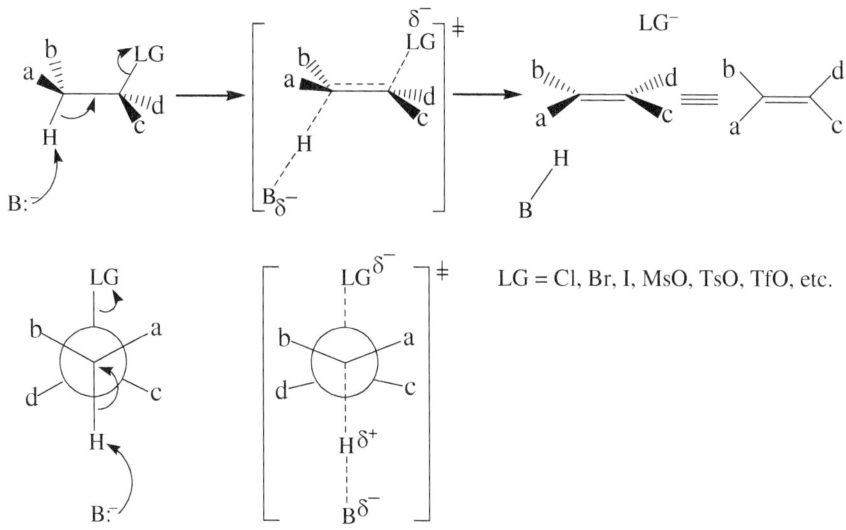

FIGURE 7.1   The E2 reaction mechanism.

alkane substrate, namely that the departing LG group and the β-hydrogen to be cleaved should be **anti-coplanar** (Fig. 7.1) in order for the reaction to be the most kinetically favorable [1–4]. This stereochemistry feature for the reaction is called **anti-elimination**. Figure 7.1 shows the general mechanism for the strong-base induced E2 reaction of a functionalized alkane. In the staggered conformation (reactive conformer), the β-hydrogen which is anti-coplanar to a functional group is activated by the LG group and becomes slightly acidic. Under certain conditions, the acidic β-hydrogen can be effectively attacked by a strong base (B:$^-$) along the direction of the H—C$_\beta$ bond to initiate a bimolecular elimination via a single transition state. In the transition state, the H—C$_\beta$ and C—LG bonds are being partially broken, coincident with the partial formation of a C=C π bond. As the transition state collapses, the alkene product is formed, together with the formation of the BH molecule and the LG$^-$ ion. Clearly, the stereochemistry of the reaction is determined by the structure of the transition state which correlates to the staggered conformation of the substrate. The relative orientations of all the groups attached to the C$_\alpha$ and C$_\beta$ (a, b, c, and d) are retained during the reaction. The reaction follows a second-order rate law (Eq. 7.3) with first order in the base (B$^-$) and first order in the substrate (RLG).

$$\text{Rate} = \text{d[alkene]}/\text{d}t = k[\text{B}^-][\text{RLG}] \tag{7.3}$$

Different from an S$_N$2 reaction which is subject to steric hindrance, the attack of a base molecule on a β-hydrogen in an E2 reaction can be free from or only have little steric hindrance as the anti-coplanar β-hydrogen is protruding, pointing away from

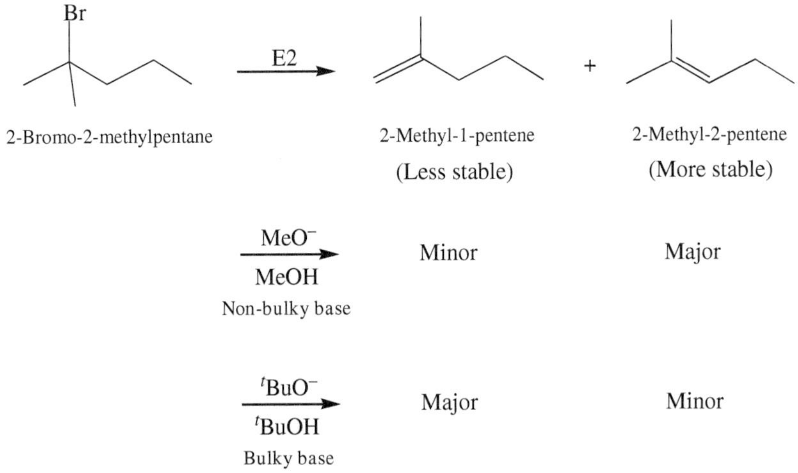

FIGURE 7.2   The E2 reaction of 2-bromo-2-methylpentane induced by different bases.

the substrate molecule and vulnerable. All types (primary, secondary, and tertiary) of functionalized alkanes (R–LG, LG = Cl, Br, I, MsO, TsO, and TfO) can undergo E2 reactions in the presence of a strong base at relatively high concentrations.

When a substrate R–LG molecule is unsymmetrical and contains two β-carbons to the functional group, its E2 reaction can be regioselective. The regiochemistry is determined by two factors, the bulkiness of the base employed in the reaction and the relative stability of structural isomers of the alkene product (Fig. 7.2).

In the example of Figure 7.2, treatment of 2-bromo-2-methylpentane by a strong base can effect an E2 reaction under certain conditions (such as heating a reaction mixture containing a high concentration of the base), and the reaction possibly gives two structural isomers of the pentene product. When methoxide (a nonbulky base) is used, the attack of the base on a β-hydrogen of neither C_1 nor C_3 carbon renders appreciable steric hindrance. In this case, the more stable isomer 2-methyl-2-pentene is the major product, and the reaction is thermodynamically controlled (Zaitsev reaction). When t-butoxide (a bulky base) is used, the attack of the base on a β-hydrogen of C_3 carbon (more sterically hindered) is relatively slow because of the steric interaction of the t-butyl group in $^tBuO^-$ with a terminal methyl group in the substrate molecule (Fig. 7.3). On the other hand, the attack of the base on a β-hydrogen of C_1 carbon (less sterically hindered) is relatively fast because of lack of steric hindrance as the t-butyl group in $^tBuO^-$ stays away from the substrate molecule (Fig. 7.3). As a result, the reaction is subject to the kinetic control (Hofmann reaction), giving the less stable 2-metyl-1-pentene as the major product.

Equation 7.4 shows another example of how the regiochemistry of an E2 reaction is dictated by the bulkiness of the base [1].

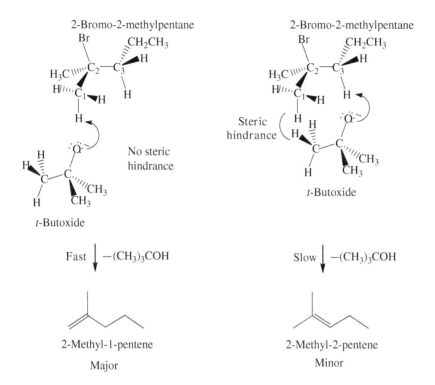

FIGURE 7.3 Steric hindrance of *t*-butoxide on the E2 reaction of 2-bromo-2-methylpentane.

$$\overset{OTs}{\text{⬡}} \quad \xrightarrow[\substack{^tBuOH \\ E2}]{^tBuO^-} \quad \text{⬡} \qquad (7.4)$$

Major (>85%)

The terminal carbon in the substrate is almost completely free from steric hindrance, while the C_3 carbon (a tertiary carbon) is much sterically hindered especially due to the aromatic ring. As a result, the bulky *t*-butoxide base attacks the hydrogen atom on the unhindered terminal carbon almost exclusively giving a 1-alkene (the kinetic product) as the major product.

In the case that an alkene isomer with its C=C bond conjugated to an unsaturated group can be formed in an elimination reaction (Reaction 7.5), this isomer is very often the sole product as it is greatly stabilized by conjugation effect. The alternative less stable isomer which contains a separated C=C π bond is not found [1].

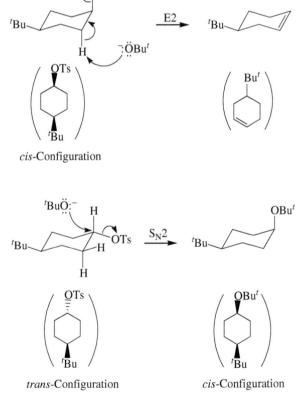

$$(7.5)$$

Sole product   Not found

(stabilized by conjugation)   (less stable)

7.1.2 E2 Eliminations of Functionalized Cycloalkanes

In cyclic systems, molecular rotations are more restricted. The outcomes of reactions for the functionalized cycloalkanes are highly affected by stereochemical environment of the substrate molecule. Figure 7.4 shows reactions of both *cis*- and *trans*-4-(*t*-butyl)cyclohexyl tosylate with *t*-butoxide [2]. Due to the bulkiness of the *t*-butyl

cis-Configuration

trans-Configuration   *cis*-Configuration

FIGURE 7.4 Reactions of *cis*- and *trans*-4-(*t*-butyl)cyclohexyl tosylate with *t*-butoxide. Modified from Bruckner [2].

group, it always stays in an equatorial position in chair conformations of any *t*-butylcyclohexane derivatives. Therefore, in *cis*-4-(*t*-butyl)cyclohexyl tosylate, the tosyl group always stays in an axial position in the chair conformation (the thermodynamically most favorable conformation) to make the two substituent groups *cis* to each other. The β-hydrogen in the axial position is anti-coplanar to the –OTs group. Thus, the substrate undergoes a fast E2 reaction with *t*-butoxide. In *trans*-4-(*t*-butyl) cyclohexyl tosylate, the tosyl group always stays in an equatorial position in the chair conformation since *t*-butyl stays in an equatorial position. None of the β-hydrogen atoms is anti-coplanar to the –OTs group. Thus, the E2 reaction is very slow (about 70 times slower than the E2 reaction for the *cis*-isomer). Instead, the major reaction is an S_N2 reaction leading to inversion of the original configuration.

Figure 7.5 shows the *t*-butoxide induced E2 reactions of *trans*- and *cis*-1-bromo-2-methylcyclohexane. For the *trans*-isomer, both the bromo and methyl groups stay in the equatorial positions in the most stable chair conformation. However, none of the β-hydrogen atoms is anti-coplanar to the equatorial –Br group. As –Br is equilibrated to the axial position though only to a small extent, the conformer becomes active toward an E2 reaction as the $C_6–H_{ax}$ hydrogen is now anti-coplanar to the axial –Br group. It can be attacked readily by $^tBuO^-$ resulting in the formation of 3-methylcyclohexene as a sole product. The reaction is regiospecific because only

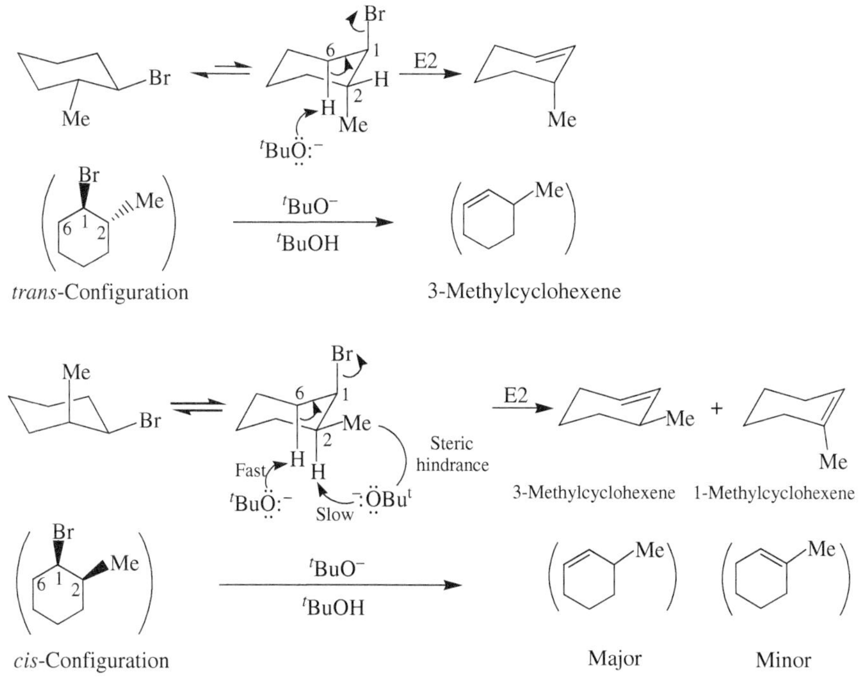

FIGURE 7.5 The E2 reactions of *trans*- and *cis*-1-bromo-2-methylcyclohexane induced by *t*-butoxide.

the C_6–H_{ax} hydrogen is anti-coplanar to the axial –Br group, but the C_2–H_{eq} hydrogen is not. For the *cis*-isomer, the reactive conformer is the one in which the bromo group stays in an axial position, while the methyl group in an equatorial position. Both C_2–H_{ax} and C_6–H_{ax} hydrogen atoms in the reactive conformer are anticoplanar to the axial –Br group. The attack of $^tBuO^-$ on the C_6–H_{ax} hydrogen is fast because the C_6-carbon is less sterically hindered. The corresponding 3-methylcyclohexene is the major product. The attack of $^tBuO^-$ on the C_2–H_{ax} hydrogen is relatively slow because of the steric interaction between the *t*-butyl group of the base and the methyl group in C_2. Thus, the corresponding 1-methylcyclohexene is only a minor product.

In the presence of *t*-butoxide or methoxide at high concentration, 1-bromo-1-methylcyclohexane undergoes E2 reaction giving methylenecyclohexane (less stable) and 1-methylcyclohexene (more stable) (Fig. 7.6). *t*-Butoxide (a bulky base) preferably attacks the less hindered methyl hydrogen to give the less stable methylenecyclohexane as the major product (Hofmann reaction). The attack of *t*-butoxide on the axial hydrogen in the ring will render substantial steric hindrance. Therefore, the corresponding more stable 1-methylcyclohexene is a minor product. On the other hand, the nonbulky methoxide base does not have appreciable steric hindrance when it attacks either β-hydrogen. Therefore, the regiochemistry for the E2 reaction is thermodynamically controlled (Zaitsev reaction) when methoxide is used, and the more stable 1-methylcyclohexene is the major product. When the reaction follows the E1 mechanism in the presence of a low concentration of the bases, the regiochemistry is

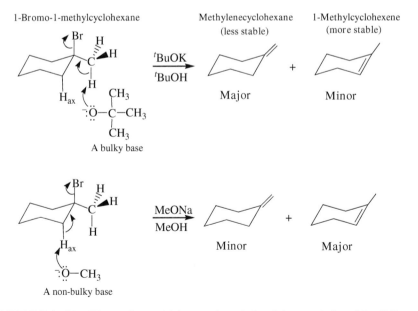

FIGURE 7.6   The E2 reactions of 1-bromo-1-methylcyclohexane induced by different bases.

only determined by relative stability of the products. It will be further addressed in Section 7.6.3.

7.1.3   Stereochemistry

Using the stereospecific 1,2-dibromo-1,2-diphenylethane substrate molecules, the anti-elimination mechanism for E2 reactions is further unambiguously demonstrated by their characteristic stereochemistry associated to the staggered reactive conformations (Fig. 7.7). The E2 reaction of a (*meso*)-compound effected by the methoxide base only gives an (*E*)-alkene product, but the (*Z*)-alkene is not found. However, the E2 reactions for both of the (1*S*,2*S*)-isomer and (1*R*,2*R*)-isomer of 1,2-dibromo-1,2-diphenylethane give the same (*Z*)-alkene product, but the (*E*)-alkene is not observed for either substrate. The results unambiguously show that each of the E2 reactions in Figure 7.7 takes place via a staggered conformation in which the β-hydrogen to be cleaved is anti-coplanar to the departing bromo group. This staggered conformation is readily formed via free rotation about the C_1—C_2 bond in each of the substrate molecules. For each of the substrates, had the reaction occurred via an eclipsed

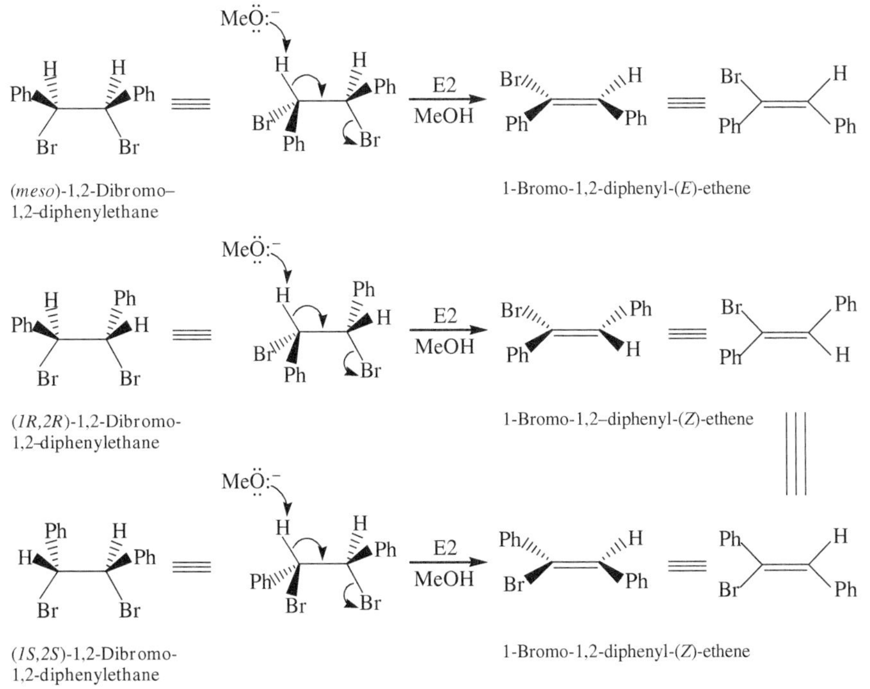

FIGURE 7.7   Stereochemistry for the E2 reactions of different stereoisomers of 1,2-dibromo-1,2-diphenylethane.

conformer with the β-hydrogen and the bromo group staying in the same side of the C—C bond, an alternative stereoisomer of the alkene to the observed would have been formed.

The anti-elimination mechanism for E2 reactions is also confirmed by a deuterium isotope labeling experiment of 2,2-dimethyl-*trans*-6-deuterocyclohexyl bromide (Reaction 7.6) [3].

$$(7.6)$$

In the reaction conditions, an E2 elimination must take place through the formation of a π bond between C_1 and C_6 carbons because of the presence of two methyl groups on C_2. Thus, either HBr or DBr will be eliminated. The actual isolated product is the one formed by loss of DBr, but not HBr. In the reactive chair conformation of the substrate, only the D atom on C_6 is anti-coplanar to the axial –Br group, but the H atom is not. The experimental results unambiguously show the necessity of the anti-coplanar arrangement for the E2 reaction, and the groups eliminated must be *trans*, but not *cis*, to each other.

The elimination of HX from a vinyl halide produces an alkyne (Reaction 7.7) [3].

$$(7.7)$$

An vinyl halide   An alkyne

Very strong bases, such as sodium amide, are generally used for the reaction. Because the product alkyne has no stereochemical features, it is difficult to obtain experimental data to support the actual mechanism. However, energetic considerations favor the concerted E2 mechanism as the E1 approach is unlikely because of instability of the vinyl cation $[R_1C^+=C(H)R_2]$.

As discussed in Section 5.8, a halobenzene undergoes analogous elimination when treated by a strong base to give a benzyne intermediate, which further reacts with the base (nucleophile) to effect an overall nucleophilic substitution.

7.2   ANALYSIS OF THE E2 MECHANISM USING SYMMETRY RULES AND MOLECULAR ORBITAL THEORY

7.2.1   Chain-Like Haloalkanes

The bimolecular β-elimination (E2) reactions of haloalkanes are among fundamental and most important organic reactions. They take place in the presence of a base. Since the halo group –X (X = Cl, Br, or I) is among the most common good leaving groups, the chain-like haloalkanes (R–X) are chosen for the mechanistic studies for E2 reactions. For chain-like haloalkanes, the E2 reaction requires a staggered conformation for the haloalkane substrate molecule, namely that the C_α–X and C_β–H bonds to be cleaved in the molecule must be anti-coplanar (Fig. 7.8). An eclipsed conformation, in which the C_α–X and C_β–H bonds stay *syn*-coplanar (see below), usually does not lead to an E2 reaction. Such a structural feature for the E2 reaction is necessary to make the β-H acidic via polarization of the C_β–H bond. This can be accounted for by a symmetry analysis on the related reacting MOs [5].

Figure 7.8a shows that when the C_α–X and C_β–H bonds in a chain-like haloalkane molecule are anti-coplanar in a staggered conformation, the C—H bonding (σ_{C-H}) orbital and the C—X antibonding ($\sigma_{C-X}*$) orbital are of symmetry-match and they partially overlap; and the C—H antibonding ($\sigma_{C-H}*$) orbital and the C—X bonding (σ_{C-X}) orbital are of symmetry-match and they partially overlap. The interaction (partial overlap) of the σ_{C-H} (filled) and $\sigma_{C-X}*$ (empty) orbitals increases the electron density in $\sigma_{C-X}*$ weakening the C—X bond and easing the departure of the halo (–X) group (Fig. 7.8b). The interaction (partial overlap) of the σ_{C-X} (filled) and $\sigma_{C-H}*$ (empty) orbitals increases the electron density in $\sigma_{C-H}*$ weakening and polarizing the C—H bond and making the β–H slightly acidic. As a result, the β–H can be effectively attacked by a base (B⁻) on the $\sigma_{C-H}*$ antibonding orbital (Fig. 7.8c), while simultaneously, the C—X bonding (σ_{C-X}) orbital partially populates $\sigma_{C-H}*$ with electrons. The combination leads to a concerted E2 reaction. The correlations of FMOs between reactants and products for the E2 reaction of a chain-like haloalkane are shown in Figure 7.9 [5].

Figure 7.10 shows [5] that when the C_α–X and C_β–H bonds in a haloalkane molecule are *syn*-coplanar in an eclipsed conformation, the σ_{C-H} and σ_{C-X} bonding orbitals (both filled) are of symmetry-match. However, the σ_{C-H} bonding orbital (filled) and the $\sigma_{C-X}*$ anti-bonding orbital (empty) are of symmetry-conflict (anti-symmetric). The $\sigma_{C-H}*$ anti-bonding orbital (empty) and the σ_{C-X} bonding orbital (filled) are also of symmetry-conflict (anti-symmetric). Neither the C_α–X bond nor the C_β–H bond can be activated by effective orbital overlap as demonstrated above. As a result, a haloalkane in an eclipsed conformation with the C_α–X and C_β–H bond *syn*-coplanar usually does not lead to an E2 reaction as the *syn*-elimination would render a high activation energy.

7.2.2 Halocyclohexane

The most stable conformation for a halocyclohexane is the chair-conformation as shown in Figure 7.11. Its E2 reaction requires that the halo (–X = –Cl, –Br, or –I) group be on an axial position, while an equatorial halo –X group cannot be eliminated readily. Only when a halo –X group is equilibrated to an axial position, can the C_α–X and C_β–H (also in axial position) bonds to be cleaved stay anti-coplanar, which makes the reacting molecular orbitals interact effectively [5].

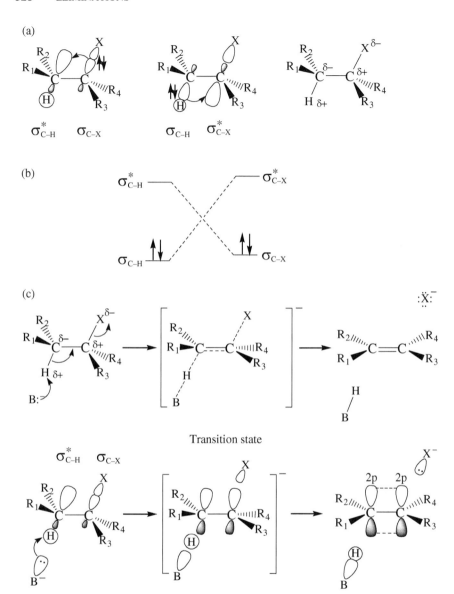

FIGURE 7.8   (a) Anti-coplanar arrangement of the C_α–X and C_β–H bonds in a haloalkane molecule (X = Cl, Br, or I; R_1, R_2, R_3, R_4 = H or alkyl) in staggered conformation. Bonding and antibonding molecular orbitals for the anti-coplanar C_α–X and C_β–H bonds. (b) Effective interactions (partial overlap) of the molecular orbitals for the C_α–X and C_β–H bonds due to symmetry-match. (c) Interactions and transformations of reacting molecular orbitals in the course of an E2 reaction.

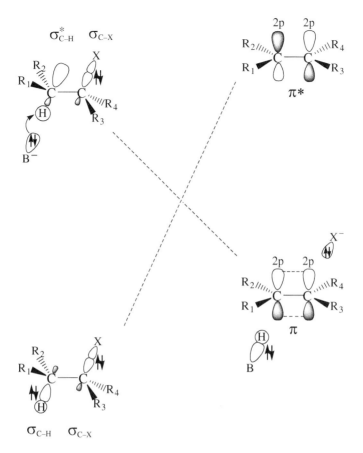

FIGURE 7.9 Correlations of frontier molecular orbitals for the E2 reaction of haloalkane. From Sun [5]. © 2003 MDPI. CC BY- 4.0. Public Domain.

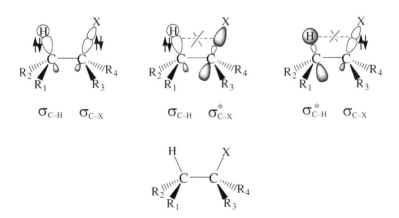

FIGURE 7.10 *Syn*-coplanar arrangement of the C_α–X and C_β–H bonds in a haloalkane molecule in eclipsed conformation. Bonding and antibonding molecular orbitals for the *syn*-coplanar C_α–X and C_β–H bonds. From Sun [5]. © 2003 MDPI. CC BY- 4.0. Public Domain.

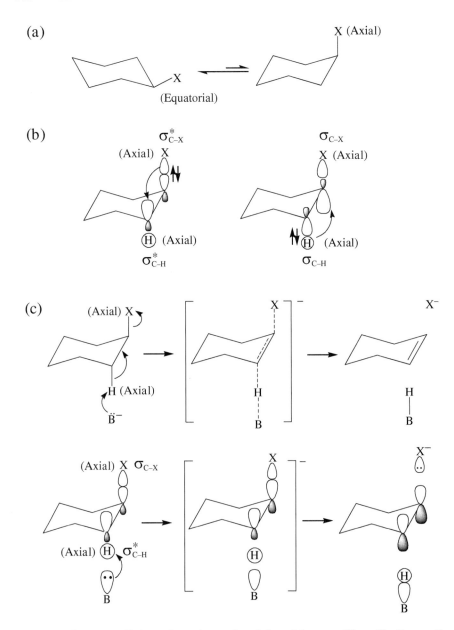

FIGURE 7.11 (a) Chair-conformations of a halocyclohexane (X = Cl, Br, or I). (b) Symmetry-match for the reacting molecular orbitals on the C_α–X (axial) and C_β–H (axial) bonds in a halocyclohexane. (c) The molecular orbital interactions in the course of an E2 reaction of a halocyclohexane.

Figure 7.11 shows that in the chair conformation of a halocyclohexane, the axial C_α–X bonding (σ_{C-X}) orbital (filled) and the axial C_β–H antibonding ($\sigma_{C-H}*$) orbital (empty) are of symmetry-match and they can partially overlap each other. The axial C_β–H bonding (σ_{C-H}) orbital (filled) and the axial C_α–X antibonding ($\sigma_{C-X}*$) orbital (empty) are of symmetry-match and they can partially overlap each other. As seen in the case of the chain-like haloalkanes, the interaction (partial overlap) of the filled σ_{C-H} and empty $\sigma_{C-X}*$ orbitals increases the electron density in $\sigma_{C-X}*$ weakening the C–X bond and easing the departure of the halo (–X) group. The interaction (partial overlap) of the filled σ_{C-X} and empty $\sigma_{C-H}*$ orbitals increases the electron density in $\sigma_{C-H}*$ weakening and polarizing the C–H bond and making the β–H slightly acidic. As a result, the β–H can be effectively attacked by a base (B⁻) on the $\sigma_{C-H}*$ antibonding orbital (Fig. 7.11c) leading to a concerted E2 elimination reaction.

The above symmetry analysis has shown that only when the C_α–X and C_β–H bonds to be cleaved in a haloalkane or a halocycloalkane molecule are anti-coplanar, can the reacting molecular orbitals be of symmetry-match. As a result, the positive overlap of σ_{C-H} and $\sigma_{C-X}*$ orbitals increases the electron density in $\sigma_{C-X}*$ weakening the C–X bond and easing the departure of the halo (–X) group. The positive overlap of σ_{C-X} and $\sigma_{C-H}*$ orbitals increases the electron density in $\sigma_{C-H}*$ weakening the C–H bond and making the β-H acidic. A proper understanding of the origin for the β-H acidity is the key to study the E2 reactions.

7.2.3   Quantitative Theoretical Studies of E2 Reactions

More quantitative theoretical studies have been performed on the fluoride ion (F⁻) catalyzed dehydrofluorination (elimination of HF) of fluoroethane (CH_3CH_2F) [6, 7]. The most favorable reaction pathway, transition state, and free energy change for the reaction are optimized and calculated. A transition state with the staggered conformation (β-H and –F being anti-coplanar) is demonstrated to possess the minimum free energy barrier ($\Delta G^{\ddagger} = 53.9$ kJ/mol) (Reaction 7.8) [7].

$$\Delta G^{\ddagger} = 53.9 \text{ kJ/mol}$$

$$(7.8)$$

However, the origin of the β-H acidity is not explained by the theoretical studies. The qualitative symmetry analysis may have provided a better understandable model to the experimental chemists and would be of much significance in guiding them to make the chemistry happen in the desired way for synthesis of useful compounds and materials.

7.3 BASICITY VERSUS NUCLEOPHILICITY FOR VARIOUS ANIONS

Basically, two factors affect basicity and nucleophilicity of a species. One is the charge density on the reactive center. The other is steric hindrance.

Usually, anions with highly concentrated charge are generally better bases than they are nucleophiles because the small size of the anion improves overlap with the small 1s orbital of a proton. Therefore, when choosing good bases for elimination reactions, emphasis is often placed on the anions with the negative charge localized on oxygen and nitrogen atoms (the elements of the second period). Examples include hydroxide (OH^-), alkoxide (OR^-), and amides (NH_2^- or NR_2^-). On the other hand, a larger anion, such as hydrogen sulfide HS^-, is a much better nucleophile than a base because of the large size of its valence orbitals which have effective overlap with a carbon valence orbital but less effective overlap with the small 1s orbital of a proton.

Basicity is relatively unaffected by steric hindrance because a proton is very small and stays in the exterior of a substrate molecule. On the other hand, as discussed in Chapter 6, steric hindrance greatly affects the rate of S_N2 reactions, and thus, it plays a major role in determining nucleophilicity of various anions. On the basis of these principles, bulky anions are usually **nonnucleophilic or weakly nucleophilic bases** [2]. For example, although t-butoxide [$(CH_3)_3CO^-$] is a much stronger base (by a factor of over 100) than methoxide (CH_3O^-), the latter is a better nucleophile because it is less sterically hindered. Due to bulkiness, t-butoxide, on the other hand, is only weakly nucleophilic although it is strongly basic. Of the extremely strong anionic nitrogen bases, $N(CHMe_2)_2^-$ and $N(SiMe_3)_2^-$ are nonnucleophilic because of the great steric hindrance exerted by the bulky $-CHMe_2$ and $-SiMe_3$ groups in the base molecules. For the same reason, the bulkiness of DBN and DBU makes each of them a nonnucleophilic base. The basicity versus nucleophilicity for various species is illustrated in Figure 7.12.

Bulky bases such as t-butoxide, $N(CHMe_2)_2^-$, and DBU are often used in E2 reactions to enhance the yield of the elimination product and to decrease or avoid the formation of the substitution product. For example, treatment of 3-methyl-4-phenyl-1-butyl tosylate (a primary alkyl tosylate) with the nonnucleophilic nitrogen base DBU induces only an E2 reaction and affords a sole alkene product (Reaction 7.9) [1].

| Major | 3-Methyl-4-phenyl-1-butyl tosylate | Sole product |

$$(7.9)$$

However, when the primary alkyl tosylate is treated with methoxide, a strongly basic and strongly nucleophilic reagent, an S_N2 reaction (Williamson ether synthesis) takes place as the major reaction. A similar situation can be seen for chloromethylcyclohexane (Reaction 7.10) [1].

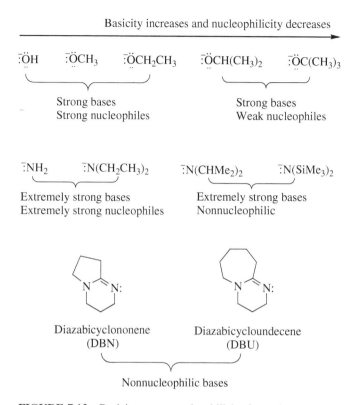

Basicity increases and nucleophilicity decreases

:ÖH   :ÖCH₃   :ÖCH₂CH₃   :ÖCH(CH₃)₂   :ÖC(CH₃)₃

Strong bases   Strong bases
Strong nucleophiles   Weak nucleophiles

:NH₂   :N(CH₂CH₃)₂   :N(CHMe₂)₂   :N(SiMe₃)₂

Extremely strong bases   Extremely strong bases
Extremely strong nucleophiles   Nonnucleophilic

Diazabicyclononene   Diazabicycloundecene
(DBN)   (DBU)

Nonnucleophilic bases

FIGURE 7.12   Basicity versus nucleophilicity for various species.

$$
\begin{array}{cc}
 & \xrightarrow[\text{MeOH}]{\text{MeO}^-\text{Na}^+} \quad\quad \xrightarrow[\text{C}_6\text{H}_6]{\substack{\text{E2}\\ \text{DBU}}} \\
\end{array}
\tag{7.10}
$$

E2   S$_N$2   Chloromethyl-   Sole product
                        cyclohexane

Utilization of the nonnucleophilic DBU base in the reaction affords a sole E2 product, while a reaction induced by methoxide results in a mixture of substitution and elimination products. Reaction of the chloroalkane with the bulky *t*-butoxide base also leads to the alkene product via an almost exclusive E2 reaction.

An epoxide contains a good leaving group (Section 6.2). In the presence of a strong nucleophile such as hydroxide, it undergoes an S$_N$2 reaction giving a diol. However, when the same epoxide is treated with a nonnucleophilic strong base, an E2 reaction takes place [2]. The mechanism is illustrated in Figure 7.13.

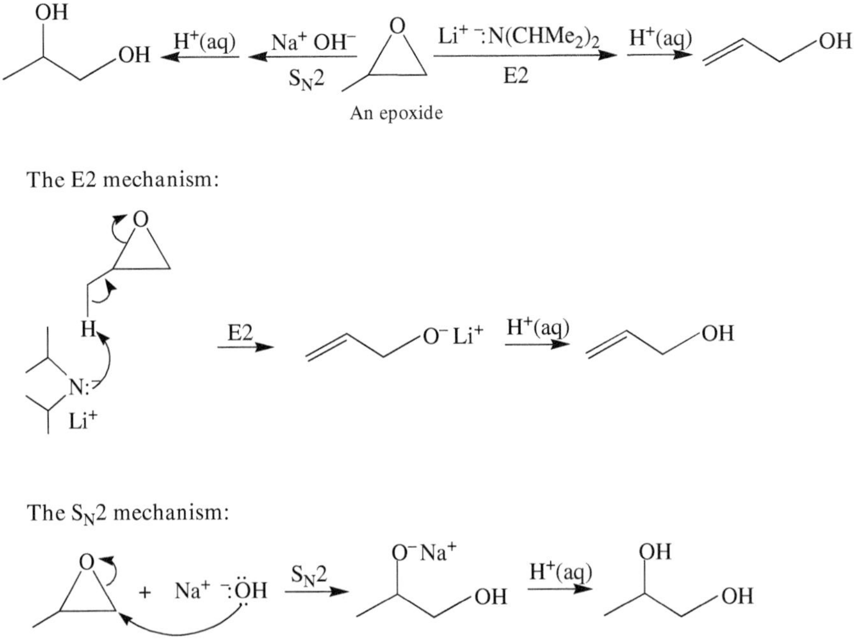

The E2 mechanism:

The S_N2 mechanism:

FIGURE 7.13   E2 and S_N2 reactions for an epoxide.

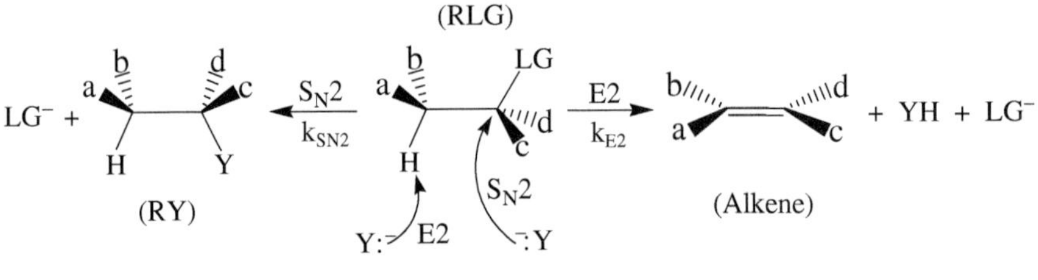

LG = Cl, Br, I, MsO, TsO, TfO, etc.

FIGURE 7.14   Competition between E2 and S_N2 reactions.

7.4   COMPETITION OF E2 AND S_N2 REACTIONS

Many bases also function as nucleophiles and vice versa. Therefore, when a functionalized alkane (R–LG) is treated by such a base (or a nucleophile, denoted as $Y{:}^-$),
both E2 and S_N2 reactions can take place concurrently (Fig. 7.14). As a base, $Y{:}^-$ can
attack an anti-coplanar β-hydrogen to the –LG group to effect an E2 reaction (with
the rate constant being k_{E2}) and lead to the formation of an alkene product. On the

other hand, Y:$^-$, as a nucleophile, can also attack the functionalized carbon to afford an S$_N$2 product RY (with the rate constant being k_{S_N2}). The rate laws for the E2 and S$_N$2 reactions are written as Equations 7.11 and 7.12, respectively.

$$d[\text{alkene}]/dt = k_{E2}[Y^-][RLG] \qquad (7.11)$$

$$d[RY]/dt = k_{S_N2}[Y^-][RLG] \qquad (7.12)$$

From Equations 7.11 and 7.12, we have

$$\frac{d[\text{alkene}]/dt}{d[RY]/dt} = \frac{k_{E2}[Y^-][RLG]}{k_{S_N2}[Y-][RLG]} = \frac{k_{E2}}{k_{S_N2}} \qquad (7.13)$$

Since the yield of a reaction product is usually directly proportional to the reaction rate, the ratio of the yield of E2 product to the yield of S$_N$2 product is equal to the ratio of the E2 reaction rate to the S$_N$2 reaction rate as shown in Equation 7.14.

$$\frac{\text{Yield of E2 product}}{\text{Yield of S}_N\text{2 product}} = \frac{d[\text{alkene}]/dt}{d[RY]/dt} = \frac{k_{E2}}{k_{S_N2}} \qquad (7.14)$$

The k_{E2}/k_{S_N2} ratio is determined by the balance of basicity and nucleophilicity of the base molecule (Y:$^-$) and by the structure and stereochemical environment of the substrate (R–LG).

As demonstrated above in Sections 7.1.2 and 7.3, when a bulky nonnucleophilic or weakly nucleophilic base (such as t-butoxide, DBU, or diisopropylamine) is used, k_{E2} will be much greater than k_{S_N2}, and as a result, the E2 reaction will be predominating. When a nonbulky nucleophilic base (such as hydroxide, methoxide, or ethoxide) is used, since both the basicity and nucleophilicity of the molecule are high, the k_{E2}/k_{S_N2} ratio will mainly depend on the structure and stereochemical environment of the substrate (R–LG). In this case, a primary substrate mainly undergoes an S$_N$2 reaction because the electrophilic carbon in the substrate molecule is only slightly sterically hindered which favors a nucleophilic attack. On the other hand, a secondary or a tertiary substrate mainly undergoes an E2 reaction as the S$_N$2 pathway is mostly or completely blocked by the steric hindrance in the substrate molecule.

Figure 7.15 shows when bromoethane (CH$_3$CH$_2$Br) is treated by sodium ethoxide NaOEt in ethanol, the observed k_{E2}/k_{S_N2} ratio is 1:100. Correspondingly, only 1% ethylene (CH$_2$=CH$_2$) product is formed. When 2-bromopropane [(CH$_3$)$_2$CHBr)] and 2-bromo-2-methylpropane [(CH$_3$)$_3$CBr)] are treated separately by sodium ethoxide NaOEt in ethanol, the observed k_{E2}/k_{S_N2} ratios are 3.5:1 and 40:1, respectively. Correspondingly, ~80% propene (CH$_3$CH=CH$_2$) product and ~100% 2-methylpropene [(CH$_3$)$_2$C=CH$_2$] product are found for the reactions of (CH$_3$)$_2$CHBr and (CH$_3$)$_3$CBr, respectively.

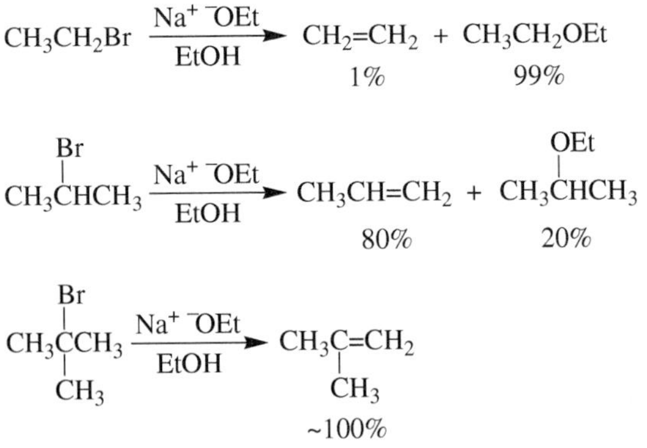

CH3CH2Br
Na+ ‾OEt
EtOH
CH2=CH2 + CH3CH2OEt
1%
99%
Br
CH3CHCH3
Na+ ‾OEt
EtOH
CH3CH=CH2 + CH3CHCH3
OEt
80%
20%
Br
CH3CCH3
CH3
Na+ ‾OEt
EtOH
CH3C=CH2
CH3
~100%

FIGURE 7.15 Ethoxide ($^{-}$OEt) induced E2 reactions versus S_N2 reactions for different bromoalkanes.

7.5 E1 ELIMINATION: STEPWISE β-ELIMINATION OF H/LG VIA AN INTERMEDIATE CARBOCATION AND ITS RATE-LAW

7.5.1 Mechanism and Rate Law

The E1 reaction is a stepwise β-elimination of a functionalized alkane (R–LG, commonly, LG = Cl, Br, I, MsO, TsO, and TfO). The first step is a unimolecular dissociation of the substrate molecule to a carbocation intermediate. Then, a β-hydrogen to the carbocation is cleaved by a base molecule resulting in the formation of an alkene product in the second step of the reaction. Its general mechanism is illustrated in Figure 7.16. In order for the elimination reaction to proceed, the $C_β$—H bond to be

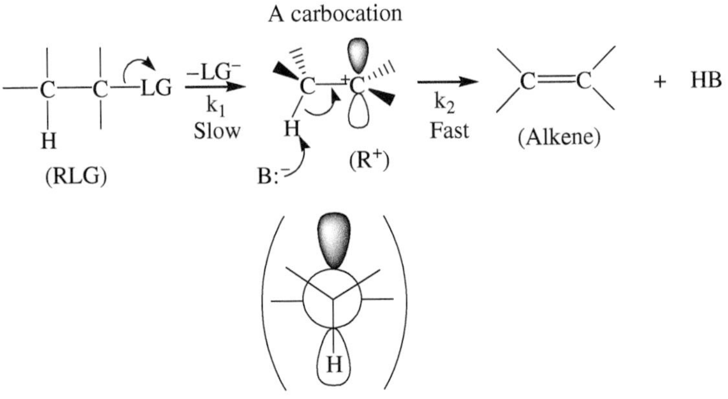

A carbocation
LG
H
(RLG)
−LG⁻
k1
Slow
H
B:
(R+)
k2
Fast
C=C
(Alkene)
+ HB
H
Overlap of Cβ-H bond
with p orbital lobe

FIGURE 7.16 The E1 reaction mechanism.

cleaved in the second step must overlap (coplanar) with one (*only one*) lobe of the empty p orbital in the carbocation, as such an overlap can have the C_β–H bonding electrons delocalized into the p orbital. As a result, the C_β–H bond is weakened and the β-hydrogen atom becomes substantially acidic. In the presence of a base mole-cule, the acidic β-hydrogen atom can be removed, eventually completing the overall elimination reaction. The activation of the β-hydrogen by the carbocation is strong, primarily owing to the positive charge held in the p orbital of the carbocation which strongly attracts the C_β–H bonding electrons. Therefore, even a weakly basic mol-ecule present in the system will be sufficient to cleave the C_β–H bond in the second step. *Different from an E2 reaction which is usually induced by a strong base, an E1 reaction does not need a strong base.*

The overall reaction rate for the E1 reaction in Figure 7.16 can be defined as d [alkene]/dt, which is directly proportional to the molar concentrations of the base (B^-) and carbocation (R^+) as shown in Equation 7.15.

$$d[\text{alkene}]/dt = k_2[B^-][R^+] \tag{7.15}$$

The rate equation in the carbocation intermediate (R^+) can be formulated as the rate of its formation $k_1[\text{RLG}]$ minus the rate of its collapse $k_2[B^-][R^+]$ (Eq. 7.16), while the steady-state assumption (d$[R^+]$/dt = 0) is applicable to the reactive carbocation.

$$d[R^+]/dt = 0 = k_1[\text{RLG}] - k_2[B^-][R^+] \tag{7.16}$$

From Equation 7.16, we have

$$[R^+] = (k_1/k_2)([\text{RLG}]/[B^-]) \tag{7.17}$$

Substituting Equation 7.17 for Equation 7.15 leads to
$$d[\text{alkene}]/dt = k_2[B^-](k_1/k_2)([\text{RLG}]/[B^-])$$
Therefore, we have

$$d[\text{alkene}]/dt = k_1[\text{RLG}] \tag{7.18}$$

Equation 7.18 is the rate law for the E1 reaction in Figure 7.16.

The rate law for an E1 reaction indicates that the reaction is first order only in the substrate, and the rate of the reaction is independent of concentration of the base. The rate law also shows that the first step of the reaction is the rate-determining step. This is consistent with the fact that the carbocation intermediate is unstable and highly reactive because it possesses a high energy level. Consequently, the first step of the reaction (the formation of the carbocation) is slow (with a high activation energy) and the second step (cleavage of the C_β–H bond) is fast (with a low activation energy).

Two important types of elimination reactions follow the E1 mechanism: (1) The acid-catalyzed dehydration reactions of secondary and tertiary alcohols and (2)

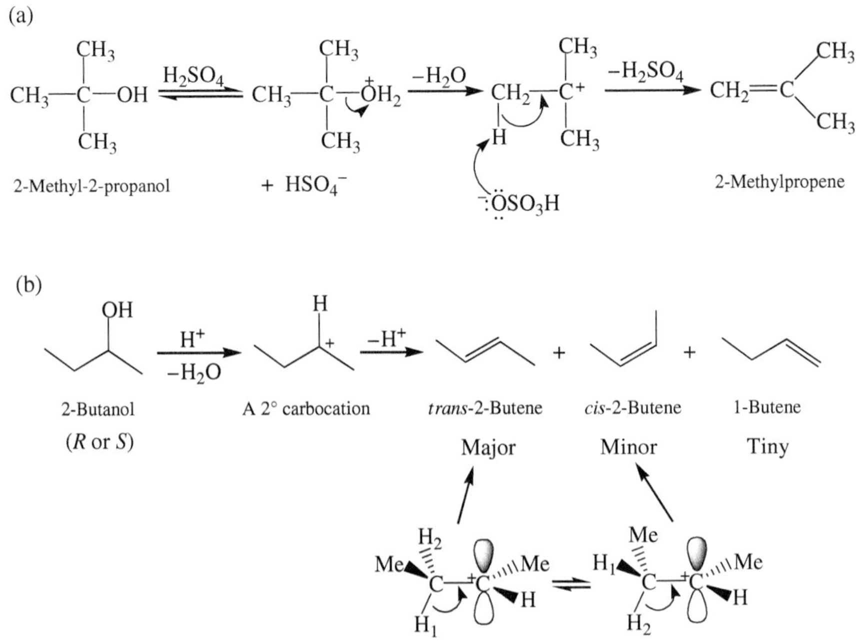

(a)

2-Methyl-2-propanol   + HSO$_4^-$ 2-Methylpropene

(b)

2-Butanol   A 2° carbocation   *trans*-2-Butene   *cis*-2-Butene   1-Butene

(R or S)   Major   Minor   Tiny

FIGURE 7.17   The acid-catalyzed dehydration of (a) 2-methyl-2-butanol and (b) 2-butanol.

β-elimination of secondary and tertiary haloalkanes and analogous functionalized alkanes in the presence of a strong base at low concentrations.

7.5.2   E1 Dehydration of Alcohols

First, let us examine the acid-catalyzed dehydration reactions of secondary and tertiary alcohols. Figure 7.17a shows the dehydration of 2-methyl-2-propanol (a tertiary alcohol) giving 2-methylpropene. In the presence of a strong acid such as sulfuric acid, the poor leaving group –OH in the alcohol molecule is protonated to –$^+$OH$_2$, a very good leaving group. The protonation is reversible and reaches equilibrium rapidly. Then, a water molecule is eliminated giving an intermediate tertiary carbocation. Finally, a C–H bond in the carbocation is cleaved readily and rapidly by HSO$_4^-$ (or H$_2$O), a weak base, producing 2-methylpropene, and the acid catalyst is regenerated. The rate-determining step is the formation of an intermediate carbocation as this step has the highest activation energy due to instability of the carbocation.

If the alcohol molecule is unsymmetrical, both the regiochemistry and stereochemistry of the E1 dehydration is thermodynamically controlled, namely that the product distribution is determined by the relative stability of different alkene isomers, with the most stable isomer being the major product. Figure 7.17b shows the acid-catalyzed dehydration of 2-butanol (R or S) (a secondary alcohol). Of the three alkene products, 1-butene is the least stable. Its quantity in the product distribution is the smallest. Free rotations about C_2—C_3 bond in the secondary carbocation intermediate results in equilibration of different reactive conformers each of which has a specifically oriented C—H bond overlap with one lobe of the carbocation p orbital. These reactive conformers allow cleavage of the activated C—H bonds in different orientations to lead to the formation of both *trans*- and *cis*-2-butene, with the more stable *trans*-2-butene being the major product.

7.6 ENERGY PROFILES FOR E1 REACTIONS

7.6.1 The Bell–Evans–Polanyi Principle

For similar concerted reactions that take place at a certain given temperature, the activation energy (E_a) can be directly correlated to the reaction enthalpy (ΔH) as follows (the Bell–Evans–Polanyi principle, Section 1.6.3) [2].

$$E_a = c_1 \Delta H + c_2 \qquad (7.19)$$

where c_1 and c_2 are positive constants.

Equation 7.19 is applicable to both endothermic ($\Delta H > 0$) and exothermic ($\Delta H < 0$) reactions and can be employed to analyze the relative activation energies for similar reactions based on their reaction enthalpies. For endothermic reactions ($\Delta H > 0$, Fig. 7.18), the greater the energy gap (ΔH) between the reactant and product (the more endothermic), the higher is the activation energy E_a. Since $\Delta H_2 > \Delta H_1$, $E_{a2} > E_{a1}$. For exothermic reactions ($\Delta H < 0$, Fig. 7.18), the greater the energy gap (the greater the absolute value of ΔH) between the reactant and product (the more exothermic is the reaction), the smaller is the activation energy E_a. The greater energy gap in an exothermic reaction makes the $c_1 \Delta H$ more negative ($\Delta H_2 < \Delta H_1$). As a result, the E_a becomes smaller ($E_{a2} < E_{a1}$) according to Equation 7.19.

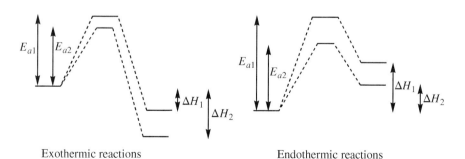

Exothermic reactions   Endothermic reactions

FIGURE 7.18   Bell–Evans–Polanyi principle: dependence of activation energy on reaction enthalpy for similar (a) exothermic and (b) endothermic reactions.

For endothermic reactions ($\Delta H > 0$), the effect of reaction enthalpy on activation energy is greater than the effect for the exothermic reactions ($\Delta H < 0$).

7.6.2 The E1 Dehydration of Alcohols (ROH)

The acid-catalyzed dehydration of secondary and tertiary alcohols follows the E1 mechanism. The formation of the carbocation intermediate is the rate-determining step. The overall reaction rate is determined by the relative stability (energy level) of the intermediate carbocation. If an alcohol molecule is unsymmetrical, both the regiochemistry and stereochemistry of the E1 dehydration of the alcohol is thermodynamically controlled, namely that the more stable is the alkene isomer, the higher is its yield.

Figure 7.19 shows the reaction profiles for acid-catalyzed dehydrations of 2-butanol (R or S, a secondary alcohol) and 2-methyl-2-propanol (a tertiary alcohol). In general, dehydration of a tertiary alcohol is faster and takes place at lower temperature than that of a secondary alcohol. For the reactions in Figure 7.19, the tertiary alcohol 2-methyl-2-propanol starts dehydrating at ~55 °C, while the secondary alcohol 2-butanol does not dehydrate until after the temperature reaches 70–80 °C. This is attributable to the difference in stabilities of the intermediate

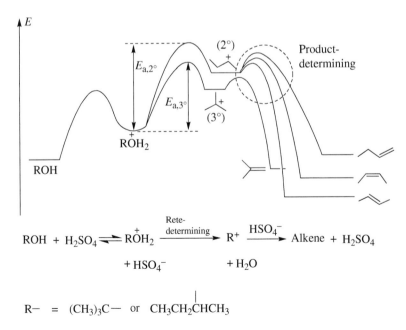

FIGURE 7.19   Reaction profiles for E1 dehydrations of 2-methyl-2-butanol and 2-butanol.

2° and 3° carbocations. Since the formation of the carbocation intermediate is the rate-determining step, the activation energies for dehydration of the alcohols ($E_{a,2}°$ and $E_{a,3}°$) are proportional to the energy levels of the carbocations (Bell–Evans–Polanyi principle) as indicated in Figure 7.19. The 3° carbocation (more stable) has a lower energy level than that of the 2° carbocation (less stable). Therefore, $E_{a3} < E_{a2}$. This makes the dehydration of the 3° alcohol faster than that for the 2° alcohol. As a result, the temperature of dehydration of the 3° alcohol is lower than that for the 2° alcohol.

While the rate of the alcohol dehydration is dictated by the stabilization (energy level) of the carbocation intermediate, the identity of the product (product distribution) is determined by the relative rate for the formation of each of the alkene isomers from dissociation of the carbocation (the product formation step). According to the Bell–Evans–Polanyi principle, the activation energy for dissociation of the carbocation to an alkene isomer is determined by the stability (energy level) of the product as indicated in Figure 7.19. The more stable is a product, the lower is the activation energy for the product formation step and the higher the yield of the product. All this explains the observed product distribution for dehydration of 2-butanol (Fig. 7.17b).

The product distribution for acid-catalyzed dehydration of 1-cyclohexylethanol (Fig. 7.20) reinforces the thermodynamic control on the product formation process as discussed above. In the course of the reaction, the initially formed secondary carbocation undergoes a facile 1,2-hydrogen rearrangement to a more stable tertiary carbocation. The subsequent cleavage of an adjacent C—H bond in the ring of the tertiary carbocation affords the major alkene product, which is thermodynamically more stable than other alkene isomers formed from the unrearranged secondary carbocation.

Acid-catalyzed dehydration of a primary alcohol follows the E2 mechanism because of the instability of a primary carbocation (Reaction 7.20).

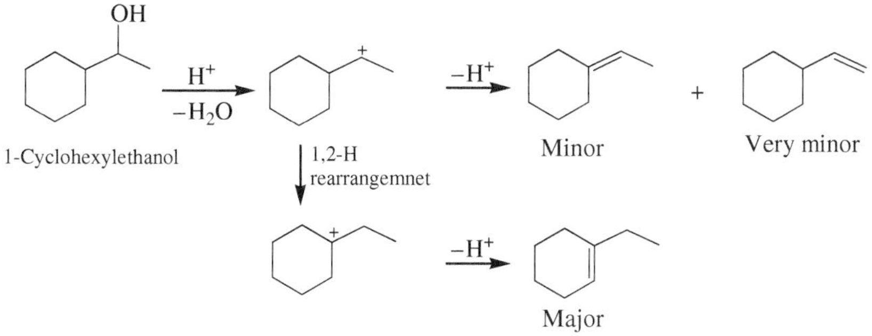

FIGURE 7.20   The acid-catalyzed E1 dehydration of 1-cyclohexylethanol.

$$
\underset{\text{A 1° alcohol}}{R\diagdown\diagup\diagdown\text{OH}} + H_3O^+ \rightleftharpoons H_2\ddot{O}\text{:} + H\underset{\overset{|}{\underset{H\ \ H}{}}}{\overset{\overset{R\ \ H}{\diagdown\ \diagup}}{C}}\overset{+}{\underset{}{OH_2}} \xrightarrow{\text{E2}} \underset{\text{A 1-alkene}}{\overset{R}{H}\diagup\diagdown\diagup\diagdown\overset{H}{H}} + H_3O^+
$$

$$(7.20)$$

As the protonated $-^+OH_2$ departs from the substrate molecule, a C—H bond in the β-carbon is cleaved simultaneously by a weakly basic water molecule. As a result, the formation of a highly unstable primary carbocation is effectively avoided, and the elimination occurs concertedly.

Due to absence of any strong bases in the system, the E2 dehydration of a primary alcohol possesses a high activation energy. Usually, the reaction requires relatively high temperatures above 100°°C. Very often, in order to make a 1-alkene from a primary alcohol, the alcohol is first converted to an alkyl tosylate or halide (RCH$_2$CH$_2$X or RCH$_2$CH$_2$OTs) instead of direct dehydration. Then an E2 elimination can be readily performed on the functionalized alkane using a bulky base to give the desired 1-alkene (referred to Reactions 7.6 and 7.9).

7.6.3   The E1 Dehydrohalogenation of Haloalkanes (RX, X = Cl, Br, or I)

Secondary and tertiary functionalized alkanes (R–LG, LG = Cl, Br, I, MsO, TsO, and TfO) can undergo E1 eliminations as major reactions when the substrates are treated with strong bases at relatively low concentrations. If the base concentration is very high, E2 reactions take place instead. When these substrates are treated with weakly basic/nucleophilic solvents, E1 reactions can only occur to a small extent. Instead, the major reactions will be the S$_N$1 reactions (referred to Chapter 6).

Similar to the E1 dehydration of alcohols, the regiochemistry and stereochemistry (product distribution) for E1 eliminations of functionalized alkanes with a good leaving group are also subject to thermodynamic control. As a result, the most stable alkene isomer will be the major product. This can be seen from E1 dehydrobromination of 2-bromobutane (R or S) (Reaction 7.21).

$$(7.21)$$

Of all the three alkene isomers, *trans*-2-butene is the most stable. It is the major product as the formation of the most stable isomer from the intermediate carbocation possesses the smallest activation energy in the product formation step.

The E1 dehydrobromination of 1-bromo-1-methylcyclohexane (Reaction 7.22) possibly gives two alkenes, with the more stable 1-methylcyclohexene being overwhelmingly major product.

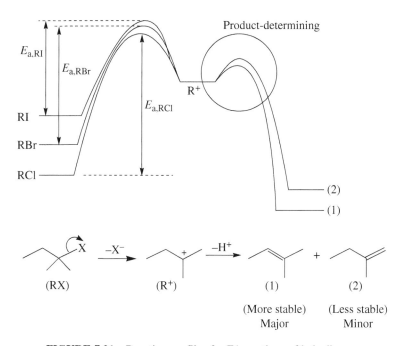

$$(7.22)$$

As described in Section 7.1.2, when 1-bromo-1-methylcyclohexane is treated by a strong base at high concentration, E2 reaction occurs. Different from the E1 reaction, the regiochemistry for E2 reaction of the substrate is determined by structure of the base (Fig. 7.6).

Figure 7.21 shows reaction profiles for E1 elimination of 2-halo-2-methylbutane (RX, X = Cl, Br, or I) [2]. The order of standard bond dissociation energies (BDEs) for C—X bonds is BDE (C–Cl, 327 kJ/mol) > BDE (C–Br, 285 kJ/mol) > BDE (C–I, 213 kJ/mol) (see Section 6.1). The relative energy levels of R–Cl, R–Br, and R–I are placed accordingly in their reaction profiles. For each of the haloalkanes, the

FIGURE 7.21   Reaction profiles for E1 reactions of haloalkanes.

rate-determining step of E1 reaction (the formation of a carbocation) has a very late transition state. Its structure greatly resembles the carbocation. Therefore, the transition states for all the haloalkanes have very similar energies. As a result, the activation energy of the rate-determining step for RX almost solely depends on BDE of the C—X bond as indicated in Figure 7.21 ($E_{a,RCl} > E_{a,RBr} > E_{a,RI}$). This means that rate of the overall reaction is determined by relative ease of departure for the functional group. *A better leaving group makes the reaction faster.* This is a general principle governing the elimination reactions. On the other hand, the regiochemistry for the E1 reactions is dictated by the relative stabilities of the alkene products. Alkene (1) is more stable and is formed as the major product, as the product formation step for the more stable alkene (which has a lower energy level) possesses a smaller activation energy and is more productive.

7.7   THE E1 ELIMINATION OF ETHERS

Analogous to acid-catalyzed dehydration of alcohols (ROH), in the presence of a strong acid, elimination can also occur to secondary or tertiary alkyl ethers (ROR′). Especially interesting is the acid-catalyzed elimination for tertiary butyl ethers (′BuOR, R is a primary or secondary alkyl group) (Fig. 7.22) [1].

The alkoxide –OR group in an ether molecule is a very poor leaving group. Upon protonation, it is converted to –$^+$O(H)R, a very good leaving group as the C—O bond is now weakened by the positive charge introduced onto the oxygen atom by protonation. Once formed, –$^+$O(H)R can readily leave as an alcohol molecule, and the positive charge is transferred to the other alkyl of the ether to form an intermediate carbocation. For the tertiary butyl ether ′BuOR, after the divalent oxygen in the molecule is protonated rapidly by a strong acid (HA), the C—O bond in the tertiary butyl group will break subsequently to give $(CH_3)_3C^+$ (a tertiary carbocation), and ROH (a primary or secondary alcohol) is generated. **The cleavage of the C—O bond is highly selective, determined by stability of the resulting carbocation.** If the C—O bond in the R group broke, a less stable primary or secondary R$^+$ carbocation would

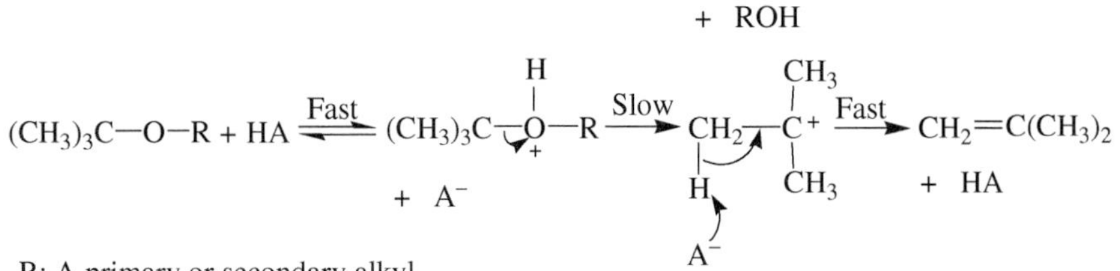

R: A primary or secondary alkyl

HA: A strong acid such as H_2SO_4, TsOH, or CF_3CO_2H

FIGURE 7.22   Mechanism for the E1 elimination of a tertiary butyl ether. Modified from Hoffman [1].

be produced, which is kinetically and thermodynamically unfavorable. Once $(CH_3)_3C^+$ is formed, a C—H bond in the carbocation will be cleaved by the conjugated base (A^-) of the acid catalyst leading to the formation of an alkene product, and the overall reaction is completed. The reaction follows an E1 mechanism with the rate-determining step being the formation of the intermediate tertiary carbocation from the initially protonated ether $^tBu-^+O(H)R$ molecule. The major outcome for the reaction lies in that a primary or secondary alcohol ROH can be readily and selectively released from a tertiary butyl ether via an acid catalyzed E1 reaction, and the reaction proceeds under relatively mild conditions. This makes the type of reactions synthetically useful.

7.8   INTRAMOLECULAR (UNIMOLECULAR) ELIMINATIONS VIA CYCLIC TRANSITION STATES

7.8.1   Concerted, *syn*-Elimination of Esters

The thermal intramolecular elimination (pyrolysis) of esters gives an alkene product, and a carboxylic acid (RCO_2H) molecule is eliminated from the ester substrate (Fig. 7.23). This is a special type of mechanistically and synthetically interesting β-elimination. The reaction is a unimolecular (concerted), *syn*-elimination that takes place in an eclipsed conformation with the two groups to be lost (a β-hydrogen and a carboxylate group RCO_2-) *cis* to each other [1]. Because the reaction is unimolecular, it has the first-order rate law, with the reaction rate proportional to the molar concentration of the ester substrate. The rate law for the reaction is related to that for an E1 mechanism. On the other hand, in terms of the concertedness, the reaction is related to an E2 mechanism. Figure 7.23 shows that the formation of a cyclic transition state from the eclipsed reactive conformation of the ester molecule only involves minimal atomic movement. This gives rise to a very small entropy $(\Delta S^{\ddagger} \sim 0)$ for the transition state. A total of six electrons participate in the concerted

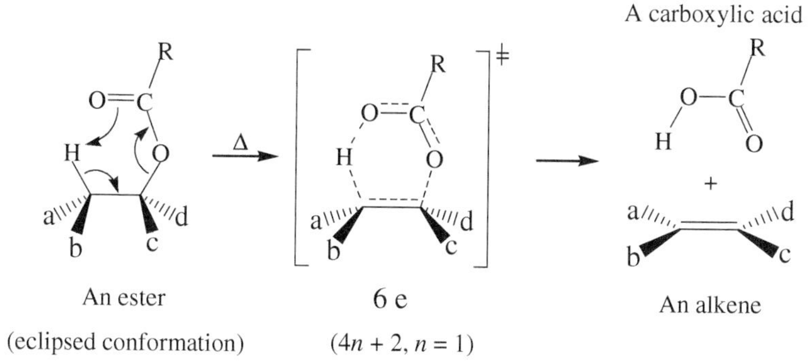

FIGURE 7.23   Mechanism for the unimolecular *syn*-elimination of an ester.

reaction process, giving a Huckel number ($4n + 2$, $n = 1$) to the transition state (refer to Section 4.3.1). Both the small entropy effect and the electronic factor favor stabilization of the cyclic transition state. As a result, the reaction possesses a low activation energy and proceeds readily.

Figure 7.24 shows examples for pyrolysis of esters. The unimolecular elimination of ethyl acetate gives ethylene and acetic acid. The reaction of 3-methyl-1-butyl acetate gives 3-methyl-1-butene. The reaction of ($1S,2R$)-3-methyl-2-hexyl acetate (a stereospecific ester) gives 3-methyl-(E)-2-hexene, but the (Z)-isomer is not formed (Fig. 7.24c). The stereochemical outcome for this reaction has confirmed the concerted, *syn*-addition mechanism for pyrolysis of esters.

(a)

Ethyl acetate   Ethylene   Acetic acid

(b)

3-Methyl-1-butyl acetate   3-Methyl-1-butene   Acetic acid

(c)

($1S,2R$)-3-Methyl-2-hexylacetate

3-Methyl-(E)-2-hexene

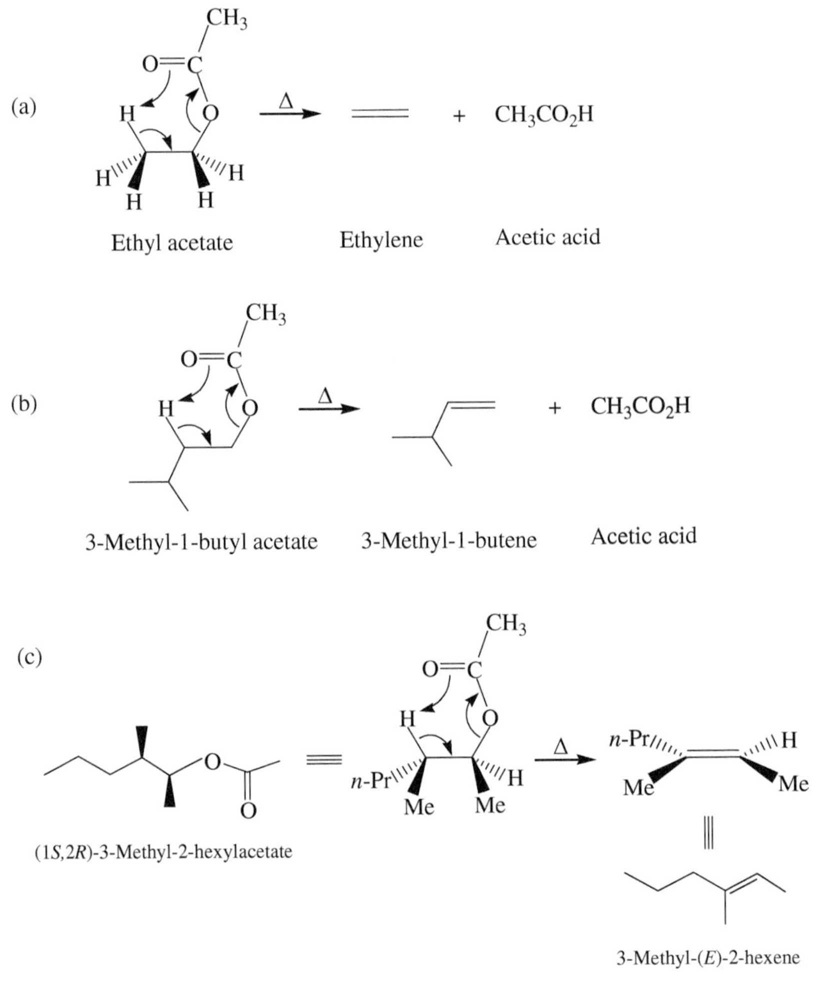

FIGURE 7.24 Examples for unimolecular *syn*-eliminations of esters: (a) ethyl acetate, (b) 3-methyl-1-butyl acetate, and (c) ($1S,2R$)-3-methyl-2-hexyl acetate.

7.8.2   Selenoxide Elimination

Selenoxide eliminations are frequently used in making α, β-unsaturated carbonyl compounds. The reaction occurs by a similar concerted, *syn*-elimination mechanism via a stabilized cyclic transition state (6e). An example is shown in Reaction 7.23 [1, 2].

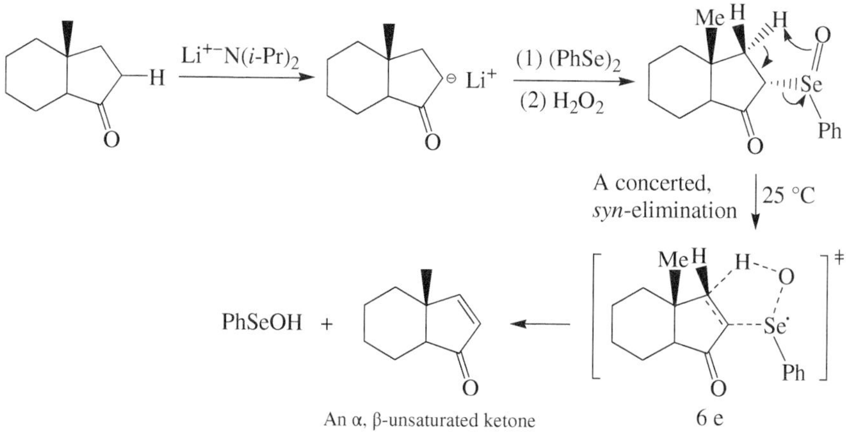

$$(7.23)$$

The electronegative carbonyl group withdraws electron density from a C_β–H bond making the β-hydrogen slightly acidic. This facilitates the formation of a carbanion when the substrate is treated with a strong base. The carbanion reacts with $(PhSe)_2$ and breaks the diselenium Se–Se bond. Consequently, a –SePh unit is introduced to the organic substrate, which can be readily oxidized to a selenoxide [–Se(O)Ph]. Since a methyl group stays in the front of the bicyclic system, the C–Se bond is formed at the back side of the ring in order to avoid possible steric hindrance by methyl. Then, the selenoxide induces a unimolecular β-elimination of the hydrogen at the back via a stabilized cyclic transition state [6e, $(4n + 2)$, $n = 1)$]. It is an effective way to install a C=C double bond in a carbonyl compound.

7.8.3   Silyloxide Elimination

Another type of eliminations that occurs concertedly via a 6e cyclic transition state is silyloxide elimination (Peterson olefination) (Fig. 7.25) [1]. When a very strong base is used to generate an oxyanion intermediate, the reaction occurs by a *syn*-elimination mechanism via an eclipsed conformation and a stabilized 6e cyclic transition state is proposed (Fig. 7.25a, each curved arrow representing two electrons). The high affinity of oxygen to silicon due to a strong O–Si bond is believed to serve as a major thermodynamic driving force for the reaction. It is an intramolecular (unimolecular) reaction and takes place readily to give a

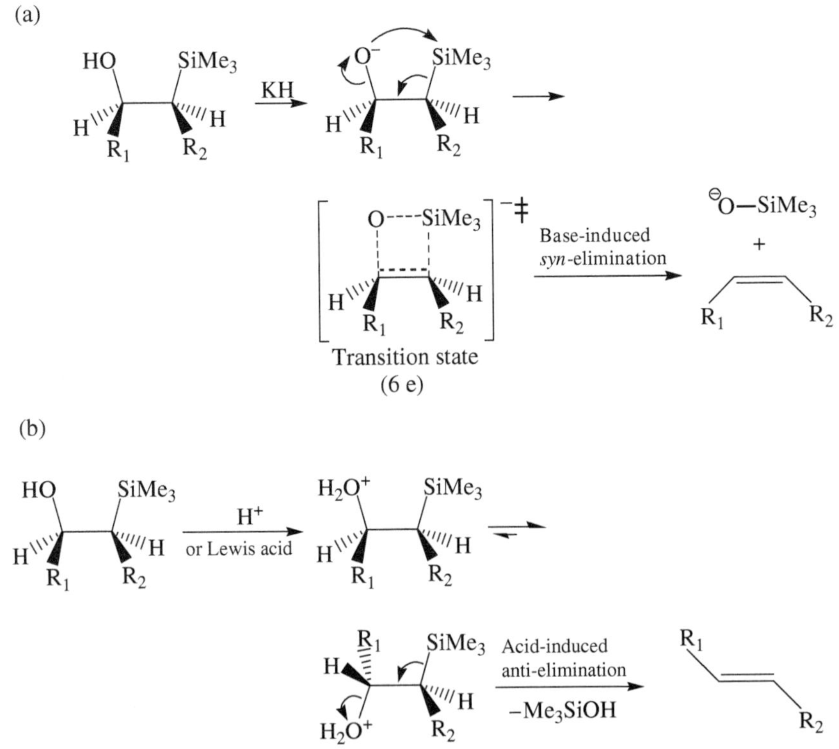

(a)

Transition state
(6 e)

(b)

FIGURE 7.25 Silyloxide elimination: (a) Base-induced *syn*-elimination and (b) acid-induced anti-elimination. Modified from Hoffman [1].

stereospecific alkene product. If a protic or Lewis acid is used to catalyze the reaction, it occurs by an anti-elimination mechanism via a staggered conformation (Fig. 7.25b). This is a very useful method to make either a (Z) or (E) alkene from the same starting material.

7.8.4   Unimolecular β-Elimination of Hydrogen Halide from Haloalkanes

Sections 7.1, 7.3, and 7.5 of this chapter indicate that dehydrohalogenation (elimination of a hydrogen halide) from a haloalkane containing one halo group usually requires a base to induce the reaction. Recent research has shown that some polyhalogenated alkanes undergo intramolecular (unimolecular) dehydrochlorination (elimination of HCl) and dehydrofluorination (elimination of HF) on photolysis with mercury emission (UV light) [8, 9]. The reactions can possibly take place via a four-centered transition state (e.g., CH_3CCl_3 in Reaction 7.24) [8].

$$H_2C-\underset{\underset{Cl}{|}}{\overset{\overset{Cl}{|}}{C}}-Cl \xrightarrow{h\nu} \left[H_2C \overset{\overset{H---Cl}{\vdots \quad \vdots}}{=\!=\!=} \underset{\underset{Cl}{|}}{C}-Cl \right]^{\ddagger} \longrightarrow H_2C=CCl_2 + H-Cl \tag{7.24}$$

A 4-center, 4-electron TS

$\Delta G^{\ddagger} = 225$ kJ/mol

Reaction 7.24 is a concerted, unimolecular β-elimination directly leading to an alkene product. In the transition state, the C—H and C—Cl bonds (containing a total of four electrons) are partially broken and the new H—Cl bond and C=C π bond (also containing a total of four electrons) are partially formed. Therefore, the cyclic transition state should contain 4e ($4n$, $n = 1$) and does not experience stabilization according to Huckle's rule. The energy of the transition state in Reaction 7.24 is determined to be 225 kJ/mol (18,800 cm^{-1}) [8]. This indicates that the transition state is not thermally accessible, and the reaction is only photochemically allowed. The theoretical analysis is consistent with the experimental conditions for the reaction. The photolysis has been used to help the reactant molecule acquire sufficient vibrational energy (activation of the reactant) to bring about the unimolecular elimination reaction.

On photolysis, $CF_3CH_2CH_2Cl$ is shown to undergo both unimolecular dehydrochlorination (β-elimination of HCl) and dehydrofluorination (β-elimination of HF) via a 4-center, 4-electron cyclic transition state to give $CF_3CH=CH_2$ and $CF_2=CHCH_2Cl$, respectively (Reaction 7.25) [9].

$$F_2C=CHCH_2Cl \overset{-HF}{\longleftarrow} \left[F_2C \overset{\overset{F---H}{\vdots \quad \vdots}}{=\!=\!=} \underset{\underset{H}{|}}{C}CH_2Cl \right]^{\ddagger} \overset{h\nu}{\longleftarrow} F_2C-\underset{\underset{H}{|}}{\overset{\overset{F}{|}}{C}}-\underset{\underset{H}{|}}{\overset{\overset{Cl}{|}}{C}}-CH_2 \overset{h\nu}{\longrightarrow} \left[F_3C\overset{\overset{H---Cl}{\vdots \quad \vdots}}{=\!=\!=} \underset{\underset{H}{|}}{C}CH_2 \right]^{\ddagger} \overset{-HCl}{\longrightarrow} F_3CCH=CH_2 \tag{7.25}$$

The rate constant for HF elimination is much smaller than for HCl elimination. The molar ratio of $CF_3CH=CH_2$ to $CF_2=CHCH_2Cl$ in the product is determined to be ~80.

7.9 MECHANISMS FOR REDUCTIVE ELIMINATION OF LG^1/LG^2 (TWO FUNCTIONAL GROUPS) ON ADJACENT CARBONS

Equation 7.26 shows a general elimination of two functional groups from two adjacent carbons by a reducing agent (electron-donor).

$$-\underset{\underset{X}{|}}{\overset{\overset{|}{|}}{C}}-\underset{\underset{Y}{|}}{\overset{\overset{|}{|}}{C}}- \xrightarrow{\text{Reductant}} \overset{}{\underset{}{C}}\!=\!\overset{}{\underset{}{C} } + X^- + Y^- \tag{7.26}$$

$(X, Y = LG^1, LG^2)$

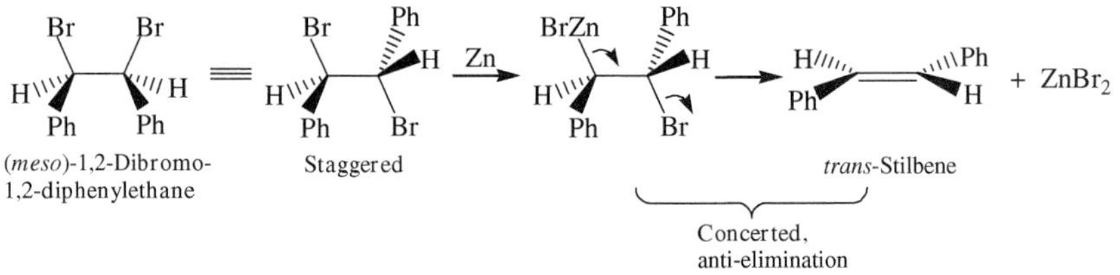

FIGURE 7.26 Mechanism for zinc-induced anti-eliminations of vicinal dihalides (dibromide).

Upon the formation of an alkene (a C=C double bond) and releasing two mono anions, the two carbon atoms are reduced. This type of elimination is called **reductive elimination** and usually follows an **anti-elimination mechanism**.

In the presence of an active metal such as zinc, tin, or magnesium, a vicinal dihaloalkane undergoes an elimination of both halo groups to form an alkene product [3]. The mechanism of this type of reactions is demonstrated using (*meso*)-1,2-dibromo-1,2-diphenylethane. Treatment of this dibromide with zinc affords only *trans*-stilbene, but not the *cis*-isomer (Fig. 7.26). The stereoselectivity shows that the reaction follows an anti-elimination mechanism via a staggered conformation. The net conversion includes the oxidation of zinc and reduction of the two adjacent functionalized carbon atoms. Further studies have shown that the formation of an alkene from a vicinal alkane dibromide occurs through an organozinc intermediate (analogous to a Grignard reagent). Similar to a Grignard reagent, the C—Zn bond in the intermediate is highly nucleophilic. The bonding electrons can be readily transferred to the C—C bond domain to form a π bond and, simultaneously, to displace a bromide. The conversion of the organozinc intermediate to a stereospecific alkene is most likely concerted.

This type of elimination of vicinal alkane dihalides can also be made by the reductive selenide (Se^{2-}) (Fig. 7.27a). Reactions of (*meso*)-2,3-dibromobutane and chiral (2R,3R)-2,3-dibromobutane with sodium selenide (Na_2Se) aqueous solution have been found to give *trans*-2-butene and *cis*-2-butene, respectively [10]. The stereoselectivity for the elimination reactions shows that both reactions are anti-eliminations and take place via staggered conformations. The mechanism for the reductive elimination by selenide is believed to be analogous to the zinc-facilitated elimination and is demonstrated by the elimination of the (*meso*)-2,3-dibromobutane (Fig. 7.27b). The reaction takes place via a postulated organoselenium intermediate in staggered conformation, with the Se–C–C–Br moiety anti-coplanar. The formation of a Se—C bond is conceivably owing to the nucleophilicity of Se^{2-}, which displaces a –Br group from the substrate readily. Very likely, the activity of the C—Se bonding electrons is comparable to that of the C–Zn electrons. They are readily transferred to the C—C bond domain to form a π bond and, simultaneously, to displace bromide.

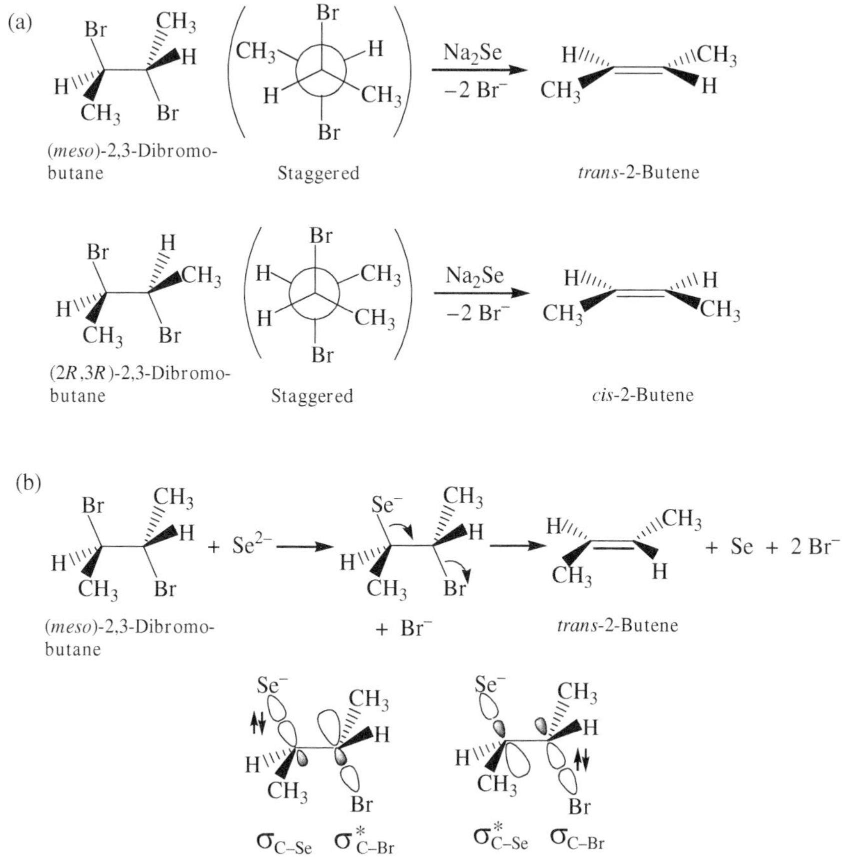

FIGURE 7.27 Mechanism for reductive anti-elimination of vicinal alkane dihalides (dibromides) by selenide (Se^{2-}). (a) Overall reactions and (b) Proposed reaction mechanism.

Figure 7.27b shows bonding and antibonding orbitals of the C—Se and C—Br bonds in the postulated organoselenium intermediate. The bonding orbital of the C—Se bond (σ_{C-Se}, occupied) and the antibonding orbital of the C—Br bond ($\sigma_{C-Br}*$, unoccupied) are of symmetry-match. They can effectively overlap in the staggered conformation. As a result, electrons flow from σ_{C-Se} to $\sigma_{C-Br}*$ to weaken the C—Br bond. Similarly, the antibonding orbital of the C—Se bond ($\sigma_{C-Se}*$, unoccupied) and the bonding orbital of the C—Br bond (σ_{C-Br}, occupied) are of symmetry-match. They can also effectively overlap in the staggered conformation to lead to transfer of electrons from σ_{C-Br} to $\sigma_{C-Se}*$ to weaken the C—Se bond. All this makes the anti-elimination occur.

It has been shown that chlorinated ethylenes (with one or more chloro –Cl groups attached to a C=C double bond, such as trans-1,2-dichloroethylene) undergoes reductive elimination by metallic zinc in the aqueous media at pH ~ 7 to give

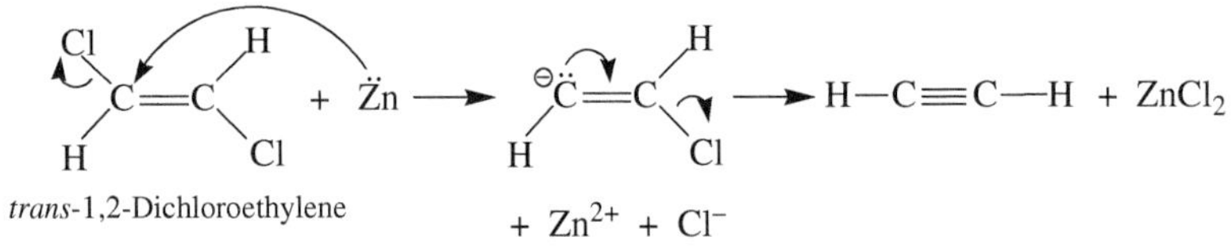

trans-1,2-Dichloroethylene

$+ Zn^{2+} + Cl^-$

FIGURE 7.28   Reductive elimination of a chlorinated ethylene by metallic zinc.

acetylene [11]. The reaction takes place via a chlorinated ethylene anion intermediate (with a negative charge formed on a doubly bonded carbon), which is formed by accepting two electrons from metallic zinc and simultaneously releasing a chloride anion from the organic substrate (Fig. 7.28). The lone pair of electrons in the anionic intermediate is highly active (nucleophilic). It is readily transferred to the C=C double bond domain to form a second π bond and to expel the chloride on the doubly bonded carbon. As a result, acetylene is formed. Alternatively, the anion intermediate may be protonated (by water in the aqueous media) to generate minor chloroethylene [11].

Research has shown that polychlorinated ethanes undergo reductive 1,2-elimination of two –Cl groups from two adjacent carbons by transition metal coenzymes, such as Vitamin B_{12} and titanium(III) citrate [12]. The reaction takes place via an intermediate radical (Reaction 7.27).

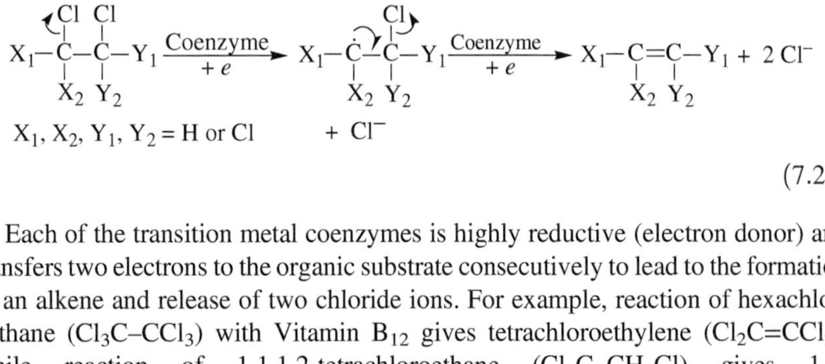

$X_1, X_2, Y_1, Y_2 = H$ or Cl $+ Cl^-$

$$(7.27)$$

Each of the transition metal coenzymes is highly reductive (electron donor) and transfers two electrons to the organic substrate consecutively to lead to the formation of an alkene and release of two chloride ions. For example, reaction of hexachloroethane ($Cl_3C–CCl_3$) with Vitamin B_{12} gives tetrachloroethylene ($Cl_2C=CCl_2$), while reaction of 1,1,1,2-tetrachloroethane ($Cl_3C–CH_2Cl$) gives 1,1-dichloroethylene ($Cl_2C=CH_2$) [12].

The reductive 1,2-elimination of halides from vicinal alkane dihalides can also be induced by low valent transition metals, such the Cr^{2+} cation (Reaction 7.28) [13].

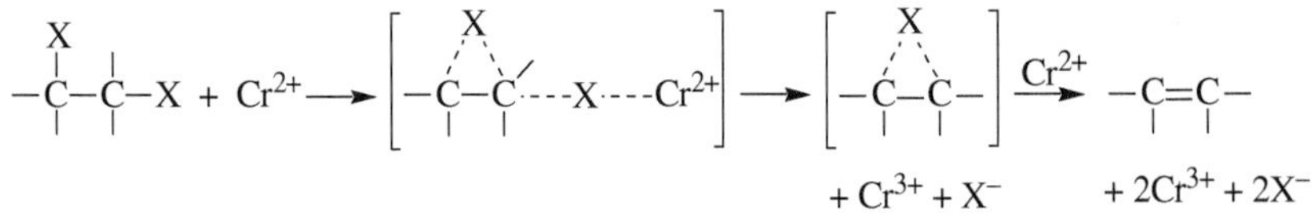

$+ Cr^{3+} + X^-$ $+ 2Cr^{3+} + 2X^-$

$$(7.28)$$

The reductive Cr^{2+} abstracts a halide. As a result, the reaction takes place via two activated intermediates (or transition states) in which the C—X bonds are partially broken. *Trans*-1,2-dibromo-1,2-diphenylethylene [*t*-Ph(Br)C=C(Br)Ph] undergoes an analogous reductive elimination by reaction with Cr^{2+} to give diphenylacetylene (PhCCPh) (Reaction 7.29) [13].

$$t - Ph(Br)C=C(Br)Ph + 2Cr^{2+} \rightarrow PhCCPh + 2Cr^{3+} + 2Br^{-} \qquad (7.29)$$

7.10   THE α-ELIMINATION GIVING A CARBENE: A MECHANISTIC ANALYSIS USING SYMMETRY RULES AND MOLECULAR ORBITAL THEORY

7.10.1   The Bimolecular α-Elimination of Trichloromethane ($CHCl_3$) Giving Dichlorocarbene (CCl_2)

If a single carbon atom contains two or more electronegative functional groups, two groups (such as H–LG) can be possibly eliminated from the carbon atom to lead to the formation of a **carbene**, a metastable species with the central carbon atom being divalent and containing a lone pair of electrons. This type of eliminations is called **α-elimination**. An important example for α-eliminations is the strong base promoted dehydrochlorination (elimination of HCl) of trichloromethane ($CHCl_3$) giving dichlorocarbene (:CCl_2) (Reaction 7.30) [2, 14].

$$CHCl_3 \overset{NaOH}{\underset{\Delta}{\rightarrow}} : CCl_2 + H_2O + NaCl \qquad (7.30)$$

The carbon atom in $CHCl_3$ contains three electronegative chloro (–Cl) groups, making the C—H bond polarized and slightly acidic. In the presence of a strong base, an HCl molecule is eliminated from $CHCl_3$ to form dichlorocarbene (:CCl_2). It has been shown in Section 4.1.2 that this α-elimination is employed for synthesizing cyclopropane rings via the reactive intermediate CCl_2. In this chapter, we present a mechanistic analysis on the α-elimination using symmetry rules and molecular orbital theory.

$CHCl_3$ possesses the C_{3v} symmetry (the same as CH_3X, Section 6.5.1) and contains a principal C_3 axis [15–17]. The 3-fold axis is along the C—H bond. Therefore, the C—H bond in this molecule is set along the *z*-axis, while the three C—Cl bonds are not covered by any of the *x*, *y*, or *z* axis (Fig. 7.29a). Each terminal atom (H or Cl) in the molecule can be regarded as a ligand coordinating to the central carbon atom. The linear combination of the four ligand orbitals (the H 1s and three Cl p orbitals) form four ligand group orbitals (LGOs) as shown in Equation 7.31 [14].

$$LGOs = 2a_1 + e \qquad (7.31)$$

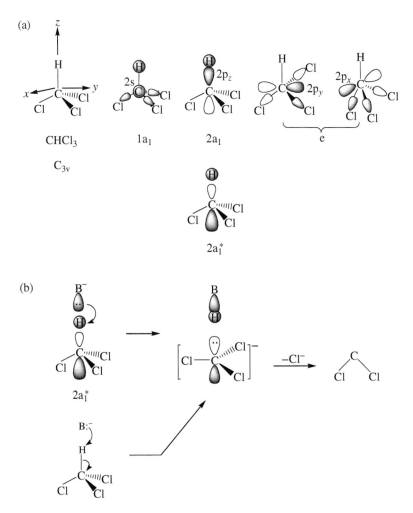

FIGURE 7.29 (a) Molecular orbitals (MOs) in $CHCl_3$ which are formed by linear combination of ligand group orbitals (LGO's $= 2a_1 + e$) and 2s, $2p_x$, $2p_y$, and $2p_z$ in the central carbon based on symmetry match. The four MO's on the top are occupied bonding orbitals. Each of them has a counterpart of an unoccupied antibonding MO. Only one antibonding MO ($2p_z$-based LUMO $2a_1^*$ along the C—H bond) is displayed. (b) Overlap of a base orbital with the $2p_z$-based LUMO $2a_1^*$ on hydrogen results in electron transfer to $2a_1^*$ breaking the C—H bond and producing a carbanion CCl_3^-.

Then, the four LGOs combine with four valence orbitals in the central carbon (2s, $2p_x$, $2p_y$, and $2p_z$) to form eight MOs, four bonding MOs and four antibonding MOs (Fig. 7.29a).

The qualitative MO treatment based on LGOs and x, y, and z basis functions in the central atom does not generate any p-orbital based bonding or antibonding MO along a C—Cl bond (Fig. 7.29a). Because the C—H bond in $CHCl_3$ is located in the z axis, the $2p_z$-based bonding ($2a_1$) [highest occupied molecular orbital (HOMO)] and antibonding ($2a_1*$) [lowest unoccupied molecular orbital (LUMO)] orbitals are formed along the C—H bond. The bonding orbital ($2a_1$) is a predominant MO responsible for the C—H bond formation and the antibonding orbital ($2a_1*$) is a predominant MO responsible for the C—H bond breaking upon accepting a pair of electrons [14]. Therefore, when the hydrogen is attacked by a strong base (e.g., OH$^-$), the base orbital can effectively overlap with $2a_1*$ (Fig. 7.29b) effecting electron transfer into $2a_1*$. This leads to dissociation of the C—H bond giving a carbanion CCl_3^-. CCl_3^- is highly unstable presumably due to strong repulsions between the axial lone-pair of electrons and the equatorial C—Cl bonding electron pairs. A chloride (Cl$^-$) is then readily eliminated by such a strong repulsion to give dichlorocarbene CCl_2 [14]. By using MO theory, the special acidity associated to $CHCl_3$ and its reactivity is properly explained.

Figure 7.30 shows the relative energy levels of the $CHCl_3$ MOs and the correlation (interaction) of its LUMO ($2a_1*$) with the base (B$^-$) orbital. Essentially, the nature of this interaction is the overlap of the B$^-$ orbital with the hydrogen 1s orbital in $2a_1*$ of $CHCl_3$. It leads to breaking of the C—H bond in $CHCl_3$ to give CCl_3^- and to form a B—H bond, which includes the filled bonding σ_{BH} and empty antibonding $\sigma_{BH}*$, simultaneously (Eq. 7.32).

$$CHCl_3 + B^- \rightarrow CCl_3^- + BH \tag{7.32}$$

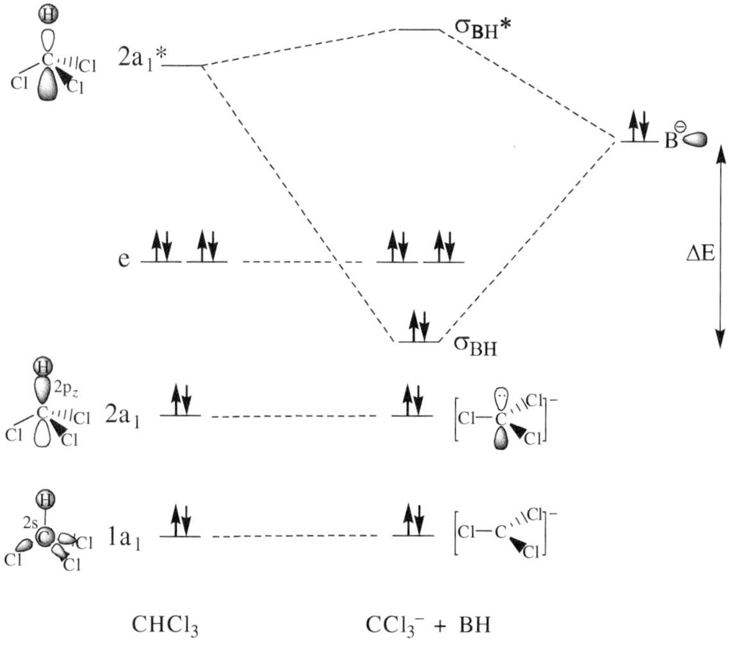

FIGURE 7.30   Molecular orbital diagram for the base induced α-elimination of $CHCl_3$ giving CCl_3^-.

The MO diagram (Fig. 7.30) indicates that the formation of the BH bond results in a substantial energy gain $(-2\Delta E)$, serving as the major thermodynamic driving force for the reaction. Basically, the reaction occurs on the LUMO $(2a_1*)$ of $CHCl_3$. On the other hand, the involvement of the $CHCl_3$ bonding MOs ($1a_1$, $2a_1$, and e) in Reaction 7.30 is minimal. The lower lying CCl_3^- MOs are correlated to the lower lying $1a_1$ and $2a_1$ MOs of $CHCl_3$. Their energies are very close.

7.10.2   Formation of a Carbene by Unimolecular α-Elimination of a Haloalkane and the Subsequent Rearrangement to an Alkene via a C—H (C—D) Bond Elimination

Recent research has shown that on photolysis some polychlorinated alkanes (e.g., $CD_3CD_2CHCl_2$) undergo unimolecular dehydrochlorination (elimination of HCl) via an α-elimination mechanism, and the reaction gives an alkene product via a 1,2-rearrangement of deuterium (Reaction 7.33) [18].

A 3-center, 4-electron TS   A carbene
$\Delta G^{\ddagger} = 276$ kJ/mol

$$(7.33)$$

The formation of $CD_3CD=CDCl$ (rather than $CD_3CD=CHCl$) unambiguously shows that a net elimination of HCl from the same carbon (an α-elimination) takes place. This way, the deuterated reagent has effectively aided establishment of the reaction mechanism.

The reaction is believed to proceed via an intermediate carbene. The formation of the carbene from the starting 1,1-dichloroalkane follows a unimolecular pathway via a 3-center, 4-electron cyclic transition state. The carbene undergoes subsequent spontaneous 1,2-rearrangement of deuterium to give the final alkene product. The transition state contains 4e $(4n, n = 1)$ and does not experience stabilization according to Huckle's rule, while its energy is determined to be 276 kJ/mol $(23,070\ cm^{-1})$ [18]. This indicates that the reaction should be only photochemically allowed, consistent with the observed experimental conditions for the reaction. The photolysis is necessary for the reaction to help the reactant molecule acquire sufficient vibrational energy (activation of the reactant) to bring about the unimolecular α-elimination reaction.

7.11   E1cb ELIMINATION

When the β-carbon to a functional group (leaving group –LG) is attached to another electronegative group, such as a carbonyl (C=O) group, the hydrogen atom on the β-carbon (β-hydrogen) becomes particularly acidic. In the presence of a strong base,

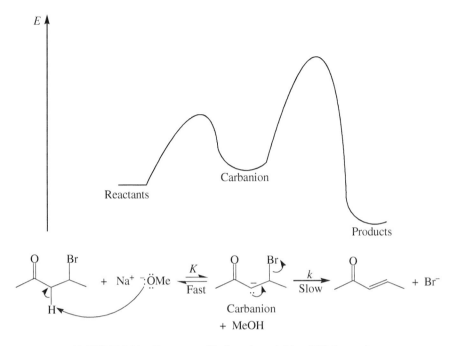

FIGURE 7.31   Energy profile for a base-initiated E1cb reaction.

the β-hydrogen can be removed via a rapid acid-base equilibrium to produce a reactive carbanion intermediate (Fig. 7.31). As discussed in Section 7.10.1 for $CCl_3{}^-$, the instability and reactivity of the carbanion intermediate are attributable to the great activity of the lone pair of electrons in the central carbon atom. It exerts a large repulsion to the C—Br bonding electrons via hyperconjugation. The repulsion force can cleave the C—Br bond spontaneously by pushing the bromo group off as a bromide Br^- anion to complete the overall elimination reaction. The elimination mechanism via a carbanion intermediate is referred to as **E1cb mechanism**. The reaction profile in Figure 7.31 shows that the second step, the cleavage of the C—Br bond is the rate-determining step. The more stable *trans*-alkene is the major product because the activation energy for its product formation step is lower than that for the less stable *cis*-isomer.

The reaction rate can be expressed as

$$d[\text{alkene}]/dt = k[\text{carbanion}] \qquad (7.34)$$

In the reaction, the carbanion reaches an equilibrium rapidly with the substrate and base added before the second step of elimination of the bromo group takes place. Therefore, we have the following equilibrium constant expression written as [16]

$$K = [\text{carbanion}]/[\text{Substrate}][\text{NaOMe}] \qquad (7.35)$$

From Equation 7.35,

$$[\text{carbanion}] = K[\text{Substrate}][[\text{NaOMe}] \qquad (7.36)$$

Substituting Equation 7.36 for Equation 7.34 leads to Equation 7.37

$$d[\text{alkene}]/dt = k \cdot K[\text{Substrate}][[\text{NaOMe}] \qquad (7.37)$$

Equation 7.37 is the rate law for the E1cb reaction in Figure 7.31. The overall reaction is the second order, first order in the organic substrate and first order in the base added. If concentration of the base is maintained constant during the reaction (usually, that is the case), the rate law will become the first order only in the substrate.

Although the hydroxyl –OH group is a poor leaving group, a β-hydroxyketone undergoes an E1cb dehydration reaction in the presence of a strong base (Reaction 7.38) presumably due to the special activity of the lone-pair of electrons in the intermediate carbanion.

Carbanion
+ HOH

$$(7.38)$$

Reaction 7.38 shows that the base molecule consumed in the first step is regenerated in the second step. As a result, its concentration remains constant during the reaction. Therefore, the rate law is the first order only in the β-hydroxyketone substrate.

Although the $-^+\text{NR}_3$ (e.g., $-^+\text{NMe}_3$) groups are poor leaving groups, quaternary amines can undergo elimination reactions in the presence of a strong base to release NMe_3 and produce an alkene (Fig. 7.32). The regiochemistry suggests that the transition state has the carbanion (E1cb) character [1]. The $-^+\text{NMe}_3$ group is electronegative due to the positive charge, and it can activate a β-hydrogen to make it acidic, vulnerable to the attack by a base. This gives rise to the carbanion (E1cb) character for the transition state of the elimination reaction. Of the two types of β-hydrogens, C_{prim}–H is more acidic than C_{sec}–H. This is because a secondary carbon contains more electron donating alkyl groups making the C_{sec}–H bond less polar than C_{prim}–H. As a result, the base preferably attacks the more acidic C_{prim}–H hydrogen to give a 1-alkene as the major product.

7.12   BIOLOGICAL APPLICATIONS: ENZYME-CATALYZED BIOLOGICAL ELIMINATION REACTIONS

7.12.1   The Enzyme-Catalyzed β-Oxidation of Fatty Acyl Coenzyme A

Elimination reactions, especially E1cb, have many biological applications as the electronegative carbonyl groups are abundant in biomolecules.

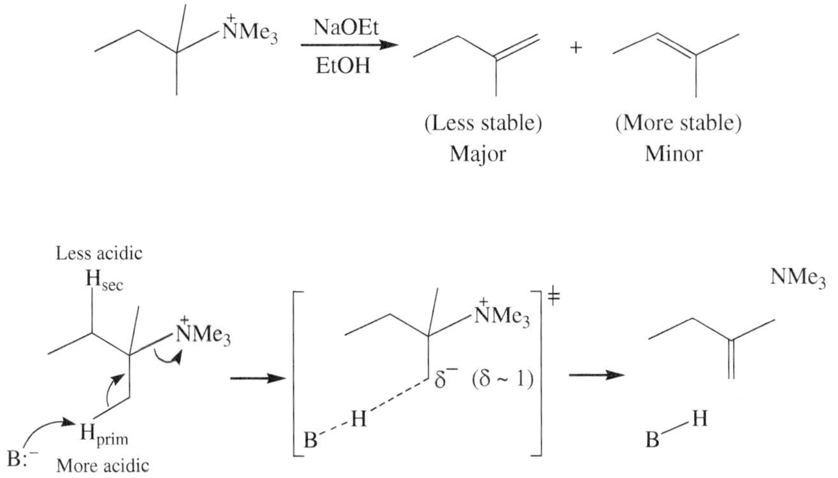

FIGURE 7.32 Elimination of NMe_3 from a quaternary amine, which possesses an E1cb character.

β-Oxidation of fatty acids is the core process in the aerobic metabolism of fatty acids. It breaks down a fatty acid by removing two carbons from the chain and produces an acetyl-coenzyme A (Reaction 7.39) [19].

$$\underset{\text{Fatty acyl CoA}}{RCH_2CH_2\overset{\overset{\displaystyle O}{\|}}{C}—SCoA} + HSCoA \xrightarrow[\text{Four steps}]{\beta\text{-oxidation}} \underset{\substack{\text{Fatty acyl CoA}\\ \text{(two carbons shorter)}}}{R\overset{\overset{\displaystyle O}{\|}}{C}—SCoA} + \underset{\text{Acetyl coenzyme A}}{CH_3\overset{\overset{\displaystyle O}{\|}}{C}—SCoA} \quad (7.39)$$

Then, the shorter fatty acyl CoA enters another cycle of the β-oxidation to further break down the carbon chain in the fatty acid until all the carbons in the fatty acid molecule is converted to acetyl coenzyme A molecules. The overall process contains four steps of pathway. The first step is removal of two hydrogen atoms from C_2 and C_3 (β-oxidation) by FAD (Reaction 7.40) [19].

$$\underset{\text{Fatty acyl CoA}}{RCH_2CH_2\overset{\overset{\displaystyle O}{\|}}{C}—SCoA} + FAD \longrightarrow \underset{\alpha,\beta\text{-Unsaturated acyl CoA}}{RCH{=}CH\overset{\overset{\displaystyle O}{\|}}{C}—SCoA} + FADH_2 \quad (7.40)$$

The mechanism in Reaction 7.40 involves a carbonyl-assisted E1cb-like elimination via a carbanion (thioester enolate) (Fig. 7.33).

The first mechanistic step is abstraction of the acidic *pro-R* hydrogen from the α-carbon to the acyl CoA to give a thioester enolate (a carbanion). Hydrogen

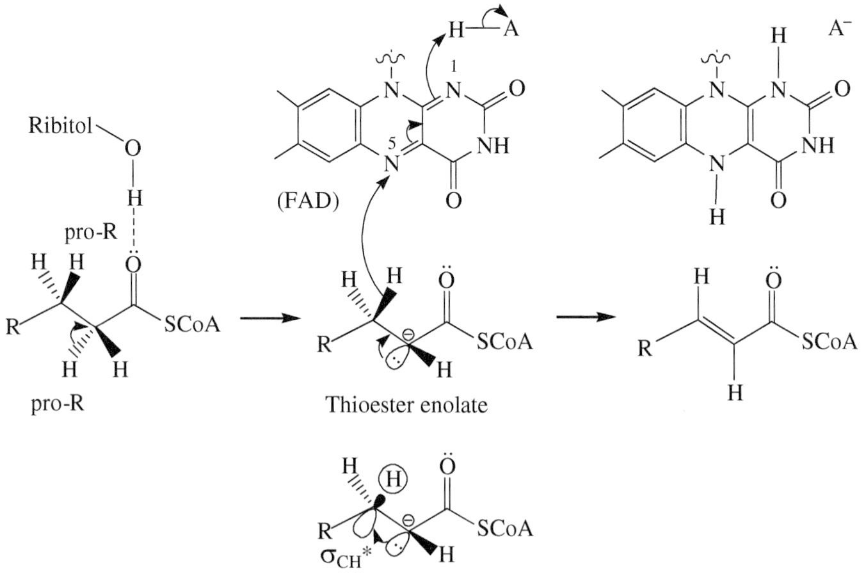

FIGURE 7.33 The FAD-facilitated E1cb-like elimination: a biological elimination involved in the process of β-oxidation of fatty acids.

bonding between the acyl carbonyl group and the ribitol hydroxyl of FAD increases the acidity of the *pro-R* hydrogen and makes the deprotonation favorable. Then, the *pro-R* hydrogen at β-carbon of the thioester enolate is transferred to FAD. Figure 7.33 shows that the highly active lone pair of electrons in the thioester enolate overlaps with the empty antibonding orbital ($\sigma_{CH}*$) of the C—H bond between the β-carbon and its pro-R hydrogen. This weakens the C—H bond. As the lone pair of electrons is transferred to the C_α—C_β bond domain, the *pro-R* hydrogen on β-carbon will be pushed off as a hydride that is simultaneously added to the doubly bonded N5 nitrogen on FAD. Concurrently, the addition is reinforced by protonation of the intermediate at N1 nitrogen. As the electron pair moves down, a C=C bond in *trans*-configuration is formed in the carbon skeleton.

7.12.2 Elimination Reactions Involved in Biosynthesis

Biosynthesis of fatty acids involves a base-catalyzed E1cb dehydration of a 3-hydroxybutyryl thioester (a β-hydroxy carbonyl compound) (Fig. 7.34). The reaction is very similar to that (Reaction 7.38) performed in laboratory. The base in this biological E1cb reaction is the imidazole ring of a histidine residue in the enzyme. In the carbanion (enolate) intermediate, a lone pair of electrons in the α-carbon is located in the opposite side of the β-hydroxyl –OH group. In the second step, as the lone pair of the electrons is transferred to the C_α—C_β bond domain to make a

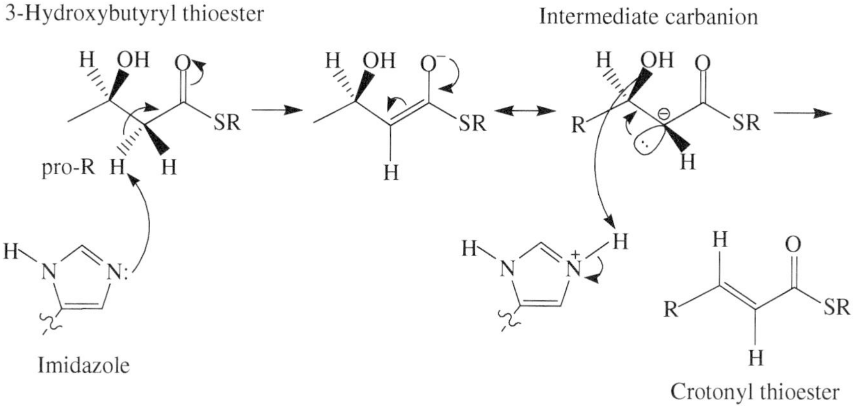

3-Hydroxybutyryl thioester   Intermediate carbanion

Imidazole

Crotonyl thioester

FIGURE 7.34   An enzyme-catalyzed E1cb elimination involved in biosynthesis of fatty acids.

π bond, the β–OH will be pushed off. The loss of hydroxide is facilitated by simultaneous protonation by an imidazole group.

Biosynthesis of limonene from linalyl diphosphate, a biological precursor, includes a comprehensive E1 reaction during which rearrangement of an intermediate carbocation takes place (Fig. 7.35) [20, 21]. The first step is unimolecular dissociation of the substrate (loss of diphosphate) to give an intermediate allyl cation. Then, a carbocation rearrangement takes place via a cyclization to form a six-carbon ring and generate a tertiary carbocation. The cyclization can also be considered a Markovnikov electrophilic alkene addition. It is followed by a C—H bond cleavage by diphosphate to give the final limonene product.

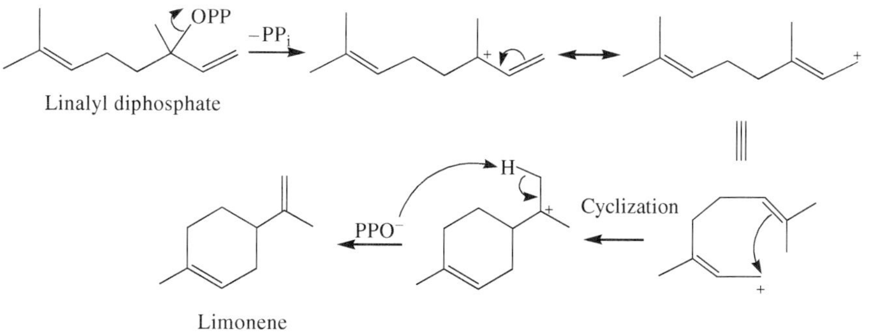

Linalyl diphosphate

Cyclization

Limonene

FIGURE 7.35   The E1 mechanism for biosynthesis of limonene from linalyl diphosphate. Carey [20] and McMurry [21].

PROBLEMS

7.1 Write out the mechanism for the following E2 reaction and account for the stereochemistry.

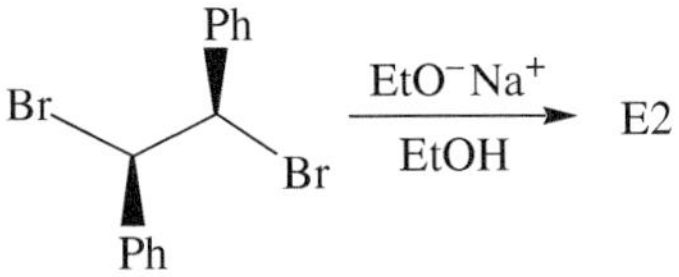

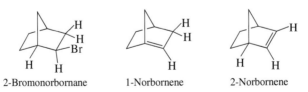

7.2 Under subtly controlled mild conditions, the following diols undergo selective monodehydration to form alkenols. Write out the mechanism for each of the reactions and account for the regioselectivity of the dehydration reactions.

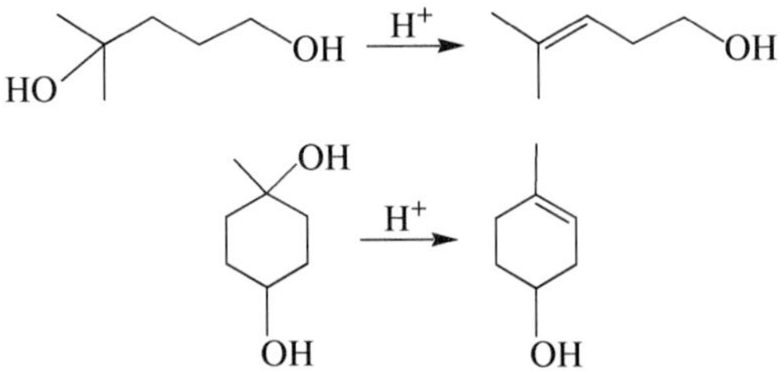

7.3 Under the same conditions, the following reactions of *cis-* and *trans*-4-chlorocyclohexanol give different products. Account for the difference.

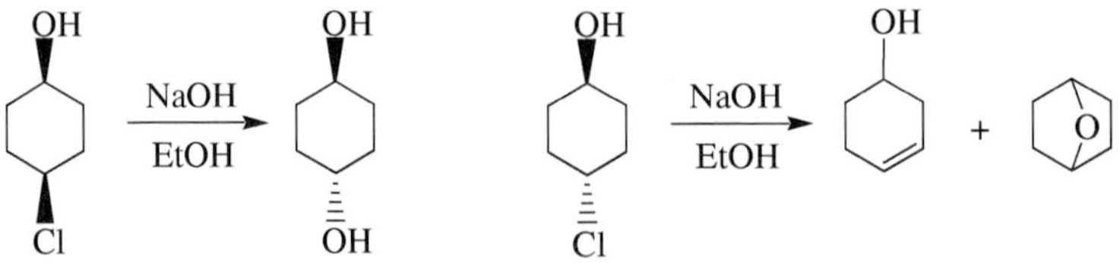

7.4 Elimination of HBr from 2-bromonorbornane gives only 2-norbornene but not 1-norbornene. Account for the regioselectivity.

2-Bromonorbornane   1-Norbornene   2-Norbornene

7.5 (1*S*,3*R*)-3-*tert*-butylcyclohexyl tosylate undergoes E2 elimination with *tert*-butoxide very slowly, while the (1*R*,3*R*)-isomer reacts much faster. Explain.

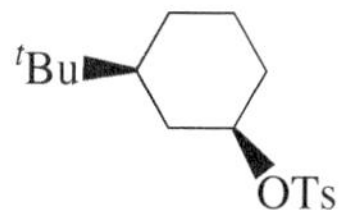

(1S,3R)-3-*Tert*-butylcyclohexyl tosylate

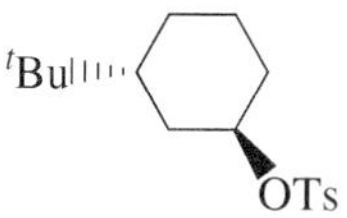

(1R,3R)-3-*Tert*-butylcyclohexyl tosylate

7.6 Show the identity of the alkene product from the following reaction. Provide a mechanism to account for the stereochemistry.

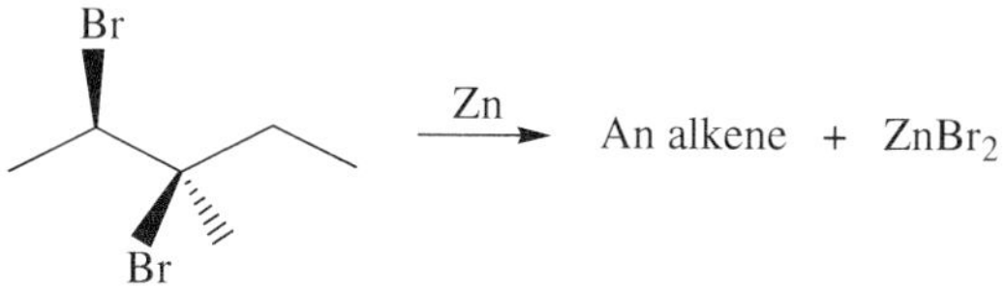

An alkene + ZnBr$_2$

7.7 Account for stereochemistry for the following E2 reaction.

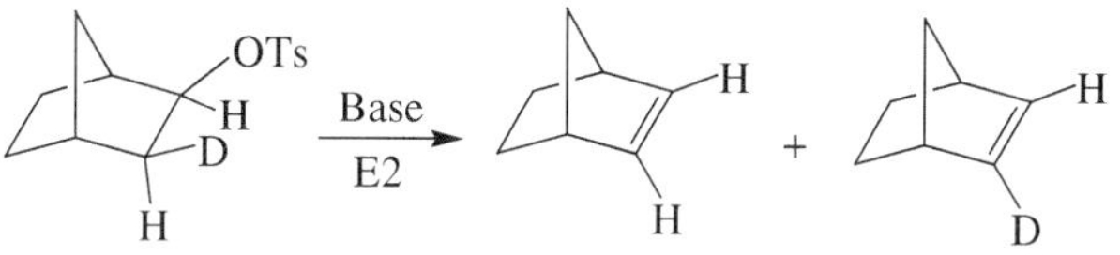

94% 6%

7.8 For the following elimination reaction of the (2-phenylethyl)trimethylammonium ion, the kinetic $^{14}N/^{15}N$ isotope effect k_{14N}/k_{15N} is 1.0133 and the deuterium isotope effect k_H/k_D is 3.2. Which of the following mechanisms does this evidence support? Explain.

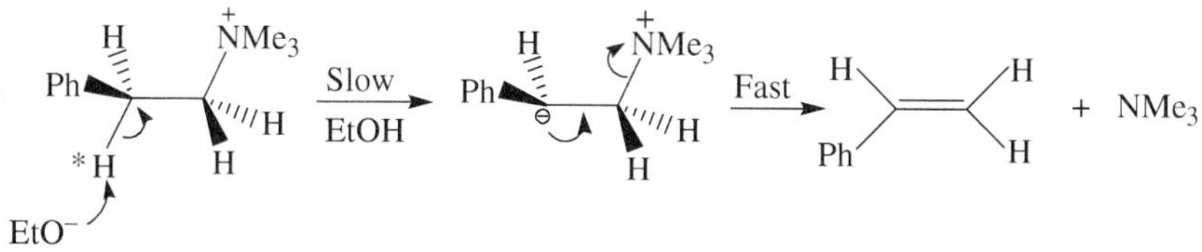

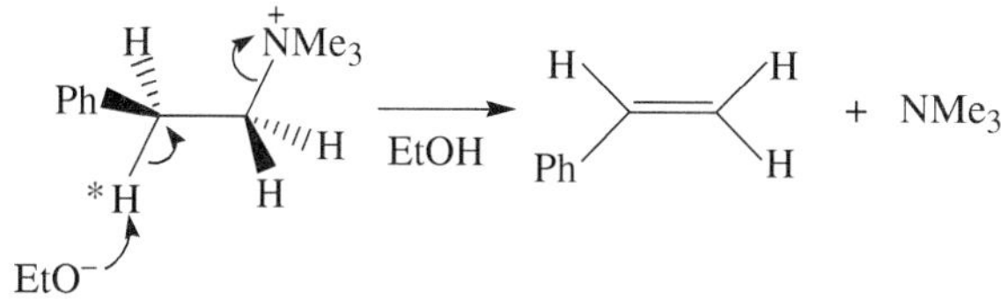

REFERENCES

1. Hoffman, R. V. *Organic Chemistry: An Intermediate Text*, 2nd ed., Wiley, Hoboken, NJ, USA (2004).

2. Bruckner, R. *Advanced Organic Chemistry: Reaction Mechanisms*, Chapter 4, pp. 129–167, Harcout/Academic Press, Orlando, FL, USA (2002).

3. Fox, M. A.; Whitesell, J. K. *Core Organic Chemistry*, 2nd ed., Jones and Bartlett, Sudbury, MA, USA (1997).

4. Brown, W. H.; Foote, C. S.; Iverson, B. L.; Anslyn, E. V. *Organic Chemistry*, 6th ed., Brooks/Cole, Belmont, CA, USA (2012).

5. Sun, X. Symmetry Analysis in Mechanistic Studies of Nucleophilic Substitution and β-Elimination Reactions, *Symmetry*, 2010, *2*, 201–212.

6. Bickelhaupt, F. M.; Baerends, E. J.; Nibbering, N. M. M.; Ziegler, T. Theoretical Investigation on Base-Induced 1,2-Eliminations in a Model System $F^- + CH_3CH_2F$. The Role of the Base as a Catalyst. *J. Am. Chem. Soc.* 1993, *115*, 9160–9173.

7. Ensing B.; Laio, A.; Gervasio, F. L.; Parrinello, M.; Klein, M. L. A Minimum Free Energy Reaction Path for the E2 Reaction between Fluoroethane and a Fluoride Ion. *J. Am. Chem. Soc.* 2004, *126*, 9492–9493.

8. Turpin, M. A.; Smith, K. C.; Heard, G. L.; Setser, D. W.; Holmes, B. E. Unimolecular Reactions of 1,1,1-Trichloropropane, and 3,3,3-Trilfluoro-1,1,1-Trichloropropane: Determination of Threshold Energies by Chemical Activation. *J. Phys. Chem. A*, 2014, *118*, 9347–9356.

9. Ferguson, J. D.; Johnson, N. L.; Kekenes-Huskey, P. M.; Everett, W. C.; Heard, G. L.; Setser, D. W.; Holmes, B. E. Unimolecular Rate Constants for HX or DX Elimination (X = F, Cl) from Chemically Activated $CF_3CH_2CH_2Cl$, $C_2H_5CH_2Cl$, and $C_2D_5CH_2Cl$: Threshold Energies for HF and HCl Elimination. *J. Phys. Chem. A*, 2005, *109*, 4540–4551.

10. Prince, M.; Bremer, B. W. Selenium Chemistry. I. Stereochemistry of Vicinal Dihalide Elimination. *J. Am. Chem. Soc.* 1967, *32*, 1655–1656.

11. Roberts, A. L.; Totten, L. A.; Arnold, W. A.; Burris, D. R.; Campbell, T. J. Reductive Elimination of Chlorinated Ethylenes by Zero-Valent Metals. *Environ. Sci. Technol.* 1996, *30*, 2654–2659.

12. Schanke, C. A.; Wackett, L. P. Environmental Reductive Elimination Reactions of Polychlorinated Ethanes Mimicked by Transition-Metal Coenzymes. *Environ. Sci. Technol.* 1992, *26*, 830–833.

13. Kray Jr. W. C.; Castro, C. E. The Cleavage of Bonds by Low-Valent Transition Metal Ions. The Homogeneous Dehalogenation of Vicinal Dihalides by Chromous Sulfate. *J. Am. Chem. Soc.* 1964, *86*, 4603–4608.

14. Sun, X. A Study of the S_N2 Mechanism by Symmetry Rules and Qualitative Molecular Orbital Theory. *Chem. Educator*, 2003, *8*, 303–306.

15. Huheey, J. E.; Keiter, E. A.; Keiter, R. L. *Inorganic Chemistry: Principles of Structure and Reactivity*, 4th ed., HarperCollins College Publishers, New York, NY, USA (1993).

16. Silbey, R. J.; Alterty, R. A.; Bawendi, M. G. *Physical Chemistry*, 4th ed., Wiley, Danvers, MA, USA (2005).

17. Pfennig, B. W. *Principles of Inorganic Chemistry*, Wiley, Hoboken, NJ, USA (2015).

18. Larkin, A. C.; Nestler, M. J.; Smith, C. A.; Heard, G. L.; Setser, D. W.; Holmes, B. E. Chemical Activation Study of the Unimolecular Reactions of $CD_3CD_2CHCl_2$ and $CHCl_2CHCl_2$ with Analysis of the 1,1-HCl Elimination Pathway. *J. Phys. Chem. A*, 2016, *120*, 8244–8253.

19. Voet, D.; Voet, J. G.; Pratt, C. W. *Fundamentals of Biochemistry*, 3rd ed., Wiley, Hoboken, NJ, USA (2008).

20. Carey, F. A. *Organic Chemistry*, 5th ed., McGraw Hill, New York, NY (2003).

21. McMurry, J. *Organic Chemistry with Biological Applications*, 2nd ed., Brooks/Cole, Belmont, CA, USA (2011).

8

NUCLEOPHILIC ADDITIONS AND SUBSTITUTIONS ON CARBONYL GROUPS

8.1 NUCLEOPHILIC ADDITIONS AND SUBSTITUTIONS OF CARBONYL COMPOUNDS

The carbonyl (C=O) group is a **carbon electrophile** due to the highly electronegative oxygen which makes the carbon atom partially positively charged. Thus, the electrophilic carbon in carbonyl can be attacked readily and effectively by strong nucleophiles such as OH⁻, OR⁻, CN⁻, or RMgX (Grignard reagent). As a new Nu—C bond is being made, the C=O π electrons are transferred to the oxygen simultaneously, resulting in a **nucleophilic addition** (Fig. 8.1a).

For ketones and aldehydes, after the nucleophilic addition takes place, the reaction will be ceased as neither alkyl nor hydrogen can be further displaced because of the strong C—C or C—H bond, and the tetrahedral adduct is relatively stable. In the presence of a strong acid (usually in a catalytic amount), a ketone or aldehyde can also undergo nucleophilic additions with weak nucleophiles [usually (but not limited to) nucleophilic protic solvents, for example, water and ethanol] (Fig. 8.1b). The protonation on the nucleophilic carbonyl oxygen takes place to create a positive charge on the oxygen atom. The positive charge is then further delocalized onto the carbon atom through the C=O π bond (refer to Fig. 1.13) to make the carbon in the carbonyl group strongly electrophilic and react with a weak nucleophile readily. The acid-catalyzed synthesis of bisphenol A from phenol and acetone discussed in Section 5.4.1 follows this mechanism. The overall reaction is initiated by addition

Organic Mechanisms: Reactions, Methodology, and Biological Applications,
Second Edition. Xiaoping Sun.
© 2021 John Wiley & Sons, Inc. Published 2021 by John Wiley & Sons, Inc.
Companion website: www.wiley.com/go/Sun/OrgMech_2e

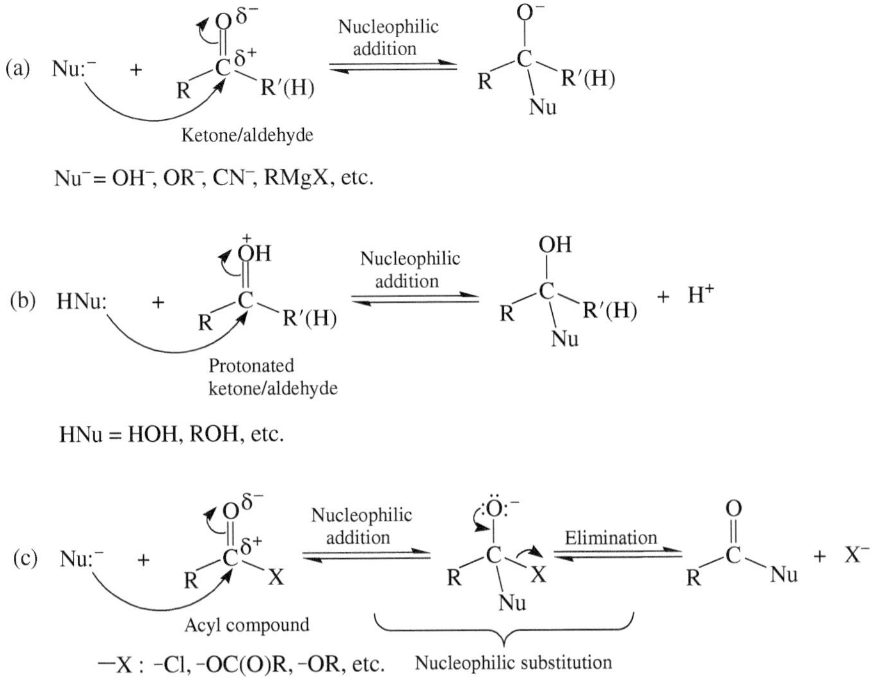

(a) Nu:⁻ + [Ketone/aldehyde] $\xrightarrow{\text{Nucleophilic addition}}$ [product]

Ketone/aldehyde

Nu⁻ = OH⁻, OR⁻, CN⁻, RMgX, etc.

(b) HNu: + [Protonated ketone/aldehyde] $\xrightarrow{\text{Nucleophilic addition}}$ [product] + H⁺

Protonated ketone/aldehyde

HNu = HOH, ROH, etc.

(c) Nu:⁻ + [Acyl compound] $\xrightarrow{\text{Nucleophilic addition}}$ [tetrahedral intermediate] $\xrightarrow{\text{Elimination}}$ [product] + X⁻

Acyl compound   Nucleophilic substitution

—X : –Cl, –OC(O)R, –OR, etc.

FIGURE 8.1 (a) Addition of a strong nucleophile to a ketone or aldehyde; (b) addition of a weak nucleophile to a protonated ketone or aldehyde; and (c) nucleophilic acyl substitution.

of weakly nucleophilic aromatic ring in a phenol molecule to the protonated acetone (refer to Fig. 5.19).

For an acyl compound bearing a good leaving group such as –Cl, –OC(O)R, or –OR on the carbonyl carbon, the tetrahedral intermediate produced from the first step of nucleophilic addition undergoes a rapid elimination to release an X⁻ anion (Fig. 8.1c). In the tetrahedral intermediate, the very active lone pair of electrons in the negatively charged oxygen exerts a strong repulsion to the C—X bond via hyperconjugation effect, which expels the –X (a good leaving group) off fairly easily. As a result, a new acyl compound with a C—Nu bond is formed. The overall net result for reaction of a nucleophile with an acyl compound is a **nucleophilic substitution**, effected by an addition first, then followed by an essentially irreversible elimination (Fig. 8.1c).

Research has shown that the nucleophilic acyl substitution reactions **do not** follow an S_N2-like mechanism as indicated in Equation 8.1.

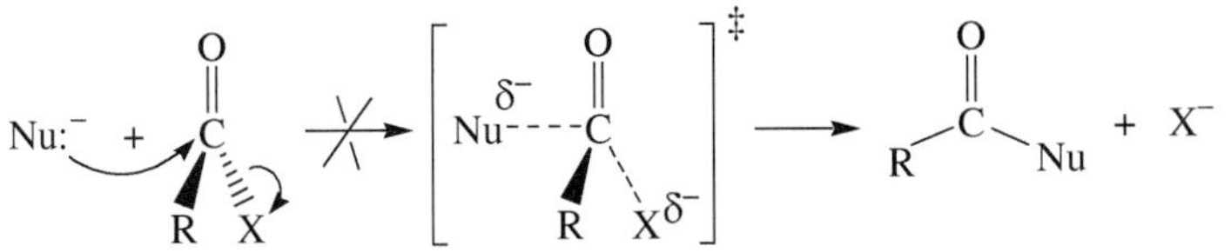

$$\text{Nu:}^- + \underset{R \quad X}{\overset{O}{\overset{\|}{C}}} \quad \cancel{\longrightarrow} \quad \left[\underset{R \quad X^{\delta-}}{\overset{O}{\overset{\|}{\text{Nu}^{\delta-}\text{---}C}}} \right]^{\ddagger} \quad \longrightarrow \quad \underset{R \quad \text{Nu}}{\overset{O}{\overset{\|}{C}}} + X^- \qquad (8.1)$$

The postulated transition state for the S_N2-like concerted acyl substitution would be very unstable because of the relatively high C—X σ bond energy and the steric interactions between the nucleophile and other groups perpendicular to it in the transition state. In contrast, the transition state for the formation of the intermediate tetrahedral adduct (Eq. 8.2) is relatively more stable.

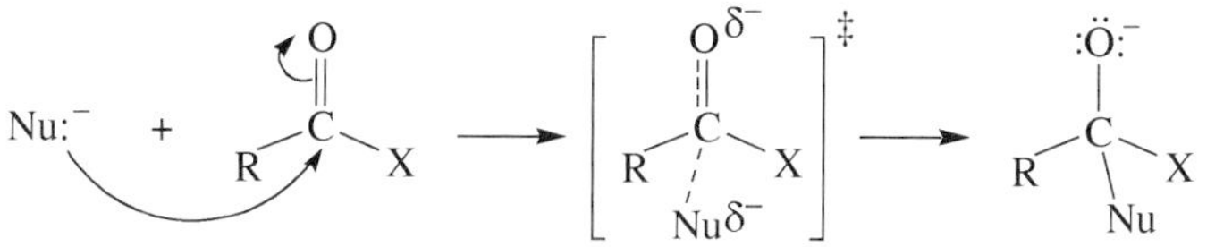

$$
\text{Nu:}^- \; + \; \underset{R}{\overset{O}{\underset{\|}{\overset{}{C}}}}\!\!-X \longrightarrow \left[\underset{Nu\delta^-}{\overset{O\delta^-}{\underset{R}{\overset{\|}{C}}}}\!\!-X \right]^{\ddagger} \longrightarrow \underset{R}{\overset{:\ddot{O}:^-}{\underset{Nu}{\overset{}{C}}}}\!\!-X \tag{8.2}
$$

This is because the transition state in Equation 8.2 contains a partially broken C=O π bond which is less energetic than a σ bond, and in addition, its structure is near tetrahedral with all the groups staying farthest apart and having small steric interactions. The relative stability of the transition states makes the acyl substitution follow a kinetically more favorable stepwise pathway as shown in Figure 8.1c.

The best evidence for the formation of a tetrahedral intermediate in an acyl substitution reaction comes from the study of the hydroxide base catalyzed hydrolysis, in normal water (H_2O^{16}), of benzoate esters [such as $PhC(O^{18})OEt$ (with $C=O^{18}$)] that has been labeled with O^{18} at the carbonyl oxygen (Fig. 8.2) [1]. The reaction is

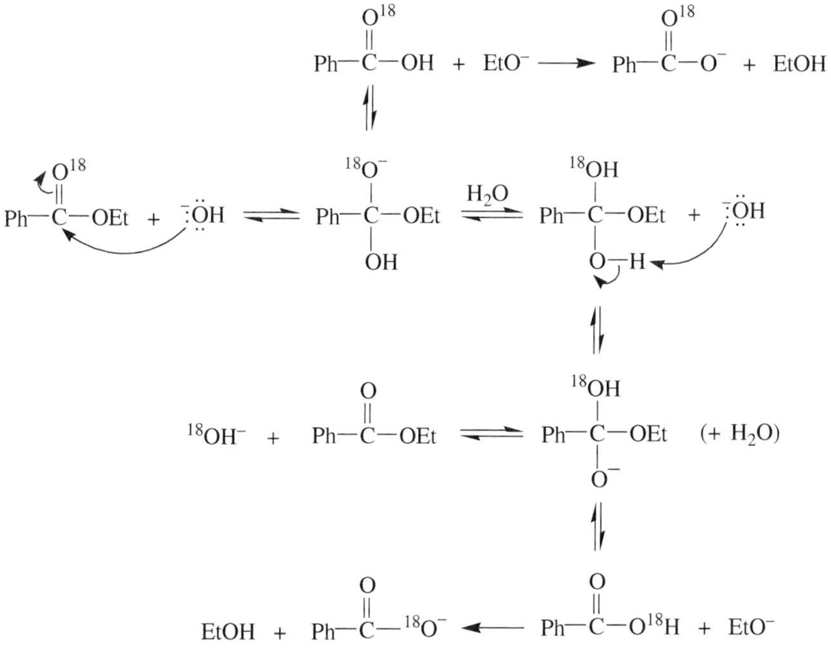

FIGURE 8.2 Mechanism for the hydroxide base catalyzed hydrolysis of ester (ethyl benzoate). Modified from Jackson [1].

not carried out to completion, and the unreacted ester is analyzed. It is found that some of O^{18} has been lost from the carbonyl group and the ester containing $C=O^{16}$ (PhCOOEt) has been formed. The ratio of the rate constant for exchange of oxygen to that for hydrolysis, k_{ex}/k_{hyd}, is determined to be 4.8, indicating that a tetrahedral adduct is formed initially (Fig. 8.2). The subsequent proton transfers make the addition reversed with loss of $^{18}OH^-$ to lead to the formation of the normal ester PhCOOEt with the carbonyl oxygen exchanged to O^{16}. Concurrently, in the course of the reaction, the tetrahedral intermediate containing either the $^{18}O^-$ or the $^{16}O^-$ anion undergoes rapid elimination of –OEt, which is facilitated by the negatively charged oxygen, to give the final hydrolysis product $PhC(O^{18})O^-$ (with $C=O^{18}$) or $PhC(O)^{18}O^-$ (with $C=O$). Since the final step of the hydrolysis reaction (the reaction of benzoic acid with ethoxide giving benzoate and ethanol) is essentially irreversible, given sufficient time, all the starting ester and tetrahedral intermediates will be converted to benzoate and ethanol in the end of the reaction. Therefore, the overall reaction for the base-catalyzed hydrolysis of ethyl benzoate (Eq. 8.3), mechanistically, follows the pattern of Figure 8.1c.

$$(8.3)$$

In principle, all the reactions in Figure 8.1 are reversible. Their reversibility depends on the natures of the nucleophile and the –R or –X group in the carbonyl compound. This will be addressed for the individual reactions of specific carbonyl compounds with various nucleophiles throughout this chapter. Many individual nucleophilic addition and substitution reactions have remarkable synthetic utility and biological applications. These aspects will be discussed in much detail along with their reaction mechanisms.

8.2   NUCLEOPHILIC ADDITIONS OF ALDEHYDES AND KETONES AND THEIR BIOLOGICAL APPLICATIONS

8.2.1   Acid and Base Catalyzed Hydration of Aldehydes and Ketones

In general, ketones and aldehydes are kinetically stable toward water because the nucleophilicity of water is not strong enough to effectively attack the weakly electrophilic carbonyl group in ketones and aldehydes. In the presence of a strong acid or base, a ketone or aldehyde undergoes reversible addition with water (hydration) to give **a germinal diol (a hydrate)** (Reaction 8.4).

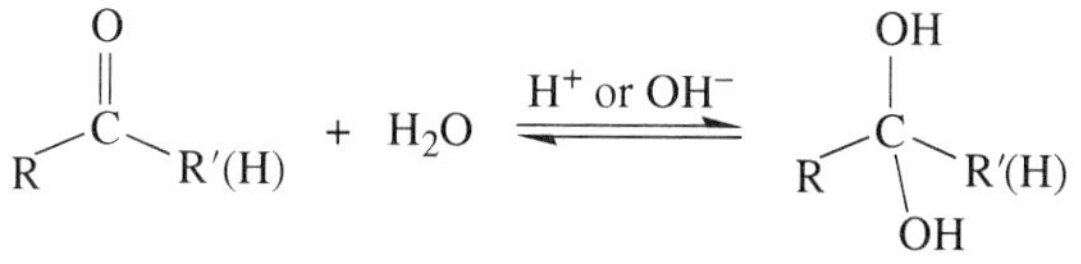

$$\text{(8.4)}$$

Ketone/aldehyde   Germinal diol (hydrate)

In the base-catalyzed hydration (Fig. 8.3a), the first step is nucleophilic addition of hydroxide (OH^-) to the carbonyl group (C=O) giving an anionic tetrahedral intermediate with the oxygen negatively charged. In the second step, the oxygen anion is protonated by water leading to the formation of a hydrate and regeneration of the hydroxide catalyst. In the acid-catalyzed hydration (Fig. 8.3b), the nucleophilic oxygen in the carbonyl group is protonated first. As a result, the electrophilicity of the carbonyl carbon is much enhanced by the positive charge. Then, it undergoes a facile addition with a water molecule to give a hydrate.

The reversibility of hydrations of ketones and aldehydes is dictated by the relative stability of the starting carbonyl substrate versus the hydrate product. In general, an electron-donating –R group stabilizes the carbonyl π bond, which disfavors the forward hydration. On the other hand, an electron-withdrawing –R group destabilizes the carbonyl π bond and makes the formation of hydrate more favorable. For this reason, hydrations of most of ketones (containing two electron-donating alkyl groups on carbonyl) are reactant-favored with a very small percentage of hydrate produced because the resulting hydrate product is less stable than the ketone reactant. Hydrations of aldehydes are more favorable than ketones because an aldehyde only contains one electron-donating alkyl group.

Hydration of trichloroacetaldehyde (Cl_3CCHO) is product-favored (essentially irreversible) presumably due to the highly electron-withdrawing –CCl_3 group attached to the carbonyl that destabilizes the carbonyl π bond (Eq. 8.5, from Fox and Whitesell [2]).

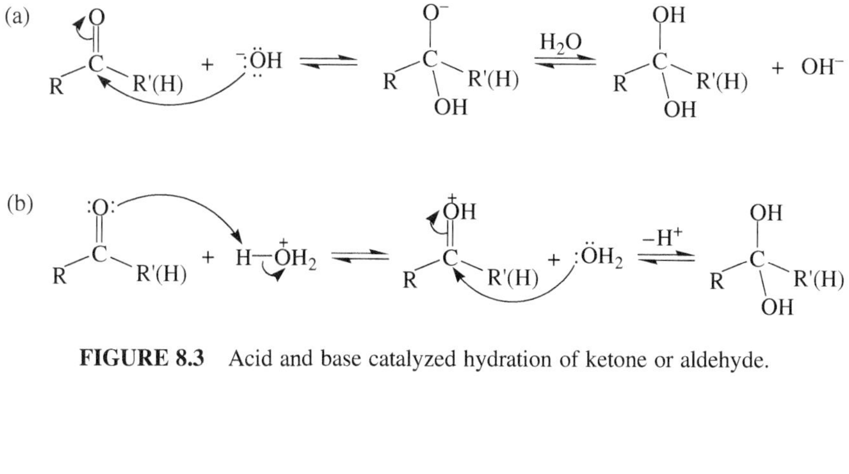

FIGURE 8.3   Acid and base catalyzed hydration of ketone or aldehyde.

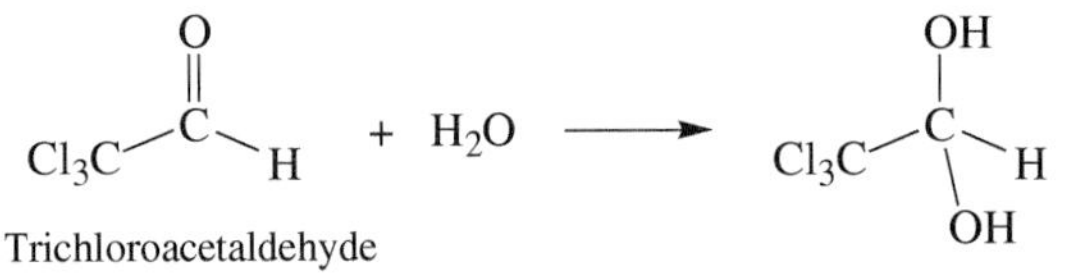

$$\underset{\text{Trichloroacetaldehyde}}{Cl_3C-\overset{\overset{\displaystyle O}{\|}}{C}-H} \quad + \ H_2O \quad \longrightarrow \quad Cl_3C-\overset{\overset{\displaystyle OH}{|}}{\underset{\underset{\displaystyle OH}{|}}{C}}-H \qquad (8.5)$$

Kinetically, this reaction is also more favorable than most of other ketones and aldehydes. It is because the carbonyl carbon in Cl_3CCHO is highly activated by the electron-withdrawing $-CCl_3$ group and becomes strongly electrophilic. As a result, the hydration proceeds rapidly even in the absence of any acid or base.

The hydration mechanism is confirmed by the outcome of the acid-catalyzed reaction of normal acetone $(CH_3)_2CO^{16}$ (in which the oxygen is O^{16}) with the O^{18}-enriched water H_2O^{18} in excess (Fig. 8.4) [2]. A net exchange of oxygen atoms between acetone and water takes place when the two substances are mixed in the presence of a strong acid (e.g., H_2SO_4). The enthalpy of the reaction is zero. However, there is an increase in entropy when H_2O^{16} is formed and subsequently mixed with excess H_2O^{18}. The overall reaction is driven thermodynamically by the H_2O^{16}/H_2O^{18} mixing entropy. Mechanistically, the O^{18}-isotope of oxygen in H_2O^{18} finds its way into the carbonyl group of $(CH_3)_2CO^{16}$ via an acid-catalyzed nucleophilic

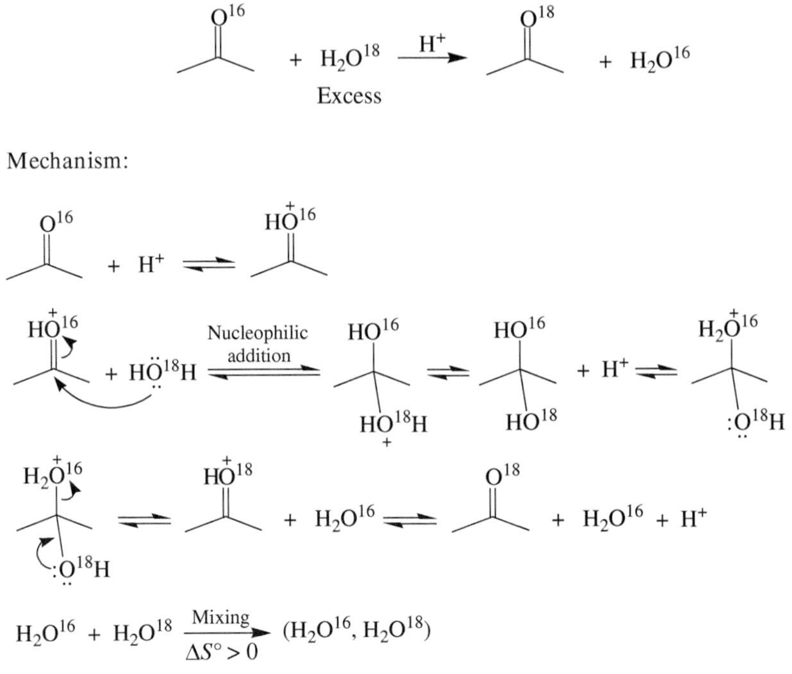

Mechanism:

$$H_2O^{16} + H_2O^{18} \xrightarrow[\Delta S° > 0]{\text{Mixing}} (H_2O^{16}, H_2O^{18})$$

FIGURE 8.4 Mechanism for oxygen exchange between acetone and water. Adapted from Fox and Whitesell [2].

hydration reaction (Fig. 8.4). Since H_2O^{18} is in excess, its initial addition to the protonated $(CH_3)_2CO^{16}$ is both thermodynamically and kinetically favorable. Following the nucleophilic addition is a heavy-water-relayed proton transfer from O^{18} to O^{16} in a hydrate. Because the intermediate hydrate is less stable than acetone and the content of H_2O^{16} in the system is low, the hydrate, once formed, is converted to acetone $(CH_3)_2CO^{18}$ favorably, releasing a H_2O^{16} molecule and regenerating H^+. The final irreversible mixing of H_2O^{16} and excess H_2O^{18} ($\Delta S° > 0$) drives the overall oxygen-exchange reaction irreversible.

8.2.2 Acid Catalyzed Nucleophilic Additions of Alcohols to Aldehydes and Ketones

Analogous to water, the hydroxyl –OH group in alcohols is weakly nucleophilic, and it can be added to the carbonyl group in ketones and aldehydes in the presence of acid.

An alkoxide base (formed by treatment of an alcohol with sodium or potassium) undergoes a reversible addition to a ketone or aldehyde giving an anionic tetrahedral adduct (Eq. 8.6, from Fox and Whitesell [2]).

$$\text{Ketone/aldehyde} \qquad (8.6)$$

However, the $C-OCH_3$ bond in the adduct is highly activated by the lone-pair of electrons in the negatively charged oxygen (Section 8.1). This gives rise to a more favorable backward elimination of –OCH_3 from the adduct giving back the original ketone or aldehyde. Therefore, the base-catalyzed addition of alcohol to ketones or aldehydes is reactant-favored and not productive.

In the presence of acid, the situation is different (Fig. 8.5). Figure 8.5 shows that an alcohol (such as methanol) undergoes reversible acid-catalyzed addition to a ketone or aldehyde to give an intermediate **hemiketal** or **hemiacetal** first. In general, hemiketals and hemiacetals are unstable with respect to the starting ketones and aldehydes. In addition, they are reactive toward an alcohol. In the presence of acid, a hemiketal or hemiacetal reacts with the second alcohol molecule spontaneously and rapidly to give a **ketal** or **acetal** and water (a by-product). The formation of a ketal or acetal from hemiketal or hemiacetal is product-favored, which drives the overall reaction forward. In addition, the overall reaction of ketone or aldehyde with alcohol can possibly become irreversible by removal of water (a by-product) from the reaction system once formed.

Mechanistically, the overall reaction in Figure 8.5 involves two nucleophilic additions of alcohol (CH_3OH) to positively charged carbonyl of a ketone or aldehyde. The first is the nucleophilic addition of CH_3OH to a protonated ketone or

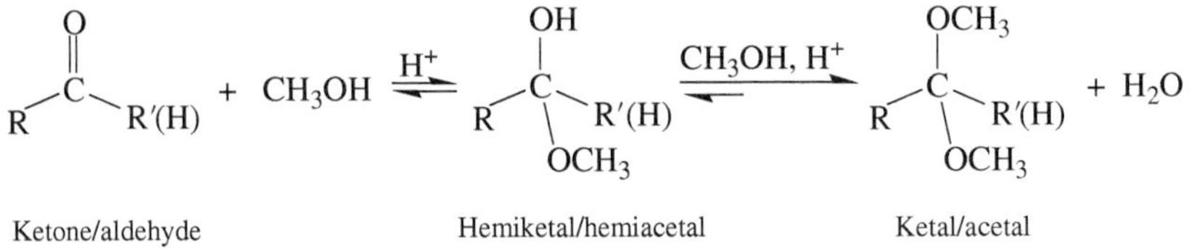

Mechanism:

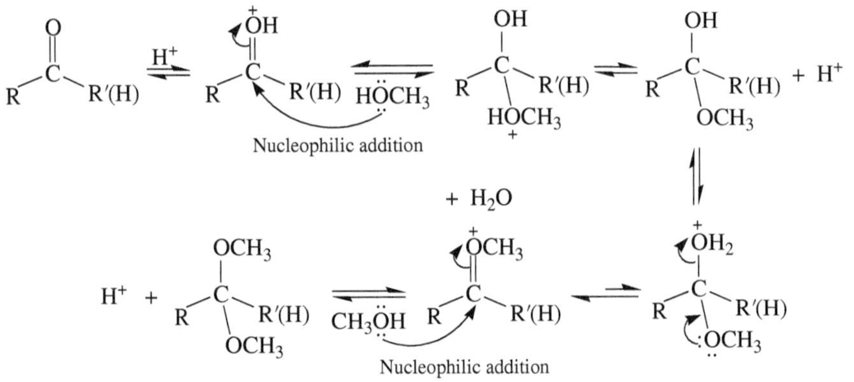

FIGURE 8.5   Mechanism for acid catalyzed nucleophilic addition of methanol to a ketone/ aldehyde giving a hemiketal/hemiacetal and a ketal/acetal.

aldehyde to give a metastable intermediate hemiketal or hemiacetal. In the following step, the hydroxyl –OH group in the hemiketal or hemiacetal intermediate is protonated. Then the active lone-pair of electrons in the $-OCH_3$ oxygen moves down to the carbon–oxygen bond domain in $C-OCH_3$ to displace a water molecule and lead to the formation of a positively charged $C=O^+CH_3$ double bond, analogous to the protonated carbonyl. Finally, the positively charged carbonyl in $C=O^+CH_3$ undergoes a facile nucleophilic addition with a second CH_3OH molecule to give a ketal or an acetal. The elimination of an H_2O molecule from the protonated hemiketal or hemiacetal serves as the major thermodynamic driving force for the overall product-favored reaction.

Figure 8.6 shows some examples for acid-catalyzed nucleophilic additions of alcohols (including diols) to ketones or aldehydes to lead to the formations of ketals or acetals via intermediate hemiketals or hemiacetals. For the reactions of diols, the two hydroxyl –OH groups in each diol are added consecutively to the carbonyl carbon in the starting ketone/aldehyde and then to the anomeric carbon of the hemiketal/hemiacetal to give the final cyclic ketal or acetal product. It is noteworthy that the conversion of hemiketal/hemiacetal to the final cyclic ketal or acetal product takes place intramolecularly with a favorable entropy. This makes the reactions of diols with ketones or aldehydes thermodynamically more favorable than the reactions of regular alcohols.

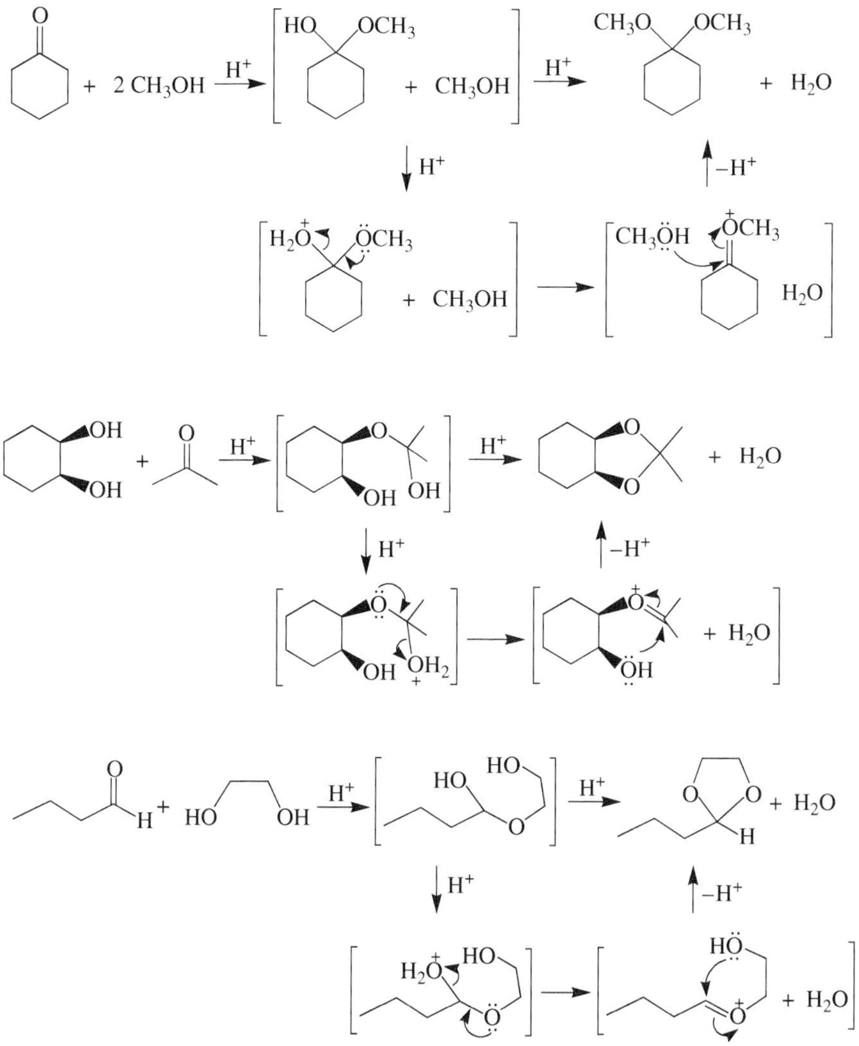

FIGURE 8.6 Acid catalyzed nucleophilic addition reactions of various alcohols with ketones and aldehydes.

In organic synthesis, the formation of a ketal or acetal can be used effectively for carbonyl group protection. The selective reduction of an ester-aldehyde in Figure 8.7 is achieved by this approach. Direct treatment of the ester-aldehyde by $LiAlH_4$ will result in reduction of both functional groups to hydroxyls. In order to only reduce the ester group selectively and maintain the identity of the aldehyde group, the ester-aldehyde substrate is first treated by CH_3OH in the presence of acid to convert the aldehyde group to an acetal. Then the intermediate is treated by $LiAlH_4$. Only

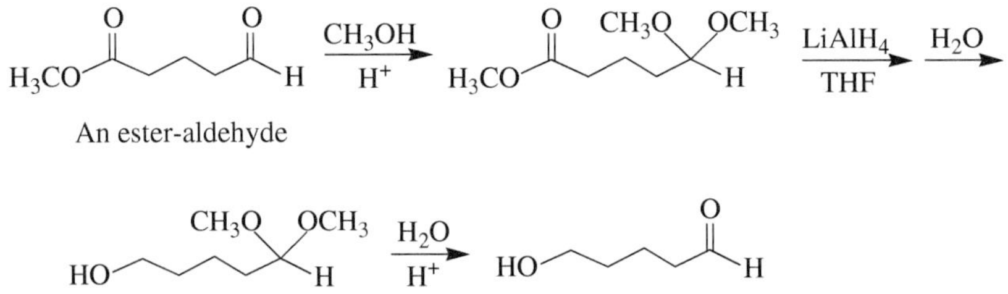

An ester-aldehyde

FIGURE 8.7 Acid catalyzed reaction of an ester-aldehyde with methanol, followed by reduction with LiAlH$_4$: The carbonyl group protection.

the ester group is reduced to a hydroxyl, but the acetal remains intact. Then acid catalyzed hydrolysis of the acetal is performed to regenerate the aldehyde. Mechanistically, the hydrolysis of an acetal to an aldehyde follows the same pathway as that for nucleophilic addition of an alcohol to an aldehyde in the reversed direction.

The carbonyl group protection by the formation of a ketal is also employed in synthesis of a specialty *cis*-unsaturated ketone by catalytic hydrogenation of an alkyne precursor (refer to Section 6.8.1 and see Fig. 6.26).

Upon heating in the presence of acid, a ketal or acetal undergoes an E1 elimination giving an alkoxy alkene (Eq. 8.7).

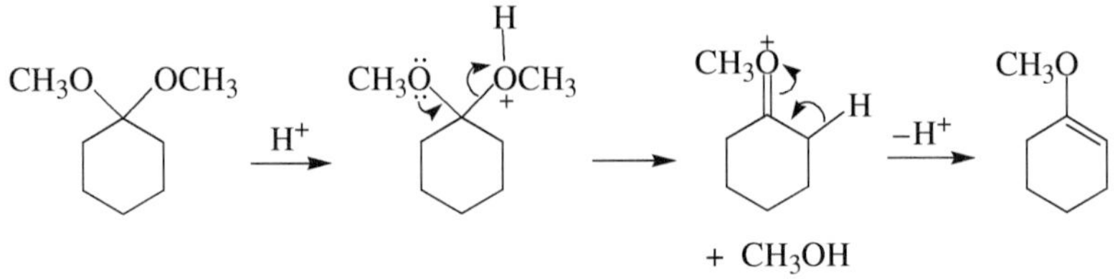

$$+ CH_3OH \tag{8.7}$$

6-Hydroxyl-2-heptanone contains both carbonyl and hydroxyl groups in the same molecule. In the presence of acid, an intramolecular nucleophilic addition of the hydroxyl to the carbonyl takes place to lead to the formation of a six-membered cyclic hemiketal (Fig. 8.8). The product is stabilized by a favorable small entropy evolved in the intramolecular reaction. In addition, the cyclic transition state for the rate-determining step (the addition of –OH to the protonated C=O) is also stabilized by a very small $\Delta S^{\ddagger}$. As a result, the formation of the six-membered cyclic hemiketal from the linear hydroxyl ketone is both thermodynamically and kinetically favorable.

8.2.3 Biological Applications: Cyclic Structures of Carbohydrates

Nucleophilic addition of the hydroxyl group (–OH) to carbonyl (C=O) giving a hemiacetal (or hemiketal) plays a critical role in carbohydrate chemistry. It is responsible for the formation of cyclic structures for many monosaccharides (simple sugars). First, let us examine the mutarotation for D-glucose.

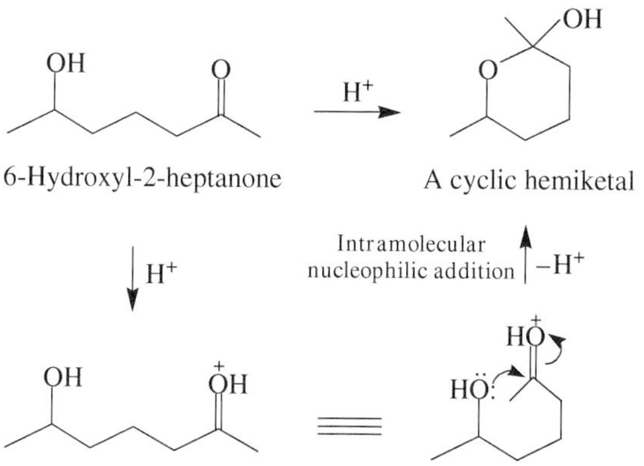

6-Hydroxyl-2-heptanone   A cyclic hemiketal

Intramolecular
nucleophilic addition $\vert -H^+$

$\vert H^+$

FIGURE 8.8   The intramolecular nucleophilic addition of 6-hydroxyl-2-heptanone giving a stable cyclic hemiketal: favorable entropy effect on an intramolecular reaction.

Figure 8.9 shows that the linear-chain form of D-glucose undergoes a reversible ring-closure and ring-opening process via an intramolecular nucleophilic attack of C_5–OH on the carbonyl (C=O) group [3]. This intramolecular nucleophilic addition is product-favored due to a favorable entropy effect. The entropy of the transition state ($\Delta S^{\ddagger}$) is also very small, making the reaction kinetically favorable as well. In an aqueous solution, 97% glucose molecules exist in the form of six-membered rings (α- and β-D-glucopyranoses). Only 3% exist in the form of the linear-chain molecules.

In the linear-chain form of D-glucose, free rotation about the C_1–C_2 bond occurs at ambient temperature. It leads to the formation of two configurations of cyclic D-glucopyranoses: When the carbonyl oxygen at the C_1-carbon of the linear-chain form is pointing downward, the nucleophilic attack of C_5–OH on the carbonyl (C=O) group gives α-D-glucopyranose, with the C_1–OH hydroxyl going downward and staying in the opposite side of the –CH_2OH of C_5-carbon. In the event that the carbonyl oxygen at the C_1-carbon of the linear-chain form is rotated upward, the nucleophilic attack of C_5–OH on the carbonyl (C=O) group gives β-D-glucopyranose, with the C_1–OH hydroxyl being positioned upward and staying in the same side of the –CH_2OH of C_5-carbon. Since the ring-opening and ring-closure effected by the nucleophilic addition is reversible and the rotation about the C_1–C_2 bond in the linear D-glucose is rapid, there is an interconversion between α-D-glucopyranose and β-D-glucopyranose via the linear-chain form of the molecule. This process is called **mutarotation**. β-D-Glucopyranose is more stable than α-D-glucopyranose because in the most favorable chair conformation, β-D-glucopyranose has all the –OH and –CH_2OH groups stay in the equatorial positions, while in α-D-glucopyranose the C_1–OH stays in an axial position. When equilibrium for mutarotation of D-glucose is reached at ambient conditions, the contents of β-D-glucopyranose and α-D-glucopyranose are 64% and 36%, respectively, owing to their relative stability.

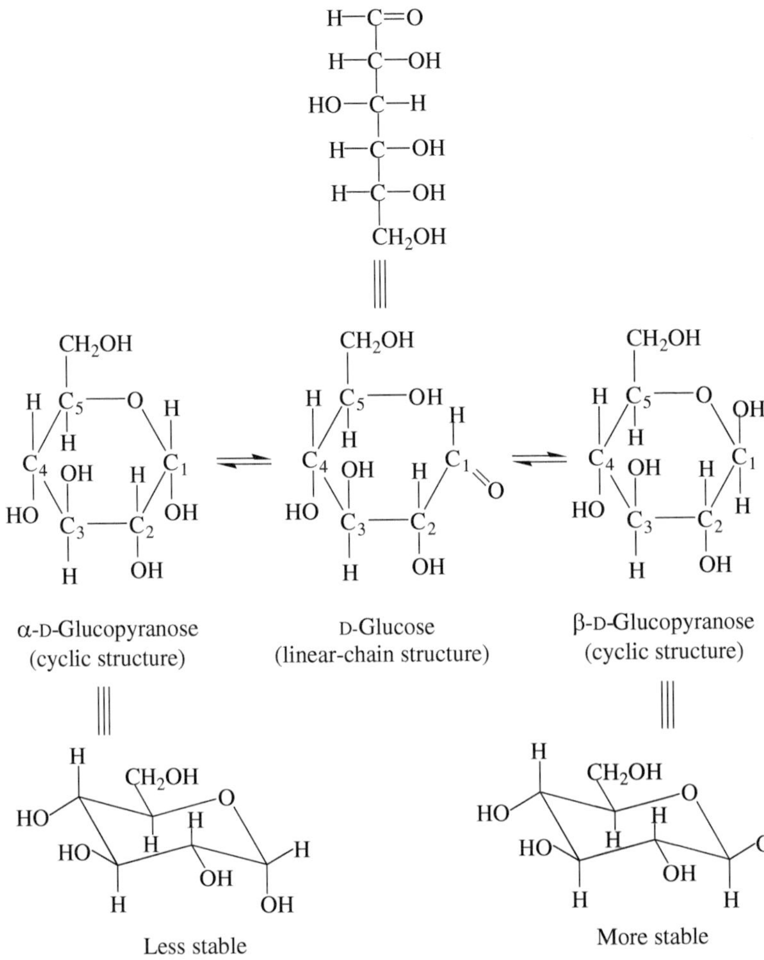

FIGURE 8.9 Cyclic structures and mutarotation of D-glucose: the intramolecular nucleophilic addition of C_5–OH to the carbonyl group of D-glucose.

D-Fructose undergoes analogous mutarotation to that of D-glucose in aqueous solution (Fig. 8.10) [3]. C_5–OH in the linear-chain form of D-fructose attacks the carbonyl (C=O) group resulting in a ring-closure and ring-opening equilibrium. The rapid free rotation about the C_2–C_3 bond moves the carbonyl oxygen at C_2-carbon of the linear-chain form downward and upward momentarily. When the carbonyl oxygen stays downward, the ring-closure via the nucleophilic attack of C_5–OH on carbonyl gives α-D-fructofuranose, with the C_2–OH hydroxyl and the –CH_2OH group of C_5-carbon staying in the opposite sides. As the carbonyl oxygen is rotated upward, the ring-closure reaction gives β-D-fructofuranose, with C_2–OH

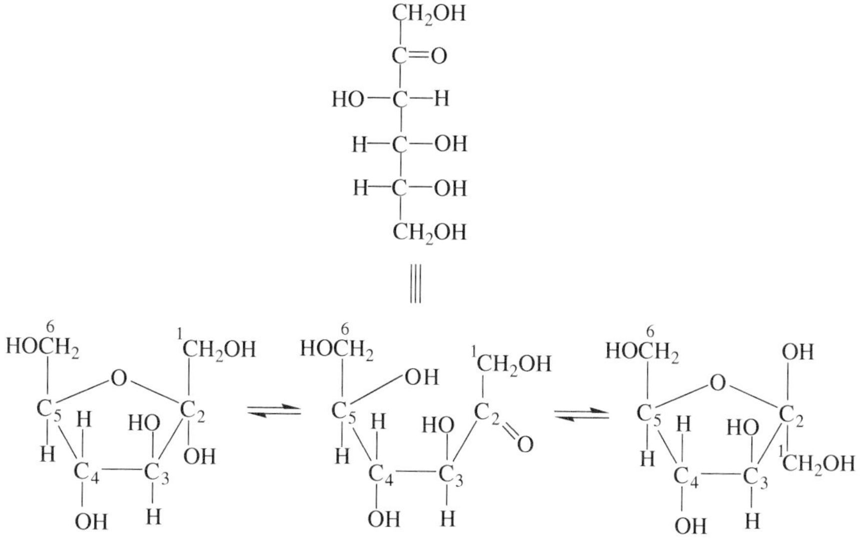

α-D-Fructofuranose (cyclic structure) | D-Fructose (linear-chain structure) | β-D-Fructofuranose (cyclic structure)

FIGURE 8.10 Cyclic structures and mutarotation of D-fructose: the intramolecular nucleophilic addition of C_5–OH to the carbonyl group of D-fructose. Modified from Voet et al. [3].

and the $-CH_2OH$ group of C_5-carbon staying in the same side. Similar to the mutarotation of D-glucose, there is a rapid interconversion between α-D-fructofuranose and β-D-fructofuranose via the linear D-fructose molecule, with the equilibrium favoring β-D-fructofuranose at ambient conditions.

All other aldohexoses (six-carbon aldehyde sugars), aldopentoses (five-carbon aldehyde sugars), ketohexoses (six-carbon ketone sugars), and ketopentoses (five-carbon ketone sugars) undergo similar ring-closure and ring-opening equilibrium and related mutarotations. They all assume cyclic structures as the major forms at ambient conditions. Some examples are shown in Figure 8.11. In particular, the cyclic D-ribose is found in RNA molecules, and the cyclic D-galactose is found in lactose (milk sugar).

8.2.4 Addition of Sulfur Nucleophile to Aldehydes

Different from hydration of aldehydes that is usually reactant-favored, and from addition of an alcohol to aldehydes that gives unstable hemiacetals, reaction of an aldehyde with $NaHSO_3$, a sulfur nucleophile, gives a stable addition product (Reaction 8.8).

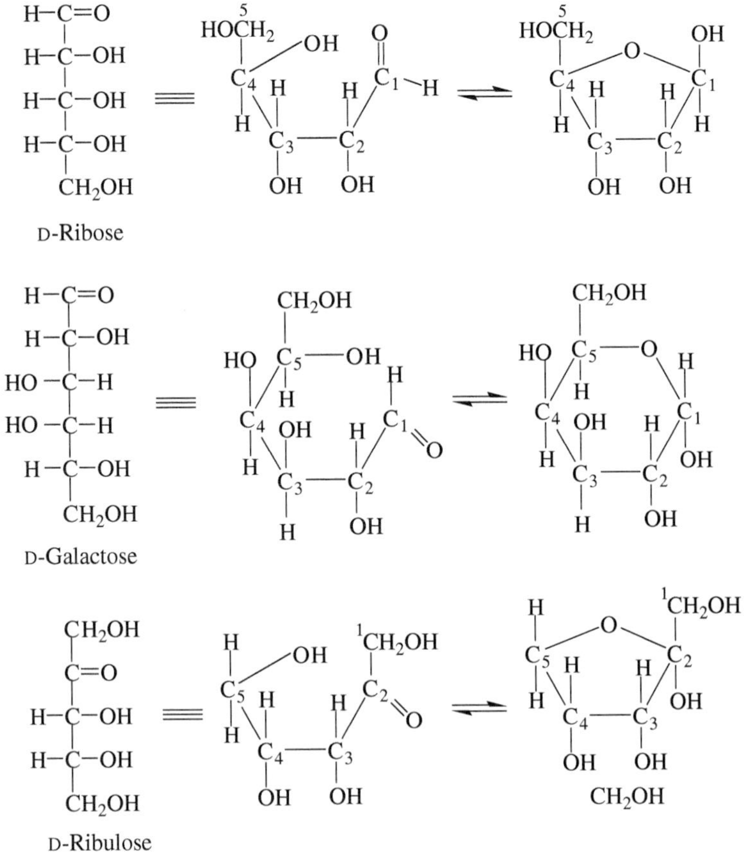

D-Ribose

D-Galactose

D-Ribulose

FIGURE 8.11 Formations of cyclic structures of five-carbon and six-carbon sugars (monosaccharides) by intramolecular nucleophilic additions of hydroxyl to carbonyl in the sugar molecules.

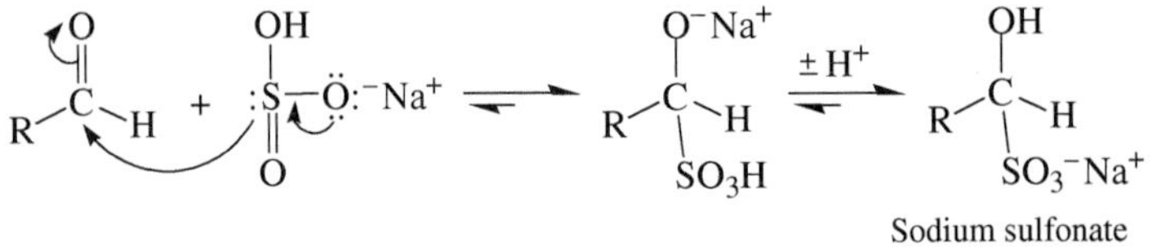

$$(8.8)$$

Sodium sulfonate

Since a sulfur nucleophile is in general stronger than oxygen nucleophiles, Reaction 8.8 does not require a catalyst. In addition, the second step of acid–base reaction makes the overall nucleophilic addition strongly product-favored (although still reversible). The sodium sulfonate product is water soluble. This reaction can be used to separate an aldehyde from other water insoluble organic compounds by converting the aldehyde to a water soluble salt. Since Reaction 8.8 is still reversible (although product-favored), treatment of the sodium sulfonate product with an acid or a base will convert it back to the original aldehyde (Reaction 8.9).

$$(8.9)$$

First, sodium sulfonate is equilibrated to an aldehyde and $NaHSO_3$ to a small extent. $NaHSO_3$ subsequently reacts with HCl or Na_2CO_3 irreversibly. It drives a complete conversion of sodium sulfonate to the original aldehyde eventually.

This reaction is only applicable to aldehydes and some methyl ketones. For other ketones, steric hindrance from the alkyl groups will prevent the nucleophilic addition reaction (Reaction 8.8) from happening.

8.2.5   Nucleophilic Addition of Amines to Ketones and Aldehydes

Primary amines (RNH_2) are moderately strong nitrogen nucleophiles owing to the particularly active lone-pair of electrons in the trivalent nitrogen. In the presence of catalytic amount of acid, they undergo nucleophilic additions to ketones and aldehydes to give imines that contain a C=N double bond (Eq. 8.10).

$$(8.10)$$

Different from the reactions of oxygen nucleophiles (water and alcohols), the reaction of a primary amine with a ketone or aldehyde has a net substitution of the oxygen by the amine nitrogen. This reactivity owes to the particular activity of the lone-pair of electrons in nitrogen. The overall reaction is reversible. If the water by-product is removed from the reaction system (by distillation or a desiccant such as molecular sieves) once formed, the reaction will become irreversible [4]. As a result, a quantitative conversion of the reactants to the imine product will be achieved.

Figure 8.12 shows the mechanism of the reaction of a primary amine with a ketone or aldehyde. The carbon–nitrogen σ bond in the imine product is formed initially by a nucleophilic addition of RNH_2 to the carbonyl group of ketone/aldehyde. This is the rate-determining step for the overall reaction and catalyzed by acid (via protonation on the carbonyl). It is followed by a proton transfer from nitrogen to oxygen, which is relayed by molecules of the protic medium. Then the active

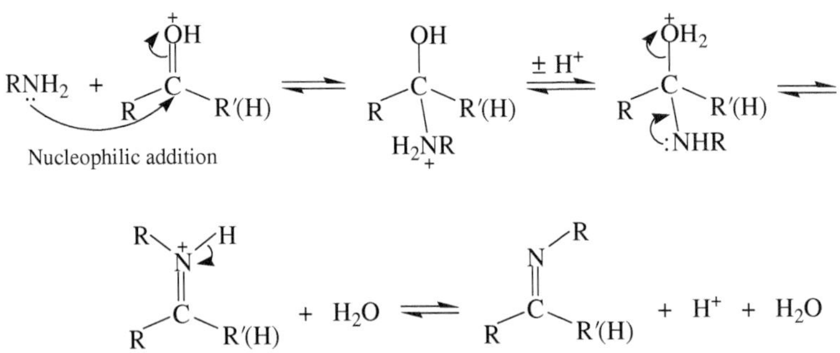

FIGURE 8.12   Mechanism for acid catalyzed nucleophilic addition of an amine to a ketone or aldehyde.

lone-pair of electrons in nitrogen moves down to the carbon—nitrogen bond domain to displace a water molecule and lead to the formation of a C=N double bond in the imine product. Since the overall reaction is reversible, treatment of an imine with excess water in the presence of acid leads to hydrolysis of the imine to a primary amine and a ketone or aldehyde. The backward hydrolysis follows the same pathway as that for the amine nucleophilic addition in the reversed direction.

This type of reactions is not restricted to the primary amines. Almost all the species containing an amino $-NH_2$ group can react with ketones and aldehydes to form a C=N double bond in the product (Fig. 8.13). All the reactions in Figure 8.13 follow the same mechanism as demonstrated in Figure 8.12. Particularly interesting is the reaction of 2,4-dinitrophenylhydrazine ($ArNHNH_2$) with a ketone or aldehyde to give a solid 2,4-dinitrophenylhydrazone ($ArNHN=CRR'$) [4]. This reaction can be used to identify aldehydes and ketones.

Reaction of a secondary amine (including acyclic and cyclic secondary amines) with a ketone or aldehyde gives an enamine (Fig. 8.14). In general, this type of reactions is also acid-catalyzed and reversible. Therefore, treatment of an enamine by excess water in acidic conditions leads to the backward hydrolysis to a secondary amine and a ketone or aldehyde. The reaction mechanism is analogous to that for the reaction of a primary amine. In the final step of the reaction, the positively charged $C=N^+$ group activates the α-hydrogen, bringing about a C—H bond cleavage (deprotonation) to lead to the formation of a C=C double bond between the α and β carbons of the enamine.

Biological application   Nucleophilic addition of amines to the ketone carbonyl group giving imines (Schiff bases) is found in conversion of fructose-1,6-bisphosphate (FBP) (a six-carbon compound) to dihydroxyacetone phosphate (DHAP) and glyceraldehyde-3-phosphate (GAP), two three-carbon compounds (Reaction 8.11) [3].

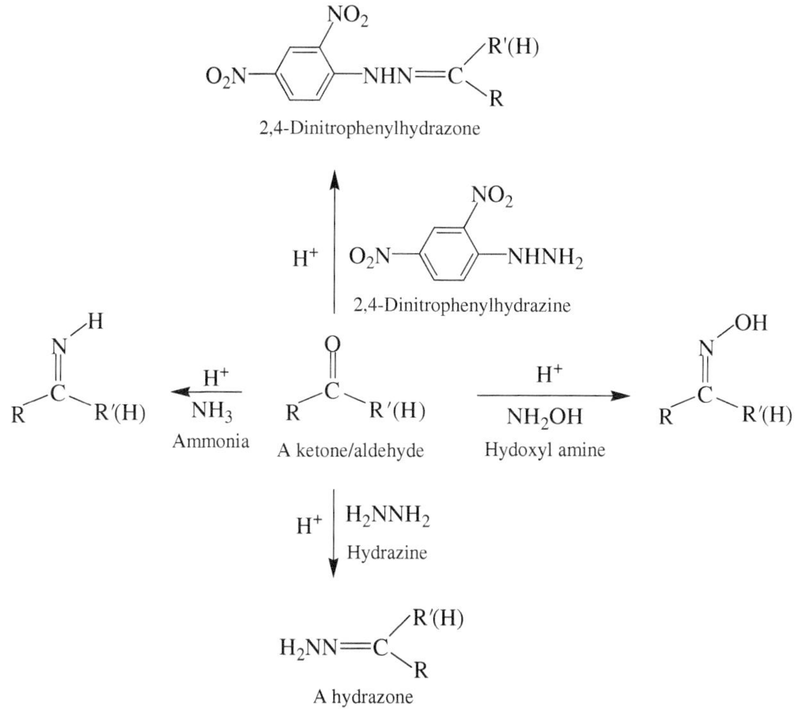

FIGURE 8.13 Reactions of a ketone or aldehyde to various compounds containing an amino ($-NH_2$) group.

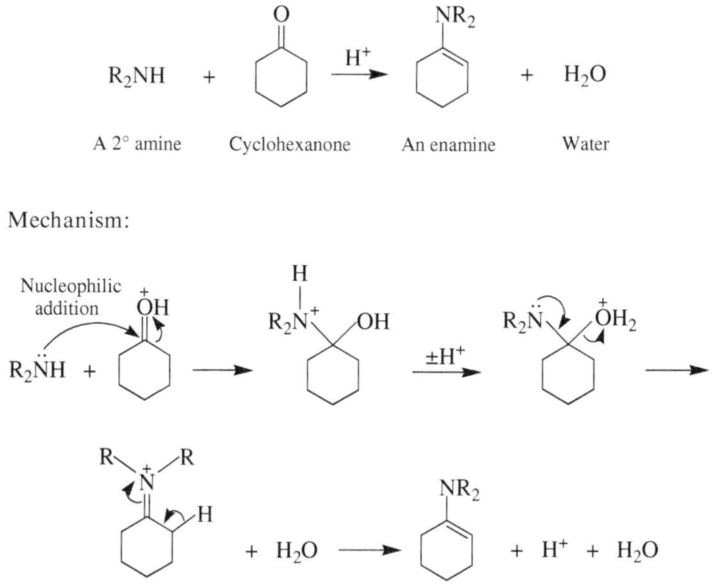

FIGURE 8.14 Mechanism for reaction of a secondary amine with a ketone (cyclohexanone) to give an enamine.

$$
\begin{array}{c}
\underset{|}{CH_2OPO_3^{2-}} \\
\underset{|}{C=O} \\
HO-\underset{|}{C}-H \\
H-\underset{|}{C}-OH \\
H-\underset{|}{C}-OH \\
CH_2OPO_3^{2-} \\
FBP
\end{array}
\quad
\overset{\text{Aldolase}}{\underset{\rightleftharpoons}{\quad\quad}}
\quad
\begin{array}{c}
\underset{|}{CH_2OPO_3^{2-}} \\
\underset{|}{C=O} \\
HO-CH_2 \\
\\
DHAP
\end{array}
\quad + \quad
\begin{array}{c}
H-\underset{|}{C}=O \\
H-\underset{|}{C}-OH \\
CH_2OPO_3^{2-} \\
\\
GAP
\end{array}
\quad (8.11)
$$

Reaction 8.11 is one of the ten metabolic steps in glycolysis (*in vivo* catabolism of glucose). It is catalyzed by aldolase. The overall enzymatic mechanism is shown in Figure 8.15 [3].

Figure 8.15 shows that the active site of the aldolase enzyme consists of the amino group ($-NH_2$) of Lys-229 side chain and the carboxylate group ($-COO^-$) of the Asp-33 side chain in the polypeptide chain of the enzyme. The first step in the reaction mechanism is that the FBP substrate binds to the enzyme in its active site to form an enzyme-substrate complex. Then, the nucleophilic Lys-229 amino group attacks the carbonyl group of FBP to effect an addition giving a protonated imine (Schiff base) which is covalently attached to the polypeptide chain of the enzyme. The nucleophilic addition follows the same mechanism as that presented in Figure 8.12. The protonated $C=N^+H$ functional group in the imine weakens the C—C bond attached to it by withdrawing the bonding electron density toward the positively charge nitrogen. As a result, the C—C bond is cleaved subsequently. A three-carbon enamine is formed and attached to the enzyme. Simultaneously, the O—H bond on the β-carbon to the C=N group is weakened by the hydrogen bond formed on the carboxylate of Asp-33, and as the proton from the O—H bond is being transferred to the Asp-33 carboxylate, its bonding electron-pair is moving to the C—O bond domain making a π bond and leading to the formation of GAP, which subsequently departs from the enzyme. Then, the C=C bond in the enamine is protonated by the Asp-33 carboxyl group (electrophilic addition) resulting in a protonated, three-carbon imine. Finally, the three-carbon imine is hydrolyzed to give DHAP and regenerate a free enzyme.

8.2.6 Nucleophilic Additions of Hydride Donors to Aldehydes and Ketones: Organic Reductions and Mechanisms

In sodium borohydride ($NaBH_4$) and lithium aluminum hydride ($LiAlH_4$), each of the B—H and Al—H bonds is nucleophilic and acts as a hydride ($H:^-$) donor. Both $NaBH_4$ and $LiAlH_4$ can react with an electrophilic ketone or aldehyde to reduce the carbonyl compound to a secondary or primary alcohol.

In general, the reaction of $NaBH_4$ is conducted in a protic medium (typically, methanol or ethanol) (Fig. 8.16). In the first step of the reaction, a nucleophilic B—H bond in BH_4^- attacks the carbonyl group in the ketone or aldehyde substrate

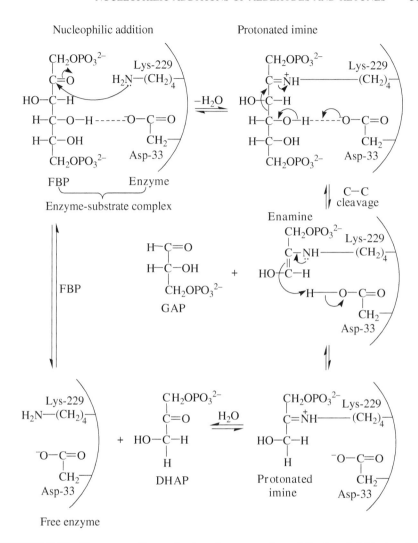

FIGURE 8.15 The enzymatic mechanism for conversion of fructose-1,6-bisphosphate (FBP) to dihydroxyacetone phosphate (DHAP) and glyceraldehyde-3-phosphate (GAP) via a protonated imine formed by nucleophilic addition of an amino (–NH$_2$) group to carbonyl. Modified from Voet et al. [3].

resulting in the formation of a C—H bond (nucleophilic addition) and breaking of the C=O π bond. In the following step, the oxygen anion in the tetrahedral intermediate is protonated by EtOH (the solvent) to give an alcohol product. The BH$_3$ produced from the first step combines with EtO$^-$ to form an [EtOBH$_3$]$^-$ adduct.

The reaction of LiAlH$_4$ has to be conducted in an unreactive aprotic solvent (typically, Et$_2$O or THF) (Fig. 8.17). The mechanism is analogous to that for the reaction

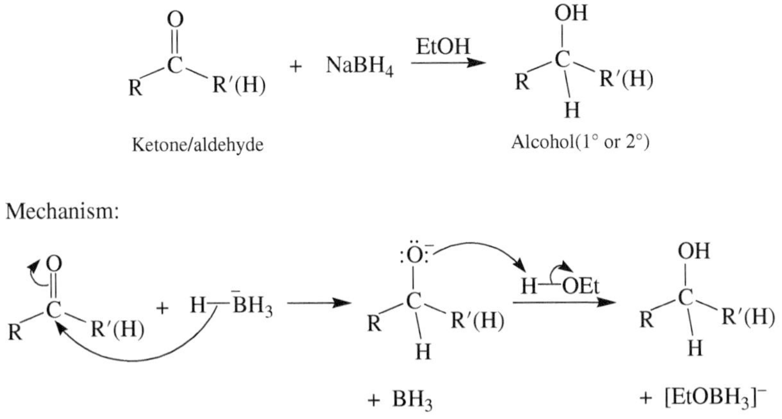

Ketone/aldehyde   Alcohol(1° or 2°)

Mechanism:

FIGURE 8.16   Mechanism for nucleophilic addition of borohydride to a ketone or aldehyde.

of NabH$_4$. It includes consecutive nucleophilic additions of four Al—H bonds from AlH$_4^-$ to the carbonyl groups of four ketone/aldehyde molecules. Finally, all the Al—O bonds are hydrolyzed in the process of aqueous work-up to give four alcohol molecules.

It is noteworthy that LiAlH$_4$ is a strong hydride donor and can reduce all types of carbonyl compounds. However, NaBH$_4$ is a relatively weak hydride donor. It can only reduce ketones and aldehydes and does not react with esters, carboxylic acids, and amides.

8.3   BIOLOGICAL HYDRIDE DONORS NAD(P)H AND FADH$_2$

Oxidation–reduction reactions in many metabolic processes of living organisms are carried out by coenzymes. The most important redox coenzymes are **nicotinamide adenine dinucleotide (phosphate) [NAD(P)$^+$, oxidized form]** and the corresponding **NAD(P)H (reduced form)**, and **flavin adenine dinucleotide (FAD, oxidized form)** and the corresponding **FADH$_2$ (reduced form)**. In the presence of suitable enzymes, NAD(P)$^+$ and FAD oxidize various biological substrates. On the other hand, the reduced coenzymes NADP(H) and FADH$_2$ act as biological hydride donors and initiate reduction.

Figure 8.18 shows structures of NAD(P)$^+$ and the corresponding reduced coenzyme NADP(H) [3]. The functioning portion of the coenzymes is the nicotinamide ring, while the large R group allows the coenzymes to bind to the active site of a suitable enzyme so that they can carry out biological oxidation–reduction reactions. In NADP(H), one of the hydrogen atoms in the *para*-carbon to nitrogen of the nicotinamide ring is activated by the nitrogen lone-pair of electrons via conjugation

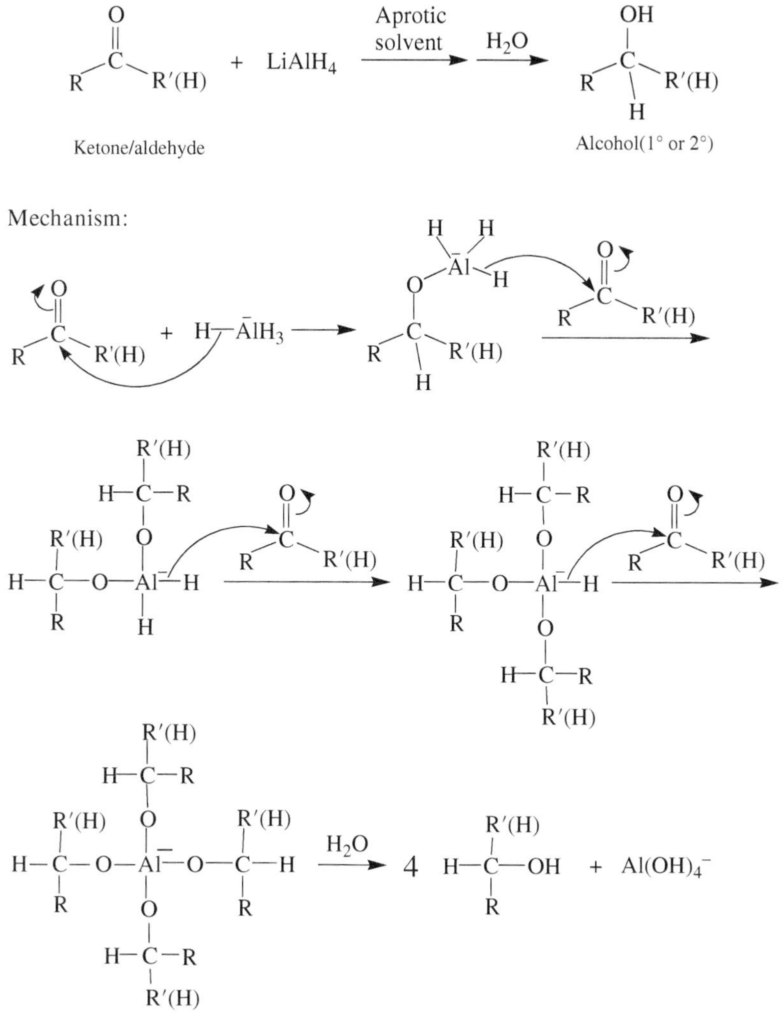

Mechanism:

FIGURE 8.17 Mechanism for nucleophilic addition of aluminum hydride to a ketone or aldehyde.

effect and is nucleophilic, readily donated as a hydride (H:$^-$) to an electrophilic center (typically, a carbonyl group) in a biological substrate. As a result, the biological substrate is reduced and NADP(H) will be in turn oxidized to NAD(P)$^+$. If the reactive center of a biological substrate is prochiral, the reduction by NADP(H) in the presence of enzyme is stereospecific. The biochemistry of NADH is illustrated by homolactic fermentation of glucose.

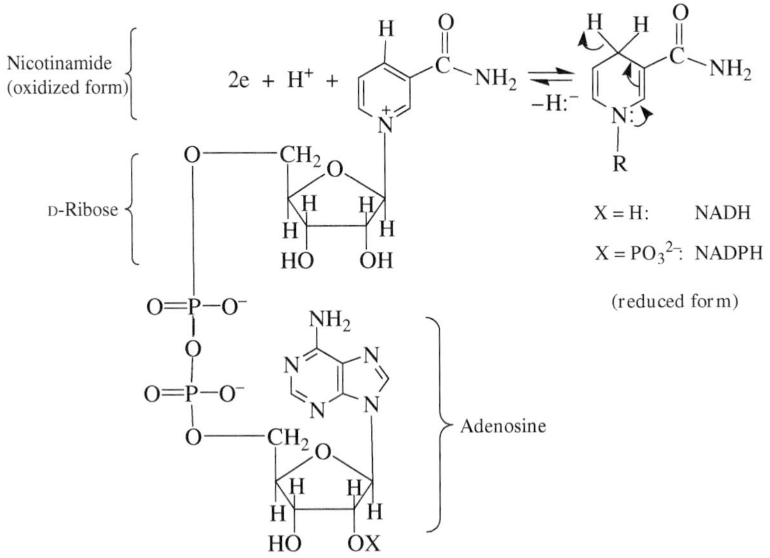

$X = H$:   Nicotinamide adenine dinucleotide (NAD^+)

$X = PO_3^{2-}$: Nicotinamide adenine dinucleotide phosphate ($NADP^+$)

(oxidized form)

FIGURE 8.18   Structures of $NAD(P)^+$ and $NAD(P)H$ (biological hydride donor).

Pyruvate (CH_3COCOO^-) is the ending product of glycolysis (anaerobic catabolism of glucose *in vivo*). In muscle, during vigorous activity, when the demand for ATP is high and oxygen is in short supply, ATP is largely synthesized via anaerobic glycolysis, together with the production of reduced coenzyme NADH. Under anaerobic conditions, the accumulated NADH in the working muscle will be oxidized back to NAD^+ by pyruvate, which is catalyzed by lactate dehydrogenase (LDH), and pyruvate is reduced to L-lactate (Reaction 8.12).

$$\underset{\text{Pyruvate}}{CH_3\overset{O}{\overset{\|}{C}}COO^-} + NADH + H^+ \underset{\text{LDH}}{\rightleftharpoons} \underset{\text{L-Lactate}}{CH_3\overset{OH}{\overset{|}{C}H}COO^-} + NAD^+ \qquad (8.12)$$

Reaction 8.12 is referred to as **homolactic fermentation** of glucose.

The mechanism for homolactic fermentation is shown in Figure 8.19 [3]. NADH is bound to the polypeptide chain of the enzyme (LDH) via the R moiety. The active site of LDH consists of the side chains of His-195 and Arg-171 residues in the polypeptide chain of the enzyme. The pyruvate substrate is bound to the active site of LDH via a hydrogen bond and ionic bonding and is held in place toward a reactive hydride in the NADH molecule. The hydrogen bond exerted by the imidazole ring of His-195 to the carbonyl oxygen in pyruvate enhances the electrophilicity of the

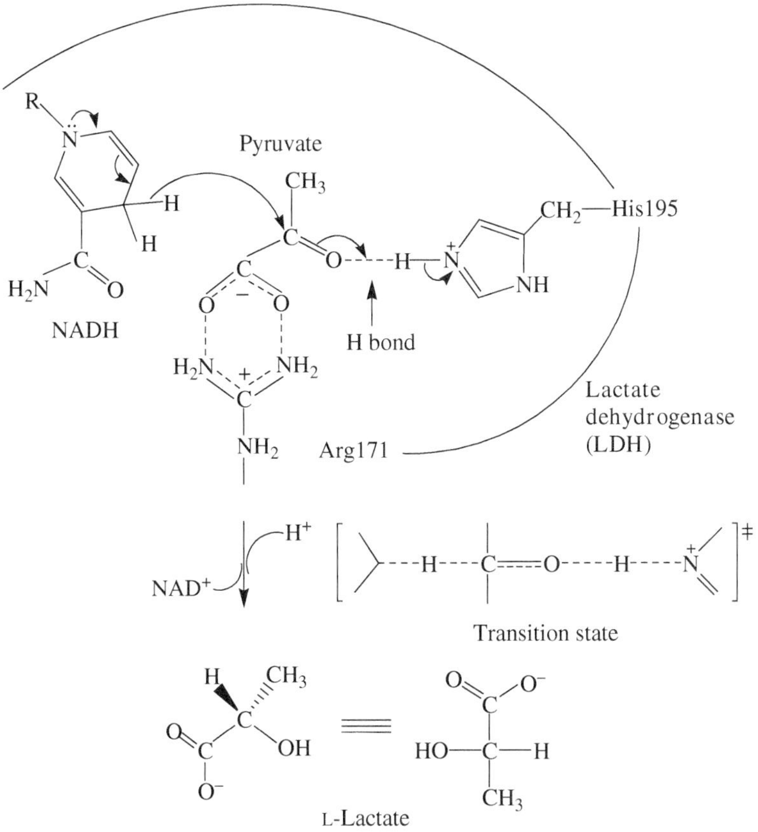

FIGURE 8.19 Mechanism for lactate dehydrogenase (LDH) catalyzed homolactic fermentation. Modified from Voet et al. [3].

carbonyl group. As a result, the carbonyl carbon can be readily and effectively attacked by the reactive hydride of NADH effecting a concerted nucleophilic addition. Pyruvate is reduced to L-lactate, and concurrently, NADH is oxidized to NAD$^+$. The attack of NADH hydride to the carbonyl group of pyruvate is face-specific: The structure of the enzyme only allows the nucleophilic attack to take place from the front, giving L-lactate. The D-configuration of lactate is not formed.

FAD and FADH$_2$ (Fig. 8.20) have similar structures to those of NAD$^+$ and NDAH. The functioning portion is the conjugated ring system. In the presence of a suitable enzyme, the reduced coenzyme FADH$_2$ donates two hydrogen atoms consecutively to electrophilic biological substrates. While the biological substrate is reduced, FADH$_2$ is oxidized to FAD via an unstable FADH˙ radical intermediate [3]. The overall half-reaction for FADH$_2$/FAD is shown in Equation 8.13.

$$FADH_2 \rightarrow FAD + 2H^+ + 2e \qquad (8.13)$$

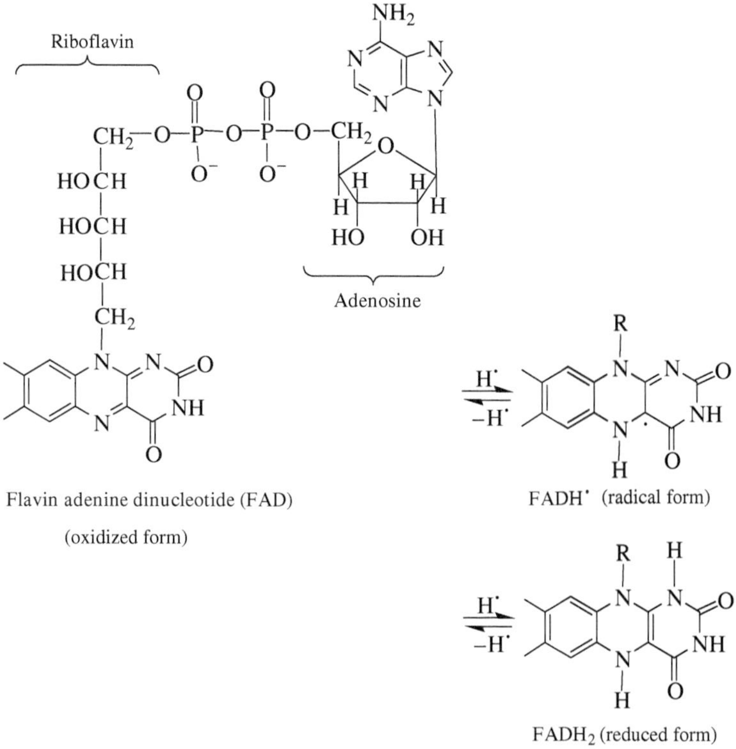

FIGURE 8.20   Structures of FAD and $FADH_2$ (biological hydride donor).

8.4   ACTIVATION OF CARBOXYLIC ACIDS VIA NUCLEOPHILIC SUBSTITUTIONS ON THE CARBONYL CARBONS

Carboxylic acids can be converted to the more reactive acyl compounds such as acyl chlorides, esters, and anhydrides. As a result, the carbonyl groups of the acids will be activated and can undergo more diversified reactions. In this section, we discuss the methodology and mechanisms for the conversions.

8.4.1   Reactions of Carboxylic Acids with Thionyl Chloride

All the carboxylic acids undergo facile reactions with thionyl chloride ($SOCl_2$) to produce acyl chlorides (Fig. 8.21). The net result for the reaction is that the hydroxyl group in the acid is replaced by a chloride from $SOCl_2$. In addition, SO_2 and HCl are formed as by-products.

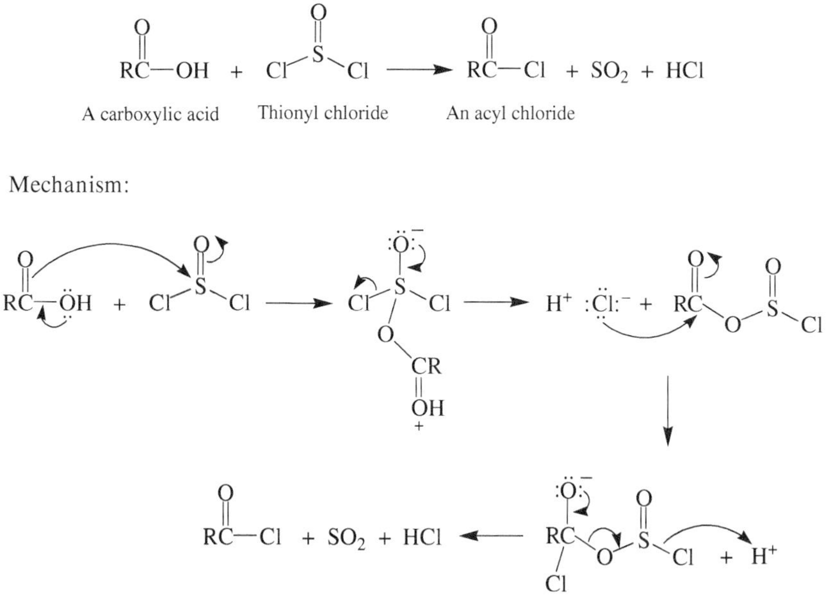

Mechanism:

FIGURE 8.21   Reaction of a carboxylic acid with thionyl chloride.

The reaction is initiated by nucleophilic attack of the carbonyl oxygen in the acid on the electrophilic sulfoxide (S=O) group in $SOCl_2$. Then a chloride anion is displaced from the first intermediate in Figure 8.21. In the next step, the carbonyl group of the second intermediate in Figure 8.21 is attacked by the nucleophilic chloride to lead to displacement of $-OS(O)Cl$ (a very good leaving group) via a tetrahedral intermediate. This is a nucleophilic acyl substitution, resulting in the formation of an acyl chloride and giving off SO_2 and HCl concurrently.

Chlorination by $SOCl_2$ takes place in various types of compounds that contain hydroxyl (–OH) by replacing the –OH with a chloride from $SOCl_2$. For example, reaction of an alcohol (ROH) with $SOCl_2$ leads to the formation of an alkyl chloride (RCl), and reaction of p-toluenesulfonic acid (TsOH) with $SOCl_2$ gives tosyl chloride (TsCl). Both reactions follow the analogous mechanism to that for chlorination of carboxylic acids by $SOCl_2$.

8.4.2   Esterification Reactions, Synthetic Applications, and Green Chemistry Methods

In the presence of a strong or moderately strong acid (such as sulfuric, p-toluenesulfonic, or phosphoric acid) in a catalytic amount, a carboxylic acid reacts with methyl, a primary or a secondary alcohol giving an ester (Fischer esterification). The overall reaction and its mechanism are shown in Figure 8.22.

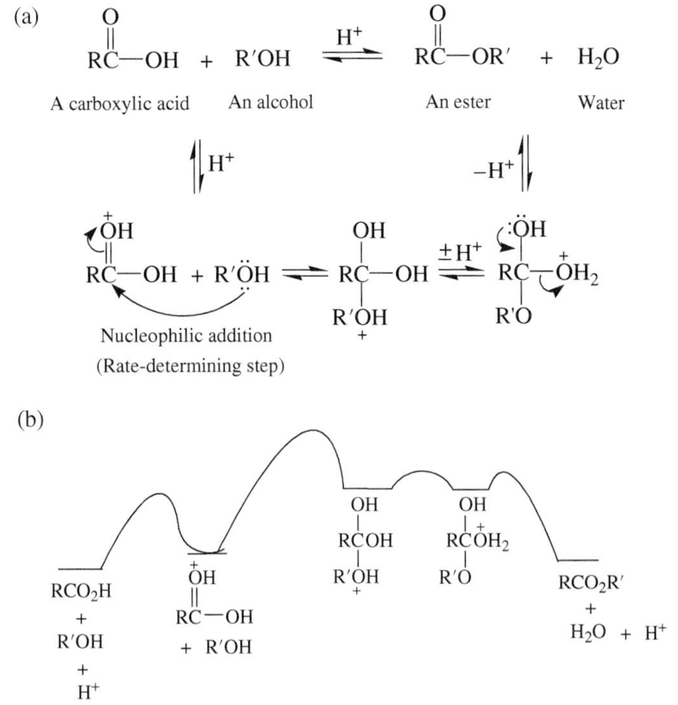

(a)

(b)

FIGURE 8.22   (a) Fischer esterification and reaction mechanism; and (b) reaction profile for the overall esterification.

Fischer esterifications are reversible with the equilibrium constants being 5–10 for most of carboxylic acids and alcohols. In order to achieve the quantitative conversion, one way is to use one of the reactants, typically the alcohol, in large excess. Alternatively, if the reactants are maintained in stoichiometric ratio, but water is removed from the system once it is formed, a quantitative conversion can also be achieved. For example, for reaction of benzoic acid ($PhCO_2H$) with methanol (MeOH) giving methyl benzoate ($PhCO_2Me$), methanol is used in more than 10-folds of excess so that this cheap alcohol can function as both a reactant and the solvent to dissolve the solid $PhCO_2H$. Because MeOH is in large excess, all the $PhCO_2H$ is consumed and converted to the ester product in the end of the reaction. For reaction of acetic acid (CH_3CO_2H) with 3-methyl-1-butanol [$(CH_3)_2CHCH_2CH_2OH$] giving 3-methyl-1-butyl acetate [$CH_3CO_2CH_2CH_2CH(CH_3)_2$], the two reactants are used in 1:1 molar ratio. The water by-product is removed from the system by azeotropic distillation immediately after it is formed. As a result, all the reactants can be converted to the ester product in the end of the reaction.

The rate-determining step for the acid-catalyzed esterification is the nucleophilic attack on the protonated carbonyl in the carboxylic acid by the alcohol (a nucleophilic addition), giving an intermediate tetrahedral adduct (Fig. 8.22). The function of the acid catalyst (such as sulfuric acid) is to introduce a positive charge onto the carbonyl group by protonation. As a result, the electrophilicity of the carbonyl is enhanced, leading to acceleration of the rate-determining nucleophilic addition step. Then a protic-medium-relayed proton transfer from the alkoxy to hydroxyl takes place. In the final step, a water molecule is eliminated, which is facilitated by the lone-pair of electrons in the –OH oxygen.

The mechanism is further confirmed by the study of acid-catalyzed hydrolysis of O^{18}-labeled ethyl acetate ($MeCOO^{18}Et$) (refer to Section 1.10 and also see Fig. 1.18) [1]. Since hydrolysis of an ester to a carboxylic acid and an alcohol is the reversal process of the Fischer esterification, according to the principle of microscopic reversibility, the esterification should follow the same mechanism as that for the hydrolysis, but in the reversed direction. The microscopic steps in the mechanism of esterification presented in Figure 8.22 are indeed the opposite steps of the hydrolysis of ester presented in Figure 1.18.

4-Hydroxypentanoic acid contains both alcoholic –OH and carboxyl (–COOH) groups in the same molecule. Treatment of this compound with sulfuric acid leads to an intramolecular esterification giving a lactone (Fig. 8.23). A positive entropy for the overall reaction makes this reaction product-favored. The entropy of the transition state ($\Delta S^{\ddagger}$) for the rate-determining step (the intramolecular nucleophilic

4-Hydroxypentanoic acid   A lactone

Intramolecular
nucleophilic addition
(rate-deter mining step)

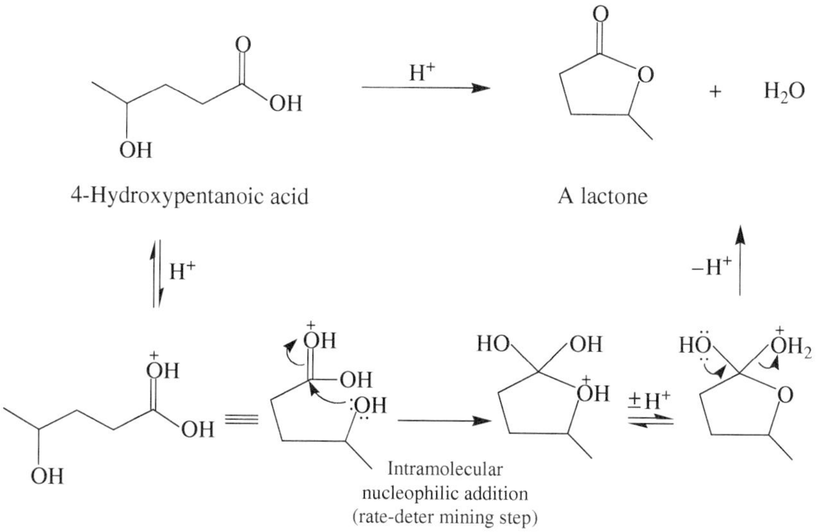

FIGURE 8.23   Formation of a lactone by the entropy-driven intramolecular esterification of 4-hydroxypentanoic acid.

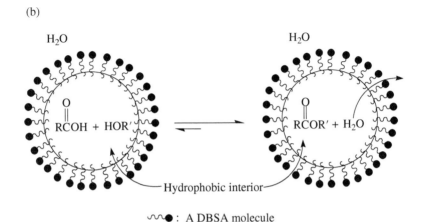

(a)

p-Dodecylbenzene sulfonic acid (DBSA)

(b)

H_2O

H_2O

RCOH + HOR′ ⇌ RCOR′ + H$_2$O

Hydrophobic interior

⌇● : A DBSA molecule

FIGURE 8.24 (a) Structure of p-dodecylbenzenesulfonic acid (DBSA); and (b) the DBSA catalyzed esterification reaction in the aqueous medium.

addition) is small, giving rise to a low activation energy for the reaction. Therefore, it is both a thermodynamically and kinetically favorable process.

Surfactant-type protic acids, such as p-dodecylbenzenesulfonic acid (DBSA), have been used to catalyze esterifications of lipophilic (water insoluble) carboxylic acids (such as lauric acid) and alcohols (such as 3-phenyl-1-propanol) in aqueous solutions [5, 6]. This has led to the development of **green chemistry methods**. DBSA contains a long hydrophobic hydrocarbon chain and a hydrophilic sulfonic acid group (Fig. 8.24a). The DBSA catalyst molecules and organic substrates (carboxylic acids and alcohols) combine in water to form emulsion droplets, which have a hydrophobic interior, through hydrophobic interactions (Fig. 8.24b). The surfactant DBSA molecules produce proton on the surface of the droplets and then enhance the rate for esterification to reach equilibrium. As illustrated in Figure 8.24b, the starting carboxylic acid and alcohol are embraced in the hydrophobic interior where they interact each other to bring about the esterification reaction. For hydrophobic substrates, the equilibrium position between the substrates and the products (esters) would lie at the ester side, because water molecules generated during the reaction will be removed from the droplets due to the hydrophobic nature of their interior. As a result, the esterification reactions can efficiently proceed even in the presence of a large amount of water as a solvent. The sulfonic acid functional group ($-\text{OSO}_2\text{OH}$) in the catalyst enhances the reaction rate by protonating the carbonyl group of the carboxylic acid, and hydrophobic interior of emulsion droplets facilitates exclusion of water molecules to shift the equilibrium to the product side and enhance the percentage of conversion.

The surfactant DBSA catalyzed esterifications work more efficiently for more lipophilic substrates as the lipophilic substrates would be more compatible with the hydrophobic interior formed by the catalyst hydrocarbon chains (Fig. 8.25) [5]. For example, reaction of 3-phenyl-1-propanol ($PhCH_2CH_2CH_2OH$) with lauric acid [$CH_3(CH_2)_{10}COOH$]) in 1:1 molar ratio in the presence of 10 mol% DBSA at 40 °C affords the corresponding ester product 3-phenyl-1-propyl laurate [$CH_3(CH_2)_{10}COOCH_2CH_2CH_2Ph$] in 63% yield. The reaction with less lipophilic butyric acid [$CH_3(CH_2)_2COOH$] gives $CH_3(CH_2)_2COOCH_2CH_2CH_2Ph$ in 31% yield. When acetic acid (hydrophilic) is used as a substrate, the yield of 3-phenylpropyl acetate $CH_3COOCH_2CH_2CH_2Ph$ is only 6%.

When a 1:1 mixture of lauric acid and acetic acid is esterified with 3-phenyl-1-propanol in the presence of DBSA in water, only $CH_3(CH_2)_{10}COOCH_2CH_2CH_2Ph$ is selectively obtained, but almost no reaction occurs to acetic acid. This selectivity is attributed to the hydrophobic nature of lauric acid and is unique. The increase in the $R'OH:RCOOH$ molar ratio to 2:1 enhances the ester product yield to more than 90% for most cases including the reaction of lauric acid and 3-phenyl-1-proponal. It should be noted that the ester formation is realized at 40 °C in water in the presence of DBSA in contrast to high temperatures that are required for conventional esterifications catalyzed by regular acids and performed in organic solvents.

It is also found that the esterification in the presence of p-toluenesulfonic acid (TsOH) as a catalyst proceeds much more slowly (Eq. 8.14) [5].

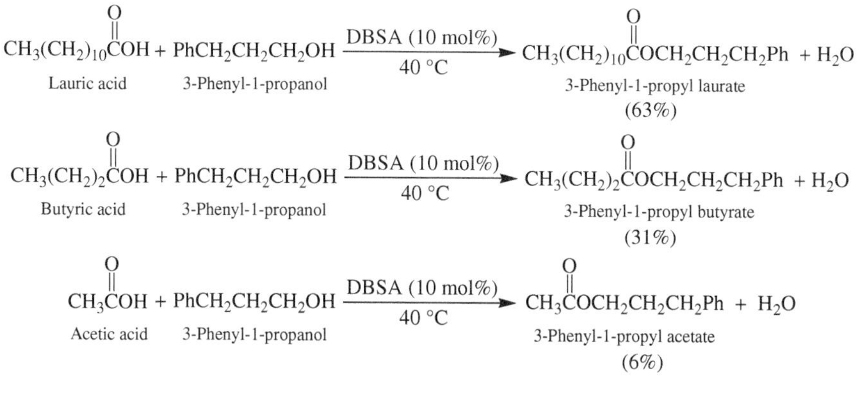

$CH_3(CH_2)_{10}\overset{\overset{\text{O}}{\|}}{C}OH + PhCH_2CH_2CH_2OH \xrightarrow[\text{40 °C}]{\text{DBSA (10 mol\%)}} CH_3(CH_2)_{10}\overset{\overset{\text{O}}{\|}}{C}OCH_2CH_2CH_2Ph + H_2O$

Lauric acid   3-Phenyl-1-propanol   3-Phenyl-1-propyl laurate
                                                                                              (63%)

$CH_3(CH_2)_2\overset{\overset{\text{O}}{\|}}{C}OH + PhCH_2CH_2CH_2OH \xrightarrow[\text{40 °C}]{\text{DBSA (10 mol\%)}} CH_3(CH_2)_2\overset{\overset{\text{O}}{\|}}{C}OCH_2CH_2CH_2Ph + H_2O$

Butyric acid   3-Phenyl-1-propanol   3-Phenyl-1-propyl butyrate
                                                                                              (31%)

$CH_3\overset{\overset{\text{O}}{\|}}{C}OH + PhCH_2CH_2CH_2OH \xrightarrow[\text{40 °C}]{\text{DBSA (10 mol\%)}} CH_3\overset{\overset{\text{O}}{\|}}{C}OCH_2CH_2CH_2Ph + H_2O$

Acetic acid   3-Phenyl-1-propanol   3-Phenyl-1-propyl acetate
                                                                                              (6%)

FIGURE 8.25 The p-dodecylbenzenesulfonic acid (DBSA) catalyzed esterification reactions of 3-phenyl-1-propanol with different carboxylic acids. Modified from Manabe et al. [5].

$$\text{CH}_3(\text{CH}_2)_{10}\overset{\overset{\displaystyle O}{\|}}{\text{C}}\text{OH} + \text{PhCH}_2\text{CH}_2\text{CH}_2\text{OH} \xrightarrow[\substack{(1)\ \text{DBSA: Fast} \\ (2)\ \text{TsOH: Slow}}]{40\ ^{\circ}\text{C}} \text{CH}_3(\text{CH}_2)_{10}\overset{\overset{\displaystyle O}{\|}}{\text{C}}\text{OCH}_2\text{CH}_2\text{CH}_2\text{Ph} + \text{H}_2\text{O}$$

Lauric acid   3-Phenyl-1-propanol   3-Phenyl-1-propyl laurate

$$(8.14)$$

According to a study on the initial rates of the esterification of lauric acid with 3-phenyl-1-propanol, DBSA catalyzes the reaction 60 times faster than does TsOH (for DBSA, the initial rate is 5.41×10^{-3} M/h; for TsOH, it is 9.05×10^{-5} M/h) [5]. These results clearly demonstrate that both the long-chain alkyl and the sulfonic acid moiety of DBSA are crucial for efficient catalysis.

Since both of the sole by-product and the solvent in the surfactant DBSA cata-lyzed esterification are water, the reactions are environmentally benign (a green chemistry method). The methodology will not only lead to a practical and greener synthetic application but also provide a new aspect of chemistry in water.

8.4.3   Formation of Anhydrides

Simple anhydrides can be made by treatment of carboxylic acids with P_2O_5, which absorbs water strongly (Eq. 8.15) [4].

$$2\ \text{R}\overset{\overset{\displaystyle O}{\|}}{\text{C}}{-}\text{OH} \xrightarrow{P_2O_5} \text{R}\overset{\overset{\displaystyle O}{\|}}{\text{C}}{-}\text{O}{-}\overset{\overset{\displaystyle O}{\|}}{\text{C}}\text{R} + \text{H}_2\text{O} \qquad (8.15)$$

A carboxylic acid   An anhydride

The reaction is a condensation between two carboxylic acid molecules. Relative to a carboxylic acid, an anhydride is much less stable. It reacts with water strongly in the presence of a catalytic amount of a strong acid to give back the starting carboxylic acid. By virtue, Reaction 8.15 is reactant-favored with a small equilibrium constant. However, P_2O_5 can remove water thoroughly from the reaction system, making the formation of an anhydride from a carboxylic acid irreversible.

Cyclic anhydrides (such as succinic anhydride) can be readily and effectively made in relatively mild conditions by ring-closure reactions of carboxylic acids (such as succinic acid) containing two carboxyl groups (Fig. 8.26). The favorable entropy effects on the overall reaction ($\Delta S > 0$) and on the transition state ($\Delta S^{\ddagger} \sim 0$) for the rate-determining step (the step of the intramolecular nucleophilic addition) make the formation of a cyclic anhydride product-favored and fast as well. Use of a catalyst is unnecessary. The first step of protonation on a carboxyl group of succinic acid is made by another succinic acid molecule.

8.4.4   Nucleophilic Addition with Alkyllithium

Carboxylic acids undergo nucleophilic additions with alkyllithium (e.g., ^{n}BuLi) giv-ing ketones (Fig. 8.27). The negative charge in the carboxylate intermediate is

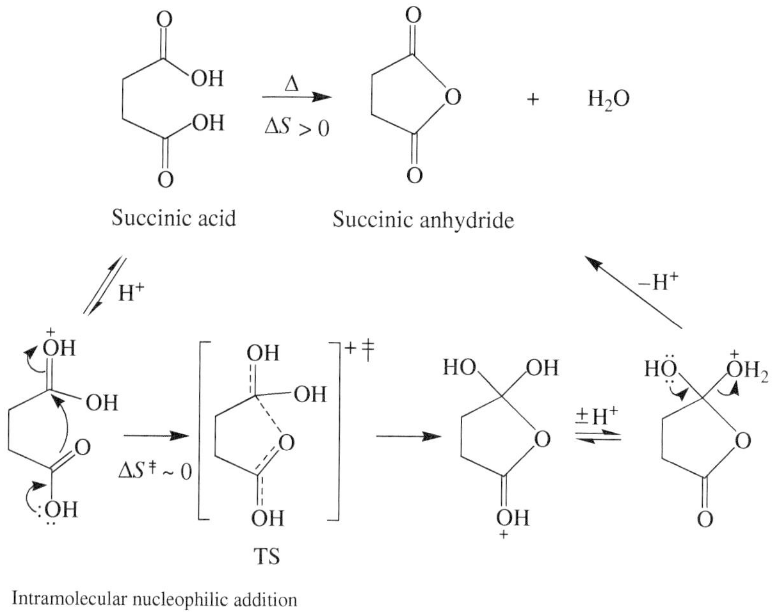

Succinic acid   Succinic anhydride

Intramolecular nucleophilic addition

FIGURE 8.26   Formation of a cyclic carboxylic anhydride (succinic anhydride) by an entropy-driven intramolecular dehydration reaction.

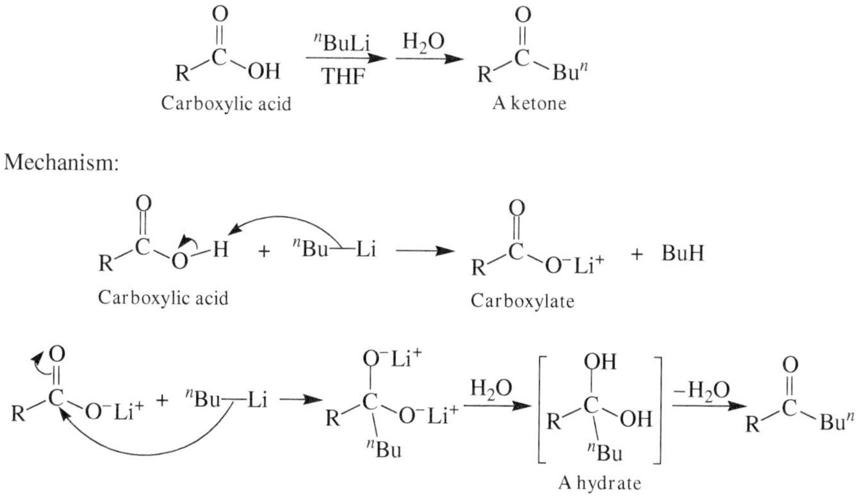

Carboxylic acid   A ketone

Mechanism:

Carboxylic acid   Carboxylate

A hydrate

FIGURE 8.27   Nucleophilic addition of ^{n}BuLi to a carboxylic acid.

strongly attracted to the very small, proton-like lithium ion. As a result, it is concentrated on one oxygen atom with very little delocalization. Thus, the carbonyl carbon in the carboxylate intermediate is electrophilic. It is effectively attacked by the strongly nucleophilic n-butyl in ^{n}BuLi to lead to the formation of an anionic tetrahedral intermediate, which contains two lithium ion stabilized negatively charged oxygen atoms. In the aqueous work-up, the anionic tetrahedral intermediate is protonated to a germinal diol (hydrate). The hydrate is unstable and converted to a ketone spontaneously (refer to Section 8.2.1 for properties of hydrates). This type of reactions has provided an effective method to synthesize various ketones.

8.5   NUCLEOPHILIC SUBSTITUTIONS OF ACYL DERIVATIVES AND THEIR BIOLOGICAL APPLICATIONS

8.5.1   Nucleophilic Substitutions of Acyl Chlorides and Anhydrides

Both acyl chlorides and anhydrides react with alcohols to lead to the formation of esters (Fig. 8.28). Since an ester is thermodynamically more stable than an acyl chloride or an anhydride, the reactions are product-favored. The reactions are initiated by a nucleophilic addition of an alcohol to the carbonyl group, which is followed by a facile and fast elimination of chloride or carboxylate giving an ester. Both the chloride and carboxylate anions are very good leaving groups.

An anhydride or acyl chloride can also react with a phenol to give an ester, while reactivity of a phenol toward a carboxylic acid is very low. Figure 8.29 shows preparation of acetyl salicylic acid (aspirin) achieved by acid-catalyzed reaction of salicylic acid (ArOH, a phenol) with acetic anhydride in mild conditions. Acetic anhydride is used in large excess so that it acts as both a reactant and the solvent, and facilitates complete conversion of salicylic acid to the acetyl salicylic acid product. The carboxyl group in salicylic acid is intact during the reaction because it is much less reactive than an anhydride and in addition, the anhydride is in excess.

A catalytic amount of sulfuric acid is used to speed up the reaction by protonating the oxygen atom of a carbonyl group in the anhydride. The protonation enhances the

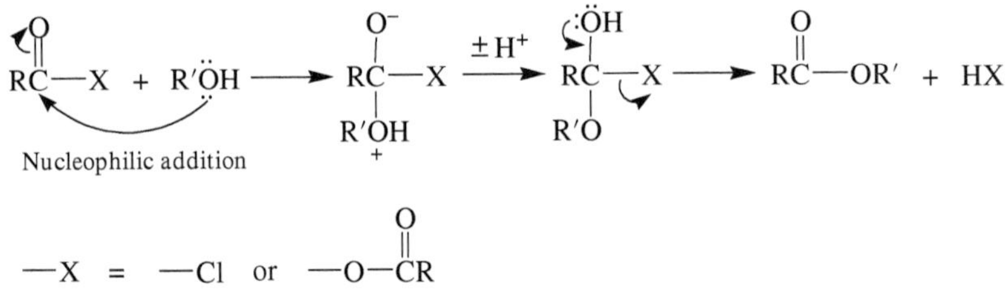

FIGURE 8.28   Nucleophilic acyl substitution reactions of an alcohol with an acyl chloride and a carboxylic anhydride.

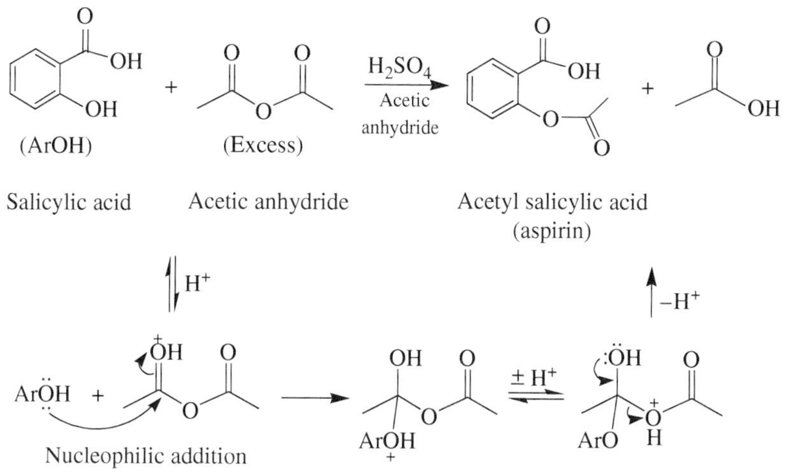

Salicylic acid   Acetic anhydride   Acetyl salicylic acid
                                                         (aspirin)

Nucleophilic addition

FIGURE 8.29 Mechanism for acid catalyzed esterification reaction of salicylic acid (a phenol) with acetic anhydride giving acetyl salicylic acid (aspirin).

Nucleophilic addition

$-X = -Cl$ or $-O-CR$ $RC-NHR' + HX$

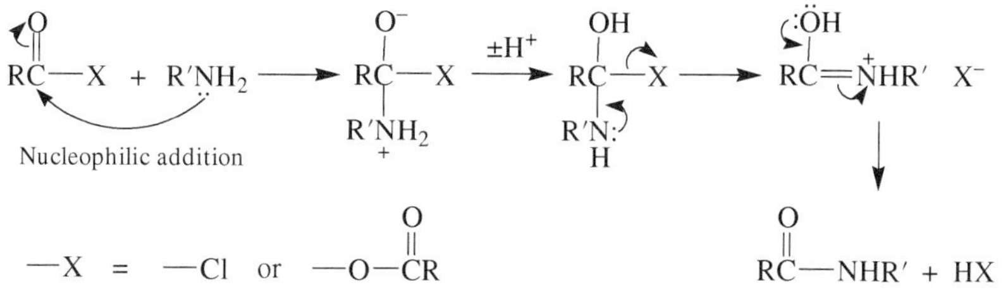

FIGURE 8.30 Nucleophilic acyl substitution reactions of a primary amine with an acyl chloride and a carboxylic anhydride.

electrophilicity of the carbonyl group and accelerates its nucleophilic addition with the hydroxyl group in salicylic acid, the rate-determining step for the overall reaction. The reaction is completed by elimination of an acetic acid molecule that is facilitated by a fast proton transfer in the intermediate adduct.

An acyl chloride or an anhydride reacts with an amine (or ammonia) to give an amide (Fig. 8.30). The overall reaction is product-favored because an amide is thermodynamically more stable than an acyl chloride or an anhydride. The first step for the reaction is a nucleophilic addition, which is the rate-determining step. Since the nucleophilicity of an amine is moderately strong, its nucleophilic addition to the acyl chloride or anhydride is fast, and there is no need to use any catalyst. The tetrahedral intermediate undergoes fast elimination of chloride or carboxylate to give the amide product.

It is noteworthy that direct reaction of a carboxylic acid with an amine (or ammonia) will not lead to the formation of an amide. Instead, a fast acid–base reaction occurs to give a carboxylate (Eq. 8.16), which is unreactive because of its low electrophilicity.

$$RC\overset{\overset{\displaystyle O}{\|}}{{}}\!\!-OH \ + \ R'NH_2 \longrightarrow RC\overset{\overset{\displaystyle O}{\|}}{{}}\!\!-O^- \ + \ R'NH_3^+$$

$$\overset{}{\nrightarrow} \ RC\overset{\overset{\displaystyle O}{\|}}{{}}\!\!-NHR' \ + \ H_2O$$

(8.16)

Thus, in order to make an amide from a carboxylic acid, the carboxylic acid would need to be first converted to an acyl chloride or an anhydride, followed by reaction with an amine (Fig. 8.30).

Acyl chlorides and carboxylic anhydrides react with Grignard reagents to give intermediate ketones first (nucleophilic acyl substitution), which undergo further nucleophilic addition reactions with Grignard reagents to give alcohols as the final products upon aqueous work-up (Eq. 8.17).

$$RC\overset{\overset{\displaystyle O}{\|}}{{}}\!\!-X \ + \ R'\!\!-MgBr \longrightarrow RC\!\!-X \longrightarrow RC\overset{\overset{\displaystyle O}{\|}}{{}} R' \xrightarrow[\text{(2) } H_2O]{\text{(1) } R'\!\!-MgBr} RC R'$$

Nucleophilic addition

$$-X = -Cl \text{ or } -O\overset{\overset{\displaystyle O}{\|}}{C}R$$

(8.17)

8.5.2   Hydrolysis and Other Nucleophilic Substitutions of Esters

As shown in Reaction 1.69 and Figure 1.18 (Section 1.10), an ester undergoes acid-catalyzed hydrolysis that follows the same mechanism as that for Fischer esterification, but in the opposite direction, to give back a carboxylic acid and an alcohol. In the reaction mechanism, water acts as a nucleophile, which attacks the protonated carbonyl group (nucleophilic addition) in the ester to effect the overall hydrolysis. In the presence of excess amount of water, the hydrolysis is complete. In the presence of the hydroxide base, the complete hydrolysis of an ester also occurs to afford an alcohol and carboxylate (Reaction 8.3 in Section 8.1).

Similar to acyl chloride and carboxylic anhydrides (Reaction 8.10), reaction of an ester with a Grignard reagent takes place readily giving a ketone intermediate, which undergoes further faster reaction with another Grignard reagent molecule to give an alcohol as the final product (Reaction 8.18).

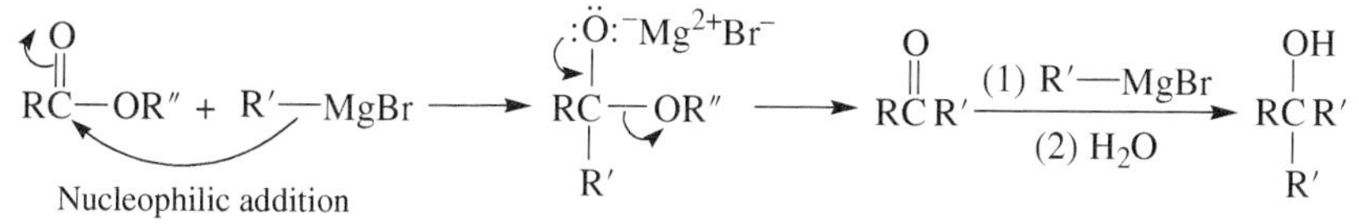

$$
\begin{array}{ccc}
\underset{\text{Nucleophilic addition}}{\overset{\overset{\displaystyle O}{\|}}{RC}-OR'' + R'\!-\!MgBr} \longrightarrow & \overset{:\ddot{O}:^-Mg^{2+}Br^-}{\underset{R'}{\overset{|}{RC}-OR''}} \longrightarrow & \overset{\overset{\displaystyle O}{\|}}{RC}R'\xrightarrow[\text{(2) H}_2\text{O}]{\text{(1) } R'\!-\!MgBr} \underset{R'}{\overset{OH}{\overset{|}{RC}R'}}
\end{array}
$$

$$(8.18)$$

An ester undergoes facile reactions with an amine (primary or secondary) or ammonia to give an amide (Fig. 8.31). The mechanism is analogous to that for the reactions with an acyl chloride or anhydride. Because an alkoxide anion is a poorer leaving group than chloride or a carboxylate anion, esters are less reactive toward an amine or ammonia than acyl chloride or anhydrides. The reaction often requires heating or high concentrations of amine.

In the presence of acid or alkoxide base, an ester undergoes nucleophilic substitution with an alcohol giving a different ester and an alcohol (Fig. 8.32). This type of reactions is called **transesterification**. The reaction is reversible. When the alcohol reactant (R_2OH) is in large excess, the starting ester (RCO_2R_1) can be completely converted to another ester RCO_2R_2. For both acid-catalyzed and base-catalyzed transesterifications, the reactions are initiated by a nucleophilic addition to carbonyl of the ester (rate-determining step), followed by a fast elimination of an alcohol or alkoxide anion. The base-catalyzed transesterification is fast, and it has been used in manufacture of biodiesel (Section 8.5.3). On the other hand, the acid-catalyzed transesterification is relatively slow and usually requires larger excess of the alcohol reactant and higher temperature. This is because the rate-determining step for the acid-catalyzed reaction involves a relatively high-energy intermediate protonated ester (Fig. 8.32), which enhances activation energy for the reaction.

8.5.3   Biodiesel Synthesis and Reaction Mechanism

Great worldwide interest has been developed in the production of methyl esters of fatty acids (FAMES) as a fuel for diesel engines or oil heaters and as a means for reducing the use of petroleum, which is entirely a fossil fuel. The bulk of fatty acid residues are embedded in triacylglycerols found in plants (vegetable oil). Therefore,

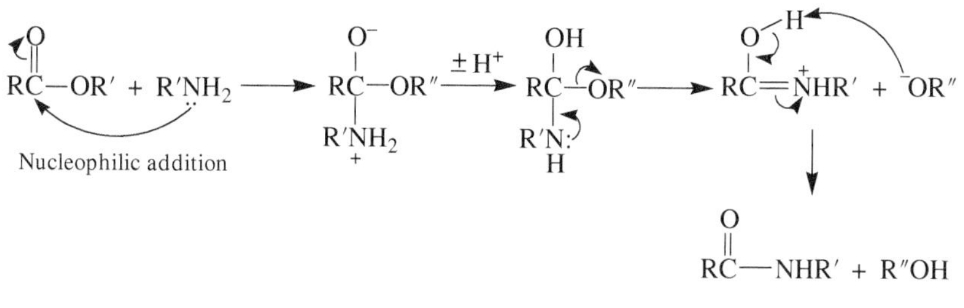

$$
\begin{array}{ccccc}
\underset{\text{Nucleophilic addition}}{\overset{\overset{\displaystyle O}{\|}}{RC}-OR' + R'NH_2} \longrightarrow & \underset{\overset{R'NH_2}{+}}{\overset{O^-}{\overset{|}{RC}-OR''}} \xrightarrow{\pm H^+} & \underset{\overset{R'N:}{\underset{H}{}}}{\overset{OH}{\overset{|}{RC}-OR''}} \longrightarrow & \overset{\overset{\displaystyle O-H}{|}}{RC}\!\!=\!\!\overset{+}{N}HR' + {}^-OR''
\end{array}
$$

$$
\underset{}{\overset{\overset{\displaystyle O}{\|}}{RC}-NHR' + R''OH}
$$

FIGURE 8.31   Nucleophilic acyl substitution reaction of a primary amine with an ester.

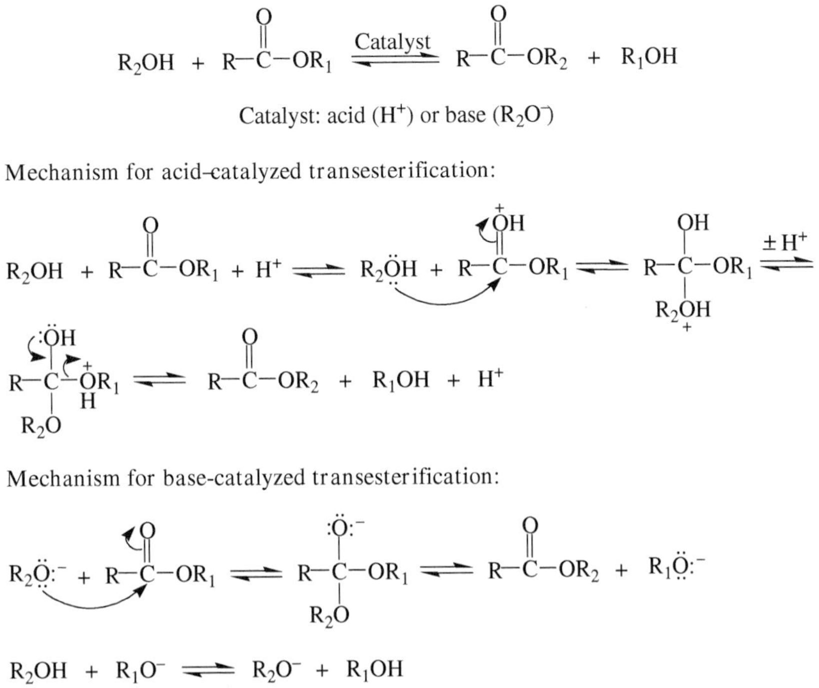

$$R_2OH + R-\overset{O}{\underset{\|}{C}}-OR_1 \underset{}{\overset{\text{Catalyst}}{\rightleftharpoons}} R-\overset{O}{\underset{\|}{C}}-OR_2 + R_1OH$$

Catalyst: acid (H^+) or base (R_2O^-)

Mechanism for acid-catalyzed transesterification:

Mechanism for base-catalyzed transesterification:

$$R_2OH + R_1O^- \rightleftharpoons R_2O^- + R_1OH$$

FIGURE 8.32 Mechanisms for acid and base catalyzed transesterification.

biodiesel (FAMES) synthesis is accomplished industrially by base-catalyzed trans-esterification of vegetable oil obtained from soybeans in United States or low-erucic acid rapeseed (canola) in Europe with excess methanol (Reaction 8.19) [7–9].

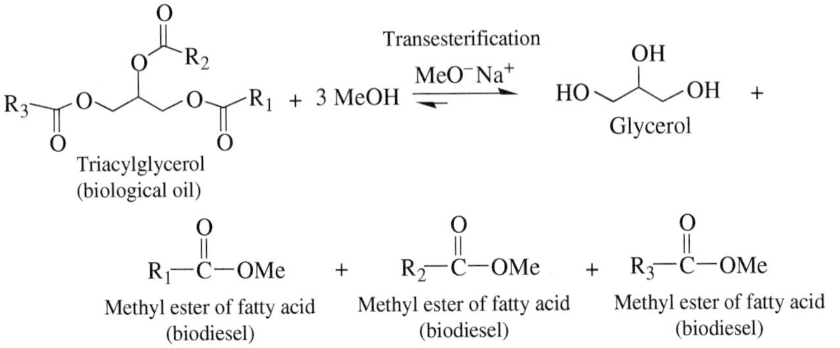

(8.19)

The R groups are long hydrocarbon chains, with the number of carbon atoms in each of the chains being 11–23 depending on the biological oil source. When the usage of methanol is in large excess, the above transesterification will be

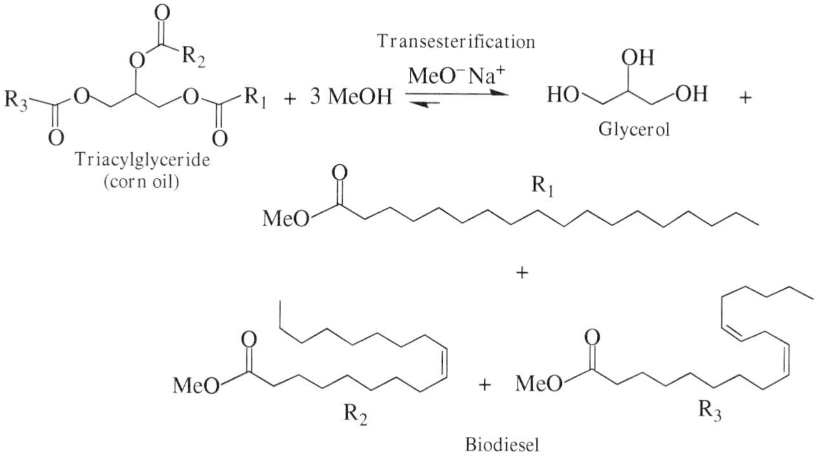

FIGURE 8.33   Synthesis of biodiesel from corn oil. Modified from Bladt et al. [8].

product-favored with almost all (99%) the vegetable oil converted to methyl esters of fatty acids (biodiesel). Figure 8.33 shows the example of biodiesel made by trans-esterification of corn oil with methanol [8].

The mechanism for the methoxide-base-catalyzed transesterification of triacyl-glycerol is shown in Figure 8.34. The reaction is initiated by nucleophilic addition of methoxide (MeO$^-$) to a carbonyl group in triacylglyceride. It is followed by elim-ination of an alkoxide anion (an anion of glycerol). Since methoxide is more basic than the anion of glycerol (The pK_a's of methanol and glycerol are 15.5 and 14.2, respectively), the transesterification to give a methyl ester and anion of glycerol is product-favored. In the last step, an anion of glycerol abstracts a proton from meth-anol to regenerate the methoxide catalyst.

8.5.4   Biological Applications: Mechanisms of Serine-Type Hydrolases

Serine proteases are a class of enzymes that catalyze hydrolysis of a peptide bond in proteins (polypeptides) (Fig. 8.35). A common example is **trypsin**. The enzyme has a single polypeptide chain, and its active site consists of functional groups of the side chains of three amino acid residues (Ser-195, His-57, and Asp-102), forming a cat-alytic triad (Fig. 8.36) [3]. Figure 8.36 shows that the enzymatic mechanism consists of consecutive nucleophilic addition and acyl substitution reactions.

The first step in the mechanism is that the protein (polypeptide) substrate (R$_1$CONHR$_2$) enters the structure of the enzyme to form an enzyme-substrate com-plex (the intermediate **1**) via hydrogen bonding and hydrophobic interactions (details omitted in the figure). The structure of the enzyme possesses an empty space (oxyanion hole) whose dimensions perfectly fit the size of an oxygen anion (–O$^-$).

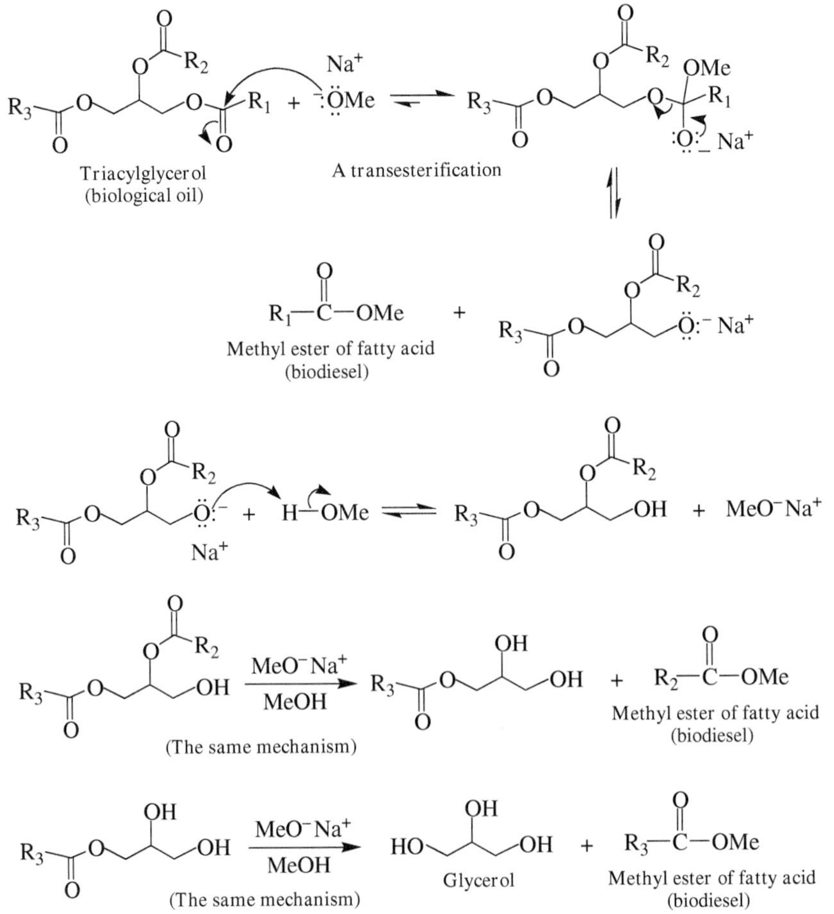

FIGURE 8.34   Mechanism for the formation of methyl ester of fatty acids (biodiesel) by the methoxide base catalyzed transesterification of triacylglycerol (biological oil).

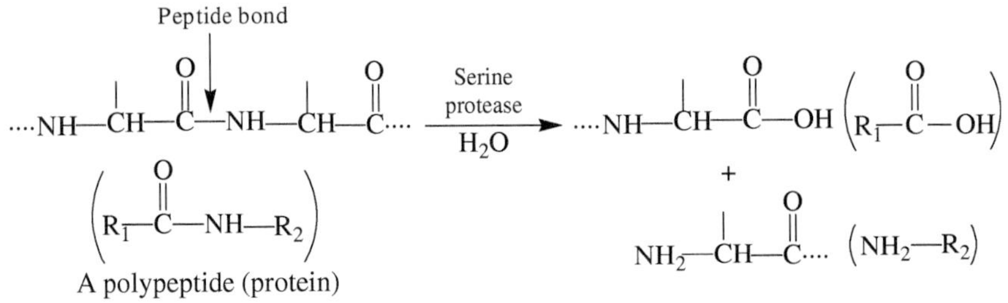

FIGURE 8.35   The hydrolytic peptide bond cleavage in proteins catalyzed by serine proteases.

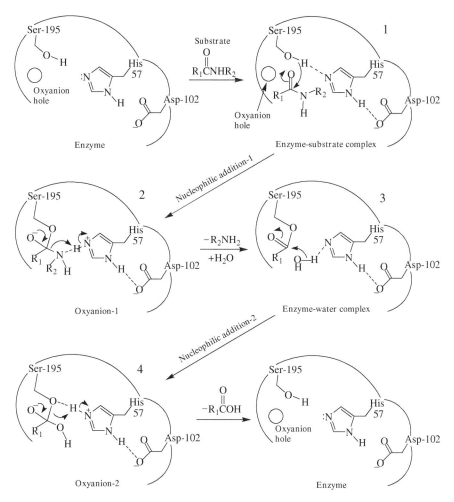

FIGURE 8.36 The catalytic mechanism for trypsin (a serine protease). Modified from Voet et al. [3].

In the enzyme-substrate complex, the hydrogen bond between the Ser-195 OH and His-57 imidazole ring in the active site weakens and polarizes the O—H bond and enhances nucleophilicity of the oxygen. The hydrogen bond between the Asp-102 carboxylate oxygen and His-57 imidazole N–H reinforces the hydrogen bond on the Ser-195 OH. As a result, the Ser OH oxygen can effectively attack the carbonyl carbon in the substrate (the first nucleophilic addition) to make the oxyanion-1 (the intermediate **2**). After it is formed, the negatively charged oxygen atom goes in the oxyanion hole in the enzyme as indicated in Figure 8.36. The formation of the oxyanion-1 (**1** → **2**) is the rate-determining step for the overall enzymatic reaction. The oxyanion is stabilized by the surrounding hydrogen bonds (not shown in the figure). This has lowered the activation energy for its formation and accelerated the overall reaction. The hydrogen bonds in oxyanion-1 (the intermediate **2**) facilitate elimination of R_2NH_2 (the cleavage of a peptide bond). As a lone-pair of electrons in the negatively charged oxygen atom is transferred to an oxygen—carbon bond domain to form a C=O π bond, the C—N bond (peptide bond) breaks simultaneously to complete a stepwise nucleophilic acyl substitution (**1** $\rightarrow$ **2** $\rightarrow$ **3**). Then R_2NH_2 departs the enzyme structure and a water molecule enters the enzyme to form an enzyme-water complex (the intermediate **3**).

In the enzyme-water complex, the OH of water is hydrogen bonded to the His-57 imidazole. This hydrogen bond weakens and polarizes the OH bond in water and enhances the nucleophilicity of its oxygen atom. Thus, the oxygen atom in water can effectively attack the carbonyl carbon to lead to the second nucleophilic addition and make oxyanion-2 (the intermediate **4**). The hydrogen bonds in oxyanion-2 (the intermediate **4**) facilitate the cleavage of a C—O bond (similar to the peptide bond cleavage in the intermediate **2**) to regenerate the OH group in Ser-195. Simultaneously, a C=O π bond is formed as a lone-pair of electrons in the negatively charged oxygen atom is transferred to an oxygen—carbon bond domain. A second stepwise nucleophilic acyl substitution (**3** $\rightarrow$ **4** $\rightarrow$ **Enzyme**) is now completed. Finally, R_1COOH departs and a free enzyme is regenerated for another cycle of catalysis.

Figure 8.37 shows a qualitative reaction profile for the tryptin catalyzed hydrolytic cleavage of a protein peptide bond (the overall reaction in Figure 8.36). The uncatalyzed hydrolytic cleavage of a peptide bond would render a high activation energy. The enzyme divides the overall reaction into two stages. The first stage includes the nucleophilic addition between the substrate and enzyme and a subsequent elimination (**1** $\rightarrow$ **2** $\rightarrow$ **3**), while water is involved in the second stage (**3** $\rightarrow$ **4** $\rightarrow$ **P + E**) for a second nucleophilic addition and acyl substitution. The nucleophilic addition of the Ser-195 OH to the substrate (**1** $\rightarrow$ **2**) has the highest activation energy and is the rate-determining step. The key to catalysis is the oxyanion hole in

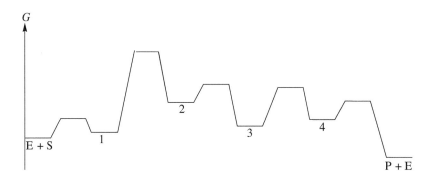

1. ES complex 2. Oxyanion-1 3. E-water complex 4. Oxyanion-2

FIGURE 8.37   The reaction profile for the trypsin catalyzed hydrolytic cleavage of a protein peptide bond.

which the oxyanion-1 (the intermediate **2**) is stabilized and as a result, the energy of the transition state is lowered. In addition, since the Ser-195 OH group is polarized by a hydrogen bond, the energy level of the ES complex (the intermediate **1**) is raised. All this results in decrease in the activation energy of the rate-determining step (**1** → **2**) and speeds up the overall reaction.

Lipases are another classes of the serine-type enzymes. In living organisms including the human body, the first step of metabolism of triacylglycerol (animal fat or plant oil) is its digestion, the stepwise hydrolysis of the ester bonds in triacylglycerol to the constituent fatty acids and glycerol. The biochemical reactions are catalyzed by a **lipase** (Fig. 8.38) and take place under mild physiological conditions. In contrast, the acid catalyzed hydrolysis of esters to carboxylic acids and alcohols usually requires heating.

A triacylglycerol lipase is a single polypeptide chain protein that contains 269 amino acid residues [10]. The active site of the enzyme is trypsin-like (serine-type) catalytic triad, which consists of side chains of Ser-144, Asp-203, and His-257 residues (Fig. 8.39).

Figure 8.39 shows that in the first step of the enzymatic mechanism, a triacylglycerol molecule binds to the enzyme in the active site to form an enzyme-substrate complex. Then, the hydroxyl group of the Ser-144 side chain attacks a carbonyl carbon in the triacylglycerol molecule (Step 1) to lead to the formation of a tetrahedral intermediate (an oxyanion), which is covalently attached to the enzyme. The intermolecular hydrogen bond between the Asp-203 carboxylate and the Ser-144 hydroxyl enhances the nucleophilicity of the –OH group and makes its attack on the carbonyl of triacylglycerol kinetically favorable. In addition, the oxygen anion of the tetrahedral intermediate is stabilized by an oxyanion hole in the enzyme structure (similar to trypsin) via hydrogen bonds formed on the negatively charged

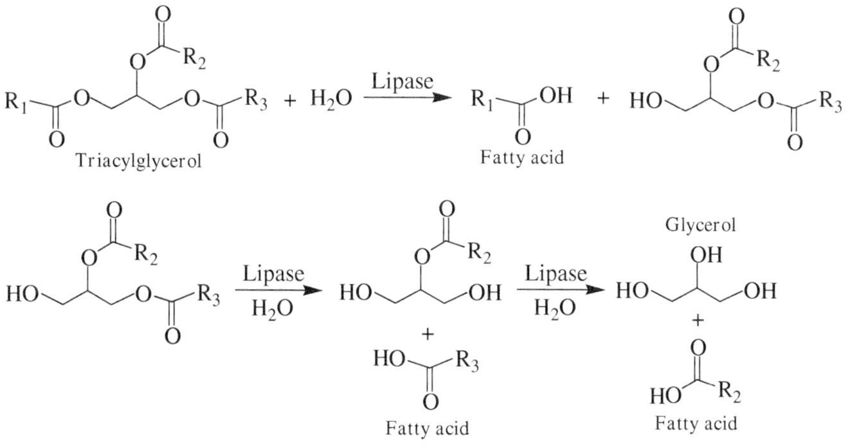

FIGURE 8.38 Lipase catalyzed hydrolysis of triacylglycerol to fatty acids.

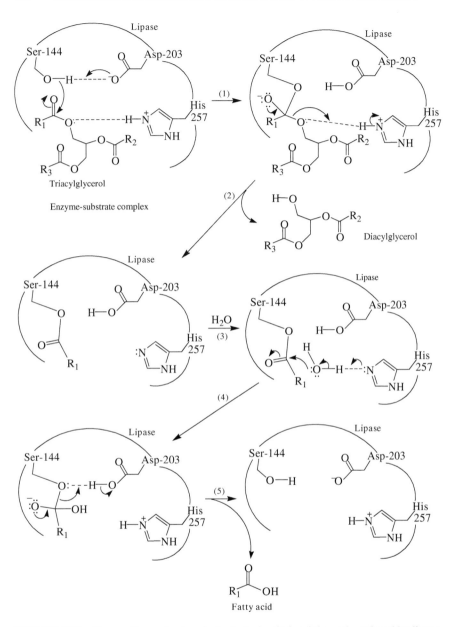

FIGURE 8.39 Enzymatic mechanism for hydrolysis of triacylglycerol catalyzed by lipase.

oxygen (omitted in Fig. 8.39). This stabilization accelerates the nucleophilic addition (Step 1), the rate-determining step of the overall hydrolysis. Then an active lone-pair of electrons in the oxyanion moves down to effect elimination of an alcohol molecule (diacylglycerol) that departs from the enzyme (Step 2), and this step is facilitated by the hydrogen bond exerted by the imidazole ring of His-257. As a dia-cylglycerol molecule is eliminated, an acyl group is formed simultaneously and attached to Ser-144 via an ester bond.

In the following step (Step 3), a water molecule enters the active site of the enzyme and effects a second nucleophilic addition (Step 4) giving another tetrahe-dral intermediate (an oxyanion). Finally, an active lone-pair of electrons in the oxy-anion of the second tetrahedral intermediate moves down to effect elimination of the Ser-144 hydroxyl group (Step 5), which is facilitated by the hydrogen bond exerted by the Asp-203 carboxyl group. This elimination step generates a fatty acid mole-cule, which leaves the enzyme subsequently, and a free enzyme is regenerated and is ready for catalysis of hydrolysis for another substrate molecule. The intermediate diacylglycerol will undergo analogous lipase catalyzed hydrolysis to release two more fatty acid molecules consecutively and glycerol eventually.

8.6   REDUCTION OF ACYL DERIVATIVES BY HYDRIDE DONORS

Acyl chlorides, carboxylic anhydrides, esters, and carboxylic acids can be reduced to primary alcohols by lithium aluminum hydride ($LiAlH_4$) (Fig. 8.40). For the first three types of acyl compounds, the reaction takes place via a nucleophilic addition

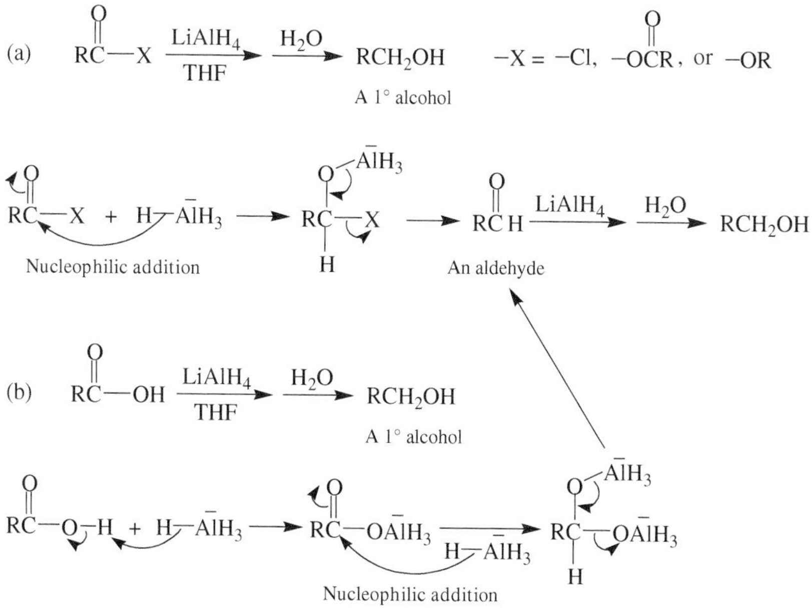

FIGURE 8.40   Reductions of (a) acyl chlorides, carboxylic anhydrides, and esters, and (b) carboxylic acids by lithium aluminum hydride.

of hydride to the carbonyl group, which is followed by elimination of chloride, carboxylate, or alkoxide anion to give an aldehyde. The intermediate aldehyde undergoes further reduction to afford a primary alcohol. The mechanism for reduction of carboxylic acids is slightly different. The reactions take place via an tetrahedral intermediate $RCH(OAlH_3)_2^{2-}$. Strong affinity of aluminum to oxygen weakens the C—O bond. As a result, a $-OAlH_3^-$ group is readily eliminated giving an intermediate aldehyde, which will be reduced to an alcohol by $LiAlH_4$.

Reduction of an amide by $LiAlH_4$ affords an amine (primary, secondary, or tertiary) (Fig. 8.41). The net result is that the divalent carbonyl oxygen in the amide is replaced by two hydrogen atoms in the end of the reaction to give $-CH_2-$ in the amine product. The overall reaction is effected by two nucleophilic hydride additions. The first hydride addition takes place on the carbonyl group of the amide giving a tetrahedral intermediate. The particularly active lone-pair of electrons in nitrogen moves down to the C—N bond domain to effectively expel the $-OAlH_3^-$ group from the intermediate, which leads to the formation of an intermediate imine. The positive charge on nitrogen of the imine makes the $C=N^+$ carbon strongly electrophilic. It is attacked readily by the AlH_4^- hydride to afford the final amine product.

8.7 KINETICS OF THE NUCLEOPHILIC ADDITION AND SUBSTITUTION OF ACYL DERIVATIVES

In general, the acyl substitution reaction is effected by a nucleophilic addition, followed by a rapid elimination of an X^- anion ($k_1 \ll k_2$) as shown in Reaction 8.20 and Figure 8.1c.

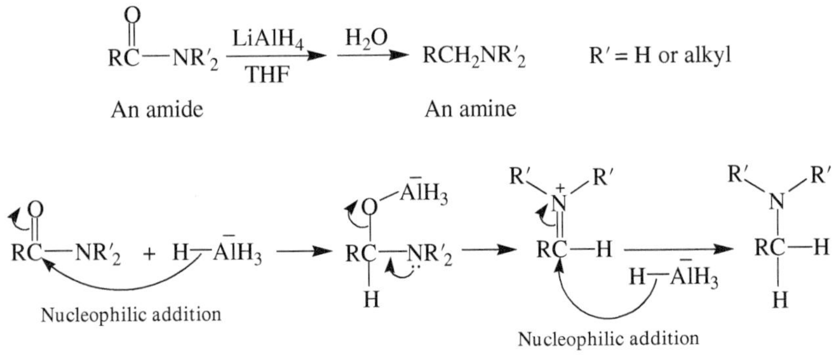

FIGURE 8.41 Mechanism for reduction of an amide to an amine by lithium aluminum hydride.

$$\text{Nu:}^- + \underset{R}{\overset{O}{\underset{\|}{C}}}{\diagdown}X \underset{k_{-1}}{\overset{k_1}{\rightleftharpoons}} \underset{\substack{R \\ \\ Nu}}{\overset{:\ddot{O}:^-}{\underset{|}{C}}}X \overset{k_2}{\longrightarrow} \underset{R}{\overset{O}{\underset{\|}{C}}}{\diagdown}\text{Nu} + X^- \qquad (8.20)$$

$$\text{Td intermediate}$$

The rate law for the overall reaction can be written as Equation 8.21.

$$\text{Rate} = k_2[\text{Td}] \qquad (8.21)$$

Due to the activity of the lone-pair of electrons in the negatively charged oxygen in the Td intermediate, the intermediate is highly reactive. Approximately, it remains in the steady-state during the reaction as formulated mathematically in Equation 8.22.

$$d[\text{Td}]/dt = 0 = k_1[\text{Nu}^-][\text{RCOX}] - k_{-1}[\text{Td}] - k_2[\text{Td}] \qquad (8.22)$$

From Equation 8.22, we have

$$[\text{Td}] = [k_1/(k_{-1} + k_2)][\text{Nu}^-][\text{RCOX}] \qquad (8.23)$$

Substituting Equation 8.23 for Equation 8.21 leads to Equation 8.24.

$$\text{Rate} = \frac{k_1}{1 + \frac{k_{-1}}{k_2}}[\text{Nu}^-][\text{RCOX}] \qquad (8.24)$$

Equation 8.24 is the rate law for Reaction 8.20.

There are two extreme situations. In the case that k_{-1} is very large, namely that $k_{-1} \gg k_1$ and $k_{-1} \gg k_2$, we have approximately $1 + k_{-1}/k_2 = k_{-1}/k_2$. Therefore, Equation 8.24 becomes

$$\text{Rate} = \frac{k_1 k_2}{k_{-1}}[\text{Nu}^-][\text{RCOX}] \qquad (8.25)$$

In this situation, pre-equilibrium for the nucleophilic addition is reached prior to the second step of elimination, with the equilibrium constant expression being $K_{eq} = k_1/k_{-1}$. Therefore, in Equation 8.25,

$$(k_1/k_{-1})[\text{Nu}^-][\text{RCOX}] = [\text{Td}] \text{ (the equilibrium concentration of the Td intermediate)}$$

It shows that Equation 8.25 is equivalent to Equation 8.21.

On the other hand, if k_{-1} is very small, namely that $k_{-1} \ll k_1$ and $k_{-1} \ll k_2$, we have approximately $1 + k_{-1}/k_2 = 1$. Therefore, Equation 8.24 becomes

$$\text{Rate} = k_1[\text{Nu}^-][\text{RCOX}] \qquad (8.26)$$

In this situation, the first step of nucleophilic addition in Reaction 8.20 becomes essentially irreversible and rate-determining step for the overall acyl substitution.

In the presence of cyanide (CN^-), two molecules of benzaldehyde combine to give benzoin (Reaction 8.27) [1].

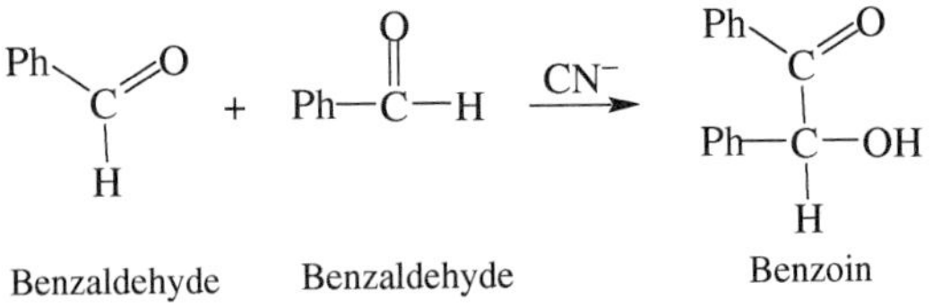

$$\text{Benzaldehyde} \qquad \text{Benzaldehyde} \qquad \text{Benzoin}$$ (8.27)

The reaction takes place via a complex mechanism (Fig. 8.42) that includes the nucleophilic addition and elimination steps.

First, the nucleophilic cyanide anion (catalyst) is added to the carbonyl group of a benzaldehyde molecule. The tetrahedral intermediate is subsequently transformed into a carbanion intermediate. The formation of the carbanion is greatly facilitated by the particular stabilization of the species by the electronegative cyanide group ($-CN$), which is accomplished via dispersion of the negative charge as demonstrated below (Eq. 8.28).

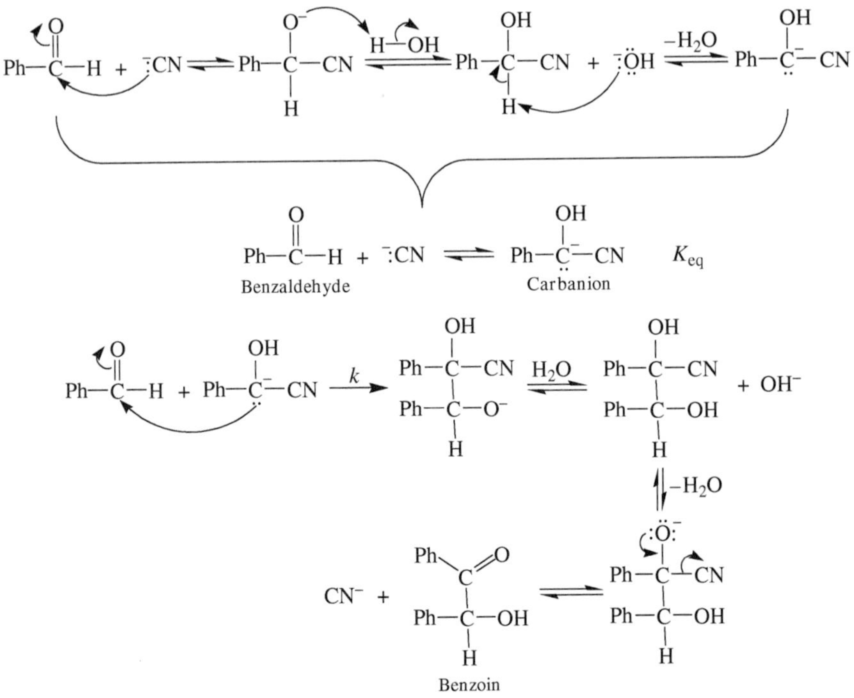

FIGURE 8.42 Mechanism for cyanide catalyzed nucleophilic addition of benzaldehyde to benzoin.

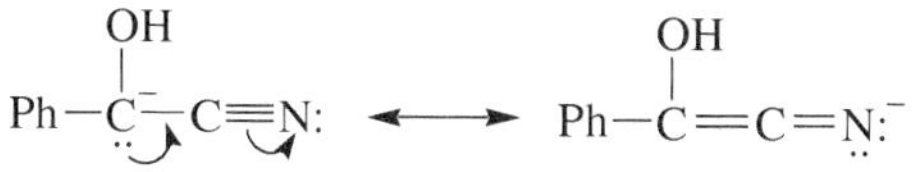

$$
\underset{\text{Ph}-\overset{\displaystyle \text{OH}}{\underset{\displaystyle }{\text{C}}}-\text{C}\!\equiv\!\text{N:}}{} \quad \longleftrightarrow \quad \underset{\text{Ph}-\overset{\displaystyle \text{OH}}{\underset{\displaystyle }{\text{C}}}\!=\!\text{C}\!=\!\text{N:}^{-}}{} \tag{8.28}
$$

As a result, there is a rapid net equilibrium between benzaldehyde and the intermediate carbanion which is initiated by cyanide as shown in Figure 8.42, with the equilibrium constant expression being

$$
K_{eq} = [\text{carbanion}]/[\text{Benzaldehyde}][\text{CN}^-] \tag{8.29}
$$

Then the carbanion strongly attacks the carbonyl group of the second benzaldehyde molecule to bring about an irreversible nucleophilic addition (with the rate constant k), which is the rate-determining step for the overall reaction. Finally, the –CN group in a tetrahedral anionic intermediate is eliminated by an active lone-pair of electrons in the oxygen anion.

Since the irreversible nucleophilic addition of the carbanion to benzaldehyde is the rate-determining step, the rate law for the overall reaction can be written as

$$
\text{Rate} = k[\text{Benzaldehyde}][\text{carbanion}] \tag{8.30}
$$

From Equation 8.29, we have

$$
[\text{carbanion}] = K_{eq}[\text{Benzaldehyde}][\text{CN}^-] \tag{8.31}
$$

Substituting Equation 8.31 for Equation 8.30 leads to

$$
\text{Rate} = kK_{eq}[\text{Benzaldehyde}]^2[\text{CN}^-] = k'[\text{Benzaldehyde}]^2[\text{CN}^-] \tag{8.32}
$$

where $k' = kK_{eq}$, the observed rate constant of the overall Reaction 8.27.

Equation 8.32 is the rate law of Reaction 8.27. The reaction rate is proportional to square of the benzaldehyde molar concentration. It is also proportional to the molar concentration of the cyanide catalyst.

PROBLEMS

8.1 Write mechanisms for the following hydrolysis of a ketal and a thioketal to acetone. Which reaction goes faster? Explain by comparison of energy profiles for the rate-determining steps of the two reactions.

A ketal   Acetone   A thioketal

8.2 Write mechanism for formation of a cyclic ketal from ethylene glycol (a diol) and acetone as shown below:

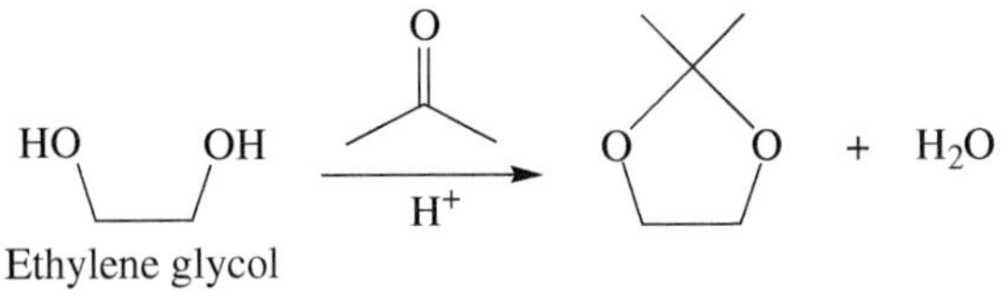

Ethylene glycol

For each of the following diols, does the reaction with acetone lead to formation of a cyclic ketal? Explain and draw the structure of the ketal formed.

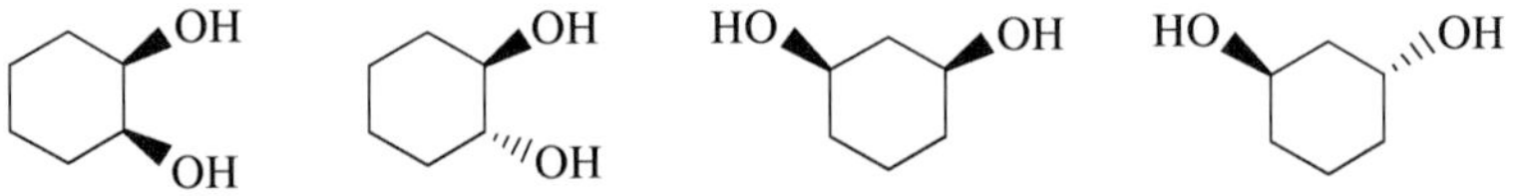

8.3 Write out detailed stepwise mechanisms for the following reactions:

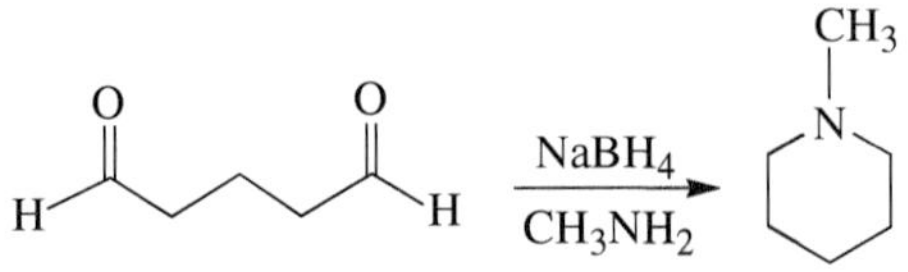

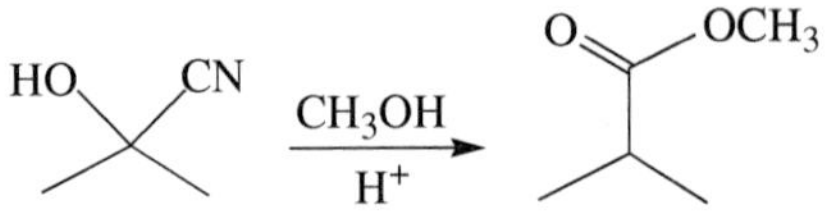

8.4 Give the product and show mechanism for the reaction of cyclohexanone with dimethylamine.

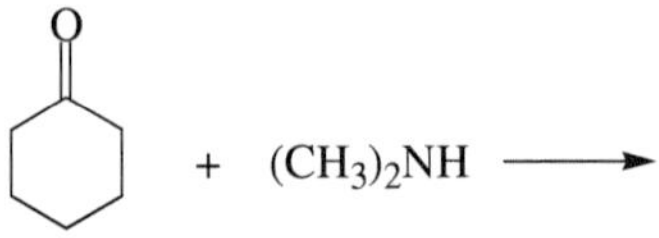

$$+ \quad (CH_3)_2NH \longrightarrow$$

8.5 Propose a mechanism for the following reaction. Explain the stereospecificity of the reaction.

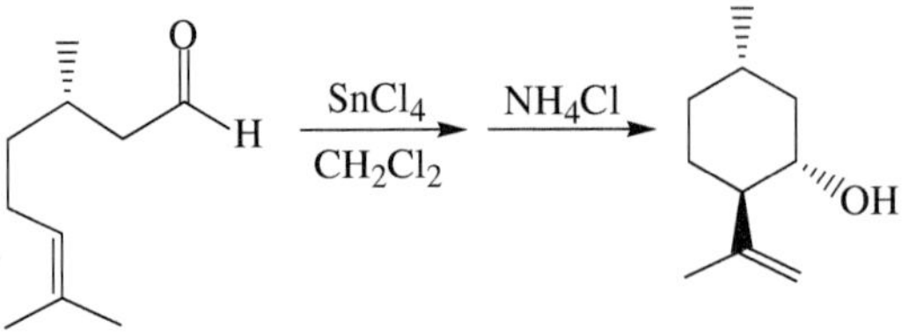

8.6 In aqueous sodium hydroxide, the keto alcohol Me₂C(OH)CH₂COMe is converted to acetone MeCOMe. Suggest a plausible mechanism and the kinetics expected.

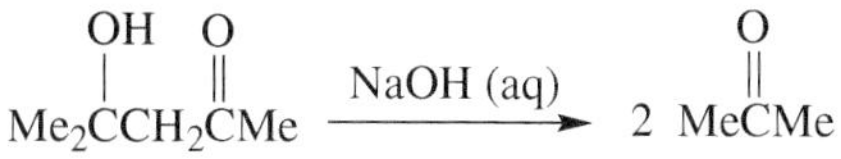

8.7 Propose a mechanism for each of the following reactions. Explain why the reactions can take place.

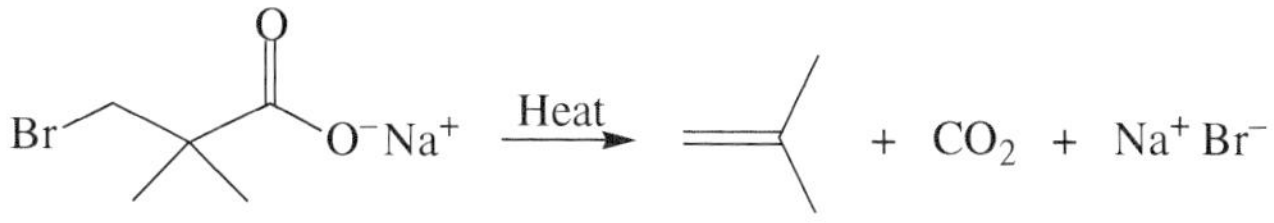

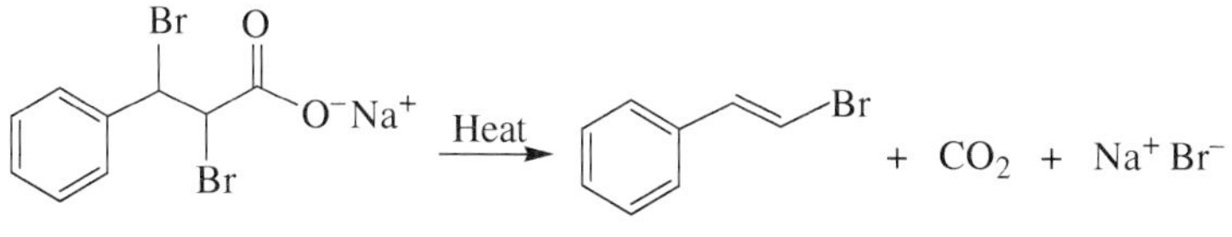

8.8 Propose a plausible mechanism to account for the following transformation in Step 1. Describe experimental conditions to bring about Step 2 and write the mechanism.

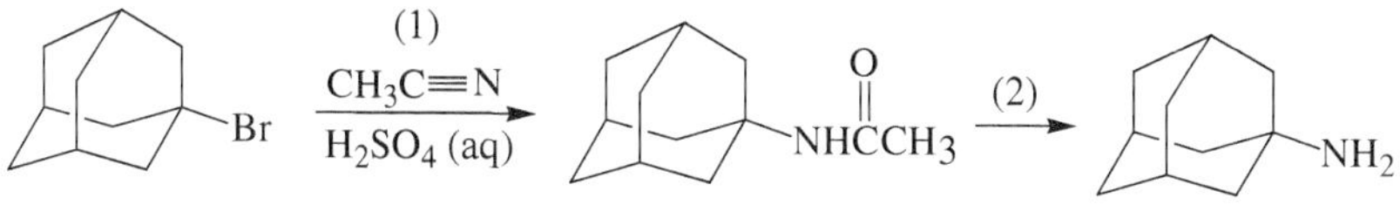

8.9 Propose necessary reagents to bring about each of the following transformations. Write mechanism for each transformation.

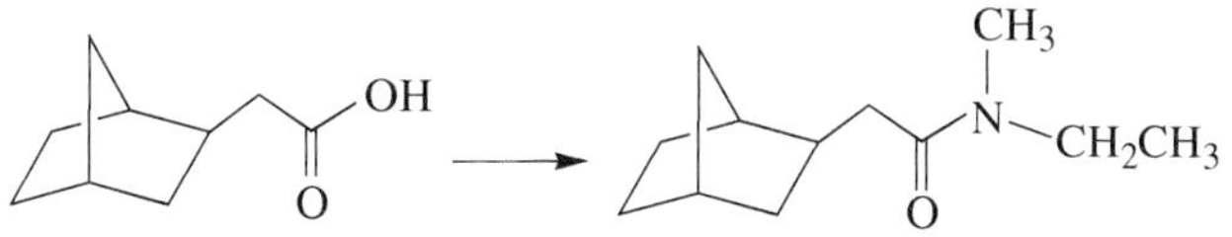

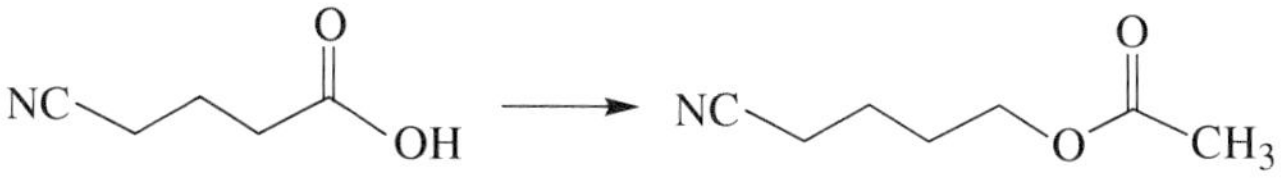

8.10 Oxalyl chloride is an effective chlorinating reagent for carboxylic acids. Write out mechanism for the following reaction:

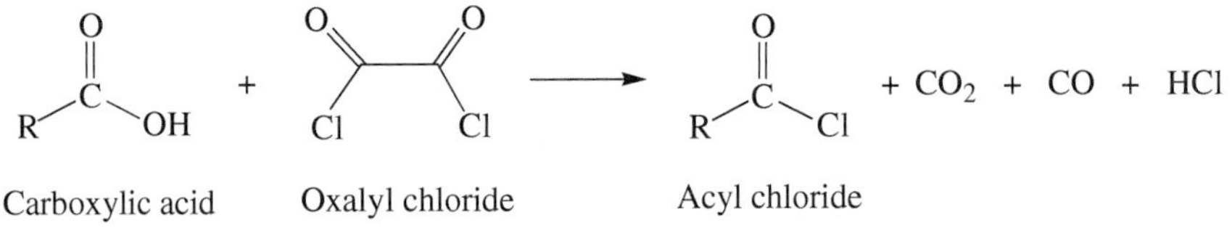

Carboxylic acid   Oxalyl chloride   Acyl chloride

REFERENCES

1. Jackson, R. A. *Mechanisms in Organic Reactions*, The Royal Society of Chemistry, Cambridge, UK (2004).

2. Fox, M. A.; Whitesell, J. K. *Organic Chemistry*, 2nd ed., Jones and Bartlett, Sudbury, MA, USA (1997).

3. Voet, D.; Voet, J. G.; Pratt, C. W. *Fundamentals of Biochemistry*, 5th ed., Wiley, Hoboken, NJ, USA (2016).

4. Hoffman, R. V. *Organic Chemistry: An Intermediate Text*, 2nd ed., Wiley, Hoboken, NJ, USA (2004).

5. Manabe, K.; Sun, X.-M.; Kobayashi, S. Dehydration Reactions in Water. Surfactant-Type Bronsted Acid-Catalyzed Direct Esterification of Carboxylic Acids with Alcohols in an Emulsion System. *J. Am. Chem. Soc.* 2001, *123*, 10101–10102.

6. Manabe, K.; Limura, S.; Sun, X.-M.; Kobayashi, S. Dehydration Reactions in Water. Bronsted Acid—Surfactant-Combined Catalyst for Ester, Ether, Thioether, and Dithioacetal Formation in Water. *J. Am. Chem. Soc.* 2002, *124*, 11971–11978.

7. Behnia, M. S.; Emerson, D. W.; Steinberg, S. M.; Alwis, R. M.; Duenas, J. A.; Serafino, J. O. A Simple, Safe, Method for Preparation of Biodiesel. *J. Chem. Educ.* 2011, *88*, 1290–1292.

8. Bladt, D.; Murray, S.; Gitch, B.; Trout, H.; Liberko, C. Acid-Catalyzed Preparation of Biodiesel from Waste Vegetable Oil: An Experiment for the Undergraduate Organic Chemistry Laboratory. *J. Chem. Educ.* 2011, *88*, 201–203.

9. Bucholtz, E. C. Biodiesel Synthesis and Evaluation: An Organic Chemistry Experiment. *J. Chem. Educ.* 2007, *84*, 296–298.

10. Brady, L.; Brzozowski, A. M.; Derewenda, Z. S.; Dodson, E.; Dodson, G.; Tolley, S.; Turkenburg, J. P.; Christiansen, L.; Huge-Jensen, B.; Norskov, L.; Thim, L.; Menge, U. A Serine Protease Triad Forms the Catalytic Centre of a Triacylglycerol Lipase. *Nature*, 1990, *343*, 767–770.

9

REACTIVITY OF THE α-HYDROGEN TO CARBONYL GROUPS

9.1 FORMATION OF ENOLATES AND THEIR NUCLEOPHILICITY

9.1.1 Formation of Enolates

Acidity of the α-hydrogen The C—H bond in the α-carbon to a carbonyl group (C=O) of aldehydes, ketones, and carboxylic acid derivatives (esters, anhydrides, amides, etc.) is slightly acidic. This is because the C_α—H bond can overlap with the C=O π bond in sideway (hyperconjugation) which results in a partial shift of electrons in the C_α—H bond domain toward the electronegative oxygen in the carbonyl group, making the C_α—H bond slightly polar. Figure 9.1 illustrates the hyperconjugation between C_α—H and C=O in a carbonyl compound. In the most favored conformation, the C_α—H σ-bond and p_π orbitals in C=O are coplanar. The sideway overlap (hyperconjugation) of the C_α—H bonding orbital with the C=O π* orbital (antibonding) results in resonance stabilization of the molecule. The ionic contributor is responsible for the acidity of the α-hydrogen. The weak acidity of the C_α—H bonds of various carbonyl compounds is indicated by comparison of their pK_a values with that of the C—H bond in alkanes: pK_a (C_α–H) = 17 (CH_3CHO), 19 (CH_3COCH_3), and 25 ($CH_3CO_2CH_3$), and pK_a = 50 (CH_3CH_3) [1].

Organic Mechanisms: Reactions, Methodology, and Biological Applications,
Second Edition. Xiaoping Sun.
© 2021 John Wiley & Sons, Inc. Published 2021 by John Wiley & Sons, Inc.
Companion website: www.wiley.com/go/Sun/OrgMech_2e

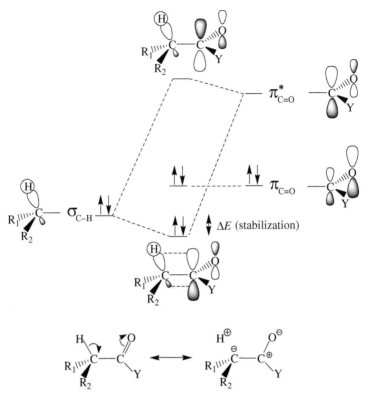

FIGURE 9.1 Molecular orbitals formed due to the hyperconjugation between the C_α–H bond and carbonyl (C=O) π bond in a carbonyl compound, and prevalent resonance structures as a result of the hyperconjugation.

Formation of enolates In the presence of a strong base such as hydroxide (OH⁻) [pK_a = 15 (HOH)], alkoxide (RO⁻) [pK_a = 16 (ROH)], amide (R₂N⁻) [pK_a = 40 (R₂NH)], or BuLi [pK_a = 50 (BuH)], the C_α–H bond of carbonyl compounds can be deprotonated to lead to the formation of an **enolate** anion (Fig. 9.2). In principle, the formation of an enolate from a carbonyl compound is reversible. The reversibility, indicated by the equilibrium constant $K = K_a(C_\alpha$–H$)/K_a$(HB), is determined by the strength of the base and the acidity (pK_a) of the C_α–H bond. When an extremely strong base, such as LiN(*i*-Pr)₂ or BuLi, is employed, the deprotonation becomes irreversible (quantitative) as the conjugate acid of each base has much greater pK_a than that of the corresponding C_α–H bond in a carbonyl compound. For example, the formation of acetone [$K_a(C_\alpha$–H)~10^{-19}] enolate by hydroxide base [K_a(HB) ~10^{-15}] has an equilibrium constant K~10^{-4} ($10^{-19}/10^{-15}$). When acetone reacts with LiN(*i*-Pr)₂ [K_a(HB)~10^{-40}] to give an enolate, the equilibrium constant for the reaction is K~10^{21} ($10^{-19}/10^{-40}$). The reaction is essentially irreversible.

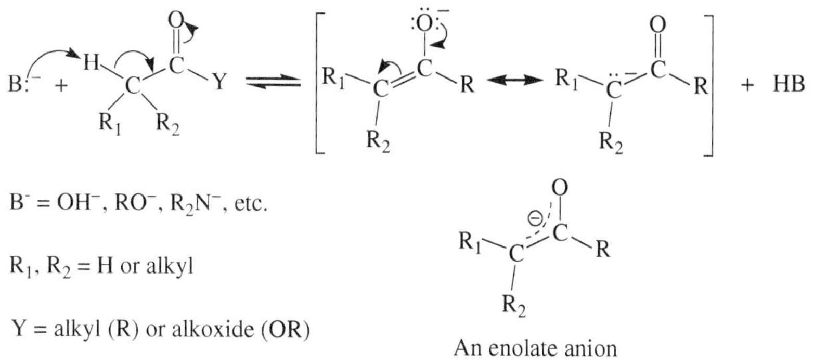

$B^- = OH^-, RO^-, R_2N^-$, etc.

R_1, R_2 = H or alkyl

Y = alkyl (R) or alkoxide (OR)

An enolate anion

Equilibrium constant: $K = \dfrac{[\text{Enolate}][\text{HB}]}{[\text{Carbonyl compound}][\text{B}^-]} = \dfrac{K_a\,(\text{C}_\alpha\text{–H})}{K_a\,(\text{HB})}$

FIGURE 9.2   Formation of the enolate from a carbonyl compound and its equilibrium constant.

The enolate anion is characterized by a *three-center, four-electron π bond*, and the negative charge is delocalized along the C_α–C–O moiety. The species experiences resonance stabilization. This also accounts for the C_α–H bond acidity. An enolate anion is a strong nucleophile with its reactive center being on the α-carbon (carbon nucleophile). Its nucleophilicity will be further addressed in Section 9.1.2.

Lithium enolates   The stability of an enolate anion is related to the nature of its counterion. Similar to proton, the lithium Li^+ ion possesses a very small radius so that its positive charge is very concentrated. In a lithium enolate, Li^+, with its concentrated positive charge, strongly attracts the negative charge on the oxygen atom greatly stabilizing the anion. In general, deprotonation of a carbonyl compound (ketone, aldehyde, or ester) by lithium diisopropyl amide [LDA, LiN (i-Pr)$_2$] leads to the formation of a stable lithium enolate in a quantitative yield (with the equilibrium constant greater than 10^{15}). Figure 9.3 shows crystal structure of the lithium enolate formed by reaction of *t*-butylmethyl ketone (3,3-dimethyl-2-butanone) with LiN(i-Pr)$_2$ in the presence of pyridine (Reaction 9.1) [1].

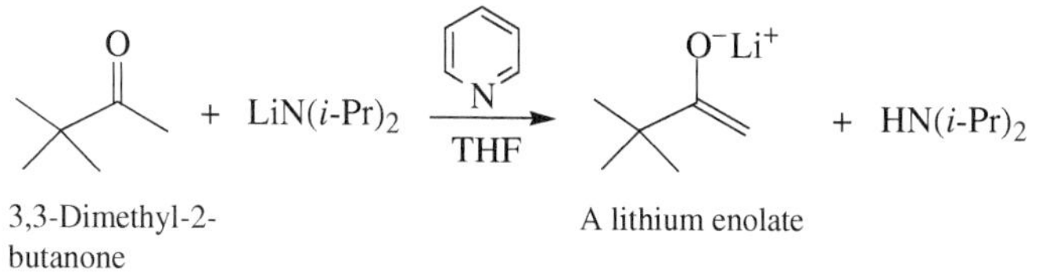

3,3-Dimethyl-2-butanone

A lithium enolate

$$+ \text{LiN}(i\text{-Pr})_2 \xrightarrow[\text{THF}]{} \quad + \text{HN}(i\text{-Pr})_2 \qquad (9.1)$$

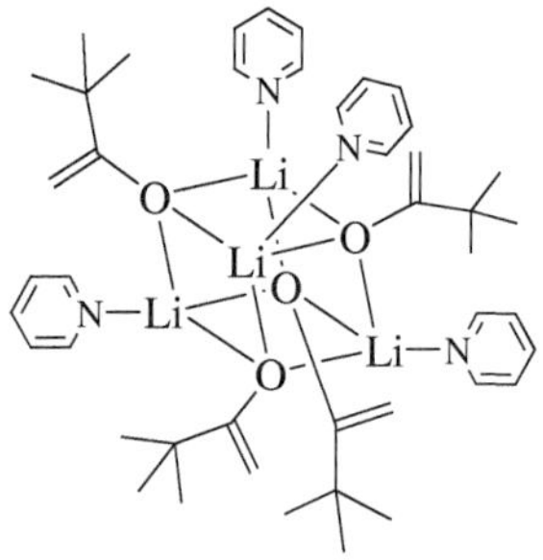

FIGURE 9.3   Crystal structure of lithium enolate of 2,2-dimethyl-2-butanone.

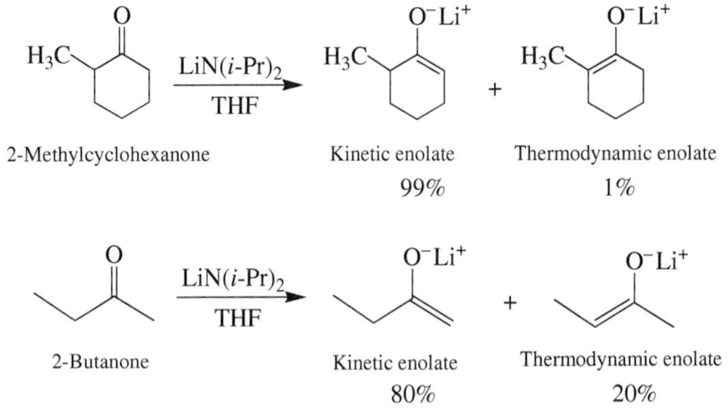

FIGURE 9.4   Regiochemistry for deprotonation of unsymmetrical ketones by LDA giving the kinetic enolates as the major products. Modified from Hoffman [2].

There are four lithium enolates (identity of which illustrated in Reaction 9.1) arranged in cubic fashion. The oxygen atoms and lithium atoms are at alternating corners of a cube. Each lithium ion is coordinated by four ligands: three oxygen atoms from enolate anions and the nitrogen atom of a pyridine molecule [1].

When an unsymmetrical ketone bearing hydrogen atoms in both α-carbons is added to LiN(i-Pr)$_2$ slowly (to maintain the base in excess in the entire reaction process), the deprotonation is regioselective, giving the less stable enolate (with the C=C carbons less substituted) as the major product (Fig. 9.4) [2]. Clearly, the deprotonation is *kinetically controlled*. The observed regioselectivity is likely due to the steric hindrance of the bulky diisopropyl amide base. When the base approaches the hydrogen on the less substituted α-carbon, the pathway is less sterically hindered, which results in a faster deprotonation and the corresponding enolate is the major product. If the ketone is in excess, the initially formed kinetic enolate (less substituted and less stable) can undergo a spontaneous isomerization to the thermodynamic enolate (more substituted and more stable). This isomerization is catalyzed by the excess ketone as shown below.

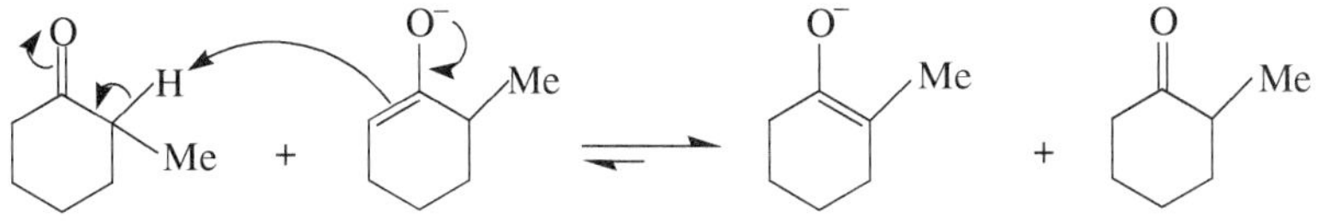

2-Methylcyclohexanone   Kinetic enolate   Thermodynamic enolate   2-Methylcyclohexanone

Net reaction:

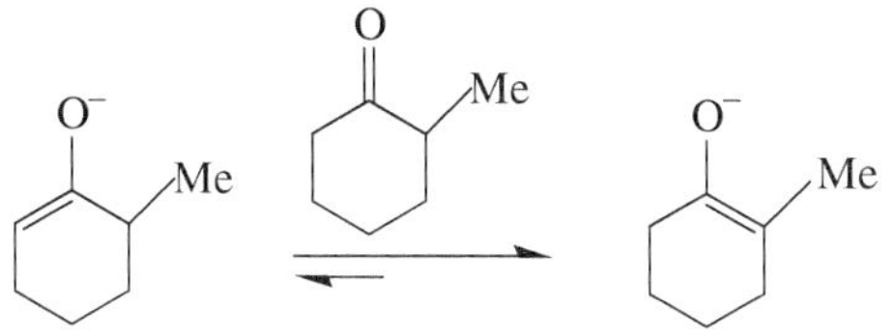

Kinetic enolate   Thermodynamic enolate

Enol   In the presence of a strong acid, a ketone or an aldehyde undergoes a reversible keto–enol tautomerization as shown in Reaction 9.2.

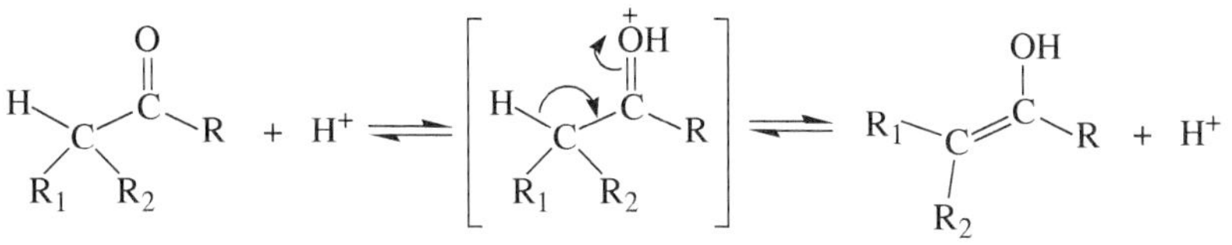

Keto   Enol

$$(9.2)$$

Similar to enolate, an enol is also a strong carbon nucleophile. The nucleophilicity of the C=C π bond is greatly reinforced by the electron donating –OH group due to the activity (hyperconjugation effect) of the oxygen lone pair of electrons. The sideway overlap of a lone-pair of electrons in the hydroxyl oxygen atom with the $\pi*$ orbital in C=C (hyperconjugation) results in resonance stabilization in enol (c.f. Figure 9.1, hyperconjugation in carbonyl compounds) and enhances the nucleophilicity of the β-carbon (Eq. 9.3).

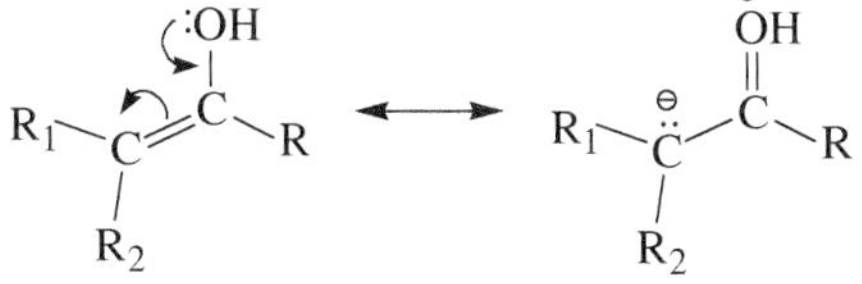

$$(9.3)$$

9.1.2   Molecular Orbitals and Nucleophilicity of Enolates

Figure 9.5 shows a molecular orbital model for the three-center, four-electron π bond formed in the C_α–C–O moiety of an enolate anion which is generated by deprotonation of a C_α–H bond in a carbonyl compound. Both the α-carbon and the carbonyl

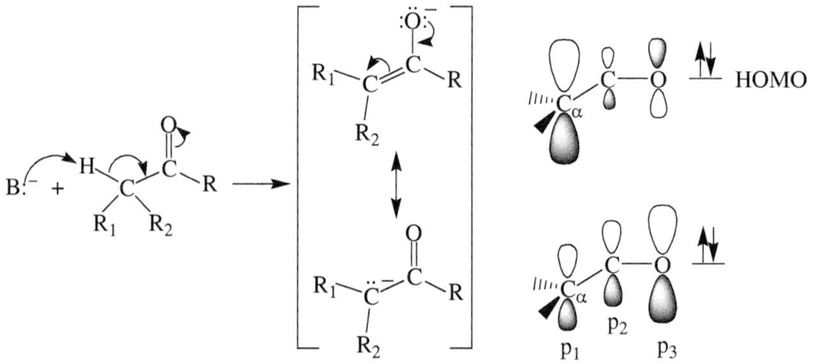

FIGURE 9.5 Occupied molecular orbitals for the three-center, four-electron bond of an enolate.

carbon in enolate are sp^2 hybridized, and each of the carbon atoms formally contributes one π electron residing in the unhybridized p_π orbital. The oxygen atom formally contributes two π electrons (a lone pair of electrons) residing in its p_π orbital. Therefore, the anion contains a total of four π electrons which are delocalized in the C_α–C–O moiety via the sideway overlap of three p orbitals (a conjugated π bond). The linear combination (overlap) of the three p orbitals in the C_α–C–O moiety (p_1, p_2, and p_3, respectively) of an enolate results in the formation of three molecular orbitals: Two occupied bonding π orbitals (Ψ_1 and Ψ_2) as shown below and an unoccupied antibonding $\pi*$ orbital (LUMO).

$$\Psi_1 = c_{11}p_1 + c_{12}p_2 + c_{13}p_3 (c_{11} \approx c_{12} < c_{13})(\text{lower-lying occupied MO})$$

$$\Psi_2 = c_{21}p_1 + c_{22}p_2 - c_{23}p_3 (c_{21} \gg c_{23} > c_{22})(\text{HOMO})$$

Only the highest occupied molecular orbital (HOMO) (Ψ_2) and lower-lying occupied MO (Ψ_1) are shown in Figure 9.5, and the lowest unoccupied molecular orbital (LUMO) is omitted. The lower-lying occupied MO is oxygen-based ($c_{11} \approx c_{12} < c_{13}$), while the HOMO is α-carbon-based ($c_{21} \gg c_{23} > c_{22}$). The relative contributions of the p_π orbitals of the carbon and oxygen atoms to HOMO and the lower-lying MO are determined by the difference in electronegativities of carbon and oxygen. Since an enolate anion is electron-rich (nucleophile), its reacting orbital toward an electrophile should be the HOMO (Ψ_2) in which the electron density is mainly concentrated on the less electronegative α-carbon [The value of the p_π eigenvector coefficient on the α-carbon (c_{21}) is much greater than those on the other atoms (c_{22} and c_{23}) as shown in Figure 9.5]. This makes the α-carbon of the enolate nucleophilic. When it reacts with an electrophile, the electrophile is usually bonded to the nucleophilic α-carbon by the end of the reaction (Reaction 9.4).

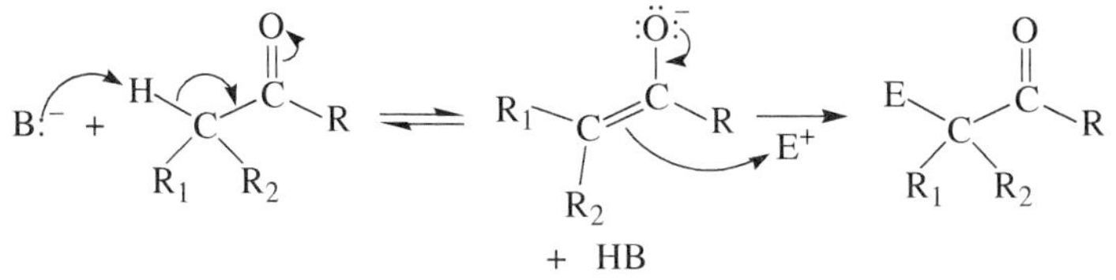

$$(9.4)$$

9.2 ALKYLATION OF CARBONYL COMPOUNDS (ALDEHYDES, KETONES, AND ESTERS) VIA ENOLATES AND HYDRAZONES

9.2.1 Alkylation via Enolates

Haloalkanes (RX, X = Cl, Br, or I) are among the most common electrophiles. Enolates generated from aldehydes, ketones, and esters undergo alkylation reactions with RX on the α-carbon of the carbonyl compounds (Fig. 9.6). The reactions follow an S_N2 mechanism on the haloalkane RX reagents (alkylating agents) [1–3]. The primary and methyl RX give the best yields essentially with no side reactions. Since the initial kinetic deprotonation for unsymmetrical ketones takes place preferably on the less substituted α-carbon, the major alkylation products for these ketones are those formed from the reactions on the less substituted α-carbons.

The reactions with secondary RX are satisfactory, but give mixtures of alkylation and elimination (E2 or E1) products as the intermediate enolates are strongly nucleophilic and strongly basic as well (Eq. 9.5) [1].

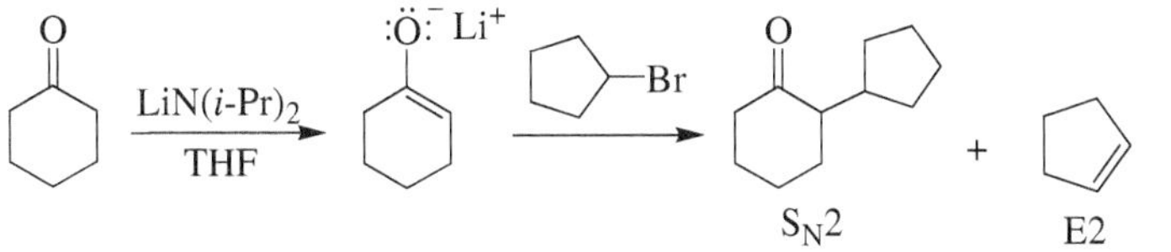

$$(9.5)$$

The alkylation reactions with tertiary RX fail completely, and the only products derived from the RX are those due to eliminations (Eq. 9.6) [3].

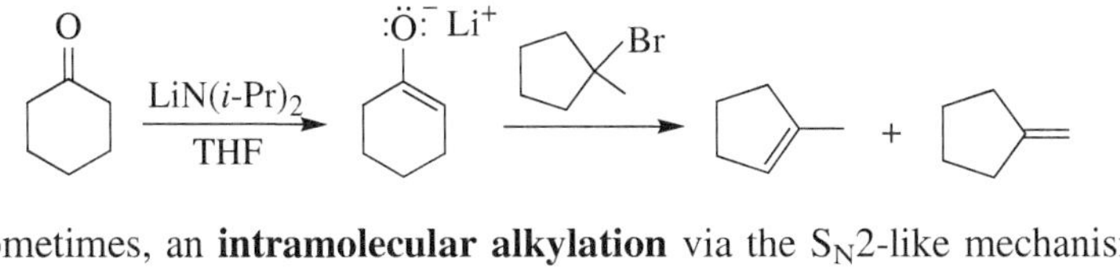

$$(9.6)$$

Sometimes, an **intramolecular alkylation** via the S_N2-like mechanism can be facilitated to introduce a cycloalkane ring to a carbonyl compound, such as an ester

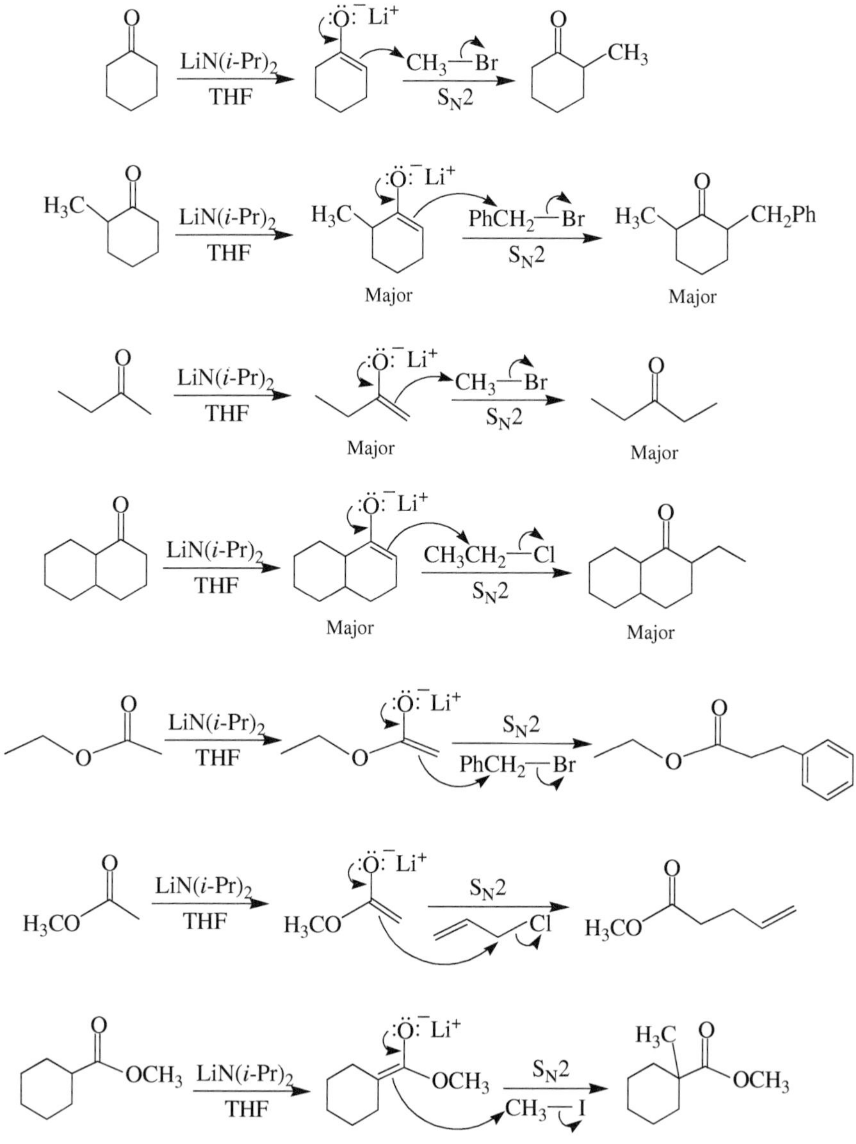

FIGURE 9.6   Alkylation of carbonyl compounds via enolates and primary alkyl halides by the S$_N$2 reactions.

(Fig. 9.7a) or a ketone (Fig. 9.7b). Reaction of 1,5-dichloropentane (a bifunctional substrate) with the enolate produced by deprotonation of ethyl acetate results in alkylation on the α-carbon of the ester. Then the intermediate ester (ethyl 7-chloroheptanoate) is treated by LiN(i-Pr)$_2$ to generate a second enolate, which undergoes

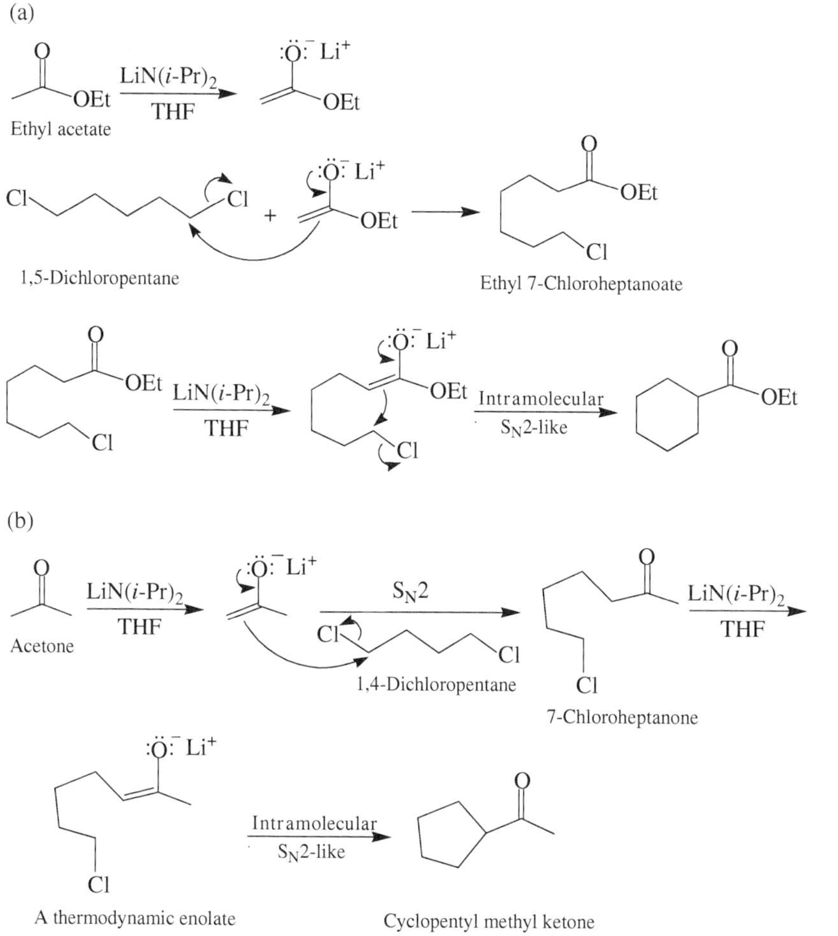

(a)

Ethyl acetate

1,5-Dichloropentane

Ethyl 7-Chloroheptanoate

Intramolecular S_N2-like

(b)

Acetone

1,4-Dichloropentane

7-Chloroheptanone

A thermodynamic enolate

Intramolecular S_N2-like

Cyclopentyl methyl ketone

FIGURE 9.7 Alkylation of carbonyl compounds via enolates and primary alkyl dihalides giving rise to ring constructions: (a) ethyl acetate/1,5-dichloropentane and (b) acetone/1,4-dichloropentane.

a subsequent intramolecular alkylation via the S_N2-like mechanism to form a cyclohexane ring on the α-carbon of the ester. Similarly, reaction of 1,4-dichlorobutane with the enolate of acetone gives 7-chloroheptanone, treatment of which by LiN(*i*-Pr)$_2$ leads to a facile intramolecular S_N2-like alkylation. It occurs via a thermodynamic enolate intermediate, giving cyclopentyl methyl ketone.

9.2.2 Alkylation via Hydrazones and Enamines

An alternative method for alkylation on the α-carbon of carbonyl compounds is accomplished via an *N,N*-dimethyl hydrazone ($R_2C=NNMe_2$) (Fig. 9.8). This

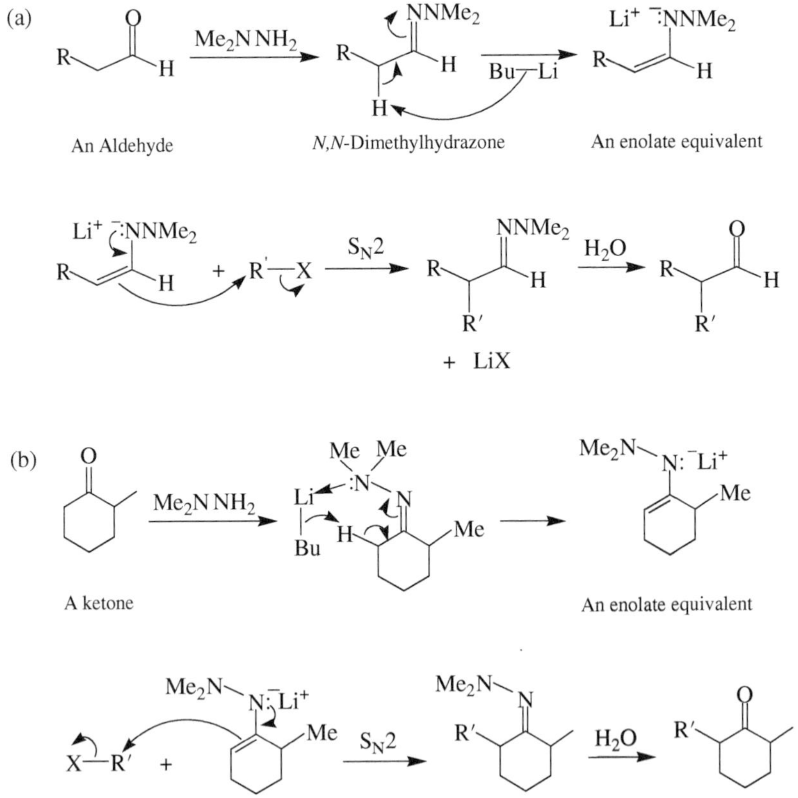

(a)

An Aldehyde   N,N-Dimethylhydrazone   An enolate equivalent

+ LiX

(b)

A ketone   An enolate equivalent

FIGURE 9.8   Alkylation of (a) an aldehyde and (b) a ketone via N,N-dimethylhydrazone. Modified from Hoffman [2].

method is particularly useful for alkylation of aldehydes. An aldehyde is more reactive (electrophilic) than a ketone or an ester. When an aldehyde is treated by a strong base, the initially formed enolate can react with the unreacted aldehyde giving the aldol side product. The problem can be overcome by converting the aldehyde to a **less reactive** N,N-dimethyl hydrazone (Fig. 9.8a). The mechanism for reaction of an aldehyde with H_2NNMe_2 giving an N,N-dimethyl hydrazone is the same as that for reactions with amines (refer to Section 8.2.5). The hydrazone is then treated by butyllithium (BuLi) [$pK_a = 50$ (BuH)], a much stronger base than $LiN(i\text{-}Pr)_2$ [$pK_a = 40$ (R_2NH)], to afford an enolate equivalent (a strong α-carbon nucleophile, analogous to an enolate). It is effectively stabilized by the lithium counterion as the negative charge on the nitrogen is strongly attracted to the small Li^+. The enolate equivalent, once formed, is allowed subsequently to react with a haloalkane ($R'X$) leading to alkylation on the α-carbon. Finally, the alkylated hydrazone product is hydrolyzed giving an alkylated aldehyde.

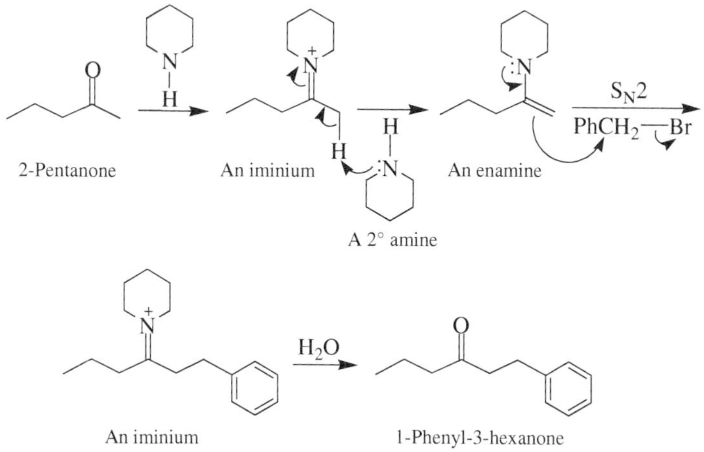

FIGURE 9.9   Alkylation of a ketone via a secondary amine.

For an unsymmetrical ketone, the alkylation through an N,N-dimethyl hydrazone preferably takes place on the less sterically hindered α-carbon (Fig. 9.8b) [2]. This regioselectivity is consistent with that of the alkylation of an enolate formed by a kinetic deprotonation of the ketone. The observed regiochemistry in Figure 9.8b is directly related to the structure of the hydrazone intermediate in which the dimethylamino Me_2N- group is pointed to the side of the less substituted α-carbon. As a result, deprotonation of the $C_α-H$ bond by Bu–Li is directed effectively by the Me_2N- group and takes place on the less substituted α-carbon, giving a kinetic enolate equivalent. Subsequent reaction of the kinetic enolate equivalent with $R'X$ results in alkylation of the less substituted α-carbon.

The alkylation of α-carbon of ketones and aldehydes can also be accomplished via a nucleophilic enamine which is generated by reaction of the carbonyl compound with a secondary amine (R_2NH) (Fig. 9.9). The enamine, once formed, is treated by a haloalkane (RX). An S_N2 reaction of RX with the enamine results in alkylation on α-carbon of the enamine. Upon hydrolysis, the intermediate iminium is converted to the alkylated ketone product (1-phenyl-3-hexanone).

9.3   ALDOL REACTIONS

9.3.1   Mechanism and Synthetic Utility

General situation   The enolate anion (nucleophile) of carbonyl compounds (aldehydes, ketones, esters, carboxylic anhydrides, or amides) undergoes addition to the electrophilic carbonyl group of aldehydes or ketones (Reaction 9.7).

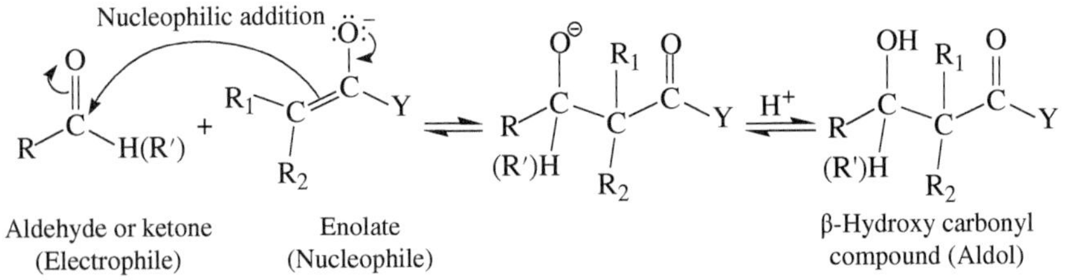

Nucleophilic addition

Aldehyde or ketone
(Electrophile)

Enolate
(Nucleophile)

β-Hydroxy carbonyl
compound (Aldol)

$$(9.7)$$

The nucleophilic addition makes a new C—C bond. The initially formed oxyanion in the reaction can be readily protonated (often by a solvent molecule). As a result, a β-hydroxy carbonyl compound is produced. This type of nucleophilic addition is referred to as the **aldol reaction (addition)**.

Aldol reaction and condensation of aldehydes and ketones   Among the common types of aldol reactions is the reaction between two molecules of the same aldehyde or ketone which is catalyzed by the hydroxide OH⁻ base (Fig. 9.10). In the presence of a catalytic amount of the hydroxide OH⁻ base, two aldehyde (or ketone) molecules combine to form a β-hydroxyaldehyde(ketone) (aldol) in which a new C—C

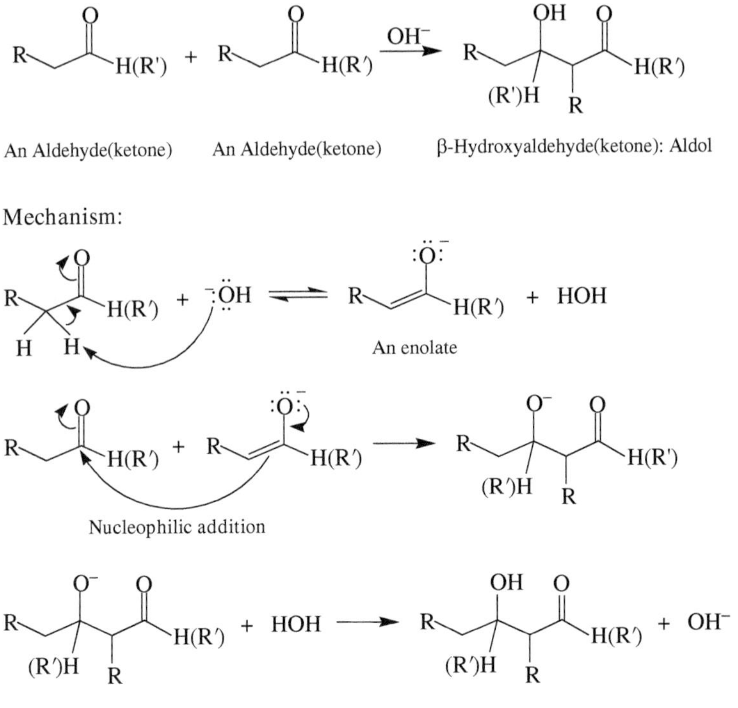

An Aldehyde(ketone)   An Aldehyde(ketone)   β-Hydroxyaldehyde(ketone): Aldol

Mechanism:

An enolate

Nucleophilic addition

FIGURE 9.10   The general mechanism for an aldol reaction of an aldehyde or ketone.

bond is made. The reaction takes place via an intermediate enolate (nucleophile) which is formed only in a small amount by a reversible deprotonation of an aldehyde (or ketone) molecule by OH^-. The enolate, once formed, attacks a second unreacted aldehyde (or ketone) molecule (electrophile) in the reaction system to effect a nucleophilic addition in which a new C—C bond is made. Finally, the intermediate oxyanion is protonated by water (solvent) giving the aldol product and regenerating OH^-. In the step of nucleophilic addition, the attack of enolate on the carbonyl group of a second aldehyde (ketone) molecule takes place equally on both faces (front and back) of the carbonyl. As a result, the aldol product is not stereospecific. Upon heating and increasing the quantity of OH^-, the initially formed β-hydroxyaldehyde (ketone) undergoes an E1cb dehydration reaction to give an α,β-unsaturated aldehyde (or ketone) (Reaction 9.8).

$$(9.8)$$

The overall reaction in Reaction 9.8 is called **aldol condensation**.
Treatment of cyclohexanone with the hydroxide base results in an aldol reaction (Reaction 9.9).

Cyclohexanone

$$(9.9)$$

The initially formed enolate of the ketone is added to the carbonyl of the second unreacted ketone molecule giving a β-hydroxyketone. Upon heating, the β-hydroxyketone undergoes an E1cb dehydration giving an α,β-unsaturated ketone, with a C=C double bond formed between the two cyclohexane rings.

Crossed aldol reactions The aldol reactions can also take place between two molecules from different aldehydes or ketones (crossed aldol reactions). A crossed aldol reaction is readily conducted when one of the carbonyl compounds is an aldehyde that lacks α-hydrogen and the other is a ketone (or an acyl derivative) [1]. In this case, it is only possible to form the enolate anion of the ketone, which preferably reacts with the more electrophilic aldehyde (rather than a second ketone molecule) to give a single aldol product. Figure 9.11a shows the hydroxide base catalyzed aldol reaction of acetone with benzaldehyde in 1:1 molar ratio. The intermediate enolate generated from acetone attacks the electrophilic carbonyl of benzaldehyde exclusively to lead to the formation of an intermediate aldol first, which undergoes subsequent E1cb

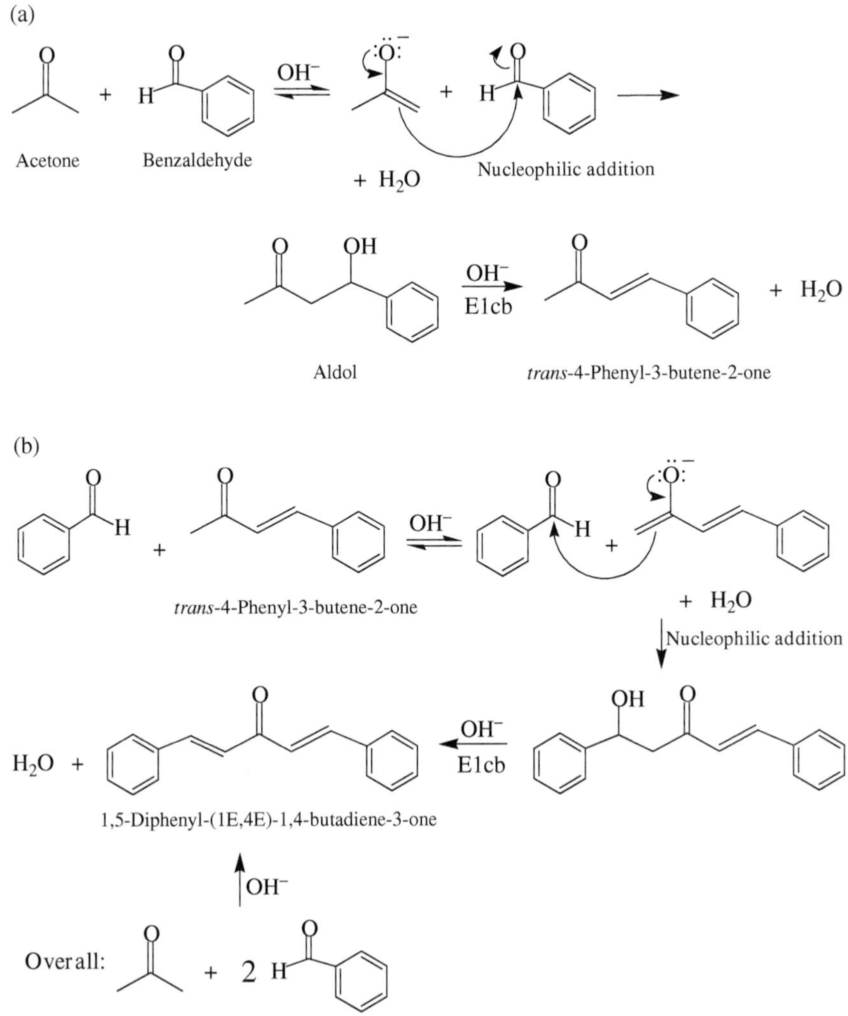

(a)

(b)

FIGURE 9.11 The crossed aldol condensation reactions between (a) benzaldehyde and acetone and (b) benzaldehyde and trans-4-phenyl-3-butene-2-one.

dehydration spontaneously at ambient temperature to give a conjugated α,β-unsaturated ketone (*trans*-4-phenyl-3-butene-2-one).

The conjugation effect of the phenyl ring greatly stabilizes the α,β-unsaturated ketone product, which makes the E1cb dehydration of the intermediate aldol favorable even at ambient temperature. If the overall reaction is performed in the molar ratio of acetone:benzaldehyde being 1:2, the *trans*-4-phenyl-3-butene-2-one product formed from the crossed aldol reaction of the first benzaldehyde molecule will undergo the aldol condensation reaction with the second benzaldyde molecule (Fig. 9.11b) to lead to the formation of a larger conjugated, more stable α,β-unsaturated ketone product [1,5-diphenyl-(1E,4E)-1,4-butadiene-3-one].

The hydroxide or alkoxide base catalyzed crossed aldol reaction between a ketone and an aldehyde bearing an α-hydrogen may lead to a mixture of products, as both ketone and aldehyde enolates can be formed in the reaction condition, and each of them reacts with an aldyhyde molecule giving a different aldol product. This difficulty can be overcome by using LDA.

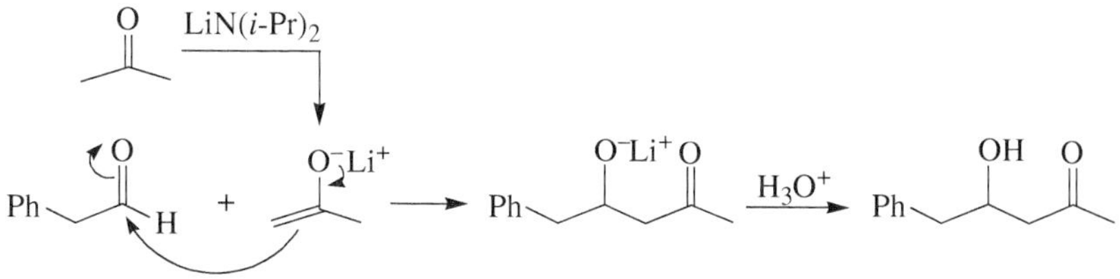

$$(9.10)$$

Reaction 9.10 shows an example of this case. In order to obtain a single product from this crossed aldol reaction, first the ketone (acetone) reactant is treated with LDA in the 1:1 molar ratio to convert all the acetone to its enolate completely. Then the aldyhyde (which contains α-hydrogens) is experimentally added to the acetone enolate to bring about the single aldol reaction as shown in Reaction 9.10, giving a sole β-hydroxyketone product.

LDA can also be used to direct a crossed aldol reaction between two different ketones. For example, in order to synthesize the β-hydroxyketone in Reaction 9.11,

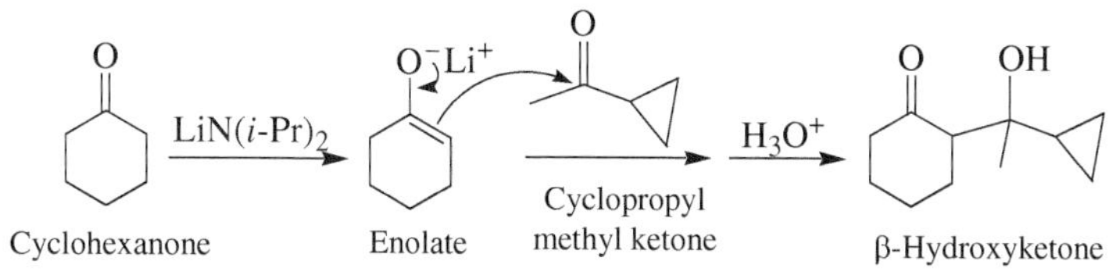

$$(9.11)$$

the crossed aldol reaction between cyclohexanone and cyclopropyl methyl ketone is conducted in such manner that the enolate of cyclohexanone is added to carbonyl of cyclopropyl methyl ketone. Experimentally, cyclohexanone is treated with LDA first to give the enolate quantitatively. Then cyclopropyl methyl ketone is added to the enolate of cyclohexanone to lead to the formation of the desired product after aqueous work-up. On the other hand, addition of the enolate of cyclopropyl methyl ketone (made by LDA quantitatively) to cyclohexanone will give a different β-hydroxyketone product (Reaction 9.12).

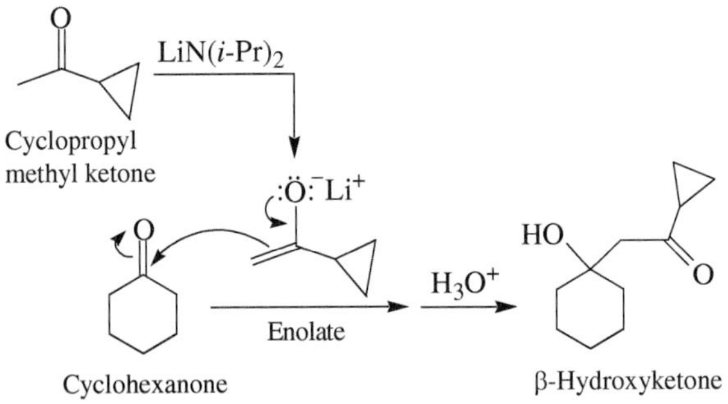

$$(9.12)$$

Research has shown that the cross aldol reactions effected by lithium enolates (generated using LDA) occur via a transition state of a six-membered ring possessing a pseudo chair-conformation, and the reactions are stereoselective (Fig. 9.12) [4]. When a ketone ($R_2CH_2COR_1$) is added to LDA to keep LDA in excess during the reaction (kinetic condition), both (Z)- and (E)-enolates (as the lithium salts) are formed with (Z)-enolate being the major product. Then the (Z)-enolate reacts with an aldehyde (R_3CHO) to give stereospecific aldols. As indicated in Figure 9.12, two possible transition states can be formed. In each of them, both the carbonyl oxygen of the aldehyde and the enolate oxygen are linked by the lithium ion and the α-carbon of the enolate approaches the carbonyl carbon of the aldehyde to have a C—C bond between the two molecules partially formed. The two transition states lead to two different 1,2-*syn* aldol products.

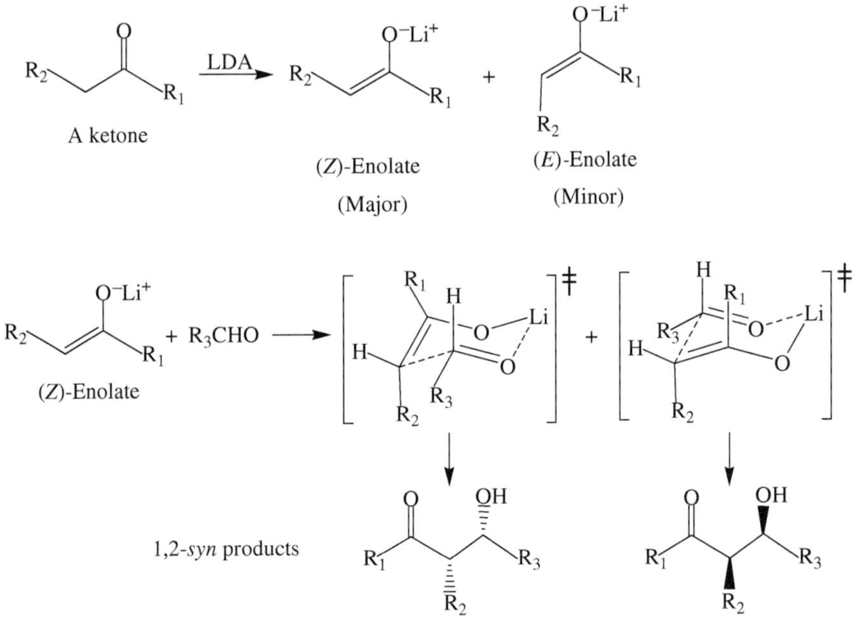

FIGURE 9.12 Transition states and stereochemistry for the aldol reaction of a lithium ketone enolate and an aldehyde. Modified from Arya and Qin [4].

Intramolecular aldol reactions   When the skeletal carbon chain of a functionalized organic compound contains two carbonyl groups (such as a diketone), a thermodynamically favorable intramolecular aldol reaction can take place via
nucleophilic addition of the enolate of one carbonyl group to the second electrophilic carbonyl (Fig. 9.13). The cyclic aldol product usually contains a five-
membered or a six-membered carbon ring. The formation of a C=C double bond
in a five-membered ring by dehydration is usually faster than that in a six-
membered ring. Therefore, for the reaction of 2,6-heptanedione, a cyclic aldol
product can be isolated at ambient temperature. Only upon heating, can an
E1cb reaction take place giving a cyclic α,β-unsaturated ketone. For reaction of
2,5-hexanedione, the intermediate cyclic aldol is unstable, and it undergoes facile
E1cb reaction even at a low temperature to give a cyclic α,β-unsaturated ketone.

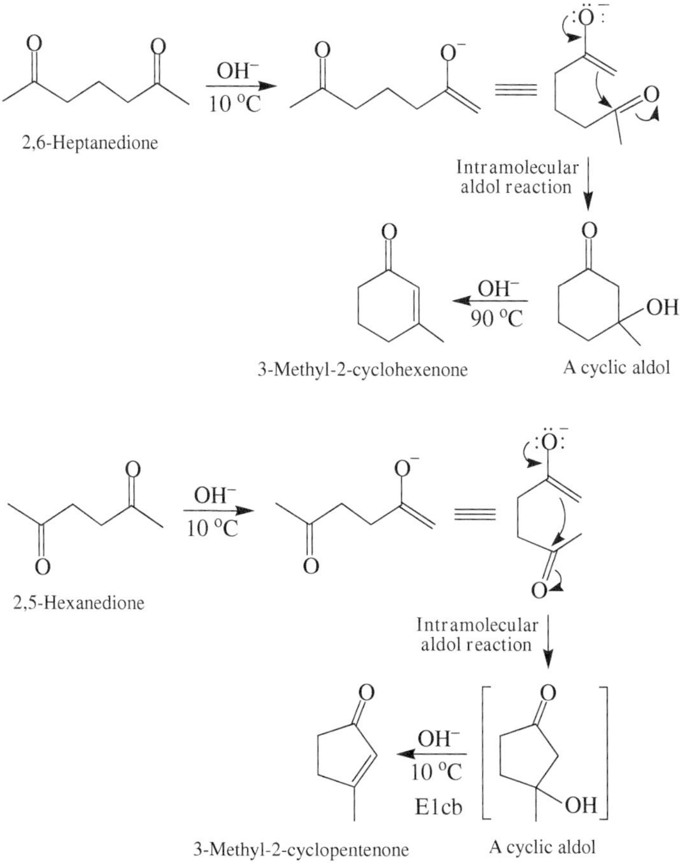

FIGURE 9.13   The intramolecular aldol condensations to form cyclic α,β-unsaturated
ketones.

The intramolecular aldol condensation of the following diketone gives a bicyclic α,β-unsaturated ketone (Eq. 9.13) [2].

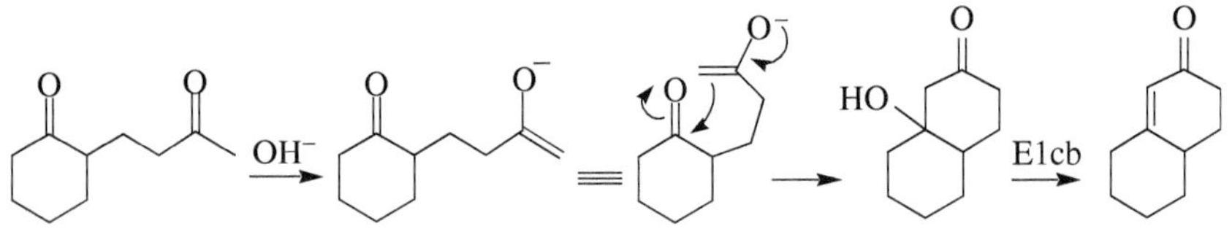

$$(9.13)$$

Other aldol reactions   Aldol condensation can also occur between an aldehyde and an enolate of an acyl derivative, such as a carboxylic anhydride (Fig. 9.14). Similar to ketones and aldehydes, the α-hydrogen of acetic anhydride is activated by an electronegative carbonyl group and is slightly acidic. In the presence of basic acetate, acetic anhydride undergoes reversible C_α–H deprotonation to give an enolate, which

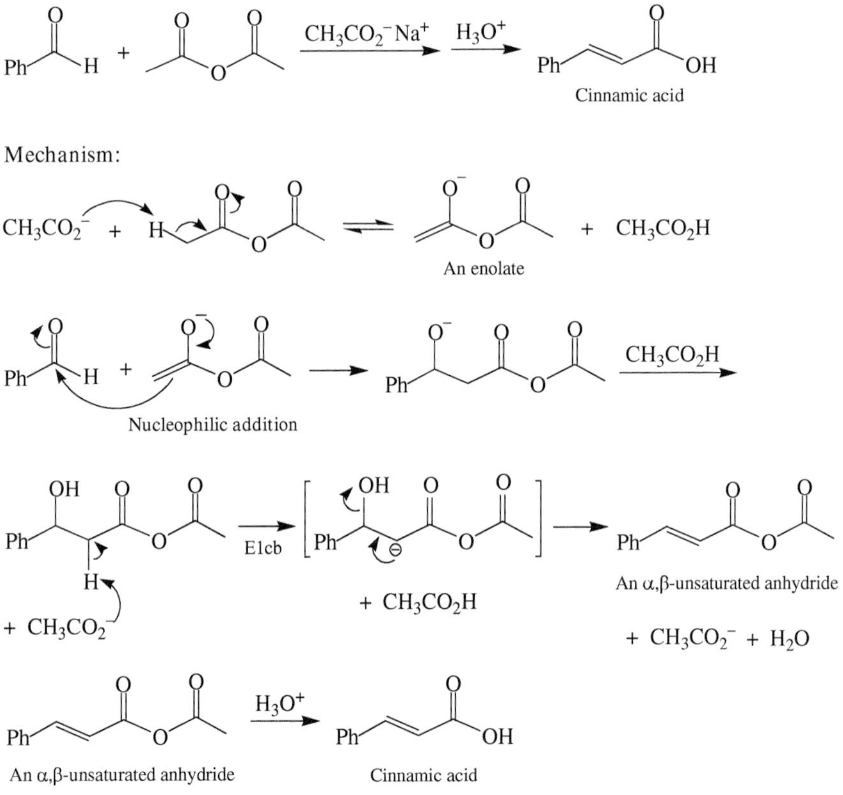

Mechanism:

FIGURE 9.14   The aldol reaction of benzaldehyde with acetic anhydride catalyzed by acetate.

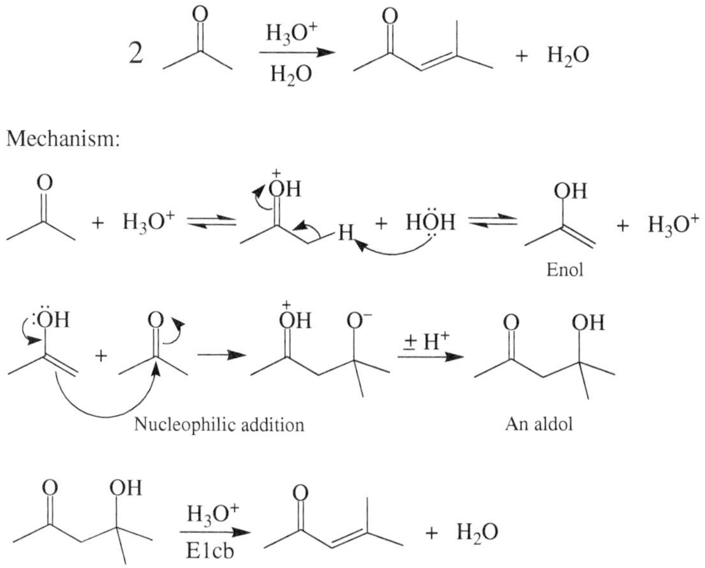

FIGURE 9.15   The acid catalyzed aldol condensation of acetone via an enol.

subsequently attacks benzaldehyde carbonyl to effect a nucleophilic addition, giving a β-hydroxyanhydride. It then undergoes an E1cb reaction to afford an α,β-unsaturated anhydride. The conjugation effect of the phenyl ring greatly enhances the stability of the α,β-unsaturated anhydride. This stabilization makes the dehydration of β-hydroxyanhydride thermodynamically favorable, and it takes place spontaneously at ambient temperature. Hydrolysis of the α,β-unsaturated anhydride gives cinnamic acid.

Figure 9.15 shows that aldol condensation reactions can also take place in the presence of a strong acid. The reaction is effected by nucleophilic addition of an enol of the ketone reactant to the carbonyl group of the second ketone molecule. Upon heating, the β-hydroxyketone undergoes E1cb reaction giving an α,β-unsaturated ketone anhydride.

9.3.2   Stereoselectivity

(S)-*Proline catalyzed aldol reaction*   Aldol reactions can be catalyzed by secondary amines (R_2NH), and the reactions take place via intermediate nucleophilic enamines. Recent research has shown that some chiral amines, typically (S)-proline and its analogues, catalyze stereoselective (asymmetric) aldol reactions with high **enantiomeric excess** (ee) [5–8]. *Enantiomeric excess (ee) is equal to the yield of the major enantiomer minus the yield of the minor enantiomer in the product whose total yield is normalized to 100%*. Figure 9.16 shows the aldol reaction of acetone (a methyl ketone) with 4-nitrobenzaldehyde which is catalyzed by 30% mol% (S)-proline (one of the naturally occurring 20 standard amino acids in living organisms). The reaction gives a stereospecific aldol product, (R)-4-aryl-4-hydroxybutanone, in 68% yield (76% ee) [5].

In the reaction mechanism, the first step is the nucleophilic attack of the amino group in (S)-proline (a secondary amine) on the carbonyl group of acetone. As a result, the carbonyl oxygen is substituted by the nitrogen of the amino group and an intermediate iminium (=$^+$NR$_2$) is formed. Deprotonation of a C$_\alpha$–H bond in the iminium affords a nucleophilic enamine. Then the enamine reacts with the

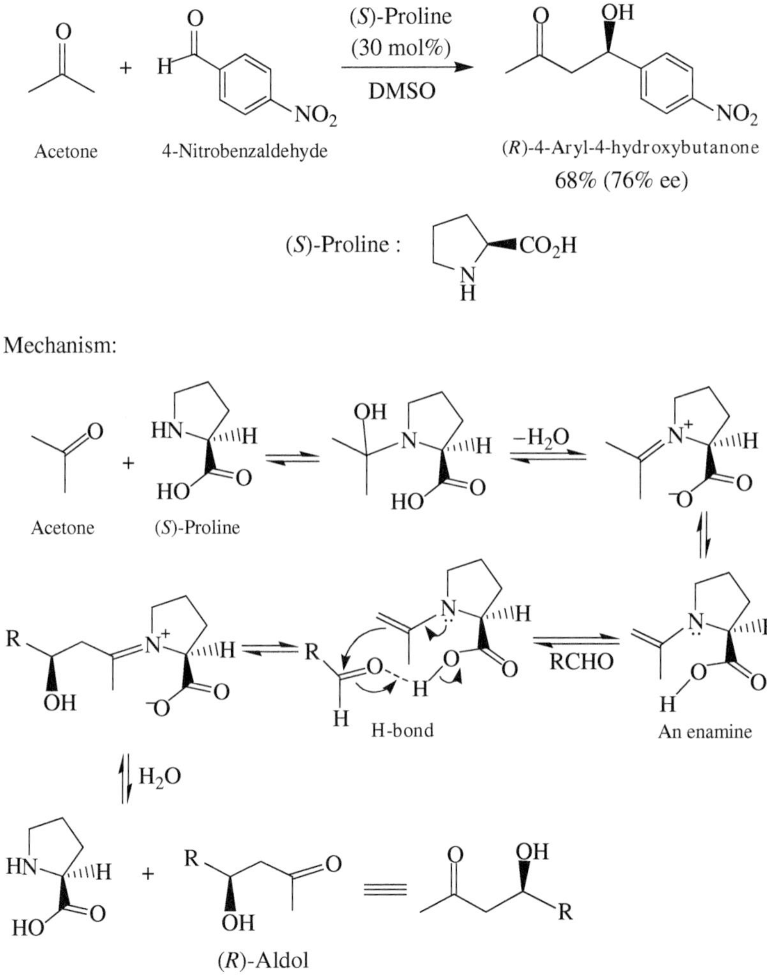

Mechanism:

FIGURE 9.16 Mechanism for the (S)-proline catalyzed enantiomerically specific aldol reaction of acetone with 4-nitrobenzaldehyde.

aldehyde substrate (nucleophilic addition). The hydrogen bond between the –OH group of (S)-proline and the carbonyl oxygen of the aldehyde plays a critical role in the observed stereoselectivity for the aldol reaction. It holds the aldehyde in place such that the nucleophilic attack of the enamine on the carbonyl group of the aldehyde mainly occurs at the back side of the carbonyl and is facial-specific, giving the (R)-enantiomer as the major enantiomer. Finally, hydrolysis of the intermediate stereospecific enamine product affords a (R)-aldol. Similarly, the (S)-proline catalyzed reaction of acetone with isobutyraldehyde (Me_2CHCHO) gives an analogous (R)-aldol, but in a higher yield (97%) with greater enantiomeric excess (96% ee) (Reaction 9.14) [5].

$$
\text{Acetone} + \text{Isobutyraldehyde} \xrightarrow[]{(S)\text{-Proline}} \xrightarrow[]{H_2O} \quad (R)\text{-4-hydroxy-5-methylhexanone}
$$

(9.14)

Acetone   Isobutyraldehyde   (R)-4-hydroxy-5-methylhexanone
                                                              97% (96% ee)

Reactions of Boc-imines   Analogous to carbonyl (C=O), the functional group of an imine (C=N) is electrophilic. It can undergo nucleophilic addition with enolate of an aldehyde or ketone [6, 7]. Figure 9.17 shows that the (S)-proline catalyzed nucleophilic addition of acetaldehyde (CH_3CHO) to an N-Boc-imine gives a specific enantiomer of β-amino aldehyde. The reaction takes place via an enamine of acetaldehyde (CH_2=CH—NR′). Similar to the (S)-proline catalyzed aldol reaction (Fig. 9.16), the intermediate enamine attacks the electrophilic N-Boc-imine from the top face giving a specific enantiomer of β-amino aldehyde after hydrolysis. This facial selectivity for the nucleophilic attack is facilitated by the hydrogen bond between the –OH group of (S)-proline and the imine functional group C=N. The bulky Boc group is preferably pointed away from the (S)-proline molecule in the hydrogen-bonded intermediate and the R group in the N-Boc-imine is *trans* to Boc.

Figure 9.18 shows that the (S)-proline catalyzed reaction of an N-Boc-imine with a ketone (or aldehyde when R_3=H) gives a specific *syn*-diasteromer of α,β-branched β-amino ketone(aldehyde). The reaction takes place via an enamine of the ketone(aldehyde) reactant, which is analogous to the enamine of acetaldehyde as shown in (Fig. 9.17). Treatment of α,β-branched β-amino aldehyde with a series of reagents affords diastereospecific β-amino acids (Reaction 9.15) [7].

$$
\xrightarrow[\substack{NaClO_2 \quad CF_3CO_2H \\ (TFA)}]{}
$$

(9.15)

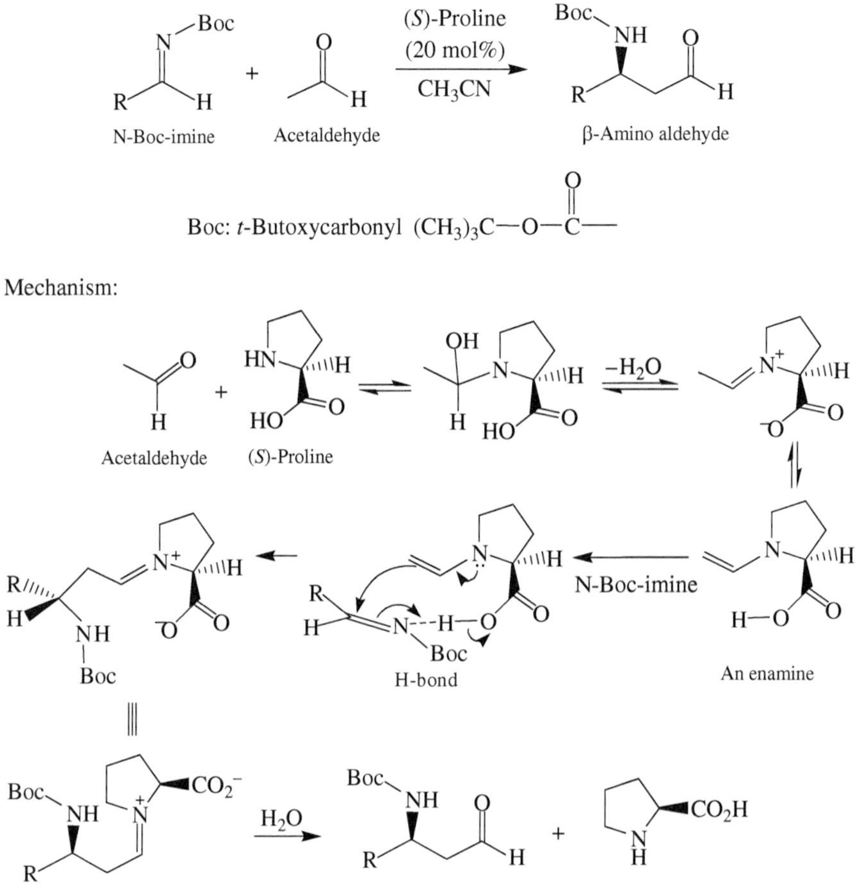

FIGURE 9.17   Mechanism for the (S)-proline catalyzed enantiomerically specific aldol reaction of an N-Boc-imine with acetaldehyde.

In Reaction 9.15, NaClO$_2$ oxidizes the aldehyde (–CHO) to carboxyl (–COOH), while CF$_3$CO$_2$H cleaves the Boc–N bond to make an amino (–NH$_2$) group. The stereospecificity of the substrate is retained in the overall treatment. The reaction is of great potential value in synthesis of peptide derivatives and related biologically active compounds.

Aldol reactions of ester enolates   According to Equation 9.7, reaction of an ester enolate (Y = OR′) with an aldehyde or ketone gives β-hydroxyester. Figure 9.19a shows **diastereoselection** in an aldol reaction effected by the (Z)-enolate of an ester. Usually, treatment of an ester by LiN(i-Pr)$_2$ gives the (Z)-enolate as the major stereoisomer [2]. The quantitatively formed, stereospecific ester enolate is then allowed

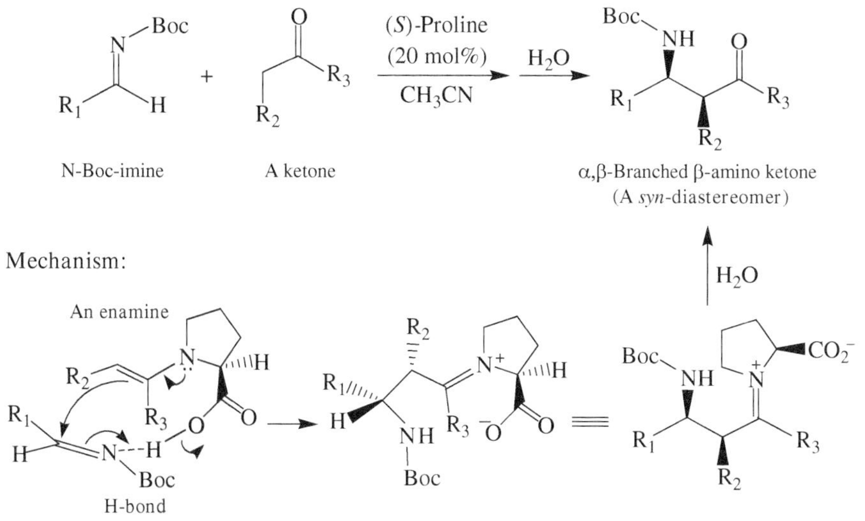

FIGURE 9.18   Mechanism for the (S)-proline catalyzed diastereoselective aldol reaction of an N-Boc-imine with a ketone.

to react with an aldehyde. In the course of the aldol reaction, the lithium counterion links the aldehyde and enolate together on the electron-donating oxygen atoms to form two possible transition states whose structures resemble the chair-conformations of cyclohexane [2, 4]. The first transition state leads to the formation of the anti-diastereomer (2R,3R) of a β-hydroxyester, and the second transition state leads to the syn-diastereomer (2R,3S) of the β-hydroxyester. In the first transition state, the isopropyl group of the aldehyde (2-methylpropanal) stays in the pseudoe-quatorial position, while in the second transition state the isopropyl group stays in the pseudoaxial position. Therefore, the first transition state is more stable than the second and is favored. *The corresponding anti-diastereomer (2R,3R) is the major product.* However, the syn-diastereomer (2R,3S) is essentially **not** formed (minor product). When the (E)-enolate of the ester (generated in a different method) is employed, the reaction gives the syn-diastereomer (2S,3R) as the major product (Fig. 9.19b). However, the syn-diastereomer (2R,3S) is essentially **not** formed (minor product).

9.3.3   Other Synthetic Applications

1,4-Addition of enolate   The aldol condensation gives rise to the formation of various α,β-unsaturated carbonyl compounds. The p_π orbitals in the alkene (C=C) portion of an α,β-unsaturated carbonyl compound can overlap with the p_π orbitals in the carbonyl (C=O) group. As a result, a conjugated π bond in the $C_\beta=C_\alpha-C=O$

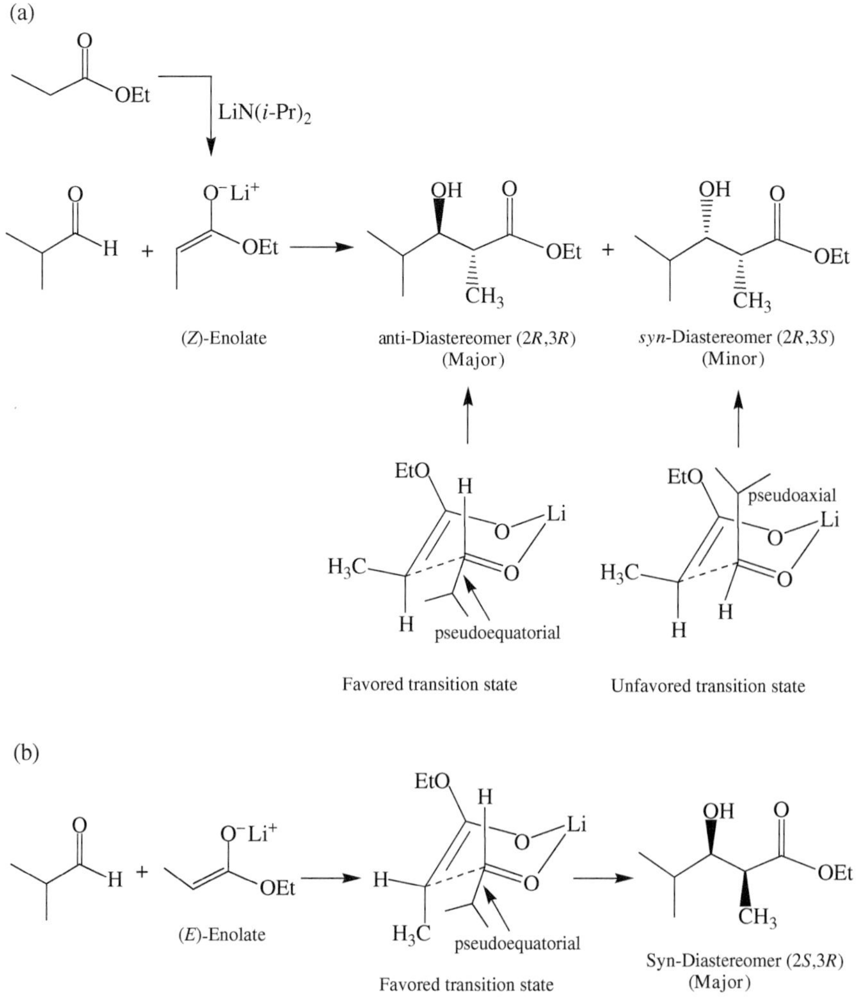

(a)

(Z)-Enolate   anti-Diastereomer (2R,3R)   syn-Diastereomer (2R,3S)
(Major)   (Minor)

Favored transition state   Unfavored transition state

(b)

(E)-Enolate   Favored transition state   Syn-Diastereomer (2S,3R)
(Major)

FIGURE 9.19   Diastereoseletive aldol reactions of (a) (Z)- and (b) (E)-enolates of esters with different aldehydes and ketones.

moiety is formed. This gives rise to two electrophilic centers in the compound, the β-carbon and the carbonyl carbon, as illustrated in Figure 9.20a. Therefore, either electrophilic carbon can be possibly attacked by a nucleophile giving a 1,2-adduct or 1,4-adduct, depending on the nature of the nucleophile. In general, a nucleophile possessing concentrated negative charge (such as alkyllithium RLi or a Grignard reagent RMgX) in a specific atom effects a 1,2-addition. On the other hand, a nucleophile in which the negative charge is dispersed into a larger space of the molecule by conjugation (such as an enolate) effects a 1,4-addition as shown in Figure 9.20b.

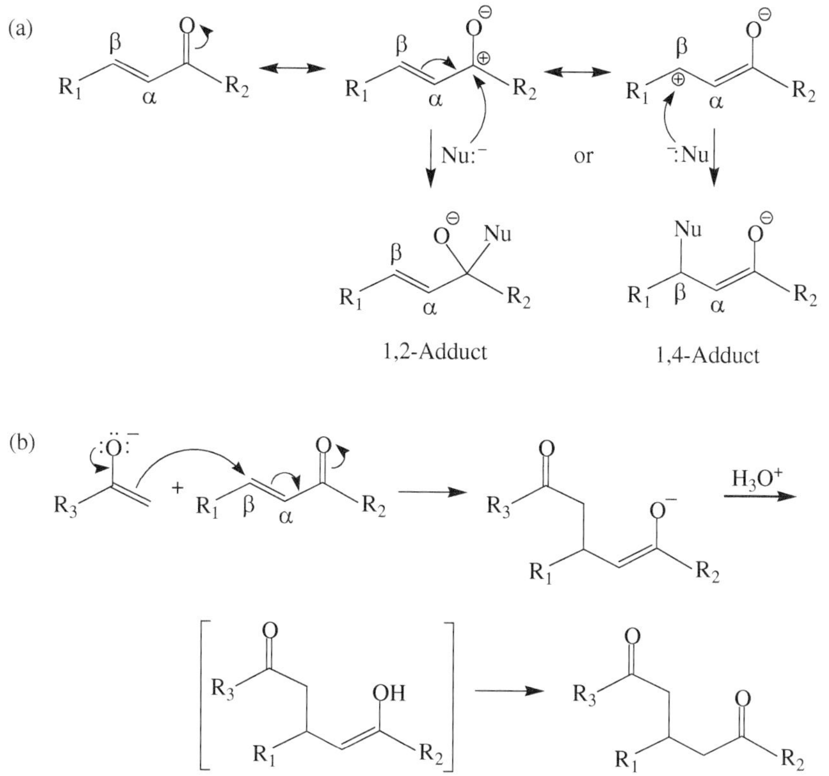

FIGURE 9.20 (a) Nucleophilic 1,2- and 1,4-additions to an α,β-unsaturated carbonyl compound; and (b) the 1,4-addition of an enolate to an α,β-unsaturated carbonyl compound.

Figure 9.20b shows that as an enolate attacks the electrophilic β-carbon of an α,β-unsaturated carbonyl compound, the C_β=C_α π electrons are shifted to the C_α—C bond domain and the C=O π electrons move to the oxygen atom, simultaneously. As a result, a new enolate is formed. Upon protonation, it is converted to an enol which undergoes rapid tautomerization to a diketone.

The α,β-unsaturated carbonyl compounds (e.g., methyl vinyl ketone) also undergoes addition to the enolate of malonate diester as shown below in Reaction 9.16.

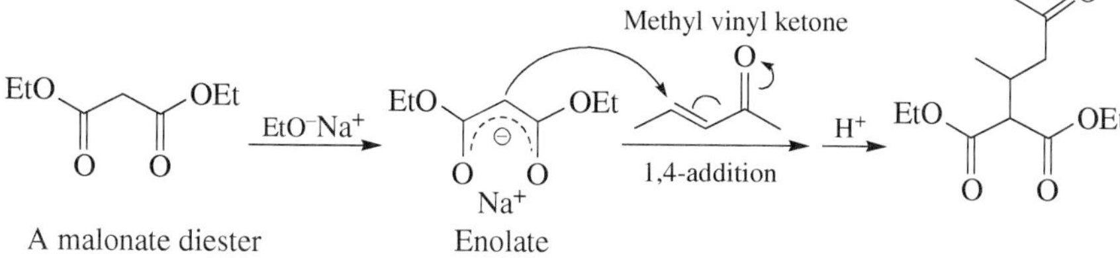

$$(9.16)$$

(a)   (b)

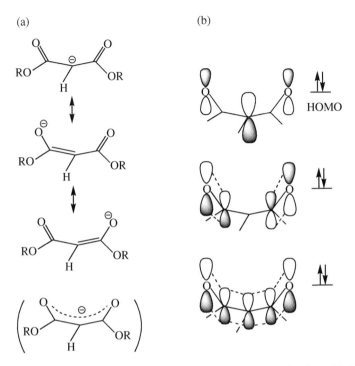

FIGURE 9.21   (a) Resonance structures and (b) the occupied molecular orbitals for the conjugated π bond of the enolate of a dicarbonyl compound.

The negative charge in the intermediate enolate is delocalized to two carbonyl groups and experiences particular resonance stabilization (Fig. 9.21a). A conjugated five-center, six-electron π bond is formed, which involves both carbonyl groups and the α-carbon in-between. Formally, each carbonyl contributes two π electrons and so does the α-carbon. Figure 9.21b shows the three occupied π molecular orbitals. In HOMO, the coefficient (contribution) of the p_π orbital of the α-carbon is much greater than those of the p_π orbitals in the carbonyl oxygen atoms. Thus, the nucleophilic center of the enolate is the α-carbon atom.

Robinson ring annulations   The combination of the 1,4-addition of enolate to α,β-unsaturated carbonyl compounds and the subsequent intramolecular aldol reaction has been effectively used in ring construction (Robinson ring annulations) in organic synthesis (Fig. 9.22). The reaction is initiated by 1,4-addition of the thermodynamic enolate of 2-methylcyclohexanone to methyl vinyl ketone (an α,β-unsaturated ketone) giving a diketone [1, 2]. Then the diketone is treated by aqueous sodium hydroxide giving a second enolate, which undergoes intramolecular aldol reaction to form a bicyclic aldol. Upon heating in the

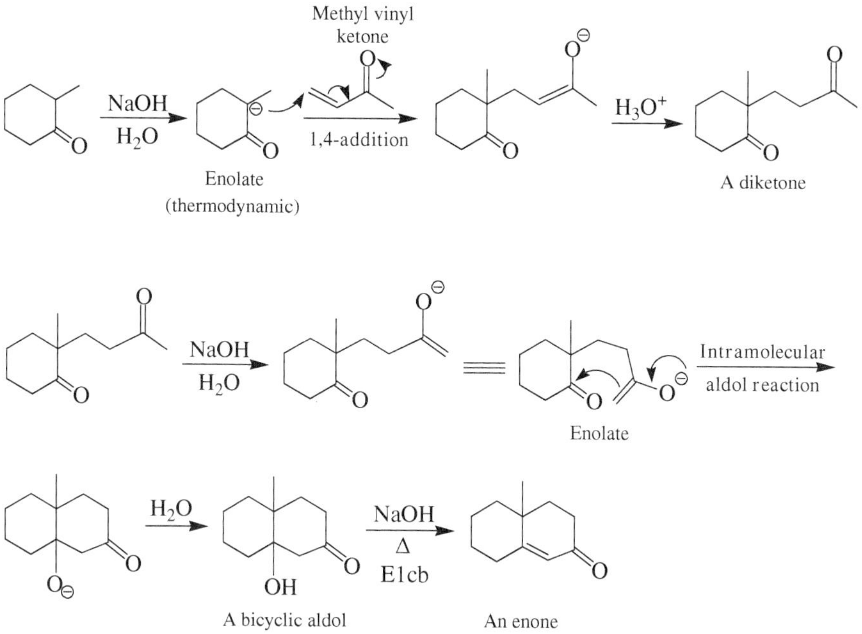

FIGURE 9.22   Mechanism for the Robinson annulations.

basic solution, the bicyclic aldol undergoes a facile E1cb reaction to produce an enone product.

Darzens condensation is the condensation of an α-haloester with a ketone or an aromatic aldehyde induced by a strong base such as *tert*-butoxide (Fig. 9.23). The

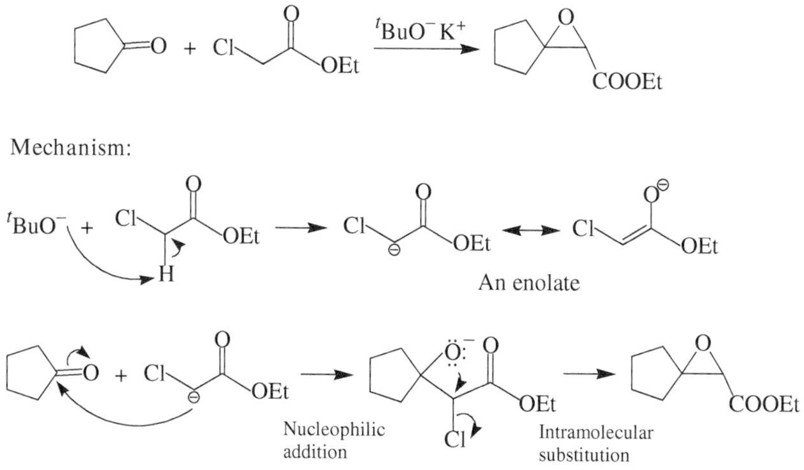

Mechanism:

FIGURE 9.23   Darzens condensation: Reaction and mechanism.

reaction takes place via the enolate of the α-haloester to give an epoxide. The α-hydrogen of the α-haloester is activated by both a carbonyl and an electronegative halo (e.g., chloro) group and becomes particularly acidic. Thus, the *tert*-butoxide base preferably deprotonates the α-haloester, rather than the ketone, to generate an ester enolate. The enolate, once formed, attacks the carbonyl of the ketone to effect a nucleophilic addition giving an intermediate oxyanion. Then the strongly nucleophilic oxyanion attacks the α-carbon bearing a chloro group to lead to an intramolecular substitution. It affords an epoxide product.

9.4   ACYLATION REACTIONS OF ESTERS VIA ENOLATES: MECHANISM AND SYNTHETIC UTILITY

In the presence of an alkoxide base, two ester molecules (often from the same ester compound) react to form a β-ketoester (Claisen condensation). The reaction takes place via an ester enolate (Fig. 9.24). Since the formation of the enolate initiated

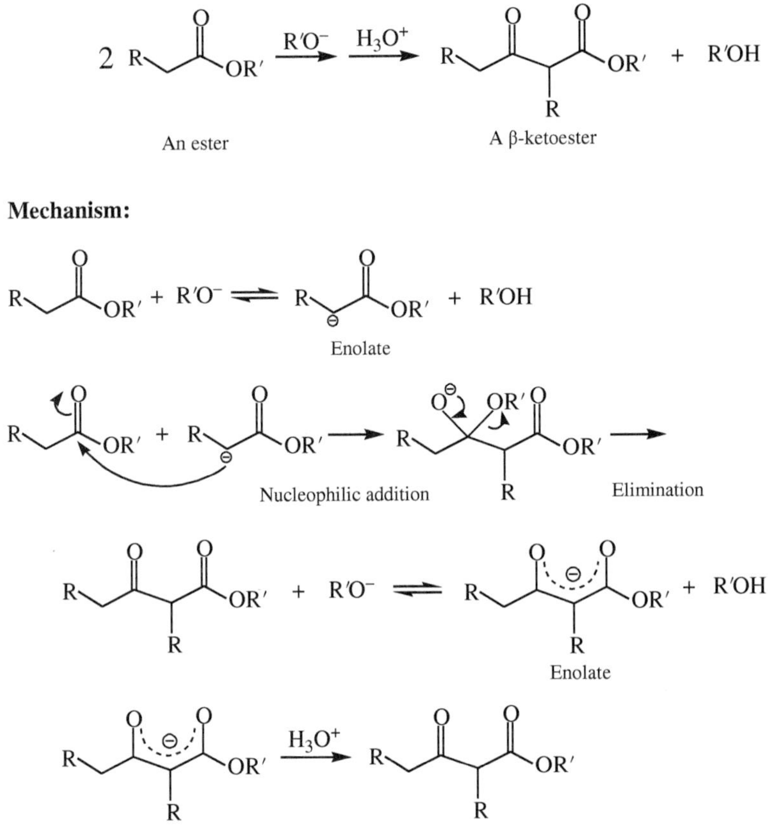

Mechanism:

FIGURE 9.24   Mechanism for Claisen condensation of an ester.

by alkoxide is reversible, only a small portion of the ester is converted to enolate and majority of the ester molecules remain unreacted. Once generated, the nucleophilic enolate is subsequently added to the electrophilic carbonyl group of an unreacted ester molecule. This is followed by elimination of an alkoxide to give a β-ketoester. The net result is that the α-carbon of the starting ester is acylated. The α-hydrogen in the β-ketoester is activated by two carbonyl groups (c.f. the enolate of malonate in Reaction 9.16) and is even more acidic than those of ketones and esters. Thus, the α-carbon in the β-ketoester product is readily deprotonated by the existing alkoxide base, and the resulting enolate is stabilized by two carbonyl groups due to delocalization of the negative charge. Upon aqueous work-up (protonation) in the end of the reaction, the enolate of β-ketoester is converted back to the β-ketoester product.

The enolate of β-ketoester undergoes facile S_N2 reaction with a primary haloalkane (such as benzyl bromide $PhCH_2Br$). This gives rise to alkylation of the α-carbon of a β-ketoester (Fig. 9.25). Upon hydrolysis, the alkylated β-ketoester experiences decarboxylation giving an unsymmetrical ketone with a branched alkyl group.

Analogous to the reactions described in Figure 9.25, the enolate of a β-ketoester can undergo consecutive concerted nucleophilic substitution reactions with a linear alkyl dihalide to form a cycloalkane ring on the α-carbon of the β-ketoester (Fig. 9.26). Upon hydrolysis, the β-ketoester possessing a cycloalkane ring is converted to a ketone. The overall reaction provides an alternative method to that in Figure 9.7b for synthesis of cyclopentyl methyl ketone.

The enolate of ethyl malonate undergoes the same type of consecutive concerted nucleophilic substitution reactions with 1,4-dibromobutane to form a cyclopentane ring on the α-carbon (Fig. 9.27). Hydrolysis of the diester bearing the ring, followed by decarboxylation, affords cyclopentylmethanoic acid.

Figure 9.28 shows that the enolate of a linear chain 1,7-diester undergoes an intramolecular Claisen condensation reaction (a nucleophilic addition to an ester carbonyl followed by elimination of EtO^-) to give a cyclic β-ketoester. In the

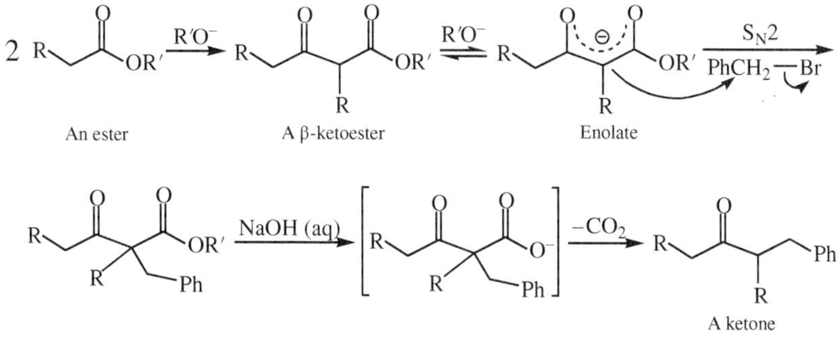

FIGURE 9.25   Alkylation of a β-ketoester via the S_N2 reaction of the enolate with a primary alkyl halide ($PhCH_2Br$).

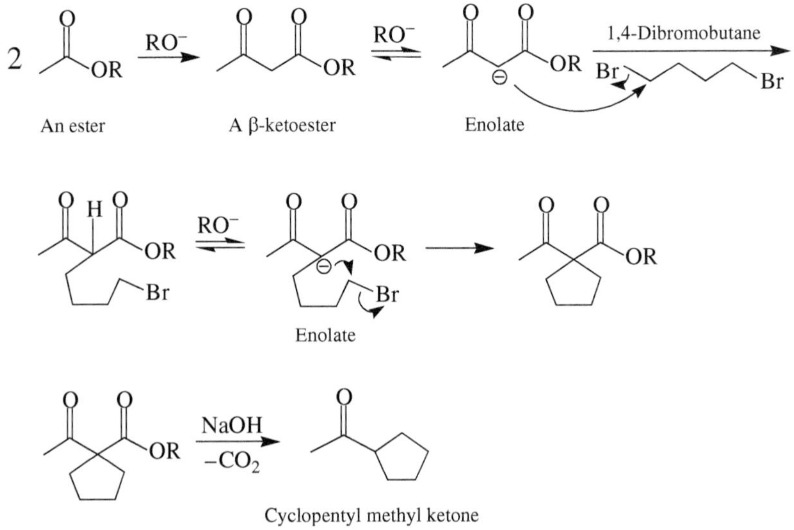

FIGURE 9.26   Alkylation of a β-ketoester via the S$_N$2 reaction of the enolate with an alkyl dihalide resulting in the formation of a cycloalkane ring on the α-carbon.

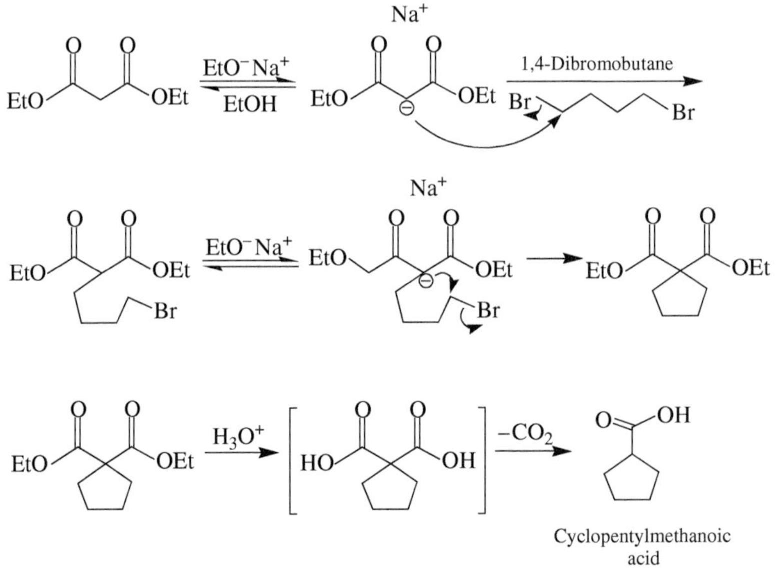

FIGURE 9.27   Alkylation of a diester by reaction of the enolate with 1,4-dibromobutane to form a cyclopentane ring on the α-carbon.

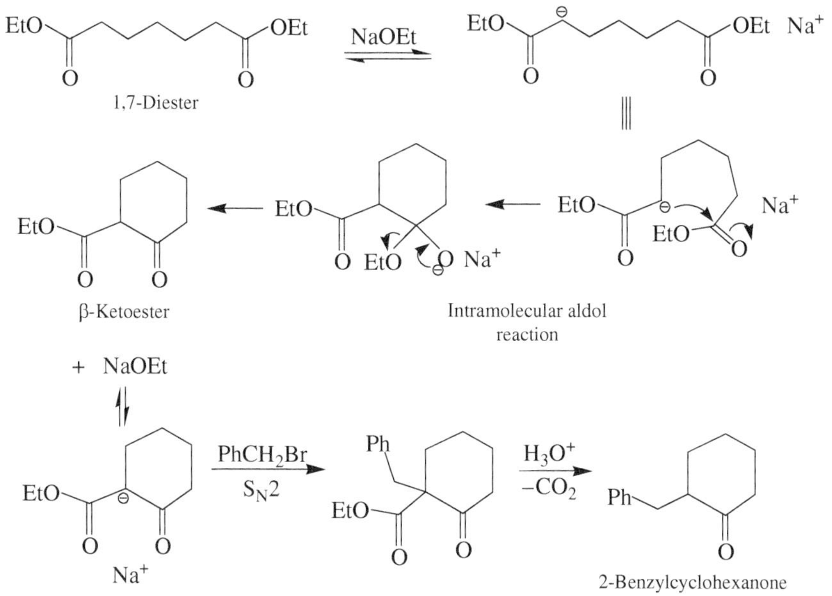

FIGURE 9.28   The intramolecular Claisen condensation of a 1,7-diester and subsequent alkylation and decarboxylation giving a cyclohexanone derivative.

presence of an alkoxide base, treatment of the β-ketoester with benzyl bromide (PhCH₂Br) gives rise to alkylation on its α-carbon. Hydrolysis of the alkylated β-ketoester, followed by decarboxylation, gives 2-benzylcyclohexanone.

The alkoxide base catalyzed Claisen condensation of two different esters (**crossed Claisen condensation**) will give a mixture of several β-ketoesters (Fig. 9.29a). Since deprotonation of the C_α—H bond of an ester by alkoxide is reversible with a small equilibrium constant ($\sim 10^{-10}$), enolates of both esters can be formed. Each reacts with a different ester giving a different product (A or B). In addition, each enolate can also react with a molecule of the same ester from which the enolate is generated (Fig. 9.24). In order for a crossed Claisen condensation to be practically meaningful and give a sole product, the reaction can be directed by LDA. Figure 9.29b shows that the reaction of methyl phenylacetate with the quantitatively pre-formed enolate of methyl acetate by LDA gives a sole β-ketoester product (A). Conversely, the reaction of methyl acetate with the quantitatively preformed (Z)-enolate of methyl phenylacetate by LDA gives a sole β-ketoester product (B) in (R)-configuration, and this reaction is stereospecific. The enantiomeric selectivity of the reaction is determined by the transition state in which the (Z)-enolate and methyl acetate are linked by the lithium ion, resembling a chair-conformation of cyclohexane. The nucleophilic attack of the enolate on the ester is facial-specific on the enolate, giving a specific enantiomer. The situation is comparable to the aldol reaction of an ester enolate with an aldehyde in Figure 9.19a, which has been discussed in Section 9.3.2.

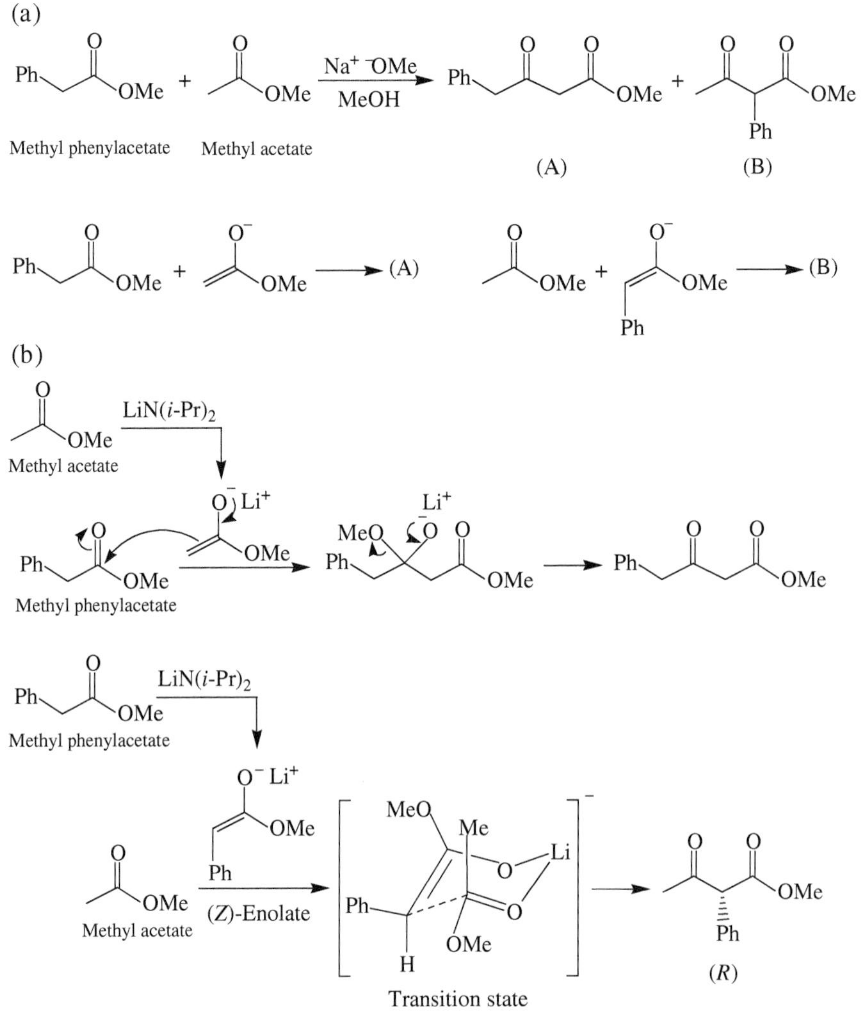

(a)

Methyl phenylacetate   Methyl acetate $\xrightarrow[\text{MeOH}]{\text{Na}^+ \ ^-\text{OMe}}$ (A) + (B)

Methyl phenylacetate + (enolate) $\longrightarrow$ (A)

(enolate) + (enolate) $\longrightarrow$ (B)

(b)

Methyl acetate $\xrightarrow{\text{LiN}(i\text{-Pr})_2}$

Methyl phenylacetate $\longrightarrow$ $\longrightarrow$

Methyl phenylacetate $\xrightarrow{\text{LiN}(i\text{-Pr})_2}$

(Z)-Enolate   Methyl acetate $\longrightarrow$ [Transition state] $\longrightarrow$ (R)

Transition state

FIGURE 9.29 (a) The alkoxide base catalyzed crossed Claisen condensation giving a mixture of different β-ketoester products; and (b) The LAD directed crossed aldol reactions giving a sole product for each reaction.

9.5 BIOLOGICAL APPLICATIONS: ROLES OF ENOLATES IN METABOLIC PROCESSES IN LIVING ORGANISMS

In living organisms, metabolism of carbohydrates (via D-glucose) and fatty acids is converged to **acetyl-coenzyme A [CH$_3$C(O)SCoA]** (Fig. 9.30), the central metabolite and carbon source [9]. By the nature, acetyl-CoA is a thioester. Its C—H bond in the acetyl group is activated. The molecule can be readily converted to a reactive enolate (nucleophile) by some basic groups in various enzymes. Thus, it can be

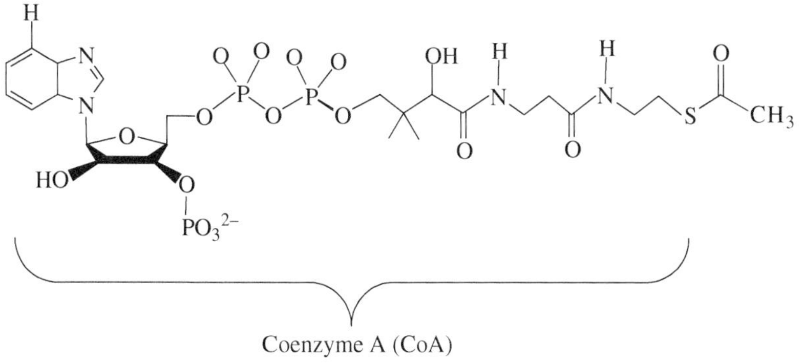

Coenzyme A (CoA)

FIGURE 9.30 Structure of acetyl-coenzyme A.

involved in aldol reactions *in vivo* leading to synthesis of various biomolecules. The portion of coenzyme A assists the binding of acetyl-CoA to various enzymes. However, its identity usually does not change in any metabolic processes.

9.5.1 The Citric Acid Cycle and Mechanism for Citrate Synthase

The citric acid cycle is a complex aerobic metabolic process of various living organisms, through which the carbon atoms in acetyl-coenzyme A [$CH_3C(O)SCoA$], initially produced from metabolism of carbohydrates and fatty acids, are fully oxidized to carbon dioxide and the chemical energy stored in acetyl-coenzyme A is completely released. The process consists of eight steps. The first step is the rate-determining step for the overall citric acid cycle. In the first step, acetyl-coenzyme A is fed into the cycle, and it combines with oxaloacetate [$^-O_2CC(O)CH_2CO_2^-$] to produce citrate [$^-O_2CCH(OH)CH_2CO_2^-$]. The overall reaction in this first step is catalyzed by the citrate synthase enzyme (Reaction 9.17) [9].

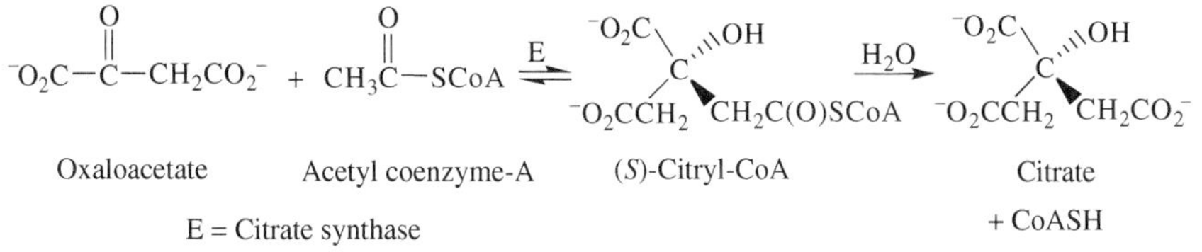

Oxaloacetate Acetyl coenzyme-A (*S*)-Citryl-CoA Citrate
+ CoASH

E = Citrate synthase

$$(9.17)$$

The irreversible hydrolysis of the intermediate (*S*)-citryl-CoA to the ending product citrate ($\Delta G° = -31.5\ kJ/mol$) makes the overall Reaction 9.17 an irreversible process, and it functions as the major rate control for the entire citric acid cycle.

The formation of the intermediate (*S*)-citryl-CoA from oxaloacetate and acetyl-coenzyme A in Reaction 9.17 (the first step of citric acid cycle) is an enzyme (citrate synthase) catalyzed aldol reaction, which occurs via the enolate of acetyl-coenzyme A (Reaction 9.18) [9].

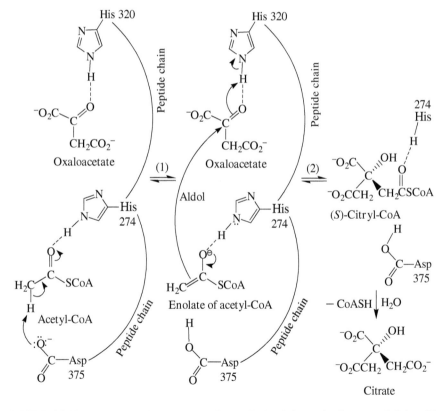

$$(9.18)$$

Generation of the enolate in Reaction 9.18 is the rate-determining step of the overall Reaction 9.17.

Figure 9.31 demonstrates the detailed enzymatic mechanism for the aldol reaction [9]. The citrate synthase enzyme contains two active sites: The oxaloacetate binding site which consists of the imidazole ring of His 320, and the acetyl-CoA binding site which consists of the imidazole ring of His 274 and the carboxylate group of Asp 375.

FIGURE 9.31   The enzymatic mechanism for synthesis of citrate, the first step of citric acid cycle. The process involves citrate synthase catalyzed aldol reaction of acetyl-CoA enolate with oxaloacetate.

In the acetyl-CoA binding site, the C—H bond of acetyl-CoA is held in place such that the α-hydrogen can be effectively attacked by the basic carboxylate of Asp 375 resulting in cleavage of the C—H bond (deprotonation) and the formation of an enolate (Step 1). The hydrogen bonding on the acetyl-CoA carbonyl by His 274 increases the acidity of the α–H, facilitating the C—H bond deprotonation by Asp 375. Then the nucleophilic enolate of acetyl-CoA attacks the carbonyl of oxaloacetate only from front (Step 2), and this facial-specificity of the aldol reaction is dictated by the enzyme structure. The hydrogen bonding on the oxaloacetate carbonyl by His 320 enhances its electrophilicity which facilitates the nucleophilic attack by the enolate of acetyl-CoA. As a new C—C bond is being formed during the nucleophilic attack, simultaneously, the π electrons in the C=O bond domain (carbonyl) are transferred to the valence shell of the oxygen atom, which attacks the N–H hydrogen in His 320 concurrently to form an O—H bond. The overall aldol reaction of the acetyl-CoA enolate and oxaloacetate (Step 2) leading to (*S*)-citryl-CoA is concerted and stereospecific. Finally, the irreversible hydrolysis of (*S*)-citryl-CoA gives citrate.

9.5.2   Ketogenesis and Thiolase

Ketogenesis   The acetyl-CoA produced by metabolism of carbohydrates (D-glucose) and fatty acids is fully oxidized to carbon dioxide in the citric acid cycle. In liver mitochondria, however, a significant fraction of acetyl-CoA has another fate. By a process known as **ketogenesis**, acetyl-CoA is converted to acetoacetate [$CH_3C(O)CH_2CO_2^-$], D-β-hydroxybutyrate [$D-CH_3C(OH)CH_2CO_2^-$] (upon NADH reduction of acetoacetate), and/or acetone [$CH_3C(O)CH_3$]. All these carbonyl compounds are called ketone bodies. They are important metabolic fuels for many muscle tissues.

Production of the ketone bodies in liver ketogenesis begins with the Claisen condensation of two acetyl-CoA molecules to acetoacetyl-CoA [$CH_3C(O)CH_2(O)SCoA$], which is catalyzed by the thiolase enzyme (Reaction 9.19).

$$2 \ CH_3 {-} \overset{\overset{\displaystyle O}{\|}}{C} {-} SCoA \xrightarrow{\text{Thiolase}} CH_3 {-} \overset{\overset{\displaystyle O}{\|}}{C} {-} CH_2 {-} \overset{\overset{\displaystyle O}{\|}}{C} {-} SCoA \ + \ CoASH \quad (9.19)$$

Acetyl-CoA   Acetoacetyl-CoA

Figure 9.32 illustrates the enzymatic mechanism for Claisen condensation in Reaction 9.19 [9, 10]. The active site of the thiolase enzyme consists of a histidine (His) imidazole ring (basic) and a cysteine (Cys) sulfide (formed by spontaneous proton transfer from the Cys–SH to the His-imidazole) [10]. The first acetyl-CoA molecule, after entering the structure of the enzyme, is effectively attacked by the nucleophilic Cys–S⁻ resulting in a tetrahedral intermediate. This is followed by

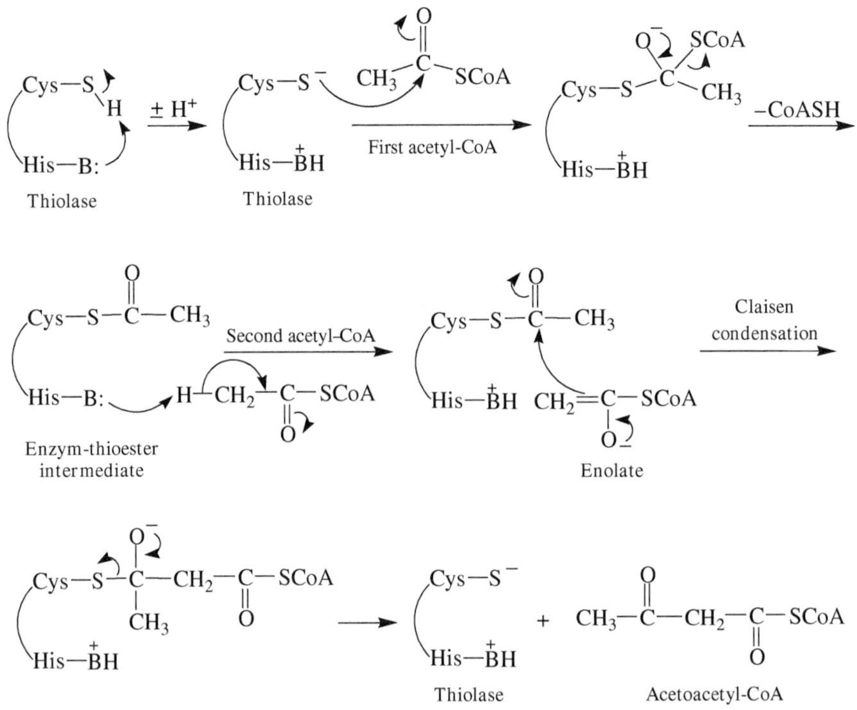

FIGURE 9.32   Mechanism for thiolase catalyzed Claisen condensation of acetyl-CoA molecules giving acetoacetyl-CoA, the precursor to ketone bodies.

elimination of CoAS⁻ (a thiolate), giving an electrophilic enzyme-thioester intermediate. CoAS⁻ is protonated to CoASH (coenzyme A) by His–B⁺H. Then the deprotonated, basic imidazole of the His residue (His–B:) attacks the acidic C–H hydrogen in the second acetyl-CoA molecule giving rise to deprotonation of the α-carbon and making an intermediate enolate (nucleophilic). The acetyl-CoA enolate, once formed, attacks the carbonyl of the acetyl-SCys thioester intermediate, leading to a Claisen condensation, and acetoacetyl-CoA is produced after elimination of the Cys–S⁻ sulfide group from a tetrahedral intermediate. The free thiolase enzyme is regenerated in the last step.

PROBLEMS

9.1   Give the product in each step and write the detailed stepwise mechanism for each of the following reactions. Indicate the nucleophile and electrophile.

(1) 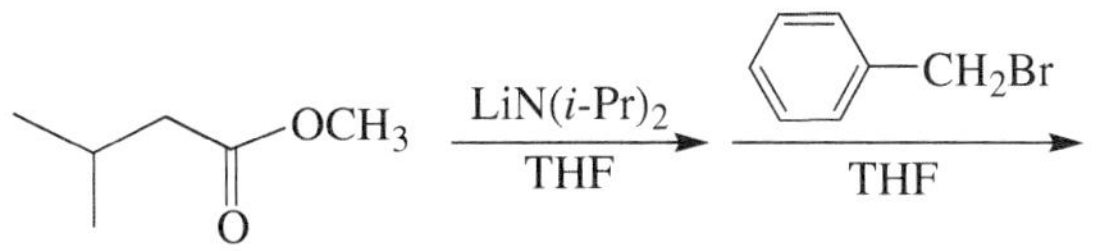   ...—OCH$_3$ $\xrightarrow[\text{THF}]{\text{LiN}(i\text{-Pr})_2}$ $\xrightarrow{\text{THF}}$ (with CH$_2$Br reagent)

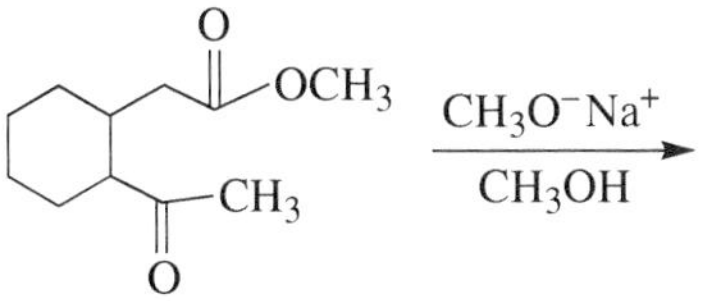

(2)   ...—OCH$_3$, —CH$_3$ $\xrightarrow[\text{CH}_3\text{OH}]{\text{CH}_3\text{O}^-\,\text{Na}^+}$

(3)   ...—OEt $\xrightarrow{\text{NaH}}$ (with Br reagent) $\xrightarrow{\text{NaH}}$ (with Br reagent)

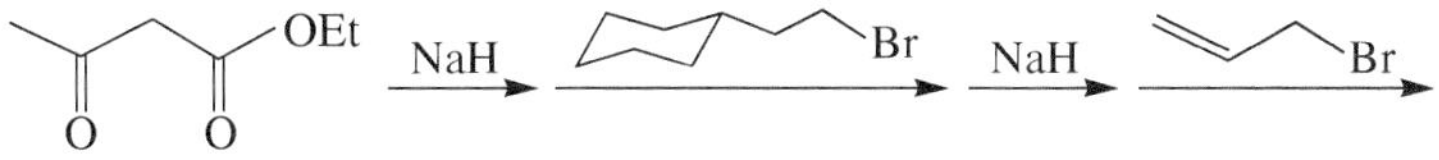

9.2 Give the product in each step and write the detailed stepwise mechanism for each of the following reactions. Account for the regiochemistry for each reaction.

(1) 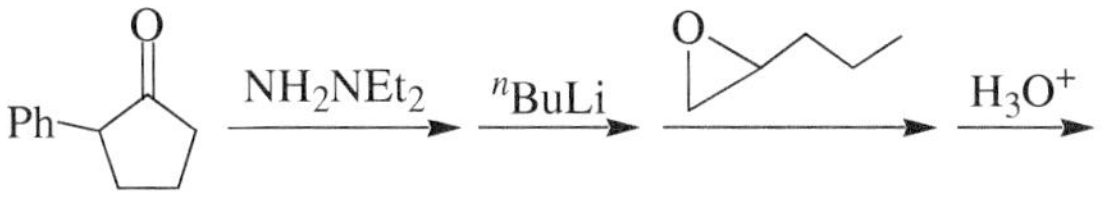   Ph— $\xrightarrow{\text{NH}_2\text{NEt}_2}$ $\xrightarrow{^n\text{BuLi}}$ (epoxide reagent) $\xrightarrow{\text{H}_3\text{O}^+}$

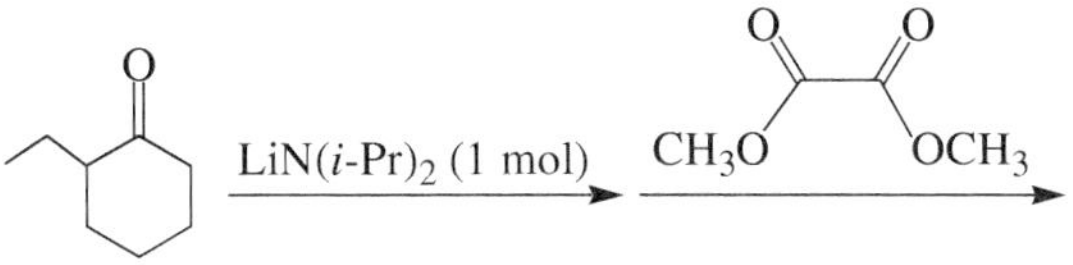

(2) $\xrightarrow{\text{LiN}(i\text{-Pr})_2\ (1\ \text{mol})}$ $\xrightarrow{\text{CH}_3\text{O}\ \ \text{OCH}_3}$

(3)   ...—OCH$_3$ $\xrightarrow{\text{NaH}}$ (with Br reagent) $\xrightarrow[\Delta]{\text{NaOH (aq)}}$

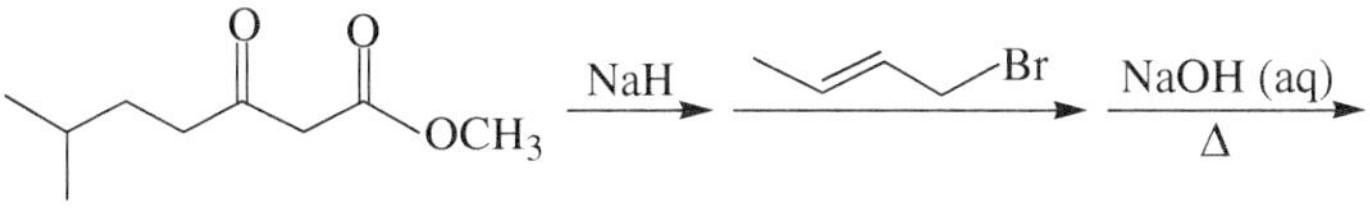

9.3 Write a detailed mechanism to account for the following synthesis. Indicate nucleophile and electrophile.

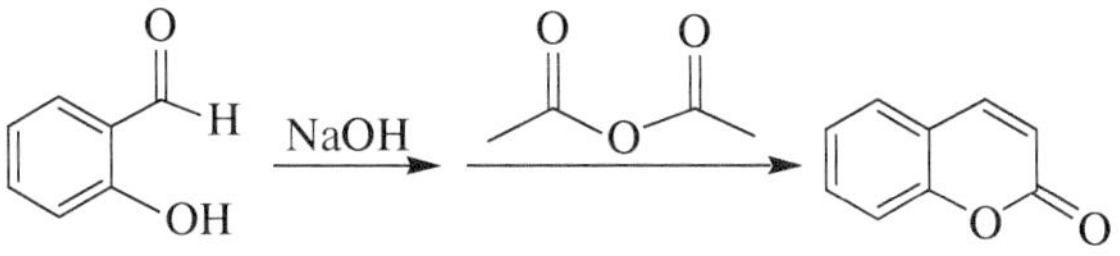

(salicylaldehyde) $\xrightarrow{\text{NaOH}}$ (acetic anhydride)   →   (coumarin)

9.4 Propose a full detailed mechanism to account for the following consecutive reactions. Write out all the intermediates.

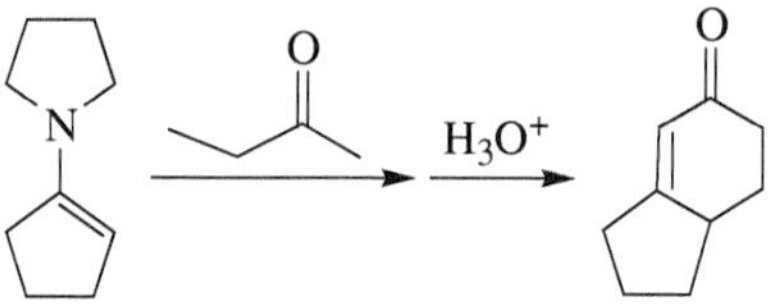

9.5 Give intermediate and final products and write the possible stepwise mechanism for the following reactions. Account for the regiochemistry and stereochemistry for the reactions.

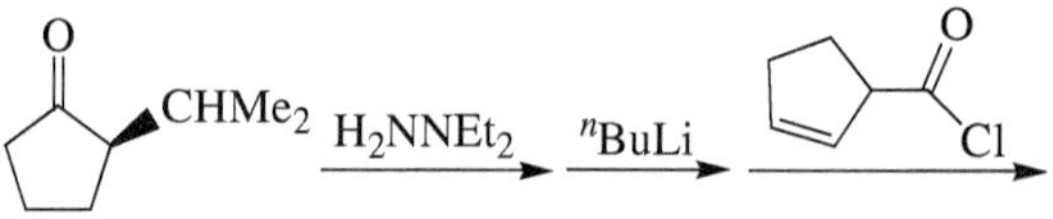

9.6 Give product and write a detailed mechanism to account for the following reaction.

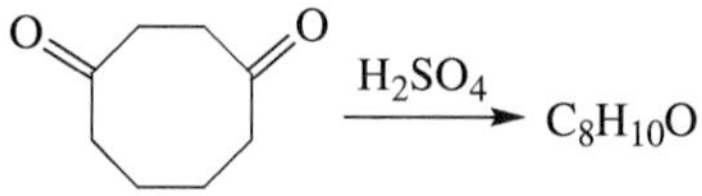

9.7 Propose a detailed mechanism to account for following synthetic reaction. Provide an explanation on how the central aromatic ring is formed.

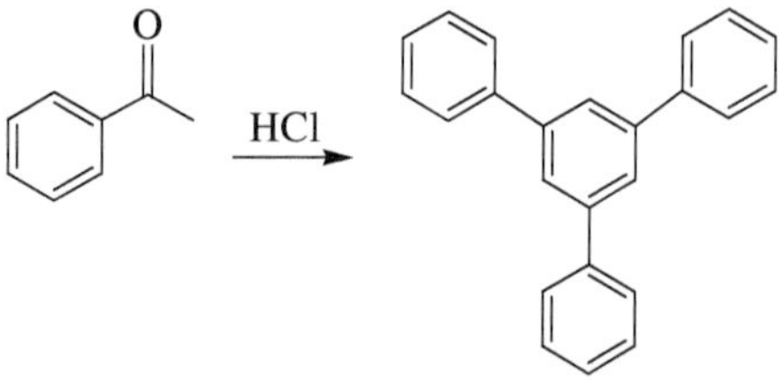

9.8 The following triketone undergoes an aldol condensation reaction in which the methyl ketone that comprises the side chain serves as the nucleophile. Write possible product(s) and show the reaction mechanism. Account for the stereochemistry of the aldol condensation reaction.

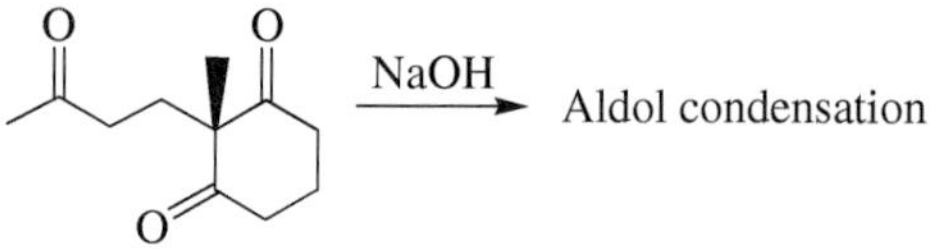

REFERENCES

1. Fox, M. A.; Whitesell, J. K. *Organic Chemistry*, 2nd ed., Jones and Bartlett, Sudbury, MA, USA (1997).
2. Hoffman, R. V. *Organic Chemistry: An Intermediate Text*, 2nd ed., Wiley, Hoboken, NJ, USA (2004).
3. Brown, W. H.; Foote, C. S.; Iverson, B. L.; Anslyn, E. V. *Organic Chemistry*, 6th ed., Brooks/Cole, Belmont, CA, USA (2012).
4. Arya, P.; Qin, H. Advances in Asymmetric Enolate Methodology. *Tetrahedron*, 2000, *56*, 917–947.
5. List, B.; Lerner, R. A.; Barbas III, C. A. Proline-Catalyzed Direct Asymmetric Aldol Reactions. *J. Am. Chem. Soc.* 2000, *122*, 2395–2396.
6. Yang, J. W.; Chandler, C.; Stadler, M.; Kampen, D.; List, B. Proline-Catalyzed Mannich Reactions of Acetadehyde. *Nature*, 2008, *452*, 453–455.
7. Yang, J. W.; Stadler, M.; List, B. Proline-Catalyzed Mannich Reaction of Aldehydes with *N*-Boc-Imines. *Angew. Chem. Int. Ed.* 2008, *46*, 609–611.
8. Schmid, M. B.; Zeitler, K.; Gschwind, R. M. NMR Investigations on the Proline-Catalyzed Aldehyde Self-Condensation: Mannich Mechanism, Dienamine Detection, and Erosion of the Aldol Addition Selectivity. *J. Org. Chem.* 2011, *76*, 3005–3015.
9. Voet, D.; Voet, J. G.; Pratt, C. W. *Fundamentals of Biochemistry*, 5th ed., Wiley, Hoboken, NJ, USA (2016).
10. Mathieu, M.; Modis, Y.; Zeelen, J. Ph.; Engel, C. K.; Abagyan, R. A.; Ahlberg, A.; Rasmussen B.; Lamzin, V. S.; Kunau, W. H.; Wierenga, R. K. The 1.8 Å Crystal Structure of the Dimeric Peroxisomal 3-Ketoacyl-CoA Thiolase of Saccharomyces Cerevisiae: Implications for Substrate Binding and Reaction Mechanism. *J. Mol. Biol.* 1997, *273*, 714–728.

10

REARRANGEMENTS

10.1 MAJOR TYPES OF REARRANGEMENTS

Many organic reactions are accompanied by **skeletal rearrangements**, processes in which migration of a certain group (usually alkyl, aryl, or a hydrogen atom) in the reactant or an intermediate occurs leading to an isomerized product. Most of rearrangement processes take place intramolecularly, while some rearrangements occur through intermolecular interactions. Very often, rearrangements are associated to transient reactive intermediates. As a result, migration of a group leads to transformation of one intermediate (initially formed) into another, the relatively thermodynamically more stable isomer. In this case, the immediate rearrangement product is not isolable, and the overall rearrangement process may be identified by comparison of structures of the starting reactant and the final product.

The **1,2-shift** (**1,2-rearrangement**) is referred to as intramolecular migration of a group or atom between two adjacent atoms in a reactant or intermediate, and this is among the most common pathway for many types of rearrangements. Depending on the types of different organic reagents, the 1,2-shifs can possibly take place between two carbon atoms (carbon–carbon rearrangement), between a carbon atom and a nitrogen atom (carbon–nitrogen rearrangement), and between a carbon atom and an oxygen atom (carbon–oxygen rearrangement) (Fig. 10.1). The carbon–carbon

Organic Mechanisms: Reactions, Methodology, and Biological Applications,
Second Edition. Xiaoping Sun.
© 2021 John Wiley & Sons, Inc. Published 2021 by John Wiley & Sons, Inc.
Companion website: www.wiley.com/go/Sun/OrgMech_2e

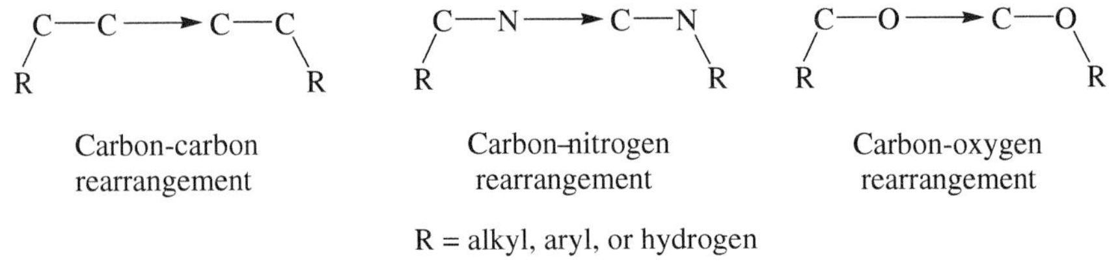

| Carbon-carbon rearrangement | Carbon-nitrogen rearrangement | Carbon-oxygen rearrangement |

R = alkyl, aryl, or hydrogen

FIGURE 10.1 Major types of rearrangements identified in organic reactions.

rearrangement usually occurs to intermediate carbocations and is referred to as **Wagner–Meerwein rearrangement**. Neither the cationic carbon nor its adjacent carbon bearing the migrating group contains a heteroatom. The carbon–nitrogen rearrangement includes **Beckmann** and **Hofmann rearrangements**. The carbon–oxygen rearrangement includes **Baeyer–Villiger oxidation** and **Claisen rearrangement**. For both carbon–nitrogen and carbon–oxygen rearrangements, commonly, the heteroatom (nitrogen or oxygen) contains a good leaving group and an active lone-pair of electrons, and a migrating group (alkyl or aryl) is attached to the adjacent carbon atom in the reactant molecule. The rearrangement of the migrating group to the adjacent heteroatom is facilitated by the good leaving group and electron pair.

The most common rearrangements are those which take place in intermediate carbocations formed in the course of various reactions. First, mechanisms of carbocation rearrangements will be discussed in this chapter. Then, detailed discussion will be presented on other important types of rearrangement reactions.

10.2 REARRANGEMENT OF CARBOCATIONS: 1,2-SHIFT

Carbocations are formed as reactive intermediates in many types of organic reactions, such as S_N1 reactions, E1 reactions, and electrophilic additions of alkenes. Frequently, facile 1,2-shifts, migration of an alkyl, aryl, or hydrogen (the R group) between the cationic carbon and its adjacent carbon atom, occur in carbocations. As a result, the initially formed, less stable carbocation is converted to a more stable carbocationic isomer (Fig. 10.2). This serves as the general thermodynamic driving force for the carbocation rearrangement (**Wagner–Meerwein rearrangement**). The transition state in a carbocation rearrangement is a delocalized three-orbital system with two electrons from the R—C covalent bond. It is stabilized because the number of electrons is a Huckel number $[(4n + 2) = 2, n = 0]$ [1]. Therefore, the rearrangement possesses a low activation energy. The stabilization of a carbocation by rearrangement can be achieved by increase in substitution of the cationic carbon for acyclic molecules or by reducing the molecular strain for cyclic molecules via ring expansion. We will demonstrate the situations using the following examples.

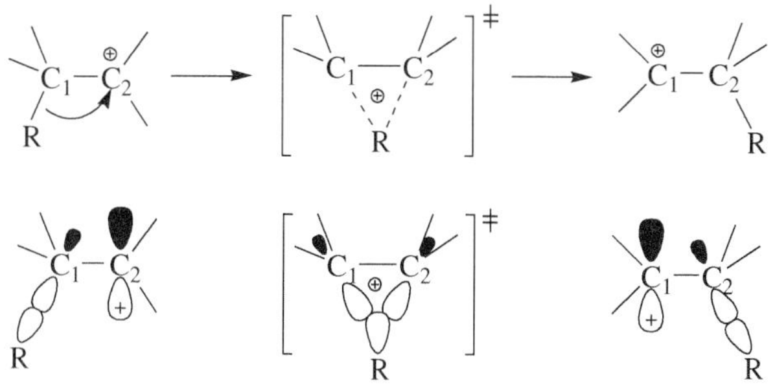

R = alkyl, aryl, or hydrogen

FIGURE 10.2   The 1,2-shift in a general carbocation.

10.2.1 1,2-Shifts in Carbocations Produced from Acyclic Molecules

In SbF$_5$, 1-fluoropropane undergoes a facile transformation to 2-propyl cation (a secondary carbocation as the SbF$_6^-$ salt) (Fig. 10.3a). The formation of the carbocation is identified by ^{1}H NMR spectroscopy, showing a characteristic septet of the CH hydrogen on the cationic carbon at 13.5 ppm [1]. Clearly, the high chemical shift is due to the strong deshielding effect of the positive charge on the CH carbon. SbF$_5$ is a very strong Lewis acid. It can abstract a fluoride (a fluorine atom with a pair of electrons) from 1-fluoropropane to form a stable SbF$_6^-$ anion (an extremely weak Lewis base), and the carbon skeleton, after loss of a fluoride, becomes a primary carbocation. The initially formed primary carbocation (as the SbF$_6^-$ salt) is then converted to a thermodynamically more stable secondary carbocation via a facile 1,2-shift of hydrogen.

The analogous SbF$_5$ initiated rearrangement of 1-fluorobutane to a *tert*-butyl cation (a tertiary carbocation as the SbF$_6^-$ salt) is demonstrated in Figure 10.3b [1]. The abstraction of a fluoride from 1-fluorobutane by SbF$_5$ gives an intermediate CH$_3$CH$_2$CH$_2$CH$_2^+$ cation. It then undergoes a thermodynamically favorable hydrogen 1,2-shift to afford a more stable secondary carbocation. Then, a methyl 1,2-shift occurs to the secondary carbocation to give a less stable primary (CH$_3$)$_2$CHCH$_2^+$ cation. Once formed, the primary (CH$_3$)$_2$CHCH$_2^+$ cation undergoes a facile 1,2-shift of hydrogen giving the thermodynamically more stable tertiary (CH$_3$)$_3$C$^+$ cation. The overall reaction involves three consecutive carbocation rearrangements, called **cascade rearrangement.** Although the second methyl 1,2-shift is endothermic (reactant-favored), both the first and third hydrogen 1,2-shifts are exothermic (product-favored), and they make the overall cascade rearrangement thermodynamically favorable.

Figure 10.4a shows electrophilic addition of HCl to 3,3,-dimethyl-1-butene giving two isomeric products, 2-chloro-3,3-dimethylbutane (17%) and 2-chloro-2,3-dimethylbutane (83%) [2]. The reaction follows Markovnikov's rule and proceeds via an initially formed secondary carbocation. The intermediate secondary

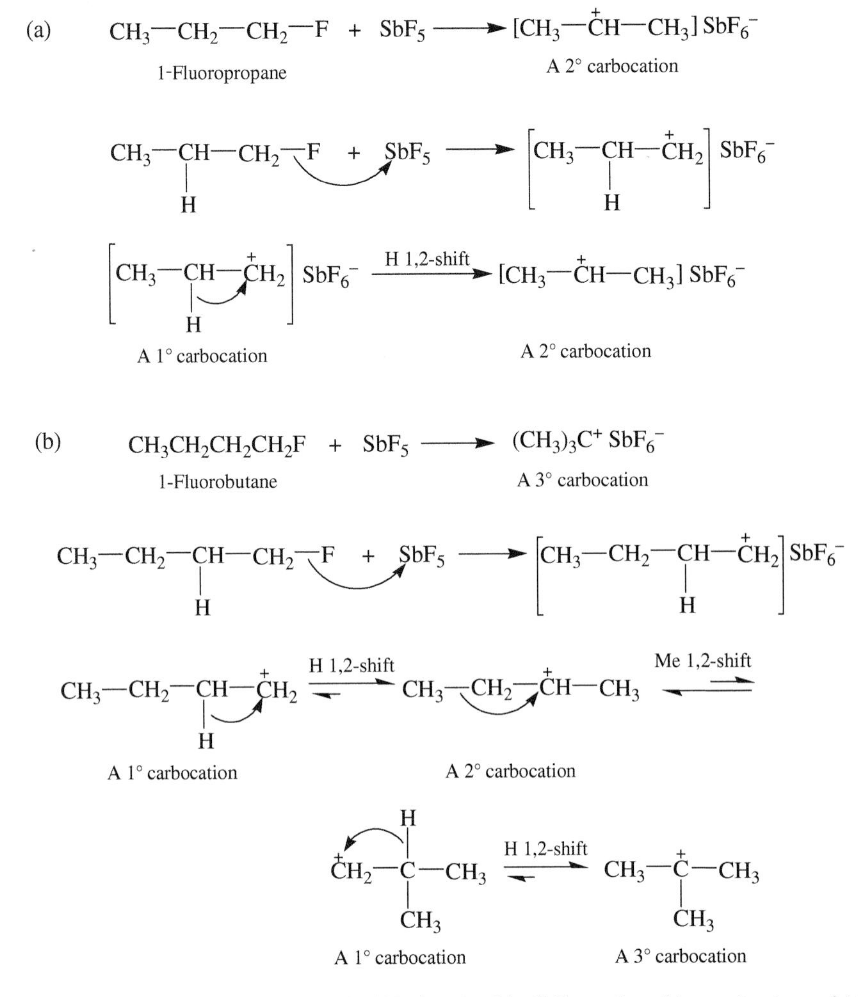

(a) $CH_3-CH_2-CH_2-F$ + SbF_5 $\longrightarrow$ $[CH_3-\overset{+}{C}H-CH_3]\,SbF_6^-$

1-Fluoropropane   A 2° carbocation

$CH_3-\underset{H}{CH}-CH_2-F$ + SbF_5 $\longrightarrow$ $\left[CH_3-\underset{H}{CH}-\overset{+}{C}H_2\right]SbF_6^-$

$\left[CH_3-\underset{H}{CH}-\overset{+}{C}H_2\right]SbF_6^-$ $\xrightarrow{\text{H 1,2-shift}}$ $[CH_3-\overset{+}{C}H-CH_3]\,SbF_6^-$

A 1° carbocation   A 2° carbocation

(b) $CH_3CH_2CH_2CH_2F$ + SbF_5 $\longrightarrow$ $(CH_3)_3C^+\,SbF_6^-$

1-Fluorobutane   A 3° carbocation

$CH_3-CH_2-\underset{H}{CH}-CH_2-F$ + SbF_5 $\longrightarrow$ $\left[CH_3-CH_2-\underset{H}{CH}-\overset{+}{C}H_2\right]SbF_6^-$

$CH_3-CH_2-\underset{H}{CH}-\overset{+}{C}H_2$ $\underset{\text{H 1,2-shift}}{\rightleftharpoons}$ $CH_3-CH_2-\overset{+}{C}H-CH_3$ $\overset{\text{Me 1,2-shift}}{\rightleftharpoons}$

A 1° carbocation   A 2° carbocation

$\overset{+}{C}H_2-\underset{\underset{CH_3}{|}}{\overset{\overset{H}{|}}{C}}-CH_3$ $\underset{\text{H 1,2-shift}}{\rightleftharpoons}$ $CH_3-\underset{\underset{CH_3}{|}}{\overset{+}{C}}-CH_3$

A 1° carbocation   A 3° carbocation

FIGURE 10.3   The carbocation 1,2-shifts involved in SbF_5 catalyzed isomerizations of (a) 1-fluoropropane to 2-propyl cation and (b) 1-fluorobutane to *tert*-butyl cation. Modified from Jackson [1].

carbocation has two fates: It can be attacked immediately by chloride to afford the unrearranged 2-chloro-3,3-dimethylbutane (the minor product). Concurrently, a methyl group on the C_3-carbon undergoes a facile 1,2-shift to lead to the formation of a thermodynamically more stable tertiary carbocation. Then, the tertiary carbocation is attacked by chloride to afford the rearranged 3-chloro-2,3-dimethylbutane (the major product).

The acid catalyzed dehydration of 3,3-dimethyl-2-butanol is accompanied by an analogous carbocation rearrangement following the pathway of 1,2-shift (Fig. 10.4b). The reaction follows E1 mechanism. Similar to the reaction in

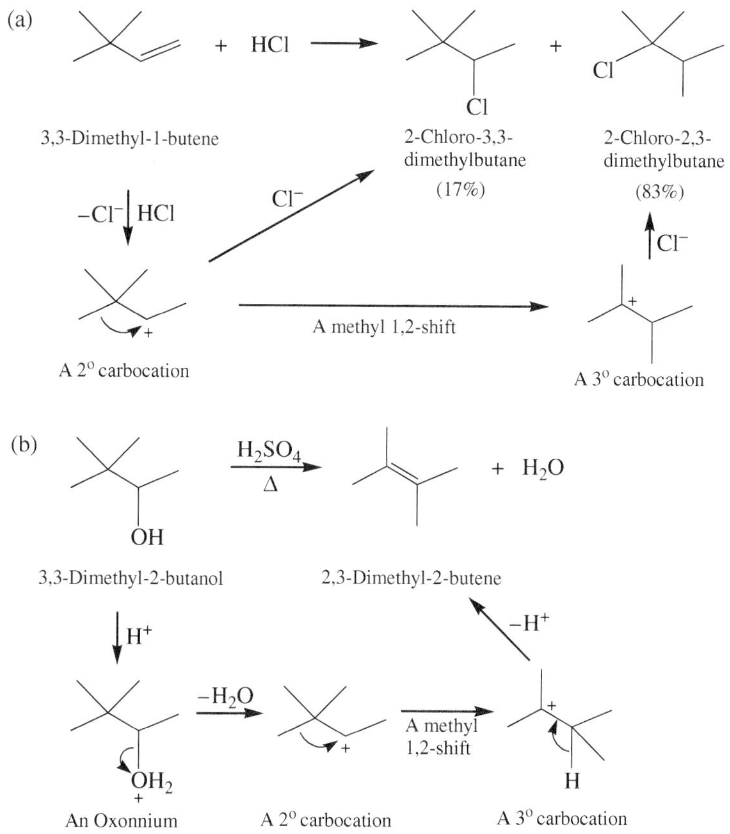

FIGURE 10.4   The carbocation rearrangements involved in (a) electrophilic addition of HCl to 3,3-dimethyl-1-butene and (b) acid catalyzed dehydration of 3,3-dimethyl-2-butanol.

Figure 10.4a, the initially formed secondary carbocation undergoes a facile 1,2-shift of a methyl group to lead to the formation of a thermodynamically more stable tertiary carbocation. Finally, the cleavage of the C−H bond adjacent to the cationic carbon gives a rearranged alkene product (2,3-dimethyl-2-butene).

10.2.2   1,2-Shifts in Carbocations Produced from Cyclic Molecules—Ring Expansion

Figure 10.5a shows nucleophilic substitution reaction (S_N1) of cyclobutylmethanol with HBr. The reaction affords bromocyclopentane as a sole product. The unrearranged substitution product bromocyclobutylmethane is not formed. As indicated in Figure 10.5a, the reaction proceeds via an intermediate oxonium ion formed by protonation of the hydroxyl group in the substrate. As a water molecule leaves from the oxonium intermediate, a carbon atom in the four-membered ring undergoes

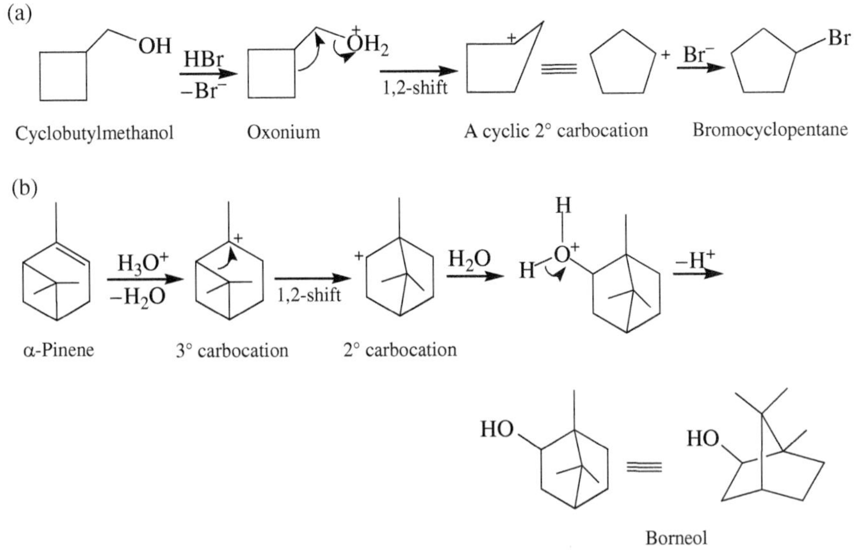

(a)

Cyclobutylmethanol   Oxonium   A cyclic 2° carbocation   Bromocyclopentane

(b)

α-Pinene   3° carbocation   2° carbocation

Borneol

FIGURE 10.5 The ring expansions involved in (a) nucleophilic substitution reaction of cyclobutylmethanol with HBr and (b) acid catalyzed hydration of α-pinene to borneol.

a simultaneous 1,2-shift (migration) to lead to a ring expansion and result in the for-mation of a cationic five-membered carbon ring. The ring expansion has relieved molecular strain pertaining to the initial four-membered ring in the substrate and led to the formation of a more stable five-member ring. In addition, the 1,2-shift of a cyclic carbon in the intermediate oxonium has effectively prevented formation of a highly unstable primary carbocation as water leaves the substrate molecule. Instead, a more stable secondary carbocation is produced as a result of the rearrange-ment. All this has made the rearrangement a thermodynamically favorable process. The attack of bromide on the cyclic secondary carbocation formed in the ring expan-sion gives the final rearranged bromocyclopentane product.

Figure 10.5b shows mechanism of acid catalyzed hydration of α-pinene, a com-plicated cyclic alkene molecule. The reaction belongs to Markovnikov alkene elec-trophilic addition, which proceeds via an initially formed tertiary carbocation (protonation of the alkene preferably occurs on the less substituted doubly bonded carbon). This carbocation intermediate contains a four-membered cyclobutane ring. A subsequent 1,2-shift of a cyclic carbon in the four-membered ring results in ring expansion and leads to the formation of a more stable five-membered ring. Although the cationic center of the rearranged carbocation intermediate (secondary) is less sta-ble than the initially formed carbocation (tertiary), the relief of the ring strain per-taining to cyclobutane in the initial tertiary carbocation through the ring expansion is predominating in stabilization of the reaction intermediate. This gives rise to an overall thermodynamically favorable carbocation rearrangement. Finally, the rearranged secondary carbocation intermediate is attacked by a water molecule resulting in the formation of borneol.

10.2.3   Resonance Stabilization of Carbocation—Pinacol Rearrangement

The acid catalyzed dehydration of 2,3-dimethyl-2,3-butadiol (pinacol) is accompanied by a skeletal methyl rearrangement giving *tert*-butylmethyl ketone (Fig. 10.6a). A 1,2-shift of a methyl group takes place in the initially formed tertiary carbocation. The resulting isomeric carbocation intermediate from the rearrangement is stabilized by the adjacent hydroxyl group through conjugation effect (resonance stabilization), giving rise to a protonated ketone (an oxonium ion). After a facile deprotonation, the

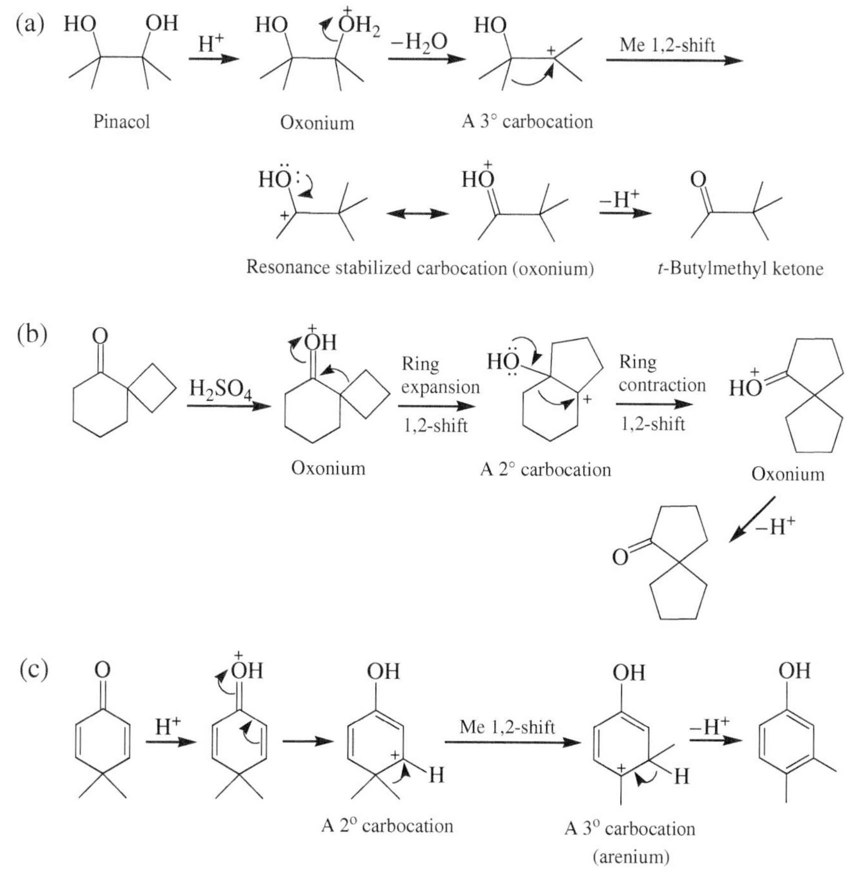

FIGURE 10.6 (a) Pinacol rearrangement, (b) acid catalyzed ring-expansion and ring-contraction in a bicyclic ketone Modified from Hoffman [3], and (c) rearrangement of a conjugated cyclic ketone to a phenol derivative.

oxonium becomes an electrically neutral ketone product. Such a rearrangement in a 1,2-diol that takes place due to resonance stabilization of the resulting carbocation (or an oxonium of doubly bonded oxygen) is called **pinacol rearrangement**.

A similar rearrangement of a cyclic ketone (**semipinacol rearrangement**) catalyzed by acid is shown in Figure 10.6b [3]. The first step in the reaction mechanism is protonation of the carbonyl group giving an oxonium intermediate. The positive charge in the protonated carbonyl oxygen drives a ring-expansion (first rearrangement via a 1,2-shift) that takes place on the cyclobutane moiety giving a five-membered ring which is fused to the cyclohexane moiety of the molecule. Although the cationic center of the rearranged carbocation intermediate (tertiary) is less stable than the initially formed oxonium ion, the relief of the ring strain pertaining to cyclobutane in the initial molecule by ring expansion is predominating in stabilization of the rearranged reaction intermediate, making the first rearrangement thermodynamically favorable. Then, a lone-pair of electrons in hydroxyl (–OH) of the tertiary carbocation intermediate drives a second rearrangement via a 1,2-shift of a carbon atom in the cyclohexane ring, and this rearrangement leads to the formation of a resonance stabilized oxonium which is more stable than the previously formed tertiary carbocation. Facile deprotonation of the second oxonium intermediate gives a rearranged ketone product, with the net expansion of a four-membered ring and contraction of a six-membered ring to two five-membered rings. The overall reaction involves two consecutive rearrangement processes of the 1,2-shifts (cascade rearrangement).

Figure 10.6c shows acid catalyzed isomerization of a conjugate cyclic ketone to a phenol derivative. A 1,2-shift of a methyl group in an intermediate secondary carbocation results in the formation of a more stable tertiary carbocation (also an arenium ion). Finally, the cleavage of an adjacent C—H bond to the cationic center affords a thermodynamically more stable aromatic ring. The major thermodynamic driving force for this rearrangement lies in the formation of a very stable aromatic system from a less stable nonaromatic system.

10.2.4 *In vivo* Cascade Carbocation Rearrangements: Biological Significance

Steroids, such as cholesterol, represent an important class of biomolecules. The *in vivo* biosynthesis of the polycyclic system starts from acetyl coenzyme A (CH$_3$CO-SCoA) and involves many sophisticated enzymatic steps [4]. First, the multistep combination of many CH$_3$CO-SCoA molecules catalyzed by various enzymes leads to the chain-like 2,3-oxidosqualene, which subsequently undergoes a concerted cyclization to lead to the formation of the polycyclic lanosterol (Fig. 10.7). Lanosterol belongs to steroids. It can be transformed subsequently to cholesterol. As can be seen in Figure 10.7, the ring-formation step for biosynthesis of steroids, the cyclization of 2,3-oxidosqualene to lanosterol, is accompanied by the concerted cascade 1,2-rearrangements of hydrogens and methyl groups in a carbocation intermediate.

First, an acidic group in the enzyme protonates the epoxide oxygen in 2,3-oxidosqualene via a hydrogen bond, triggering concerted cascade electron transfer processes, as indicated by the arrows in 2,3-oxidosqualene, toward the enzyme. All

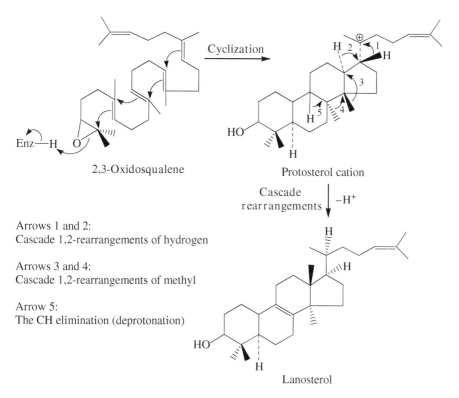

FIGURE 10.7 The concerted cascade carbocation 1,2-rearrangements in a biochemical process: The *in vivo* biosynthesis of steroids (cholesterol).

this has led to concerted cyclizations giving an intermediate fused polycyclic system (protosterol cation) with a positive charge introduced onto an acyclic carbon (a tertiary carbocation). The cationic center in protosterol activates the C—H bond on its adjacent cyclic carbon resulting in a 1,2-rearrangement of hydrogen (Arrow 1 in protosterol). This hydrogen rearrangement relays the positive charge to the adjacent cyclic carbon and triggers a second concerted 1,2-rearrangement of hydrogen (Arrow 2). The second hydrogen rearrangement further relays the positive charge downward, driving a concerted 1,2-rearrangement of methyl group (Arrow 3). For the same reason, the methyl rearrangement drives a second concerted 1,2-rearrangement of methyl group (Arrow 4). Finally, the positive charge is relayed by the cascade rearrangements to a C–H hydrogen giving rise to the C—H bond elimination (deprotonation) (Arrow 5), and a π bond is formed in the polycyclic lanosterol product. The concerted cascade carbocation rearrangements are illustrated in Figure 10.7. All the rearrangement steps are stereospecific. For example, the first methyl rearrangement (Arrow 3) results in addition of a methyl group to the front of the adjacent cyclic carbon, while the second methyl rearrangement (Arrow 4) transfers the methyl group to the back side of the ring. The stereochemical feature of the rearrangements is determined by the relative positions (configurations) of the hydrogens and methyls which undergo the rearrangements and leads to stereospecific product.

10.2.5   Acid Catalyzed 1,2-Shift in Epoxides

In the presence of a protic or Lewis acid, an epoxide can undergo a 1,2-shift of alkyl or hydrogen giving a ketone product (Fig. 10.8) [3]. The reaction is initiated by the formation of an oxonium intermediate through interaction of a lone-pair of electrons in the epoxide oxygen and the acid center. Then, a C—O bond in the more substituted carbon (containing fewer H atoms if the epoxide is unsymmetrical) breaks to give a relatively more stable carbocation (a tertiary carbocation). In the next step, a hydrogen 1,2-shift takes place in the carbocation which is facilitated by an active lone-pair of electrons in the oxygen. As a result, the initially formed tertiary carbocation is converted to a more stable oxonium ion owing to resonance stabilization. Due to the stereospecificity of the epoxide reactant as indicated in Figure 10.8 (with oxygen staying backside of the ring), the 1,2-shift of hydrogen (in the front of the ring) is stereospecific and only occurs in the front face of the cyclopentane ring, which gives the *trans*-configuration of a ring structure (with the two methyl groups staying in different sides of the ring in the second oxonium intermediate). Departure of the BF₃ molecule from the carbonyl oxygen affords a cyclopentanone derivative in *trans*-configuration.

10.2.6   Anion Initiated 1,2-Shift

Sometimes, a 1,2-shift between two adjacent carbon atoms can also be initiated by an anion. Figure 10.9a shows that treatment of a 1,2-diketone with the hydroxide anion results in a 1,2-shift of a phenyl (–Ph) group during the overall reaction [3]. The reaction is initiated by a nucleophilic addition of OH⁻ to a carbonyl in the diketone

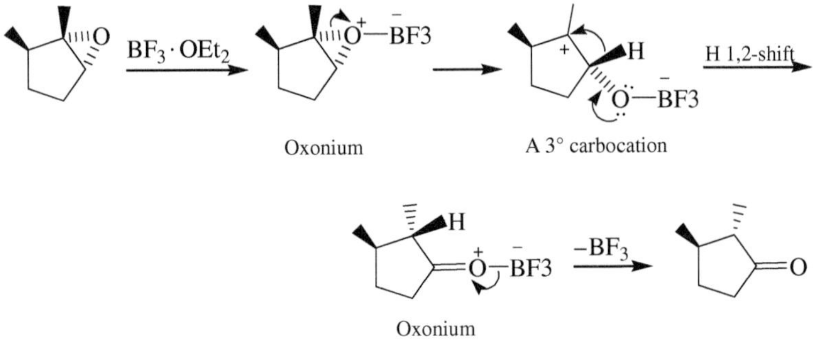

FIGURE 10.8   The 1,2-shift in an epoxide. Modified from Hoffman [3].

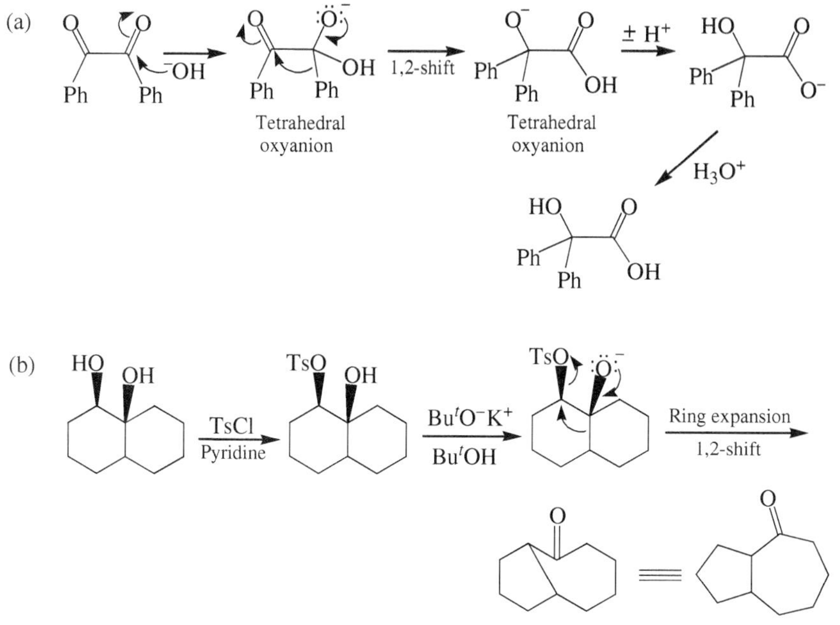

FIGURE 10.9 Anion-initiated 1,2-shifts: (a) Isomerization of a 1,2-diketone to an α-hydroxycarboxylic acid and (b) ring-expansion and ring-contraction of a bicyclic diol.

substrate giving a tetrahedral oxyanion intermediate. In the following step, an active lone-pair of electrons in the oxyanion and the adjacent electronegative carbonyl group work together to drive a 1,2-shift of a phenyl group: A lone-pair of electrons in the negative oxygen acts as a pushing force against the phenyl group, while simultaneously, the electronegative oxygen in the adjacent carbonyl withdraws the Ph—C bonding electron pair toward the carbonyl carbon. As a result, the phenyl group in the tetrahedral carbon is eliminated and simultaneously, migrates to the adjacent carbonyl carbon. A new carbonyl group and oxyanion are formed. A subsequent net spontaneous proton transfer from the carboxyl group to the second oxyanion intermediate gives a thermodynamically more stable carboxylate. Upon aqueous workup using an acid, an α-hydroxycarboxylic acid is produced. It is noteworthy that the basicity of the initial reaction product (a carboxylate) is lower than that of the reactant (hydroxide). This difference in basicity functions as the major thermodynamic driving force for the rearrangement. In addition, repulsive force between the positive carbonyl carbons in the 1,2-diketone substrate destabilizes the reactant, making the rearranged product thermodynamically more favorable.

The reaction in Figure 10.9b involves an anion initiated, a good leaving group assisted 1,2-shift of carbon. First, one of the hydroxyl groups in the 1,2-diol substrate is tosylated, followed by deprotonation of the other hydroxyl with *t*-butoxide. Then, an active lone-pair of electrons in the oxyanion moves down and drives a 1,2-shift of a carbon in a six-membered ring. This 1,2-shift is directly facilitated by simultaneous departure of tosylate (a very good leaving group).

10.3   NEIGHBORING LEAVING GROUP FACILITATED 1,2-REARRANGEMENT

This type of 1,2-rearrangement usually takes place between adjacent carbon and nitrogen atoms (carbon–nitrogen rearrangement) or adjacent carbon and oxygen atoms (carbon–oxygen rearrangement) (Fig. 10.10). The heteroatom Y (nitrogen or oxygen) bears a good leaving group (LG) and contains an active lone pair of electrons. The adjacent carbon atom has a migrating group R (alkyl, aryl, or hydrogen). The lone pair of electrons in the heteroatom acts as a pushing force, and it moves down to the C—Y bond domain expelling off the R group from carbon. As the R—C bond breaks, the R group migrates with an electron pair to its adjacent heteroatom Y. The formation of the new R—Y bond is facilitated and accompanied by concurrent departure of the LG group. All the electron transfer processes illustrated in Figure 10.10 occur concertedly, resulting in an overall 1,2-rearrangement. The neighboring leaving group facilitated 1,2-shift is found in Beckmann rearrangement, Hofmann rearrangement, and Baeyer–Villiger oxidation (rearrangement).

10.3.1   Beckmann Rearrangement

In this type of the rearrangement, an oxime, formed by reaction of a ketone with NH_2OH, is converted to an amide (Fig. 10.11). The reaction can be initiated by a protic or Lewis acid, which interacts with the hydroxyl group in the starting oxime to form a good leaving group (such as water, formed by protonation of –OH) [5]. In addition, an alternative good leaving group such as tosylate (TsO–) may be formed by reaction of the oxime with TsCl. Then, a good leaving group (such as water or tosylate) facilitates a 1,2-rearrangement that takes place spontaneously in the doubly bonded C=N moiety, resulting in the formation of a resonance stabilized intermediate **nitrilium ion**. The R_1 group which is *trans* to –OH of the oxime migrates to the adjacent nitrogen, while the R_2 group (*cis* to –OH) is not rearranged. In the

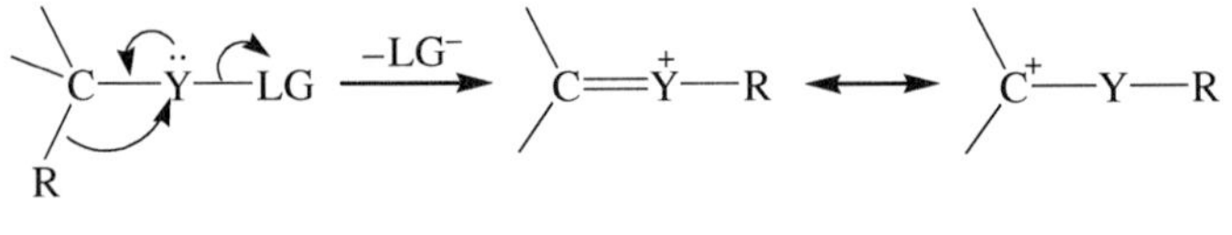

Y = nitrogen or oxygen

R = alkyl, aryl, or hydrogen

FIGURE 10.10   The general mechanism for neighboring leaving group facilitated 1,2-rearrangement in functionalized organic compounds.

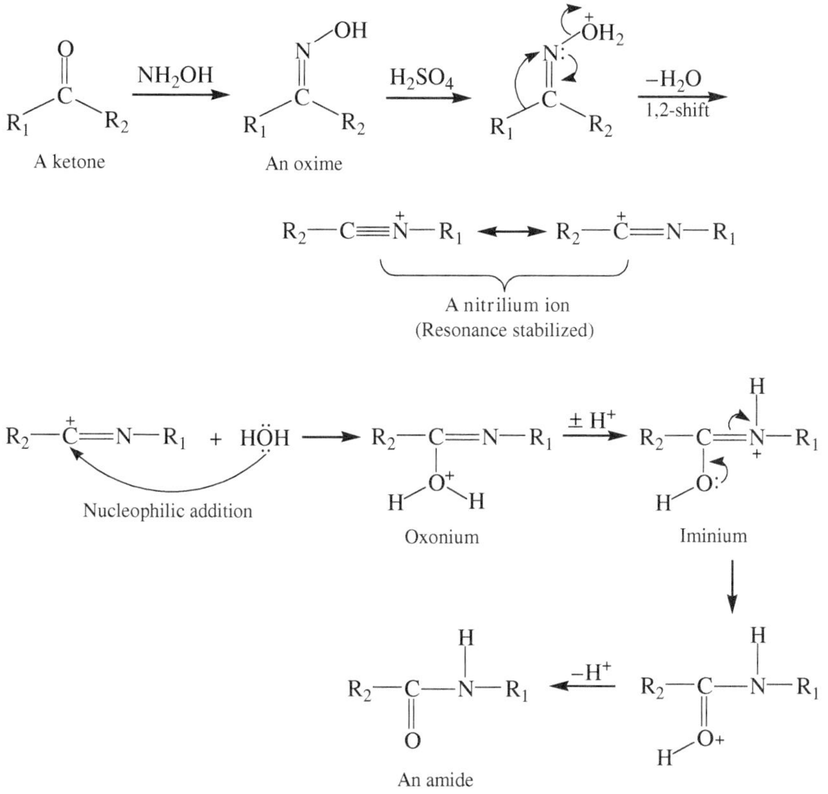

FIGURE 10.11   Mechanism of Beckmann rearrangement, a 1,2-shift in a nitrilium ion which results in transformation of an oxime to an amide.

following step, the strongly electrophilic carbon atom in the nitrilium intermediate is attacked by a water molecule giving an oxonium intermediate, which is transformed subsequently to an iminium intermediate via facile proton transfer processes. The iminium intermediate undergoes a thermodynamically favorable tautomerization to give the final amide product.

The Beckmann rearrangement can also take place in cyclic oximes. For example, in concentrated sulfuric acid, the oxime of cyclohexanone is converted to caprolactam (a cyclic amide) via a 1,2-shift of a cyclic carbon in the $C=N$ domain of the protonated oxime (Fig. 10.12). The Beckmann rearrangement leads to a ring expansion giving a seven-membered cyclic nitrilium intermediate. The angular $-^+C=N-$ double bond should be the predominately major resonance contributor to the cyclic nitrilium intermediate as a $-C\equiv N^+-$ triple bond in a ring structure would be expected unstable due to the molecular strain. In the next steps, similar to the reaction of an acyclic oxime (Fig. 10.11), the strongly electrophilic carbon in the cyclic nitrilium is attacked by a water molecule, which eventually leads to the formation of

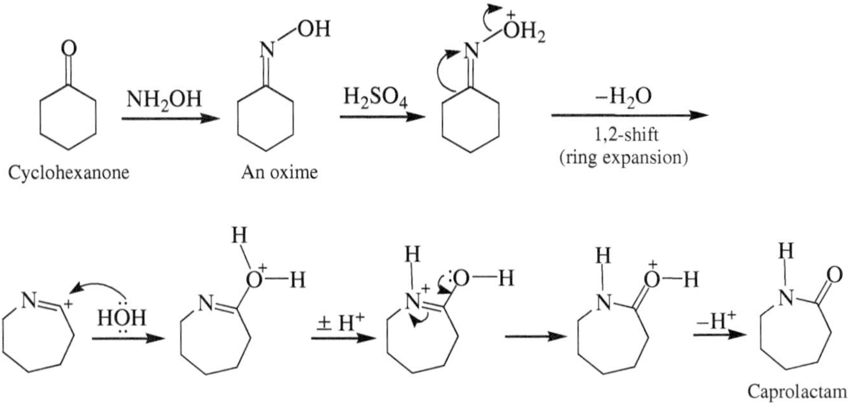

FIGURE 10.12   Mechanism of Beckmann rearrangement involved in transformation of oxime of cyclohexanone to the cyclic amide caprolactam.

caprolactam, a cyclic amide, after proton transfers and a tautomerization. The conversion of oxime of cyclohexanone to caprolactam via Beckmann rearrangement is an industrially useful process, as the caprolactam product is the direct precursor of nylon 6, a versatile polymer that has many applications—among them, the manufacture of fibers for carpeting and other textiles.

10.3.2   Hofmann Rearrangement

In the presence of hydroxide (OH⁻) and elemental bromine (Br_2), an amide can be converted to an amine with one less carbon atom (Fig. 10.13). Clearly, this conversion involves a 1,2-rearrangement of the R group in the C—N bond of the amide substrate, called **Hofmann rearrangement**. The electronegative carbonyl group in the amide can activate its adjacent N—H bond, making the hydrogen slightly acidic. In the presence of the strongly basic OH⁻, the N—H bond can be deprotonated giving a highly nucleophilic nitrogen anion. Then, the nucleophilic nitrogen anion attacks Br_2, resulting in the introduction of a good leaving group –Br to the amide nitrogen. In the following step, the leaving group –Br and lone pair of electrons in the C–N nitrogen drive a 1,2-shift of the –R group to make a cationic O=C=⁺N(H)R intermediate. The electrophilic carbon in the cationic intermediate is subsequently attacked by water, and the resulting intermediate undergoes proton transfers and decarboxylation to afford an amine product.

In general, Hofmann rearrangement of cyclic amides results in the formation of cyclic amines which is accompanied by ring contraction by one carbon atom (Fig. 10.14). The first two steps in the mechanism of the reaction in Figure 10.14 are the same as those for reaction of an acyclic amide in Figure 10.13. Then, the leaving group –Br facilitated ring contraction (1,2-shift of a cyclic carbon) occurs. As a result, the initial six-membered ring is converted to a five-membered cyclic

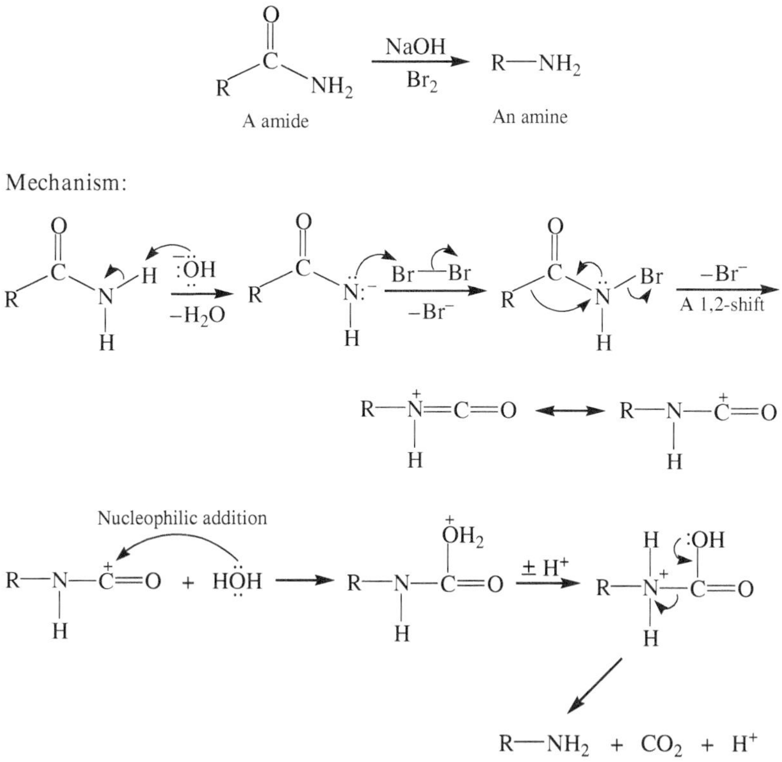

Mechanism:

FIGURE 10.13   Mechanism of Hofmann rearrangement involved in transformation of an acyclic amide to an amine.

cationic intermediate. Nucleophilic attack on the intermediate by water eventually leads to the formation of a five-membered cyclic amine.

10.3.3   Baeyer–Villiger Oxidation (Rearrangement)

A ketone can be oxidized to an ester by a percarboxylic acid ($R'CO_3H$). The reaction involves a 1,2-shift in a C–O moiety. The carbon–oxygen rearrangement associated to the oxidation of a ketone by a percarboxylic acid is called **Baeyer–Villiger rearrangement**. Figure 10.15 shows the reaction mechanism. First, the carbonyl group in the ketone substrate is attacked by the nucleophilic OH oxygen in $R'CO_3H$ giving an intermediate tetrahedral oxyanion. By the nature, this step is a nucleophilic addition to carbonyl. Then, facile proton transfers take place leading to an analogous neutral tetrahedral intermediate. The electronegative carbonyl withdraws the electron density from the peroxy O–O bond, greatly weakening the bond and making the carboxylate ($R'COO–$) group a very good leaving group. Departure of the carboxylate facilitates a 1,2-shift of an R group in the C–O moiety of the intermediate, and a

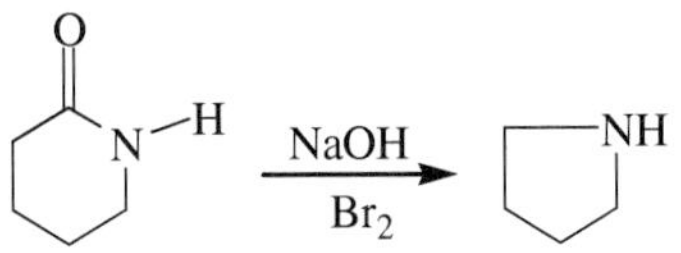

Mechanism:

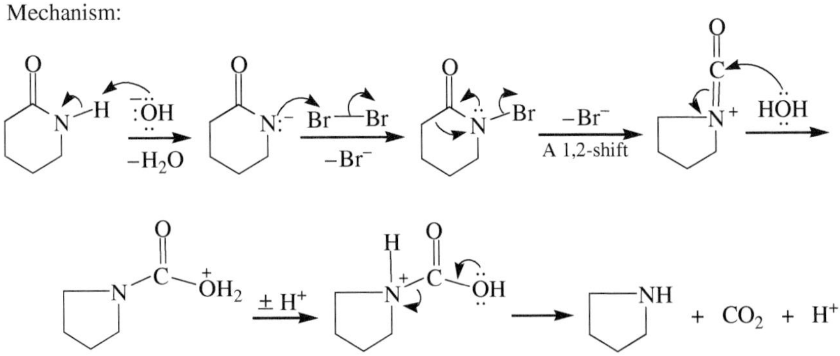

FIGURE 10.14 Mechanism of Hofmann rearrangement for a cyclic amide.

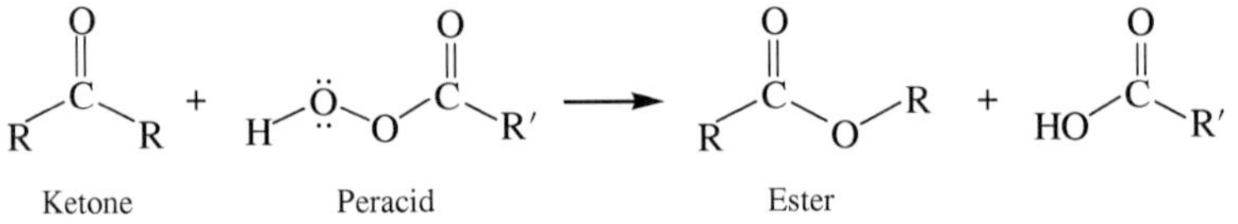

Ketone   Peracid   Ester

Mechanism:

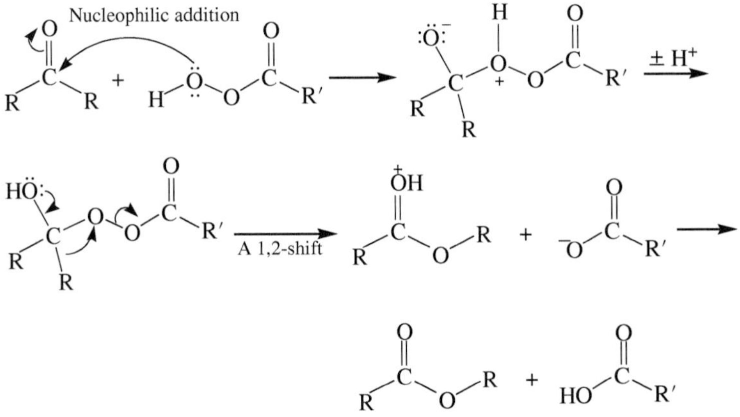

FIGURE 10.15 Mechanism of Baeyer–Villiger oxidation (rearrangement): Oxidation of a ketone to an ester by percarboxylic acid.

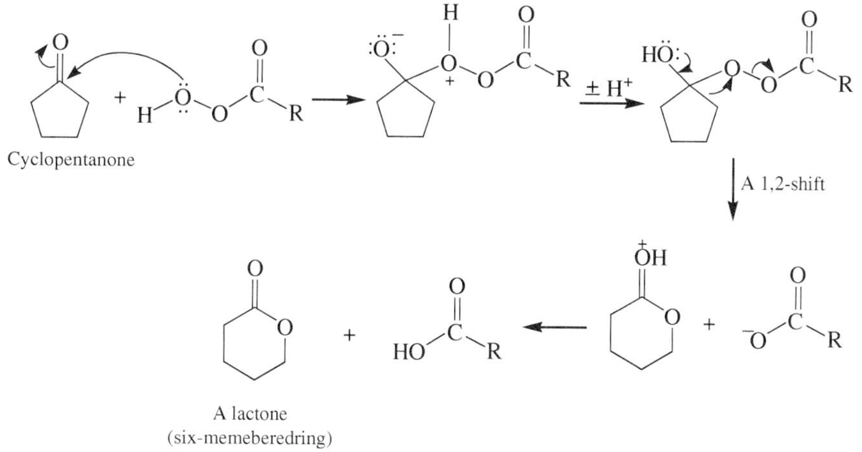

Cyclopentanone

A 1,2-shift

A lactone
(six-memeberedring)

FIGURE 10.16   Mechanism for Baeyer–Villiger oxidation of cyclopentanone to a six-membered lactone.

protonated ester is produced after the carbon–oxygen rearrangement. Finally, a thermodynamically favorable proton transfer between the protonated ester and the carboxylate takes place (an acid-base reaction) to afford an ester and R'COOH.

The Baeyer–Villiger rearrangement has found synthetic applications in making lactones from cyclic ketones (Fig. 10.16). The example in Figure 10.16 shows that oxidation of cyclopentanone by a percarboxylic acid (RCO_3H) results in the net insertion of an oxygen atom in the C_α—CO bond of the cyclopentane ring to give a lactone (a six-membered ring). In the reaction mechanism, departure of a carboxylate (RCOO–) group initiates a 1,2-shift of a ring carbon in the C–O domain of the neutral tetrahedral intermediate formed by a nucleophilic addition. This results in a ring expansion giving the six-membered lactone.

For unsymmetrical ketones, the Baeyer–Villiger rearrangement is regioselective, preferably with the more stable group (more substituted carbon) migrating to the C–O oxygen atom [3]. In other words, the relative capability of migration for different groups follows the order of aryl > tertiary carbon > secondary carbon > primary carbon. Figure 10.17 shows examples of this situation. In the first example, oxidation of acetophenone by m-chlorobenzoic peracid (MCBPA) affords phenyl acetate. The isomeric methyl benzoate is not observed. The regiochemistry originates from the regioselective Baeyer–Villiger rearrangement: Preferably, the phenyl group in the tetrahedral intermediate undergoes 1,2-shift giving the observed phenyl acetate product. However, the less stable methyl is not rearranged. Therefore, the isomeric methyl benzoate is not formed. In the second example, oxidation of methyl(iso)propyl ketone by MCBPA affords isopropyl acetate. The isomeric methyl isoproponoate is not observed. Similarly, the regiochemistry originates from the regioselectivity associated to the Baeyer–Villiger rearrangement: It is the more

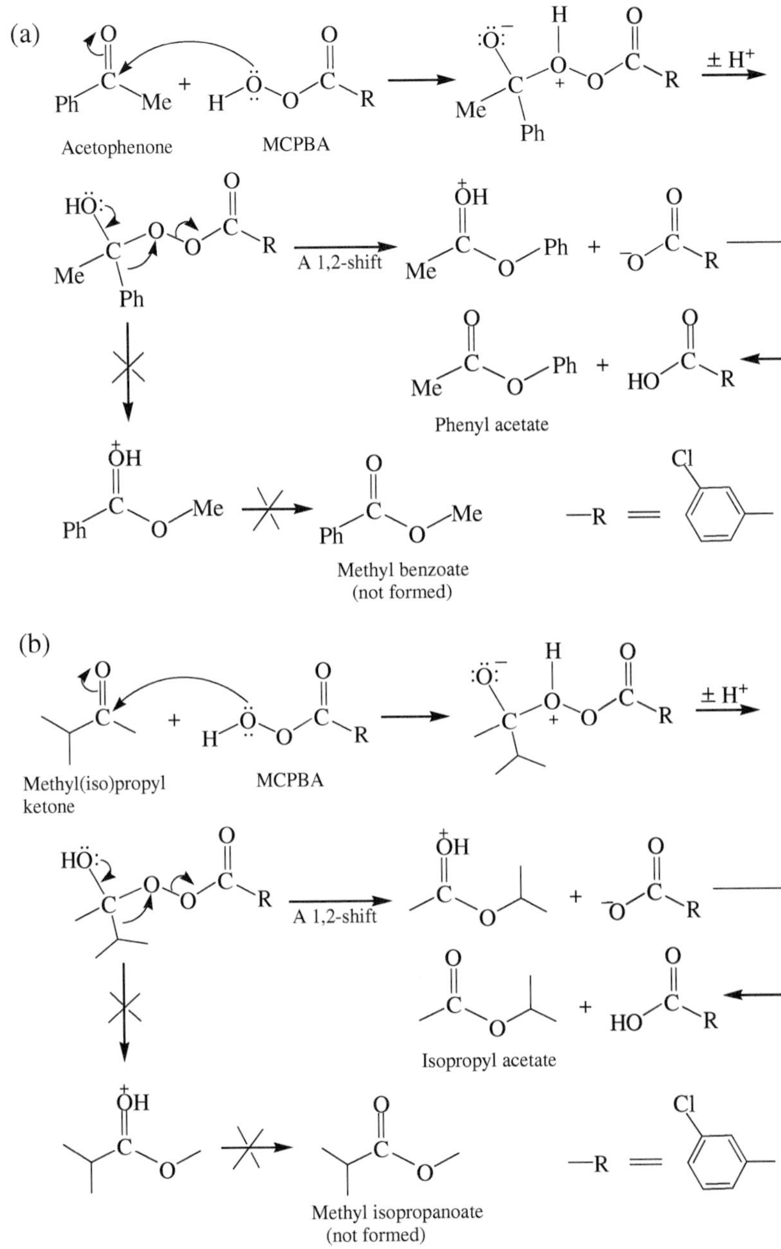

FIGURE 10.17 The regioselectivity for the Baeyer–Villiger oxidation of unsymmetrical ketones.

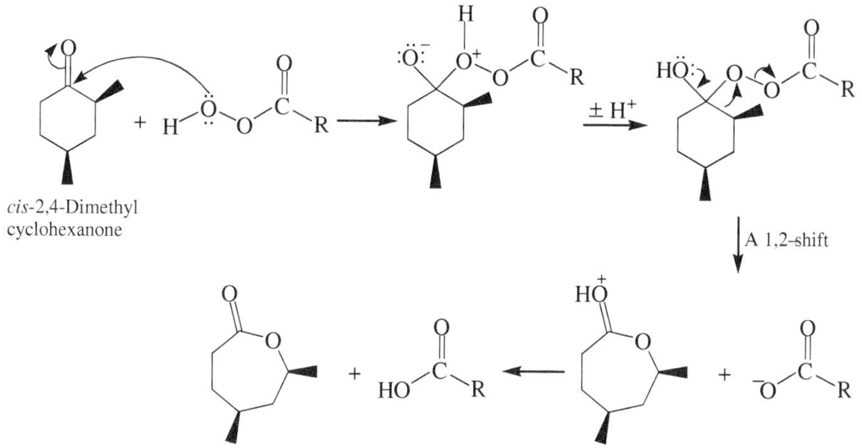

cis-2,4-Dimethyl
cyclohexanone

A 1,2-shift

A lactone
(A *cis*-seven-membered ring)

FIGURE 10.18   The regio- and stereo-chemistry for the Baeyer–Villiger oxidation of cis-2,4-dimethylcyclohexanone.

stable isopropyl group (but methyl) that rearranges to give the observed isopropyl acetate. The isomeric methyl isoproponoate is not observed because the less stable methyl is unrearranged.

Figure 10.18 shows mechanism and regio- and stereo-chemistry for the Baeyer–Villiger oxidation of *cis*-2,4-dimethylcyclohexanone. In the tetrahedral intermediate, it is a tertiary carbon that undergoes 1,2-rearrangement to lead to the formation of the observed lactone product. The secondary carbon in the ring is unrearranged. In addition, the configuration of the tertiary carbon in the ring is retained in the rearrangement, giving an observed *cis*-lactone.

10.3.4   Acid Catalyzed Rearrangement of Organic Peroxides

Autoxidation of cumene ($PhCHMe_2$) affords cumyl peroxide (see Section 2.5 and Figure 2.15). In the presence of a strong acid (such as H_2SO_4), cumyl peroxide undergoes a carbon–oxygen rearrangement to give acetone and phenol (Fig. 10.19). The function of the acid is to protonate the hydroxyl group in cumyl peroxide, making a good leaving group H_2O on the C–O oxygen. As a water molecule leaves and an electron pair in the C–O oxygen moves down to the C—O bond domain to form a π bond, the phenyl group undergoes a 1,2-rearrangement to give a resonance stabilized carbocation intermediate. Similar to the Baeyer–Villiger rearrangement, this carbon–oxygen rearrangement is regioselective, preferably with the more stable phenyl migrating to the C–O oxygen while the methyl groups remaining unrearranged. Then, the electrophilic carbon in the cationic intermediate is attacked

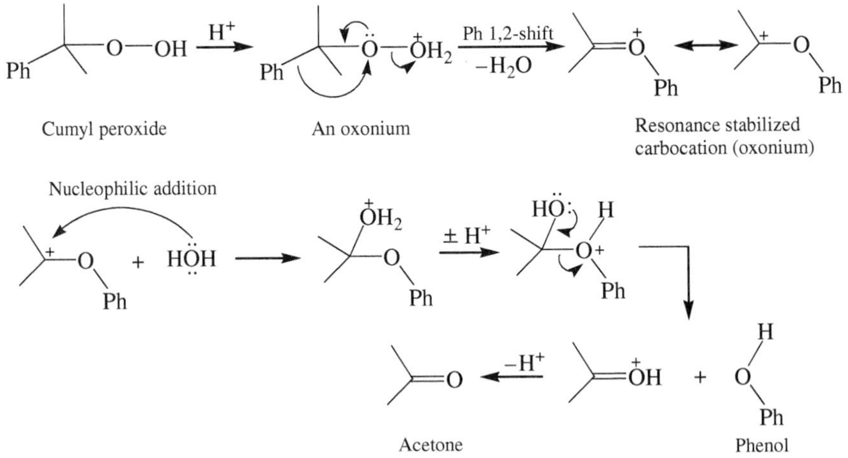

FIGURE 10.19 Mechanism for the acid catalyzed carbon–oxygen rearrangement of cumyl peroxide to give acetone and phenol.

by H_2O to afford an oxonium intermediate. It undergoes subsequent proton transfers to lead to the second oxonium intermediate. An electron pair in the hydroxyl oxygen of the second oxonium intermediate moves down to the C—O bond domain, resulting in the formation of a C=O π bond and simultaneously, pushing off a phenol PhOH molecule. A facile deprotonation of the protonated carbonyl group affords acetone. As mentioned in Chapter 2, this carbon–oxygen rearrangement has been employed in industry for manufacture of acetone and phenol, two very useful fundamental organic chemicals.

10.4   CARBENE REARRANGEMENT: 1,2-REARRANGEMENT OF HYDROGEN FACILITATED BY A LONE PAIR OF ELECTRONS

A **carbene, R_2C:,** is a neutral molecule in which a carbon atom is divalent (possessing two covalent bonds) and surrounded by only six valence electrons [2]. In the lowest electronic state of most carbenes, the divalent carbon is sp^2 hybridized with a lone pair of electrons occupying a sp^2 orbital. The unhybridized 2p orbital is vacant and perpendicular to the plane defined by the sp^2 orbitals (Fig. 10.20). Because of the vacant 2p orbital, a carbene is highly electrophilic on the divalent carbon. In some carbene molecules, the lone pair of electrons in the central divalent carbon facilitates a 1,2-rearrangement of hydrogen such that a C—H bond in an adjacent carbon is cleaved by the electrophilic vacant p orbital which initially overlaps with the bond, with the bonding electrons flowing into the vacant p orbital and the hydrogen migrating to the divalent carbon (1,2-shift). Simultaneously, the lone pair of electrons in the sp^2 orbital of the divalent carbon moves down to the C—C bond domain to form a π

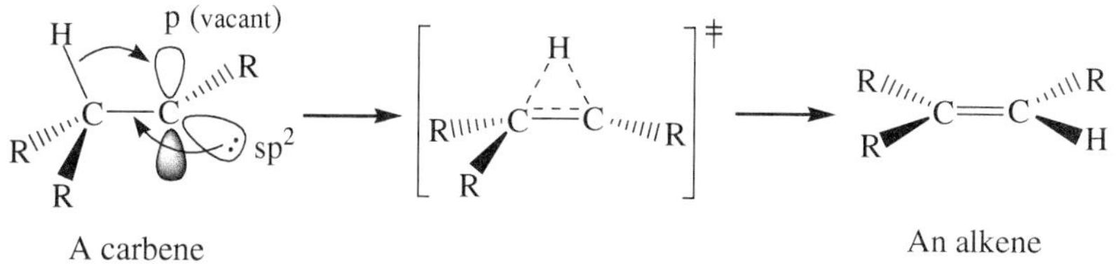

A carbene   An alkene

FIGURE 10.20   Electron-pair facilitated 1,2-rearrangement in carbene.

bond. The overall concerted 1,2-rearrangement of hydrogen affords a thermodynamically more stable alkene (Fig. 10.20).

A carbene can be made by reaction of a ketone ($R_2C=O$) with tosylhydrazine ($TsNHNH_2$), followed by treating the intermediate ($R_2C=NNHTs$) with ^nBuLi to give a tosylhydrozone [$R_2C=NN^-(Li^+)Ts$]. Upon heating, the tosylhydrozone is transformed into a carbene (Fig. 10.21) [6]. This synthetic method is referred to as *Bamford–Steven synthesis*. The cyclic carbene synthesized in Figure 10.21 undergoes a spontaneous 1,2-rearrangement of hydrogen to give a thermodynamically more stable alkene, and the rearrangement is subject to stereoelectronic control [7].

Figure 10.21 shows that the lone pair of electrons in the conformationally rigid carbene stays in the equatorial position, and the vacant unhybridized p orbital is located approximately in the axial orientation [7]. The axial $C-R_1$ (R_1 = H or D) bond is only ~10° away from alignment with the vacant p orbital, thus the $C-R_1$ bond and p orbital have good overlap. On the other hand, the equatorial $C-R_2$ (R_2 = H or D) bond is ~10° away from alignment with the sp^2 orbital holding a lone of electrons, and the $C-R_2$ bond and p orbital is roughly perpendicular to each other with essentially no overlap. As a result, R_1 undergoes the 1,2-rearrangement, but R_2 does not. This stereoelectronic control on the carbene 1,2-rearrangement has been confirmed by deuterium isotope labeling experiments as illustrated in Figure 10.21.

The stereochemistry for the 1,2-rearrangement of carbenes is further demonstrated by the reaction in Figure 10.22. The deuterium isotope labeling experiment shows that the axial H (which has a good overlap with the carbene p orbital) undergoes a 1,2-rearrangement, but the equatorial D remains intact because the $C-D$ bond does not overlap with the p orbital.

The carbene species in Figure 10.23 does not undergo a hydrogen 1,2-rearrangement. Instead, a facile intramolecular cyclization takes place via a 1,3-rearrangement of hydrogen [8, 9].

Under certain conditions, some alkenes undergo the reversal photochemical 1,2-rearrangement of hydrogen to afford a thermodynamically less stable isomeric carbene molecule. Figure 10.24 shows an example of such nonspontaneous (endergonic $\Delta G > 0$) photochemical rearrangement [10]. As the intermediate carbene is produced, a spontaneous intramolecular cyclization occurs. The cyclization follows the same 1,3-rearrangement mechanism as that described in Figure 10.23.

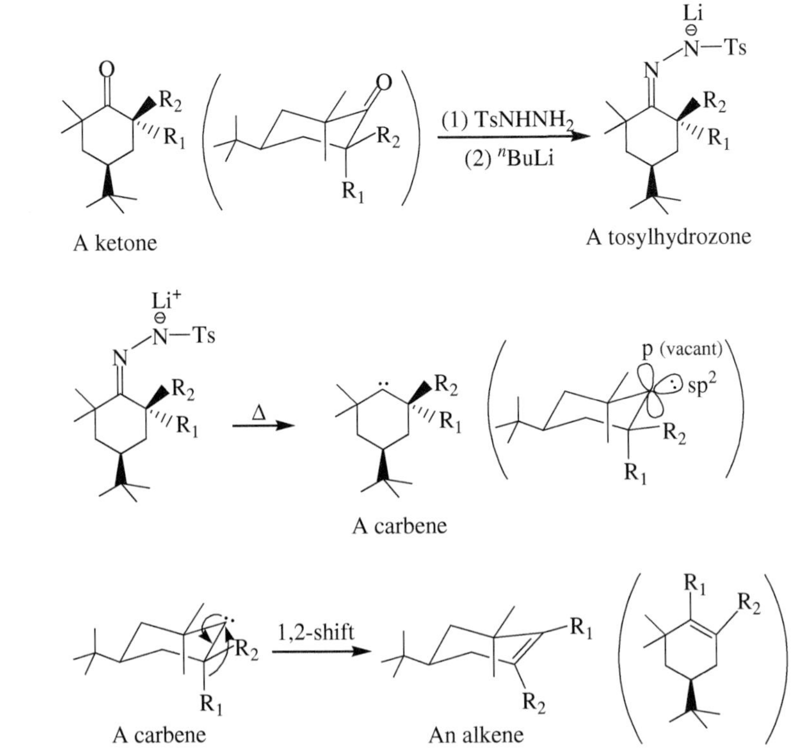

A ketone   A tosylhydrozone

A carbene

A carbene   An alkene

(A) $R_1 = H, R_2 = H$ (C) $R_1 = H, R_2 = D$

(B) $R_1 = D, R_2 = H$ (B) $R_1 = D, R_2 = D$

FIGURE 10.21 Synthesis of carbene from a ketone and stereoelectronic control on the carbene 1,2-rearrangement. Kyba and Hudson [6].

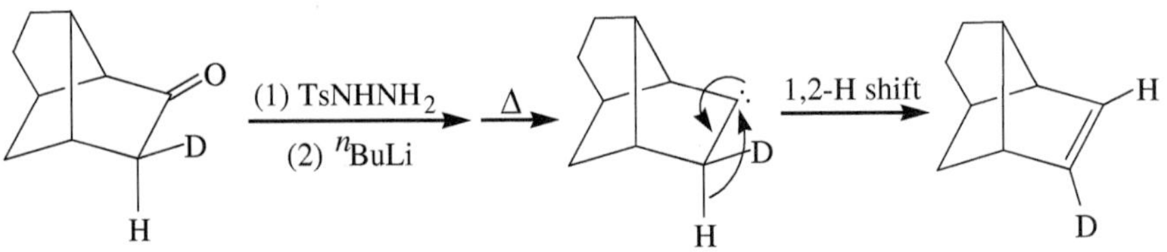

FIGURE 10.22 Stereoselective 1,2-rearrangement of a carbene.

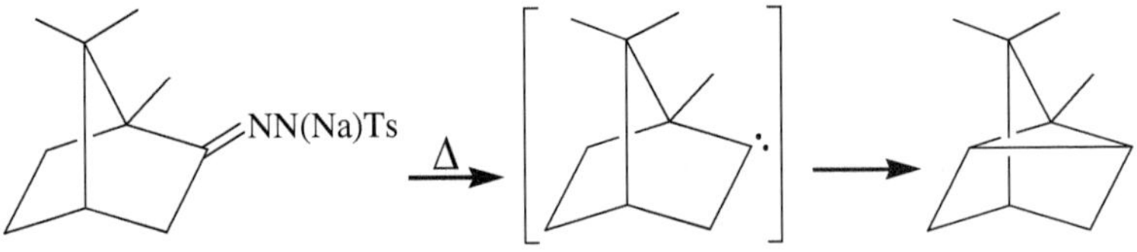

FIGURE 10.23 Cyclization of a carbene via a 1,3-shift.

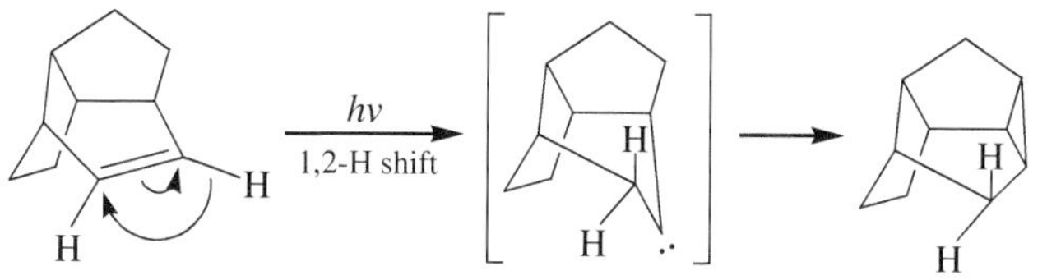

FIGURE 10.24   Photochemical 1,2-rearrangement of an alkene to a carbene, the backward process of the spontaneous carbene rearrangement, followed by a carbene cyclization through a 1,3-shift.

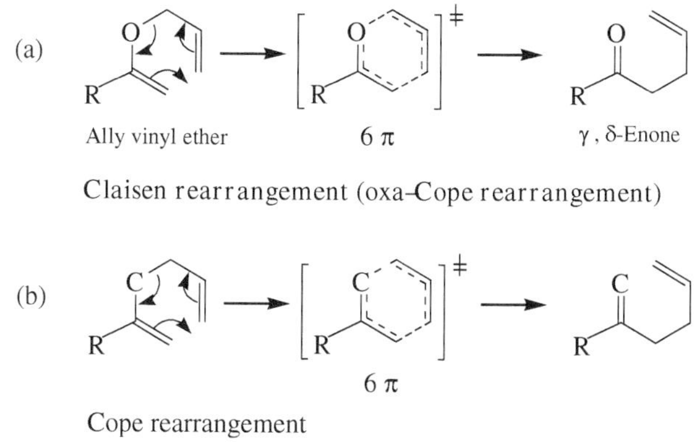

(a)

Ally vinyl ether   6 π   γ , δ-Enone

Claisen rearrangement (oxa–Cope rearrangement)

(b)

6 π

Cope rearrangement

FIGURE 10.25   Claisen rearrangement (a), an analogous process to the Cope rearrangement (b).

10.5   CLAISEN REARRANGEMENT

The **Claisen rearrangement** (Fig. 10.25) is a pericyclic reaction very similar to the Cope rearrangement, which has been discussed in Section 4.5. It takes place via a 6π $[(4n + 2)$, $n = 1$, aromatic] cyclic transition state, analogous to that for a thermally symmetry allowed Diels–Alder cycloaddition [1, 5]. During the rearrangement, a carbon–oxygen σ-bond and a carbon–carbon π-bond are broken, coincident with the formations of a carbon–oxygen π-bond and a carbon–carbon σ-bond. As a result, an allyl vinyl ether (a separated diene) is isomerized to a γ,δ-enone. The formation of a C=O π-bond in Claisen rearrangement makes the reaction highly exothermic. In addition, because the C=O π-bond energy (441 kJ/mol) is much higher than the C=C π-bond energy (256 kJ/mol), the transition state for Claisen rearrangement possesses lower energy (more stable) than does the transition state for the Cope rearrangement. Therefore, the temperature required for the Claisen rearrangement is lower than that for the Cope rearrangement.

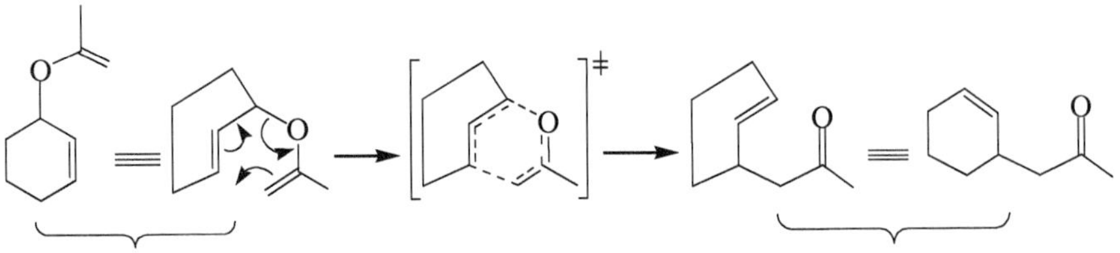

Allyl vinyl ether   γ, δ-Enone

FIGURE 10.26   Claisen rearrangement of a cyclic allyl vinyl ether to a γ,δ-enone.

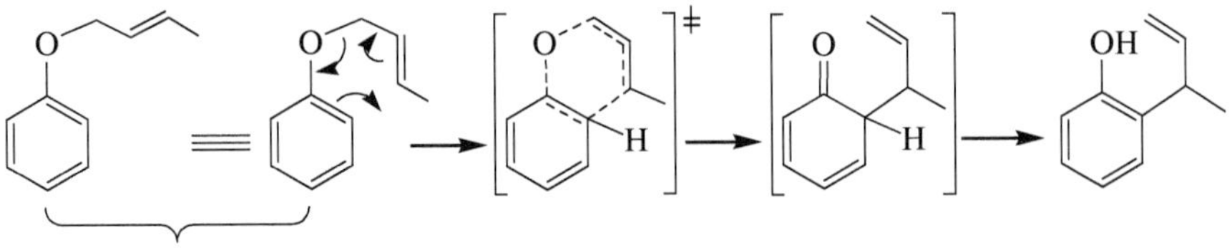

trans-2-Butenyl phenyl ether   A cyclohexadienone   *o*-Allylphenol
                                                                  intermediate   (racemic)

FIGURE 10.27   Claisen rearrangement of *trans*-butenyl phenyl ether to an *o*-allylphenol.

Figure 10.26 shows an example for synthesis of a γ,δ-enone by Claisen rearrangement. The overall reaction is accompanied by a rearrangement of a π bond in a cyclohexene ring.

The Claisen rearrangement also takes place for allyl aromatic ether (Fig. 10.27) following a similar mechanism. In the six-membered cyclic transition state for this rearrangement, the *o*-carbon in the aromatic ring is connected to a carbon of the allyl group. A cyclohexadienyl intermediate is metastable due to disruption of the initial aromatic system. It undergoes subsequent facile tautomerization to afford an *ortho*-substituted phenol.

In the presence of a strong base, the Claisen-rearrangement-like pericyclic reaction occurs to an allyl ester to lead to the formation of a γ, δ-unsaturated carboxylate anion (Fig. 10.28). First, deprotonation of the ester occurs on its α-carbon giving an intermediate enolate. It is analogous to an allyl vinyl ether and will experience a facile Claisen-rearrangement-like process to afford a γ, δ-unsaturated carboxylate. Upon protonation, the carboxylate is converted to a γ, δ-unsaturated carboxylic acid.

10.6   CLAISEN REARRANGEMENT IN WATER: THE GREEN CHEMISTRY METHODS

Analogous to some Diels–Alder cycloadditions which take place much faster in water than in organic solvents or as neat reactants (Section 4.6), certain Claisen rearrangement reactions also occur much faster in water than in organic media. In

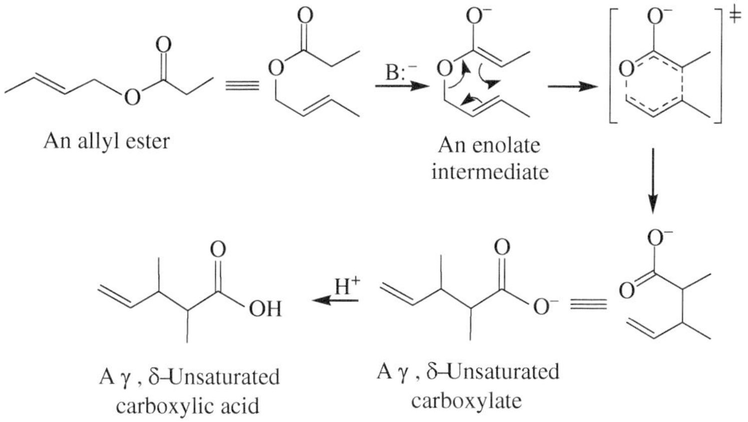

FIGURE 10.28   A strong base induced Claisen rearrangement of an allyl ester to a γ,δ-unsaturated carboxylate.

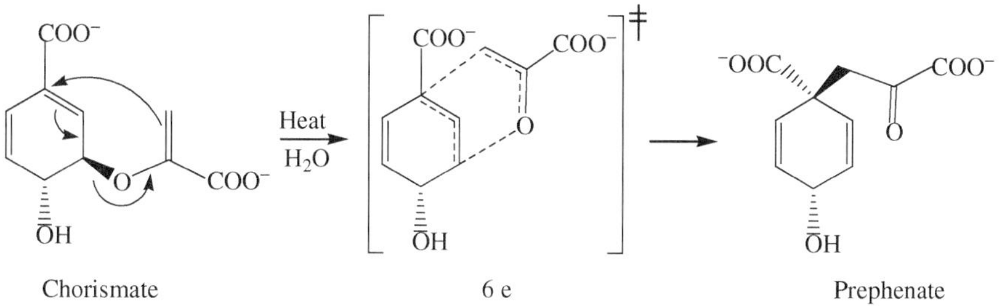

FIGURE 10.29   The Claisen rearrangement of chorismate to prephenate in water via a 6 e cyclic transition state.

general, as discussed in Section 1.12, the molecules of organic substrates (reactants) aggregate in water to increase their surface area and thus their energy is enhanced relative to their energy levels in organic solvents or in the neat states. In addition, the water molecules can be hydrogen bonded to the transition state of the reaction in the surface to stabilize it. As a result, the activation energy for the reaction in water is substantially lowered as illustrated in Figure 1.24.

The conversion of chorismate (equivalent to an allyl vinyl ether) to prephenate (Fig. 10.29) is apparently a Claisen rearrangement via a 6 e cyclic transition state, and the reaction possesses significant biological relevance [11, 12]. It is catalyzed *in vivo* by the enzyme chorismate mutase and is the first step in the conversion of chorismate to phenylalanine and tyrosine in bacteria and low plants [12]. Research has shown that this rearrangement is unusually facile even *in vitro* in the absence of any catalyst. The concerted, uncatalyzed Claisen rearrangement in Figure 10.29 has a half-life of 10 min at 75 °C in water at pH 5, which is 4200 times faster than the rearrangement of the allyl vinyl ether in di-*n*-butyl ether at the same temperature [12].

Figure 10.30 shows the Claisen rearrangements of 6-β-glycosylallyl vinyl ether and 6-α-glycosylallyl vinyl ether in water and toluene [11]. The reactions are performed in the presence of NaBH₄. Each concerted rearrangement gives an aldehyde intermediate, which is subsequently reduced by NaBH₄ to an alcohol. For the rearrangement of 6-β-glycosylallyl vinyl ether (Fig. 10.30a), it only takes 1 h to complete when the reaction proceeds in water at 80 °C. When the reaction is performed in toluene at the same temperature, it takes 13 d to complete. For the rearrangement of 6-α-glycosylallyl vinyl ether (Fig. 10.30b), it takes 1.5 h to complete

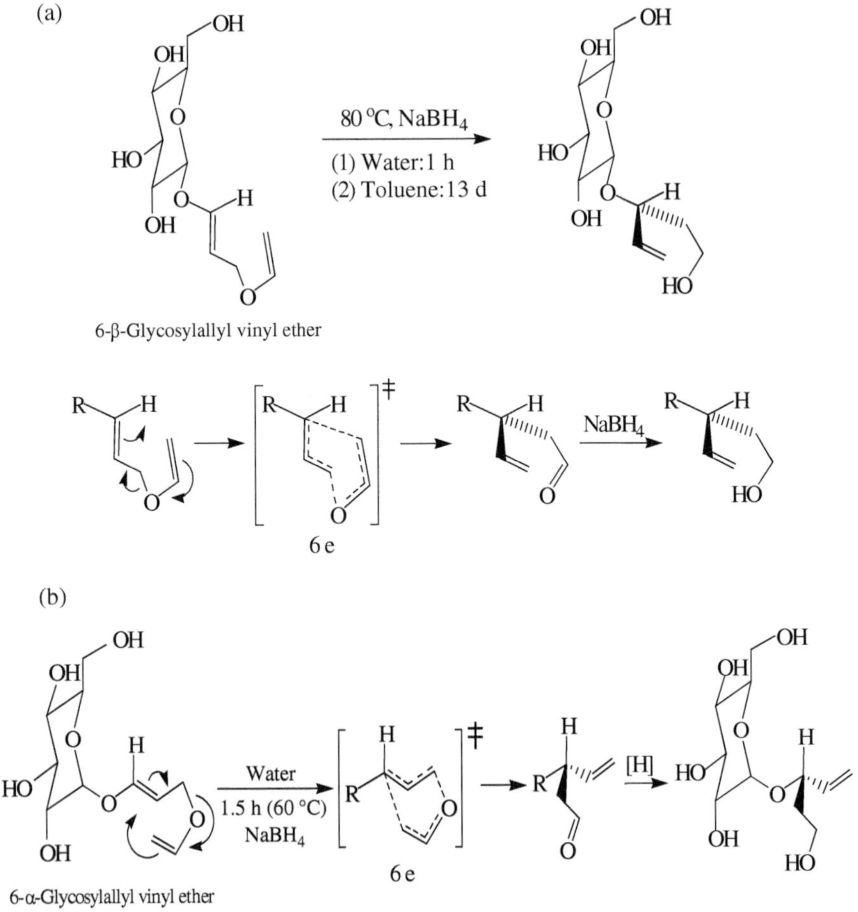

FIGURE 10.30 The Claisen rearrangement of (a) 6-β-glycosylallyl vinyl ether and (b) 6-α-glycosylallyl vinyl ether in water and toluene in the presence of NaBH₄.

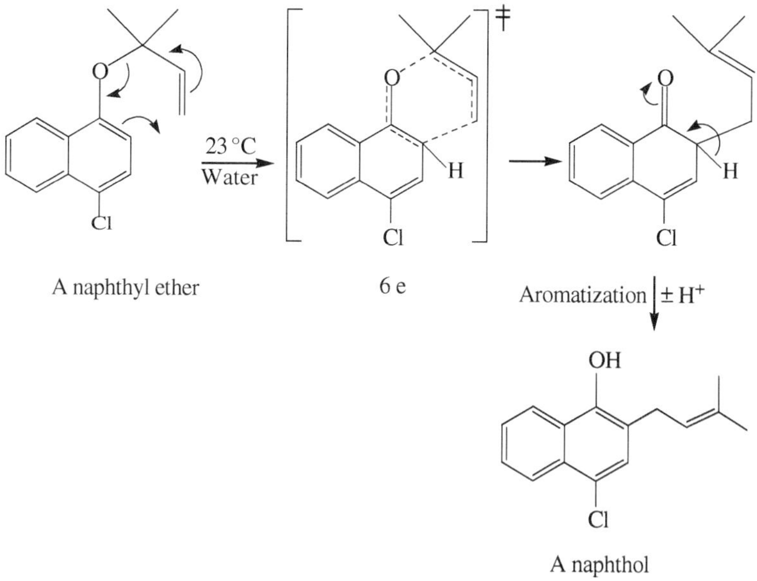

FIGURE 10.31   The Claisen rearrangement of a naphthyl ether in water.

when the reaction proceeds in water at 65 °C. However, no product is formed when the reaction is conducted in toluene at 110 °C.

   Research also shows that the Claisen rearrangement of a naphthyl ether to a naphthol (Fig. 10.31) is much more efficient when the reaction is performed in water than the reactions performed in organic solvents [11, 13, 14]. First, the concerted Claisen rearrangement gives a ketone intermediate, which undergoes a facile aromatization to the final naphthol product. The reaction performed in water for 5 d at 23 °C leads to complete conversion of the naphthyl ether to a naphthol with 100% yield, while the yields for the reactions conducted in toluene and methanol at the same temperature and length are 16% and 56%, respectively. The reaction with the neat reactant has a yield of 73%.

10.7   PHOTOCHEMICAL ISOMERIZATION OF ALKENES AND ITS BIOLOGICAL APPLICATIONS

The rotation about a C=C double bond in an alkene is hindered at typical reaction temperatures (<300 °C) because of a high activation energy required to break the π bond. Therefore, the isomerization of a *cis*-alkene to its *trans*-isomer does not occur thermally at ambient temperature although the *trans*-isomer is thermodynamically more stable. When an alkene molecule absorbs a photon from ultraviolet radiation, an electron can be excited from the bonding HOMO orbital (π) to the antibonding

LUMO orbital ($\pi*$), resulting in breaking of the π bond. Then, a rotation about the C—C σ bond takes place readily leading to a geometric isomerization.

10.7.1   Photochemical Isomerization

The photochemically induced interconversion between the *cis*- and *trans*-configurations of an alkene can be demonstrated using *trans*- and *cis*-2-stilbenes as an example (Fig. 10.32). *Trans*-2-stilbene exhibits two maxima (296 and 305 nm) in the ultraviolet absorption due to the π–$\pi*$ transition (excitation of an electron from HOMO to LUMO) in the alkene molecule [5]. The energy of the photon at these wavelengths matches the π–$\pi*$ gap in the molecule. For *cis*-2-stilbene, steric interactions of the two phenyl rings prevent full planarity of the π system. This causes decrease in conjugation, and as a result, the π–$\pi*$ gap in the alkene molecule becomes larger. The ultraviolet light with a shorter wavelength (higher energy) is required to excite an electron from the HOMO to LUMO for *cis*-2-stilbene.

Figure 10.32 shows when *trans*-2-stilbene is irradiated at 296 nm (which matches π–$\pi*$ gap), an electron is excited from HOMO (π) to LUMO ($\pi*$). Now, the energy decrease of the electron in HOMO is offset by the energy increase of the electron in LUMO. This causes breaking of the π bond and the formation of an unstable diradical intermediate in *trans*-configuration. Free rotation about the two carbon atoms bearing a single electron becomes kinetically favorable, and it can take place rapidly even at very low temperatures to give another diradical intermediate in *cis*-configuration. Ultimately, the electron in the high-energy antibonding orbital

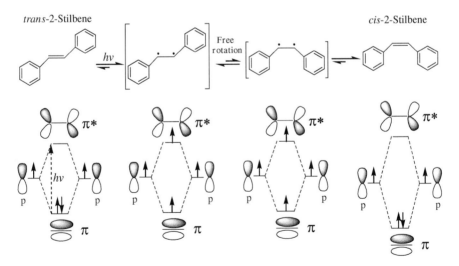

FIGURE 10.32   Mechanism for the photochemical isomerization of *trans*-2-stilbene to *cis*-2-stilbene. The HOMO (π)–LUMO ($\pi*$) gap in *cis*-2-stilbene is greater than that in *trans*-2-stilbene.

of the *cis*-diradical returns to the bonding orbital, leading to the formation of a C=C π bond. This results in *cis*-2-stilbene and releases most of the absorbed energy as heat. Since the irradiation wavelength does not match the π–π* gap in *cis*-2-stilbene, the excitation of an electron from HOMO to LUMO in *cis*-2-stilbene by photon of 296 nm would be much slower than that for *trans*-2-stilbene. Therefore, the π–π* transition in *trans*-2-stilbene will be predominating when irradiated by the 296 nm ultraviolet light, and most of the *trans*-2-stilbene molecules will be converted to *cis*-2-stilbene.

This example has demonstrated a general photochemical isomerization of a *trans*-alkene to its thermodynamically less stable *cis*-isomer. On the other hand, when ultraviolet light with a shorter wavelength which matches the π–π* gap in *cis*-2-stilbene is employed, *cis*-2-stilbene will be converted to the thermodynamically more stable *trans*-2-stilbene in the opposite direction of the photochemical pathway for *trans*-to-*cis* transformation as shown in Figure 10.32.

10.7.2   Biological Relevance

The photochemically induced *cis*–*trans* isomerization of alkenes has significant biological consequences. It forms the chemical basis for mammalian vision. Light-sensitive receptor cells in the eye contain 11-*cis*-retinol, which is chemically bound to the protein opsin through an amino (–NH_2) group, forming rhodopsin (Fig. 10.33) [5]. The π–π* gap of the conjugated alkene matches the wavelengths of visible light. Absorption of a photon from the visible light induces the π–π* transition similar to that for 2-stilbenes as described in Figure 10.32. This can cause a free rotation about the C_{11}—C_{12} bond in the 11-*cis*-retinal molecule, giving rise to the geometric isomerization as indicated in Figure 10.33. The structure of the resulting 11-*trans*-retinal molecule is more extended than that of the *cis*-isomer. Therefore, the *cis*-to-*trans* isomerization directly causes a structural change in the opsin protein, which results in release of calcium ions. The increase in concentration of Ca^{2+} triggers a nerve impulse that is interpreted by the brain as vision.

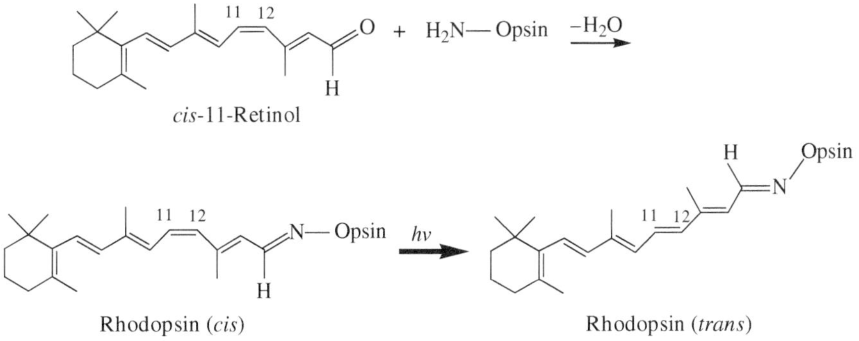

FIGURE 10.33   The photochemical rearrangement of *cis*-11-retinol to its *trans*-isomer: The origin of vision.

10.8   REARRANGEMENT OF CARBON–NITROGEN–SULFUR CONTAINING HETEROCYCLES

It has been shown (see Section 4.4.3) that the dithionitronium NS_2^+ cation undergoes concerted symmetry-allowed cycloaddition reactions with nitriles to give the 6π 5-organo-1,3,2,4-dithiadiazolium heterocycles (Fig. 10.34a) [15, 16]. Upon reduction, 5-organo-1,3,2,4-dithiadiazolium is transformed into stable 7π 5-organo-1,3,2,4-dithiadiazolyl radical (Fig. 10.34b) [16, 17]. Some derivatives of this radical system, such as those with R = Ph and Bu^t (electron donating groups), have been isolated as pure compounds in the dark [16]. In the normal room light, these derivatives of 5-organo-1,3,2,4-dithiadiazolyl radical have been found to undergo facile rearrangement in liquid solution to the isomeric 7π 4-organo-1,2,3,5-dithiadiazolyl (Fig. 10.35) [15, 17]. Apparently, the rearrangement is effected by the net exchange of sulfur (S1) and nitrogen (N2) atoms within the five-membered ring of 5-organo-1,3,2,4-dithiadiazolyl. *In the dark, the rearrangement of the tert-butyl and phenyl derivatives of 5-organo-1,3,2,4-dithiadiazolyl does not occur in solution* [16].

The unrearranged 5-phenyl-1,3,2,4-dithiadiazolyl (R = Ph) radical exhibits two UV–Vis absorption bands centered at 376 and 480 nm, respectively. The absorbances of the radical at both wavelengths are determined to be directly proportional to *the square of molar concentration of the radical*, unambiguously showing a dimerization equilibrium (Fig. 10.35), and the photon absorptions are due to a radical dimer [17]. The standard enthalpy and entropy changes for the dimerization are determined by variable temperature ESR spectroscopy to be $\Delta H_d° = -19.0$ kJ/mol and $\Delta S_d° = -66.5$ J/mol, respectively. On this basis, the dimerization equilibrium constant at room temperature (298 K) is $K_{298} = 0.7$ (R = Ph) [17].

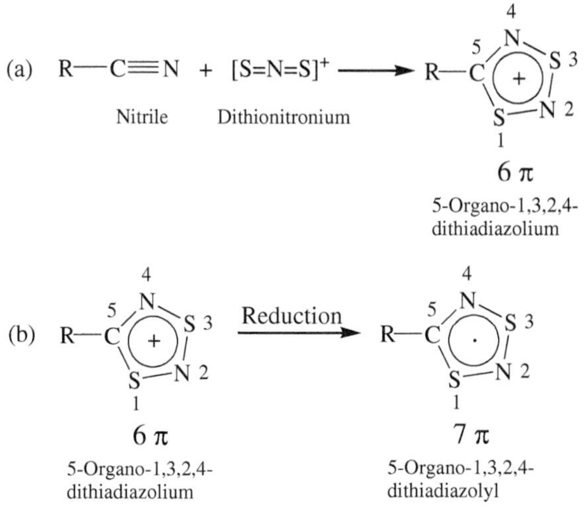

FIGURE 10.34 Synthesis of (a) the 6π 5-organo-1,3,2,4-dithiadiazolium heterocycle. Parsons and Passmore [15] and Passmore et al. [16], and (b) the 7π 5-organo-1,3,2,4-dithiadiazolyl heterocyclic radical. Passmore et al. [16] and Passmore and Sun [17].

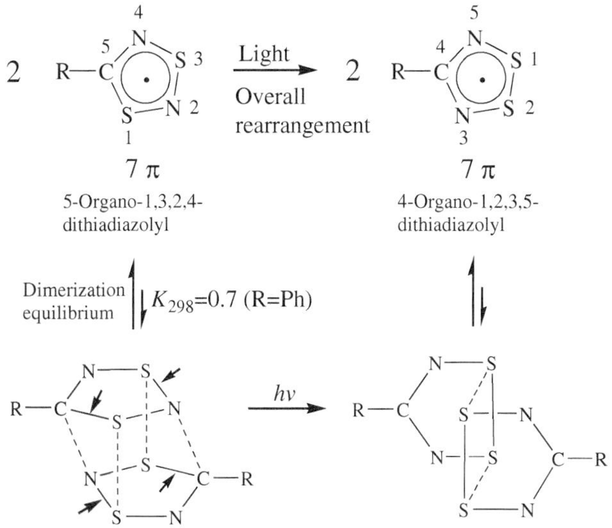

FIGURE 10.35 The photochemical rearrangement of 5-organo-1,3,2,4-dithiadiazolyl to the isomeric 5-organo-1,2,3,5-dithiadiazolyl through an intermediate dimer. Parsons and Passmore [15] and Passmore and Sun [17].

Further photochemical studies show that upon irradiations at 376 and 480 nm, the 5-phenyl-1,3,2,4-dithiadiazolyl (R = Ph) radical rapidly and quantitatively rearranges to the isomeric 4-phenyl-1,2,3,5-dithiadiazolyl (R = Ph) radical [17]. Since the absorptions of photons at 376 and 480 nm are due to the dimer of the unrearranged 5-phenyl-1,3,2,4-dithiadiazolyl radical as demonstrated above, the photochemical rearrangement has been determined to take place via an intermediate radical dimer as shown in Figure 10.35 as the consequence of its photon absorption at 376 or 480 nm. Structural characterization of related species suggests that in solution, the unrearranged 5-organo-1,3,2,4-dithiadiazolyl radical dimer may possess an energetically favorable *trans*-configuration with the symmetry i (center of symmetry) as shown in Figures 10.35 and 10.36. The two monomers are weakly linked together by their $\pi*-\pi*$ (SOMO–SOMO) interactions on the sulfur atoms (two S…S contacts). Upon irradiation of the *trans*-dimer of 5-phenyl-1,3,2,4-dithiadiazolyl (R = Ph) at 376 or 480 nm, the dimer allows a concerted rearrangement via only slight movement of all the atoms by the net cleavage of two C—S and N—S bonds (indicated by arrows in Figure 10.35) within the two five-membered rings and the concurrent formation of two C—N and S—S bonds (indicated by dotted lines of the first dimer in Figure 10.35) between two monomers. As a result, a thermodynamically more stable ($\sim$316 kJ/mol) $\pi*-\pi*$ (SOMO–SOMO) dimer of 4-phenyl-1,2,3,5-dithiadiazolyl (R = Ph) is formed. Spontaneous dissociation of the 4-phenyl-1,2,3,5-dithiadiazolyl (R = Ph) dimer leads to the rearranged monomeric 4-phenyl-1,2,3,5-dithiadiazolyl radical.

In the process of the concerted bimolecular (dimeric) rearrangement, an inversion center i of the dimer is retained. Therefore, the molecular orbitals of the starting unrearranged 1,3,2,4-dithiadiazolyl dimer can be correlated to those of the resulting rearranged 1,2,3,5-dithiadiazolyl dimer with respect to i (Fig. 10.36) [17]. Theoretical calculations indicate that each of the unrearranged 1,3,2,4-dithiadiazolyl (R=H) and rearranged 1,2,3,5-dithiadiazolyl (R=H) radical dimers has 27 occupied

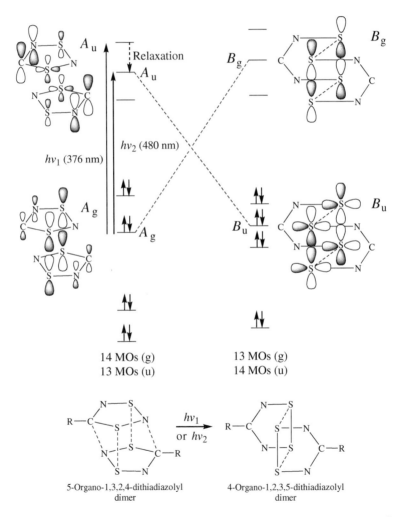

FIGURE 10.36 Mechanism for the concerted photochemically symmetry allowed bimolecular (dimeric) rearrangement of 5-organo-1,3,2,4-dithiadiazolyl to the isomeric 5-organo-1,2,3,5-dithiadiazolyl. From Passmore and Sun [17].

molecular orbitals in the valence shell. Of the 27 occupied MOs, the unrearranged 1,3,2,4-dithiadiazolyl dimer has 14 symmetric (g) and 13 antisymmetric (u) with respect to i (the center of symmetry), while the rearranged 1,2,3,5-dithiadiazolyl dimer has only 13 occupied MOs that are symmetric (g) and 14 antisymmetric (u). As a result, correlations of MOs of the two dimers involve a crossover correlation between an occupied antibonding π–π MO (A_g) of the 1,3,2,4-dithiadiazolyl dimer and a high-energy unoccupied antibonding $\sigma*$ S–S MO (B_g) of the 1,2,3,5-dithiadiazolyl dimer and between a high-energy unoccupied $\pi*$–$\pi*$ MO (A_u) of the 1,3,2,4-dithiadiazolyl dimer and an occupied σ S–S MO (B_u) of the 1,2,3,5-dithiadiazolyl dimer (Fig. 10.36). Thus, the rearrangement is *photochemically symmetry allowed but thermally symmetry forbidden*, consistent with the observation that the rearrangement is photochemically induced.

The rearrangement is believed to proceed by the allowed electronic transition from an occupied MO (A_g) to an unoccupied MO (A_u) in the 1,3,2,4-dithiadiazolyl dimer, which may be the origin of the absorption band of the dimer at 480 nm ($h\nu_2$). The absorbed photon energy leads to breaking of two C—S and N—S bonds and concurrent formation of two C—N and S—S bonds in the dimer. The other dimer absorption band at 376 nm ($h\nu_1$) may originate from an electronic transition from occupied MO (A_g) to a higher energy unoccupied MO, which then relaxes to the MO (A_u) to bring about the rearrangement. The overall mechanism for the concerted photochemically symmetry allowed bimolecular (dimeric) rearrangement of 5-organo-1,3,2,4-dithiadiazolyl to 4-organo-1,2,3,5-dithiadiazolyl is illustrated in Figure 10.36.

When the five-membered ring of 5-organo-1,3,2,4-dithiadiazolyl is attached to an electronegative group such as R = CF_3, 4-$O_2NC_6H_4$, or 3,5-$(O_2N)_2C_6H_3$, it can also undergo rearrangement to the isomeric 4-organo-1,2,3,5-dithiadiazolyl thermally *in the dark* at room temperature. However, the rearrangement in the dark has been found 5–10 times slower than that in the normal room light [16]. This low-energy thermal rearrangement in the dark may also proceed via the formation of an intermediate 5-organo-1,3,2,4-dithiadiazolyl *trans*-dimer in a concerted, but symmetry disallowed mechanism, consistent with the observed qualitatively slower rearranging rates in the dark.

PROBLEMS

10.1 Propose a detailed stepwise mechanism to account for ring expansion involved in the following reaction.

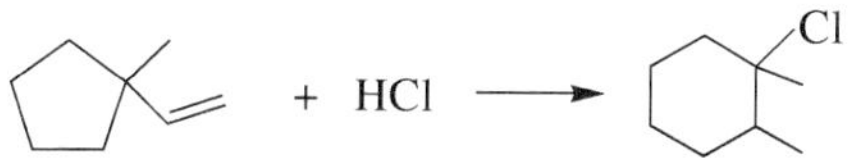

10.2 Propose a detailed stepwise mechanism to account for the following reaction. Explain the stereochemistry for the reaction.

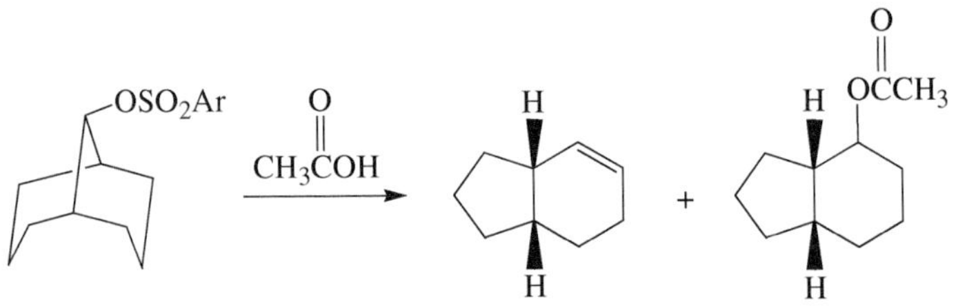

Arenesulfonic acid (ArSO₃H) is a strong acid, and its conjugate base $ArSO_3^-$ is a weak base and a good leaving group.

10.3 Write out mechanism for the following reaction and account for the ring expansion involved in the reaction.

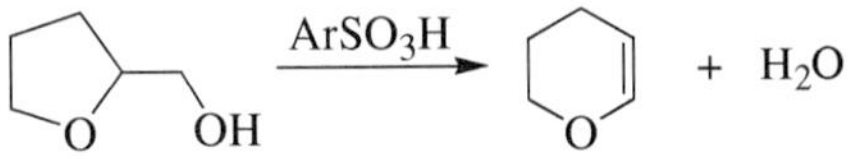

10.4 Treatment of an α-halocyclohexanone with sodium hydroxide results in a ring contraction. The reaction takes place via a proposed cyclopropanone interme-diate. Propose a detailed stepwise mechanism to account for formations of the cyclopropanone intermediate and subsequent conversion of the unstable inter-mediate to a carboxylate containing a cyclopentane ring. What factors have facilitated formation of the unstable cyclopropanone intermediate?

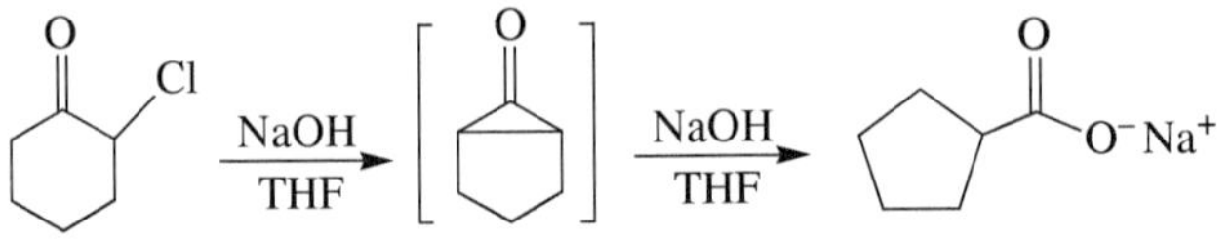

10.5 Suggest a stepwise mechanism to explain the following reaction. Account for the driving force for the rearrangement process.

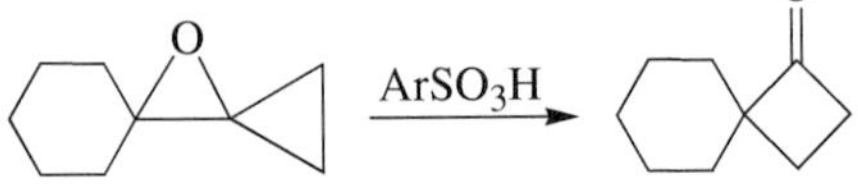

10.6 Usually, a tertiary alcohol is stable toward an oxidant under mild conditions. However, a tertiary allyl alcohol can be oxidized by chromic acid (H_2CrO_4) to an α,β-unsaturated ketone as shown below. Suggest a mechanism to account for the reaction.

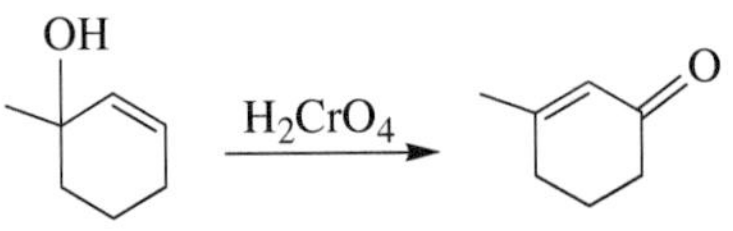

10.7 Treatment of phenol with sodium hydroxide followed by a reaction of an allyl bromide at elevated temperature results in an alkylation on the aromatic ring as shown below. Write a mechanism to account for the reaction. What type of rearrangement is involved in this reaction?

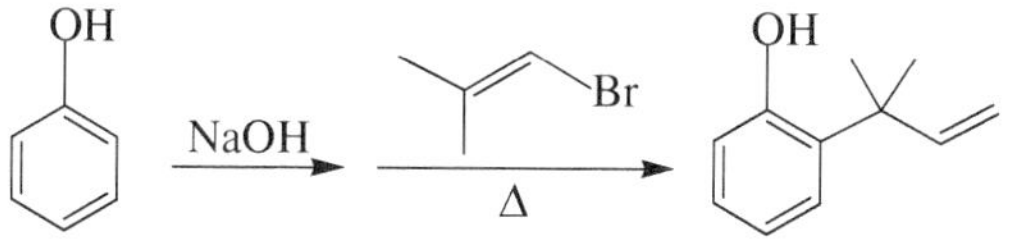

10.8 For each of the following reactions, give the necessary reagents for the conversion. Show the possible reaction mechanism and account for the rearrangement involved in the reaction.

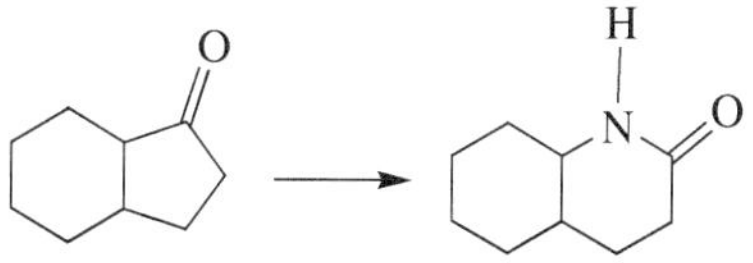

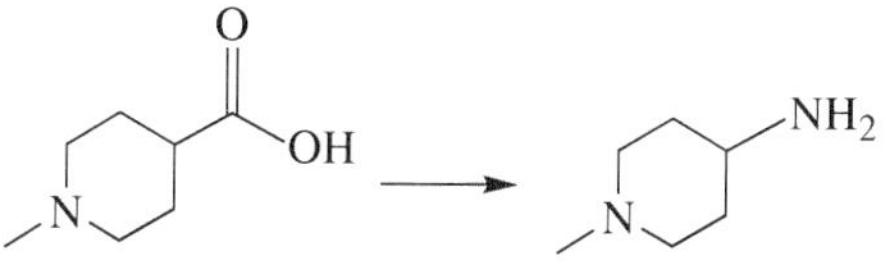

REFERENCES

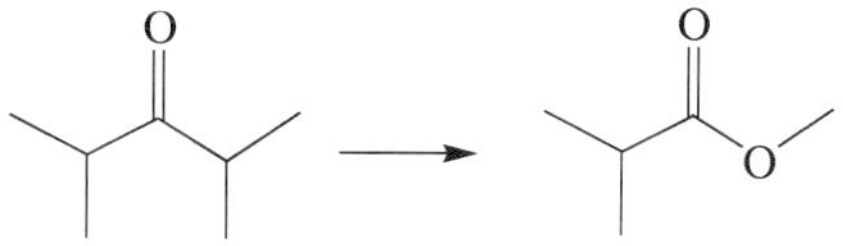

1. Jackson, R. A. *Mechanisms in Organic Reactions*, The Royal Society of Chemistry, Cambridge, UK (2004).

2. Brown, W. H.; Foote, C. S.; Iverson, B. L.; Anslyn, E. V. *Organic Chemistry*, 6th ed., Brooks/Cole, Belmont, CA, USA (2012).

3. Hoffman, R. V. *Organic Chemistry: An Intermediate Text*, 2nd ed., Wiley, Hoboken, NJ, USA (2004).

4. Voet, D.; Voet, J. G.; Pratt, C. W. *Fundamentals of Biochemistry*, 5th ed., Wiley, Hoboken, NJ, USA (2016).

5. Fox, M. A.; Whitesell, J. K. *Organic Chemistry*, 2nd ed., Jones and Bartlett, Sudbury, MA, USA (1997).

6. Kyba, E. P.; Hudson, C. W. 1,2-Hydrogen Shifts in Carbenes. The Question of Stereoelectronic Control of Migration to an Alkylcarbene. *J. Am. Chem. Soc.* 1976, *98*, 5696–5697.

7. Kyba, E. P.; John, A. M. 4-*tert*-Butyl-2,2-Dimethylcyclohexylidene. A Surprising Lack of Stereoselectivity in a 1,2-Hydrogen Shift to an Alkylcarbene. *J. Am. Chem. Soc.* 1977, *99*, 8329–8330.

8. Freeman, P. K.; Hardy, T. A.; Balyeat, J. R.; Wescott, Jr., L. D. Hydrogen Migration in 2-Carbena-6,6-dimethylnorbornane. *J. Org. Chem.* 1977, *42*, 3356–3359.

9. Kyba, E. P.; Hudson, C. W. 1,2-H Shifts in Carbenes. The Benzonorbornenylidene System. *J. Org. Chem.* 1977, *42*, 1935–1939.

10. Nickon, A.; Ilao, M. C.; Stern, A. G.; Summers, M. F. Hydrogen Trajectories in Alkene to Carbene Rearrangement. Unequal Deuterium Isotope Effects for the axial and Equatorial Paths. *J. Am. Chem. Soc.* 1992, *114*, 9230–9232.

11. Butler, R. N.; Coyne, A. G. Water: Nature's Reaction Enforcer–Comparative Effects for Organic Synthesis "In–Water" and "On–Water". *Chem. Rev.* 2010, *110*, 6302–6337.

12. Gajewski, J. J.; Jurayj, J.; Kimbrough, D. R.; Gande, M. E.; Ganem, B.; Carpenter, B. K. On the Mechanism of Rearrangement of Chorismic Acid and Related Compounds. *J. Am. Chem. Soc.* 1987, *109*, 1170–1186.

13. Narayan, S.; Muldoon, J.; Finn, M. G.; Fokin, V. V.; Kolb, H. C.; Sharpless, K. B. "On Water": Unique Reactivity of Organic Compounds in Aqueous Suspension. *Angew. Chem. Int. Ed.* 2005, *44*, 3275–3279.

14. Nicolaou, K. C.; Xu, H.; Wartmann, M. Biomimetic Total Synthesis of Gambogin and Rate Acceleration of Pericyclic Reactions in Aqueous Media. *Angew. Chem.* 2005, *117*, 766–771.

15. Parsons, S.; Passmore, J. Rings, Radicals, and Synthetic Metals: The Chemistry of SNS^+. *Acc. Chem. Res.* 1994, *27*, 101–108.

16. Passmore, J.; Sun, X.; Parsons, S. Cycloaddition Reactions of $SNSAsF_6$ with Aryl Nitiles and Diphenylacetylene; the Preparation and Characterization of Aryl 1,3,2,4- and 1,2,3,5-Dithiadiazolyl Radicals. *Can. J. Chem.* 1992, *70*, 2972–2979.

17. Passmore, J.; Sun, X. Identification of the 1,3,2,4-Dithiadiazolyl RCNSNS. Radical Dimers in Solution, Their Dimeric Concerted Photochemically Symmetry-Allowed Rearrangement to 1,2,3,5-Dithiadiazolyl RCNSSN. By the Net Exchange of Adjacent Cyclic Sulfur and Nitrogen Atoms, and the Photolysis of RCNSNS. *Inorg. Chem.* 1996, *35*, 1313–1320.

INDEX

Organic Mechanisms: Reactions, Methodology, and Biological Applications,
Second Edition. Xiaoping Sun.
© 2021 John Wiley & Sons, Inc. Published 2021 by John Wiley & Sons, Inc.
Companion website: www.wiley.com/go/Sun/OrgMech_2e